Studies in Systems, Decision and Control

Volume 585

The series "Studies in Systems, Decision and Control" (SSDC) covers both new developments and advances, as well as the state of the art, in the various areas of broadly perceived systems, decision making and control–quickly, up to date and with a high quality. The intent is to cover the theory, applications, and perspectives on the state of the art and future developments relevant to systems, decision making, control, complex processes and related areas, as embedded in the fields of engineering, computer science, physics, economics, social and life sciences, as well as the paradigms and methodologies behind them. The series contains monographs, textbooks, lecture notes and edited volumes in systems, decision making and control spanning the areas of Cyber-Physical Systems, Autonomous Systems, Sensor Networks, Control Systems, Energy Systems, Automotive Systems, Biological Systems, Vehicular Networking and Connected Vehicles, Aerospace Systems, Automation, Manufacturing, Smart Grids, Nonlinear Systems, Power Systems, Robotics, Social Systems, Economic Systems and other. Of particular value to both the contributors and the readership are the short publication timeframe and the worldwide distribution and exposure which enable both a wide and rapid dissemination of research output.

Indexed by SCOPUS, DBLP, WTI Frankfurt eG, zbMATH, SCImago.

All books published in the series are submitted for consideration in Web of Science.

Jyotindra Narayan · Hassène Gritli
Editors

Design and Control of Rehabilitation Robots

From Concept to Therapy

Springer

Editors
Jyotindra Narayan
Department of Mechanical Engineering
Indian Institute of Technology Patna
Patna, Bihar, India

Department of Computing
Imperial College London
London, UK

Hassène Gritli
Laboratory of Robotics, Informatics
and Complex Systems (RISC Lab)
University of Tunis El Manar
Tunis, Tunisia

Higher Institute of Information
and Communication Technologies
University of Carthage
Tunis, Tunisia

ISSN 2198-4182 ISSN 2198-4190 (electronic)
Studies in Systems, Decision and Control
ISBN 978-3-031-86976-1 ISBN 978-3-031-86977-8 (eBook)
https://doi.org/10.1007/978-3-031-86977-8

This Springer imprint is published by the registered company Springer Nature Switzerland AG
The registered company address is: Gewerbestrasse 11, 6330 Cham, Switzerland

To our family for unwavering love and support

Preface

Rehabilitation robotics stands at the intersection of advanced technology and compassionate care, revolutionizing the way physical impairments are addressed. This book on design and control of rehabilitation robots offers a comprehensive exploration of the field, starting with a historical overview and progressing through key areas such as human-robot interaction, sensing and perception technologies, path planning, and assist-as-need control schemes. These topics are carefully examined to demonstrate their practical applications in enhancing therapy outcomes and fostering patient autonomy. Notably, the book emphasizes designing rehabilitation robots that seamlessly integrate into clinical settings while prioritizing patient safety and comfort.

With real-world case studies and a focus on ethical and regulatory considerations, the book bridges the gap between theory and practice, providing invaluable insights for researchers, clinicians, engineers, and policymakers. The concluding chapter synthesizes key lessons and outlines future strategies for advancing the field. By highlighting both the opportunities and challenges in deploying rehabilitation robots, this book serves as a vital resource for driving innovation in patient-centric care, ultimately aiming to improve the quality of life for individuals with physical impairments.

Patna, India Jyotindra Narayan
Tunis, Tunisia Hassène Gritli

Introduction

Rehabilitation robotics is transforming healthcare by integrating advanced technological innovations with patient-centered approaches to improve therapy outcomes and foster autonomy for individuals with physical impairments. This book offers a thorough exploration of the field, beginning with its historical evolution and progressing through key areas such as human-robot interaction, sensing technologies, path planning, and assist-as-need control schemes, providing practical insights into their clinical applications. It showcases real-world case studies that highlight the impact of these technologies across various medical domains, emphasizing both their successes and the challenges encountered in deployment. Ethical and regulatory considerations are also addressed, stressing the importance of responsible innovation to ensure patient safety and well-being. By synthesizing key insights and proposing future directions, the book equips researchers, clinicians, engineers, and policymakers with the necessary tools to advance rehabilitation robotics. Ultimately, it underscores the transformative potential of these technologies in reshaping patient care, improving therapy efficiency, and enhancing the quality of life for individuals with mobility impairments.

The book has been organized into 18 chapters that highlights different studies and topics related to the design and control of rehabilitation robots for healthcare solutions.

Chapter "Assistive and Rehabilitative Brain–Computer Interfaces: Innovations in Patient Care" This chapter delves into Brain-Computer Interfaces (BCIs), a transformative technology that enables direct communication between the brain and external devices, bypassing traditional muscle-based control. It explores the evolution, technical foundations, and applications of BCIs in assistive and rehabilitative contexts. The chapter covers the distinction between invasive and non-invasive BCIs, highlighting their role in empowering individuals with disabilities to regain autonomy and enhancing neurorehabilitation by promoting brain plasticity. Additionally, it discusses the integration of BCIs with complementary technologies such as virtual reality, robotics, and machine learning to improve their precision and effectiveness. By examining these advancements, the chapter underscores the growing impact of BCIs in improving the quality of life for individuals with limited mobility and contributing to innovative neurorehabilitation approaches.

Chapter "Assessing Performance and Mechanics of the PUCP IArm Upper Limb Prosthesis" addresses the challenges in accessing advanced upper limb prostheses in developing countries, with a focus on Peru, where high costs often lead individuals to choose basic or aesthetic prostheses over functional ones. It highlights the pressing need for affordable prosthetic solutions that retain essential functionality to improve users' daily lives. The chapter introduces the PUCP IArm prosthesis, a cost-effective, functional hand prosthesis developed at the Pontificia Universidad Católica del Perú using additive manufacturing. The study details the mechatronic characterization and performance evaluation of this prosthetic hand, including tests on finger force, tensile strength, grip capacity, and response latency. With a weight of 230 g and strong performance in pinch and cylindrical grips, the PUCP IArm demonstrates promising results compared to other myoelectric prostheses, offering an accessible alternative for users in resource-limited settings.

Chapter "Generative Design of Wearable Assistive Devices: Strategies and Selection" explores the application of generative design methods in creating wearable assistive devices (WADs) such as orthoses, prostheses, and exoskeletons. It provides an overview of the evolution of WAD design and examines how generative methods address key challenges like reducing weight, enhancing actuator efficiency, improving adjustability, and increasing responsiveness. The chapter analyzes various generative strategies and their design objectives, presenting a structured framework to guide the selection of design methods based on specific needs. To illustrate these concepts, a case study involving the development of an ankle-foot orthotic for short-term injury support is discussed, showcasing the practical application of generative design principles in assistive hardware development.

Chapter "Comparative Evaluation of Analytical Techniques for Estimating Ground Reaction Force in Human Walking" focuses on the estimation of Ground Reaction Force (GRF), a crucial factor in understanding human gait dynamics and designing effective assistive devices. The study explores various analytical techniques for GRF estimation using a publicly available gait dataset, beginning with the Smooth Transition Approach and progressing to an inverted pendulum model for double stance phases. Different estimation methods are evaluated across gait phases, with a particular focus on improving a single model capable of predicting GRF throughout the entire gait cycle. The chapter compares methods using both visual analysis and RMSE values to assess accuracy, highlighting the trade-offs between phase-specific and unified models. The insights gained from this study have significant implications for assistive device design and the development of physics-informed learning models to enhance gait analysis and rehabilitation strategies.

Chapter "A Comprehensive Comparative Analysis of Gait Phases Using Machine Learning: Binary and Ternary Classifications with Dual Training Methods and Scaling Techniques" delves into the classification of human gait phases using advanced machine learning techniques to aid in diagnosing and treating gait abnormalities and musculoskeletal disorders. It explores binary and ternary phase classifications based on data from 100 individuals sourced from an open-access platform, utilizing two distinct training methodologies: stratified random sampling and participant-based splits. The chapter details the preprocessing steps applied to the dataset, including scaling methods like Min-Max Scaling, Standard Scaling, and Principal Component Analysis. It then assesses the performance of various machine learning models, such as k-Nearest Neighbors, Logistic Regression, Decision Tree, Random Forest, Support Vector Machine, Naive Bayes, Linear Discriminant Analysis, and Quadratic Discriminant Analysis, based on accuracy, cross-validation scores, Mean Squared Error, Root Mean Squared Error, and R2 scores. Random Forest emerged as the most effective model, achieving the highest accuracy for both binary and ternary classifications, demonstrating its efficacy in accurately predicting gait phases.

Chapter "Solving Inverse Kinematics of a 2-DoF Upper Limb Exoskeleton Robot Using a Hybrid Approach Combining Metaheuristic Algorithms and Artificial Neural Networks" explores advanced methods for solving the Inverse Kinematics (IK) problem in upper limb exoskeletons, focusing on overcoming the challenges posed by the non-linearity of traditional analytical approaches. The study begins by applying metaheuristic algorithms, including Particle Swarm Optimization, Genetic Algorithm, BAT Algorithm, and Grey Wolf Optimizer, to address the IK model for two-degree-of-freedom (2-DoF) exoskeletons. The chapter then introduces a hybrid approach by integrating Artificial Neural Networks (ANNs), which are trained using random step-size and sinusoidal-signal datasets. By optimizing ANN hyperparameters, such as hidden layers and activation functions, through novel techniques, the methodology achieves improved accuracy and lower error rates. The experimental results validate the effectiveness of this hybrid model and demonstrate its potential scalability to more complex robotic systems, making a significant contribution to the field of exoskeleton control.

Chapter "Association of Heart Rate Variability with Cardiovascular Disease Risk in Healthy Adults" investigates the relationship between heart rate variability (HRV) and cardiovascular risk factors in healthy adults using the Framingham Risk Assessment (FRA) tool. It investigates whether low HRV can serve as an early indicator of cardiovascular disease (CVD) risk by analyzing time-domain and frequency-domain HRV variables, including SDNN, RMSSD, and LF/HF ratio. The study collected HRV data from 38 healthy participants using a Hexoskin smart shirt and correlated it with FRA variables such as CVD risk score, heart age, and hypertension risk. Results revealed that lower HRV was associated with higher CVD risk, hypertension risk, and heart age, as well as with higher age, BMI, resting heart rate, and waist circumference. Additionally, the LF/HF ratio showed a moderate correlation with walking speed and prefrontal cortical activation, suggesting a link between autonomic function and physical activity.

Chapter "Telerehabilitation and Remote Monitoring: Improving Efficiency and Security in Healthcare" delves into the transformative impact of telerehabilitation and remote monitoring in modern healthcare, emphasizing the advantages of dedicated medical servers over third-party systems. It highlights how dedicated servers offer faster data transmission, reduced network interruptions, and enhanced security, leading to more accurate patient monitoring. The research findings demonstrate that dedicated servers achieve up to 50% faster transmission speeds, a 95% reduction in network interruptions, and an 85% decrease in data breaches, ensuring patient confidentiality and trust. With a 30% improvement in patient monitoring accuracy, this chapter underscores the critical role of dedicated servers in optimizing medical response times and enhancing clinical decision-making in telerehabilitation, paving the way for more reliable and secure remote healthcare solutions.

Chapter "AI-Driven Gait Classification Using Portable Wearable Sensors: Advances and Case Study" discusses into the advancements in AI-driven gait classification, focusing on the use of wearable portable sensor systems to characterize and measure human gait. It highlights the application of machine learning (ML) and deep learning (DL) models to accurately classify various gait activities, demonstrating their potential for gait rehabilitation applications. The chapter includes a review of recent progress in gait classification techniques and presents a case study where gait data was collected using an IMU sensor attached to the right shank of 20 young subjects performing six different gait activities: walking uphill, downhill, on level ground, climbing stairs, descending stairs, and standing. Through the training and evaluation of various ML and DL models, the chapter provides comprehensive experimental results, showcasing the practical implications of these technologies in enhancing gait analysis and rehabilitation outcomes.

Chapter "Data-Driven Grasp Posture Recognition with Data Gloves and Machine Learning Models" explores the application of machine learning algorithms for classifying various hand grasp postures using data collected from a sensor-equipped data glove. The study evaluates five models—Support Vector Machine (SVM), Decision Tree (DT), Random Forest (RF), k-Nearest Neighbors (KNN), and Logistic Regression—based on their accuracy, precision, recall, and computational efficiency. KNN and SVM achieved the highest classification accuracy, particularly for common grasp types like Cylindrical Grip and Lateral Pinch, though certain grips, such as the Hook Grip, posed challenges across models. The chapter highlights the trade-offs between accuracy and computational time, noting that RF offers robust accuracy but slower computation, while DT are faster but less accurate. The research emphasizes the importance of selecting suitable algorithms for specific tasks and contributes to the advancement of gesture recognition systems in robotics, neurorehabilitation, and assistive technologies, where precise, real-time hand posture recognition is essential.

Chapter "Predicting Lower Limb Joint Movements from Muscle Signals Using Neural Networks: A Comparative Study of Training Optimizers" This chapter delves into the estimation of lower limb joint angles from muscle signals in able-bodied adults across different locomotion modes using an open-access dataset. The study focuses on motion data and surface EMG signals from 11 right lower limb muscles of 22 subjects under various conditions, particularly analyzing treadmill walking at a speed of 1.3 m/s. The chapter aims to achieve two key objectives: first, to preprocess and prepare the dataset for analysis and model training, and second, to train backpropagation neural network models using three optimization techniques—Levenberg-Marquardt (LM), Bayesian Regularization (BR), and Scaled Conjugate Gradient (SCG). The performance of these models is assessed by comparing their predictions with actual joint angle measurements using metrics like mean squared error and correlation coefficients. The LM method demonstrated the best performance with the lowest error and highest correlation, while BR and SCG also yielded promising results. This study offers valuable insights into the relationship between muscle activity and joint movement, contributing to the development of effective control strategies for lower-limb rehabilitation robots.

Chapter "Advanced Impedance Electromyographic-Based Control for an Upper Limb Exoskeleton Robot" introduces an adaptive impedance control strategy designed to enhance active rehabilitation using a 7-DOF wearable robotic exoskeleton with an unknown dynamic model. The core challenge addressed is enabling the exoskeleton to accurately interpret the user's movement intentions, despite the nonlinear and time-varying nature of human motion. To achieve this, the chapter employs Hill's model to estimate human forces through the use of gravitational dynamics, force sensors, and surface electromyogram (sEMG) signals. A Radial Basis Function Neural Network (RBFNN) integrated with a sliding mode estimator is then utilized to predict the user's desired motion intention (DMI) in real time. The estimated DMI is incorporated into the adaptive impedance control framework, allowing the exoskeleton to provide precise and comfortable assistance by tracking an impedance target. Experimental trials with healthy participants validate the effectiveness of this approach, demonstrating smooth and adaptive interaction between the robot and its wearer across various rehabilitation scenarios.

Chapter "Computer-Aided Design and Control of an Integrated Exoskeleton-Upper Limb System" outlines the various stages involved in constructing Computer-Aided Design (CAD) simulation models using the combined environments of SolidWorks and Matlab/Simulink. The study specifically focuses on developing an exoskeleton model designed for rehabilitation purposes, coupled with a human upper limb model. These simulation models are essential for visualizing the system's ability to track desired trajectories during movement, providing a virtual testbed for evaluating control strategies and mechanical behavior. The first part of the research is dedicated to the design and analysis of the exoskeleton-upper limb system's CAD model. This stage involved creating a detailed 3D mechanical representation of the exoskeleton and human arm in SolidWorks, where joint configurations, mechanical constraints, and movement capabilities were defined. The next phase of the research focused on extracting the Electromyography (EMG) signals generated during elbow flexion and extension movements. To control the exoskeleton-upper limb system, a Model-Free Adaptive Terminal Sliding Mode Controller with Gravity Compensation was implemented. This advanced control algorithm was chosen for its robustness in handling the complex, nonlinear dynamics of the exoskeleton interacting with the human. The effectiveness of both the models and the control algorithm were validated through simulation results.

Chapter "Robust Stabilization of a 1-DoF Knee Rehabilitation Exoskeleton Using an Enhanced LMI-Based Composite Nonlinear Controller" focuses on a control strategy designed to robustly stabilize a one-degree-of-freedom (1-DoF) knee rehabilitation exoskeleton robot by addressing the system's nonlinear dynamics and uncertainties. The proposed approach integrates a linear state-feedback controller with nonlinear control laws to manage parameter uncertainties and disturbances effectively. The nonlinear dynamics are approximated through bounded functions, leading to the development of Linear Matrix Inequality (LMI) conditions using techniques such as Young's inequality, the S-procedure, the Schur complement, and the matrix inversion lemma. The chapter presents three LMI design approaches and employs a quadratic Lyapunov function to ensure stability and robustness. The control strategy's effectiveness is validated through simulations under external disturbances, and a comparative analysis highlights the robustness of the proposed methods in maintaining the stability of the knee-joint exoskeleton system.

Chapter "Experimental Investigation of Robotic Ankle Foot Orthosis with Different Actuators for Foot Drop Rehabilitation" focuses on the development of a robotic ankle-foot orthosis (RAFO) aimed at improving mobility for patients with foot drop by addressing the limitations of conventional ankle-foot orthoses in providing sufficient ankle range of motion (ROM). It presents the design and integration of two distinct actuators—the Series Elastic Actuator (SEA) and Scissor Actuator (SA)—into the RAFO, along with the implementation of a fuzzy logic controller in MATLAB Simulink to manage plantarflexion and dorsiflexion movements. The chapter details the experimental comparison of both actuators using bench tests with an Arduino Mega microcontroller, rotary angle sensor, and ultrasonic sensor. Key findings reveal that while SA offers greater ROM and torque generation, the SEA excels in energy efficiency. The chapter concludes by proposing a hybrid actuator design combining both mechanisms to optimize power consumption, ROM, and torque, offering a promising solution for enhancing mobility and quality of life for individuals with foot drop.

Chapter "Influence of Lazy Zone on Orthotropic Bone Adaptation in Human Femur: Pre- and Post-surgery Insights" explores the impact of the lazy zone threshold on orthotropic bone remodeling in the context of peri-prosthetic bone adaptation following hip replacement surgery. Using a finite element model, the study simulates the elastic material property distribution across an implanted femur subjected to various physiological loading conditions, incorporating eight distinct lazy zone thresholds. The findings reveal that variations in lazy zone size significantly influence bone remodeling, with lower thresholds leading to increased bone resorption around the hip stem and higher thresholds reducing bone loss. The study highlights the importance of selecting an appropriate lazy zone threshold in biomechanical simulations to better predict post-surgery bone adaptation and inform rehabilitation strategies.

Chapter "Ethical and Regulatory Considerations in Rehabilitation Robotics: Ensuring Safe and Equitable Care" delves into the critical ethical and regulatory considerations surrounding the integration of robotics in rehabilitation therapies. It begins by outlining the ethical foundations essential to healthcare technologies, such as patient autonomy, consent, and beneficence, while also examining global regulatory standards like FDA guidelines and CE marking. The chapter emphasizes the importance of adhering to safety standards and risk management protocols to mitigate potential harm, alongside addressing privacy concerns in line with regulations such as HIPAA and GDPR. Furthermore, it highlights the need for equitable access to rehabilitation robotics, advocating for inclusive practices that overcome socioeconomic barriers. Through case studies and best practices, the chapter provides valuable insights into navigating the evolving ethical and regulatory landscape, offering guidance for ensuring safe, ethical, and patient-centered care in rehabilitation robotics.

Chapter "Advancement and Challenges in Rehabilitation Robotics: Insights from Case Studies" explores the emergence of robotic technologies in healthcare, focusing on rehabilitation robotics designed to assist patients with physical disabilities. It provides an in-depth analysis of real-life case studies, addressing the practical applications, technological advancements, and user experiences of various rehabilitation robots like Burt and Myopro. The chapter also examines the integration of AI, ML, DL, and AR/VR technologies in enhancing rehabilitation robotics and discusses future directions and solutions to existing challenges, offering valuable insights into improving the accessibility and efficiency of robotic rehabilitation technologies.

Contents

Editors and Contributors

About the Editors

Jyotindra Narayan serves as an Assistant Professor in the Department of Mechanical Engineering at the Indian Institute of Technology Patna, starting from December 2024. He is actively engaged in healthcare robotics, control theory, and artificial intelligence and is also associated with the Department of Computing at Imperial College London as a Honorary Research Associate. He completed his Ph.D. in January 2023 from the Mechanical Engineering Department at IIT Guwahati, focusing on designing, developing, and controlling passive-assist and active-assist control schemes for pediatric lower-limb exoskeleton systems. With over 80 publications in reputed journals and conferences, including more than 30 Q1/Q2-indexed journals, Dr. Narayan is an academic editor at PLOS ONE, an associate editor at *Measurement and Control Journal* (Sage Publications), and a guest editor at the *International Journal of Advanced Robotic Systems*. He has chaired special sessions at prestigious international conferences such as CoDIT 2024 and IC_ASET 2024 and recently received the S. K. Mitra Memorial Best Research Paper Award (August 2024) by IETE, recognized by DSIR, Government of India. He has recently co-authored a contributed book *Intelligent Cyber-Physical Systems for Healthcare Solutions: From Theory to Practice* published by Springer Nature. Additionally, he is a lead reviewer for top-tier journals from Springer, IEEE, Elsevier, MDPI, and international conferences ICRA 2024 and BioRob 2024.

Hassène Gritli obtained the National Engineering Diploma and the Research Master's degree in Industrial Informatics and Automation from the National Institute of Applied Sciences and Technology (INSAT), respectively in February 2006 and July 2007. He obtained his Ph.D. in Electrical Engineering from the National Engineering School of Tunis (ENIT), Tunisia, in January 2014, followed by the HDR degree in June 2020. His academic career began as an Assistant at the Higher Institute of Applied Sciences and Technology (ISSAT) of Mateur from 2008 to 2012. He then served as a Technologist Teacher at the Higher Institute of Technological

Studies (ISET) of Kélibia from 2012 to 2014. Between 2014 and 2020, he was an Associate Professor at the Higher Institute of Information and Communication Technologies (ISTIC), Tunisia, before being appointed as Professor-Lecturer in Automation and Industrial Informatics at ISTIC, a position he holds to date. As a tenured research scientist at the Laboratory of Robotics, Informatics, and Complex Systems (RISC Lab) at ENIT, his research focuses on biped and exoskeleton robots, hybrid dynamical systems, chaos theory, robust control, and machine learning applications in real-world models. He has organized special sessions at international conferences, served as a (Lead) Guest Editor for special journal issues, and delivered a keynote lecture on robust control at IEEE SSD'20. Additionally, he was part of the organizing committees for IC_ASET conferences in 2020, 2022, and 2023. With over 150 publications in leading journals and conferences, he regularly reviews numerous international publications in Robotics and Control Theory.

Contributors

Victoria E. Abarca Pontificia Universidad Católica del Perú, Lima, Peru

Azamat Ali School of Liberal and Creative Arts (Journalism and Mass Communication), Lovely Professional University, Phagwara, Punjab, India

Safya Belghith RISC Laboratory, National Engineering School of Tunis, University of Tunis El-Manar, Tunis, Tunisia

Sana Bembli RISC Laboratory, National Engineering School of Tunis, University of Tunis El-Manar, Tunis, Tunisia

Santushti Betgeri Department of Computer Engineering, Vishwakarma Institute of Technology, Pune, Maharashtra, India

Alka Bishnoi Department of Physical Therapy, College of Health Professions and Human Services, Kean University, Union, NJ, USA

Rania Bouzid Polytechnic School of Tunisia, University of Carthage, La Marsa, Tunisia;
Laboratory of Robotics, Informatics and Complex Systems (RISC Lab—LR16ES07), National Engineering School of Tunis, University of Tunis El Manar, Tunis, Tunisia

Brahim Brahmi Department of Electrical Engineering Center for Interdisciplinary Research Center for Intelligent Manufacturing & Robotics (IRC-IMR), King Fahd University of Petroleum and Minerals, Dhahran, Eastern Province, Saudi Arabia

Debabrata Chakraborty Fracture Mechanics and Composite Structures Laboratory, Department of Mechanical Engineering, Indian Institute of Technology Guwahati, Guwahati, Assam, India

Tara Chand Jindal Institute of Behavioural Sciences, O.P. Jindal Global University, Sonipat, India

Souptick Chanda Biomechanics and Simulations Laboratory, Department of Biosciences and Bioengineering, Indian Institute of Technology Guwahati, Guwahati, Assam, India

Sekar Anup Chander CSIR-Central Scientific Instruments Organisation, Sector-30C, Chandigarh, India;
Academy of Scientific and Innovative Research (AcSIR), Ghaziabad, Uttar Pradesh, India

Arockia Selvakumar Arockia Doss Vellore Institute of Technology, Chennai, Tamil Nadu, India

Dante A. Elias Pontificia Universidad Católica del Perú, Lima, Peru

Siddharth Garg Netaji Subhas University of Technology, Delhi, India

Shiv Manjaree Gopaliya School of Mechanical Engineering, VIT Bhopal University, Bhopal, India

Gowrishankar Govindaraj Vellore Institute of Technology, Chennai, Tamil Nadu, India

Hassène Gritli Laboratory of Robotics, Informatics and Complex Systems (RISC Lab—LR16ES07), National Engineering School of Tunis, University of Tunis El Manar, Tunis, Tunisia;
Higher Institute of Information and Communication Technologies, University of Carthage, Technopole of Borj Cédria, Hammam Chatt, Ben Arous, Tunisia

Ayush Gupta Department of Mechanical Engineering, Indian Institute of Technology Guwahati, Guwahati, Assam, India

Rakhi Gupta Jindal Institute of Behavioural Sciences, O.P. Jindal Global University, Sonipat, India

Shyamanta M. Hazarika Department of Mechanical Engineering, Indian Institute of Technology Guwahati, Guwahati, Assam, India

Ankan Hazra Fracture Mechanics and Composite Structures Laboratory, Department of Mechanical Engineering, Indian Institute of Technology Guwahati, Guwahati, Assam, India

Manuel E. Hernandez Department of Biomedical and Translational Sciences, Carle Illinois College of Medicine, University of Illinois Urbana-Champaign, Urbana, IL, USA;
Department of Kinesiology and Community Health, College of Applied Health Sciences, University of Illinois Urbana-Champaign, Urbana, IL, USA

Kazuaki Ito Department of Mechanical Engineering, Gifu University, Gifu, Japan

Sahar Jenhani Laboratory of Robotics, Informatics and Complex Systems (RISC Lab - LR16ES07), National Engineering School of Tunis, University of Tunis El Manar, Tunis, Tunisia

Sanchit Jhunjhunwala Translead Medtech, Indian Institute of Science, Bengaluru, Karnataka, India

Navkiran Kalsi Jindal School of Psychology and Counselling, O.P. Jindal Global University, Sonipat, India

V. S. R. Kamaleshwaran Gnanamani College of Technology, Namakkal, Tamilnadu, India

Nahla Khraief Haddad RISC Laboratory, National Engineering School of Tunis, University of Tunis El-Manar, Tunis, Tunisia

L. Kiruthika Dhanalakshmi College of Engineering, Chennai, Tamilnadu, India

Estefany Macedo Pontificia Universidad Católica del Perú, Lima, Peru

Amal Mekni Laboratory of Robotics, Informatics and Complex Systems (RISC Lab—LR16ES07), National Engineering School of Tunis, University of Tunis El Manar, Tunis, Tunisia

Monisha Mohan Department of Computer Engineering, Marwadi University, Rajkot, Gujarat, India

Megha Mudholkar Department of Computer Engineering, Marwadi University, Rajkot, Gujarat, India

Pankaj Mudholkar Faculty of Computer Applications, Marwadi University, Rajkot, Gujarat, India

Sivakumar Murugesan Dhanalakshmi College of Engineering, Chennai, Tamilnadu, India

Sonam Nahar Department of Computer Science and Engineering, School of Technology, Pandit Deendayal Energy University, Gandhinagar, Gujarat, India

Jyotindra Narayan Department of Mechanical Engineering, Indian Institute of Technology Patna, Patna, Bihar, India;
Department of Computing, Imperial College London, London, UK

Subhash Pratap Department of Mechanical Engineering, Indian Institute of Technology Guwahati, Guwahati, Assam, India;
Department of Mechanical Engineering, Gifu University, Gifu, Japan

Mohammad Habibur Rahman Department of Mechanical Engineering, University of Wisconsin-Milwaukee, Milwaukee, WI, USA

Enzo Romero Pontificia Universidad Católica del Perú, Lima, Peru

Maarouf Saad Department of Electrical Department, Ècole de technologie supèrieure, Montrèal, QC, Canada

Vhatkar Dattatraya Shivling CSIR-Central Scientific Instruments Organisation, Sector-30C, Chandigarh, India;
Academy of Scientific and Innovative Research (AcSIR), Ghaziabad, Uttar Pradesh, India

Ravjeet Singh Thapar Institute of Engineering and Technology, Patiala, Punjab, India

Ashish Singla Thapar Institute of Engineering and Technology, Patiala, Punjab, India

Preet Sojitra Department of Computer Science and Engineering, School of Technology, Pandit Deendayal Energy University, Gandhinagar, Gujarat, India

Dinesh Subbiah Dhanalakshmi College of Engineering, Chennai, Tamilnadu, India

Anugya Tripathi School of Mechanical Engineering, VIT Bhopal University, Bhopal, India

I. Vandhana Dhanalakshmi College of Engineering, Chennai, Tamilnadu, India

Vineet Vashista Department of Mechanical Engineering, Indian Institute of Technology Gandhinagar, Gandhinagar, Gujarat, India

Kedarmal Verma Cognitive Experimental Laboratory, School of Humanities and Social Sciences, Indian Institute of Technology Indore, Indore, India

Assistive and Rehabilitative Brain–Computer Interfaces: Innovations in Patient Care

Tara Chand, Kedarmal Verma, Rakhi Gupta, and Navkiran Kalsi

Abstract Brain-Computer Interfaces (BCIs) have emerged as a groundbreaking technology that connects the brain directly to external devices, bypassing traditional muscle and peripheral control. The advancements in neuroscience and machine learning allow users to control computers, artificial limbs, and other assistive devices using brain signals alone. By decoding brain activity, BCIs provide new communication and control methods for individuals with limited mobility and paralyzed, allowing them to speak, type, or manipulate objects in their environment. BCIs also play a vital role in neurorehabilitation, where they promote neural recovery by enhancing the brain's plasticity. This dual application of BCIs, both in assistive technology and rehabilitation, makes them a powerful tool for enhancing patients' quality of life and independence. The chapter provides a detailed overview of BCIs, discussing their history, development, and the technical foundations that enable them to translate neural signals into commands for external devices. It also covers the differences between invasive and non-invasive BCIs, highlighting their respective strengths. While assistive BCIs give individuals with disabilities newfound autonomy, rehabilitative BCIs work in synergy with physical therapy to restore motor skills and cognitive functions. The chapter also highlights the importance of the multimodal approaches that integrate additional technologies, such as virtual reality, robotics, and machine learning. These hybrid systems show promise in enhancing the precision, accuracy, and overall effectiveness of BCIs in both assistive and rehabilitative contexts. As these technologies evolve, BCIs are expected to become even more robust, offering better support for patients with disabilities and contributing to more effective neurorehabilitation strategies.

Tara Chand and Navkiran Kalsi are contributed equally.

T. Chand · R. Gupta
Jindal Institute of Behavioural Sciences, O.P. Jindal Global University, Sonipat, India

K. Verma
Cognitive Experimental Laboratory, School of Humanities and Social Sciences, Indian Institute of Technology Indore, Indore, India

N. Kalsi (✉)
Jindal School of Psychology and Counselling, O.P. Jindal Global University, Sonipat, India
e-mail: navkiran.kalsi@jgu.edu.in

J. Narayan and H. Gritli (eds.), *Design and Control of Rehabilitation Robots*, Studies in Systems, Decision and Control 585, https://doi.org/10.1007/978-3-031-86977-8_1

Keywords Brain-computer interfaces · Rehabilitation · Assistive technology

1 Introduction

Over the few decades, due to the advancements of neuroscience, and the development in machine learning techniques have given birth to the new technology known as Brain-Computer Interfaces (BCIs). This cutting-edge technology provides a communication and control channels to directly link the brain to external devices that do not depends on the brain's normal output for the peripheral and muscles control [1]. It allows us to control computers, robotic limbs, and other assistive external devices using only brain activity. BCIs in helping patients with severe motor disabilities is becoming a powerful tool as an assistive technology [2]. The use of BCI mediates the brain and the outside world and provides a new way of interacting and controlling the environment previously unavailable to individuals with quite severe physical disabilities. Through this technology, individuals who are unable to move can control computers, prosthetic limbs, or even communicate via speech-generating devices [2, 3]. It helps paralyzed patients to speak or control artificial limbs and other external devices. These systems offer a basic avenue through which these patients with conditions like ALS or spinal cord injuries can communicate. For example, the use of EEG has helped some of the paralyzed patients to speak or write by interpreting electrical impulses in the brain [4].

The significance of BCIs is not limited to assistive technology. In addition to helping patients control their environment, BCIs are increasingly being used in neurorehabilitation. it has been possible to apply EEG-based BCIs in synchronization with behavioral physiotherapy in order to enhance neuroplasticity to restore the function of the stroke victim [5]. This combined approach builds on the plasticity of the brain and works with the understanding that serious injury invokes the remainder of the person's life potential to heal. By leveraging the brain's neuroplasticity, BCIs enable patients to regain lost motor functions, furthering their recovery and improving their long-term prognosis [6].

This chapter provides a comprehensive overview of BCIs, focusing on their development, technological aspects, and clinical applications. We begin by exploring the history and evolution of BCIs, from early neurophysiological research to modern innovations in neural signal processing. Then, we delve into the technical underpinnings of BCIs, explaining how these systems work by interpreting brain signals and translating them into meaningful commands. Following this, we examine the differences between invasive and non-invasive BCIs, highlighting their respective advantages, disadvantages, and real-world applications. The latter half of the chapter discusses two of the primary uses of BCIs: assistive and rehabilitative technology. Assistive BCIs provide a critical means of Communication and control for patients with severe disabilities, offering them independence. Rehabilitative BCIs, on the other hand, aid in neural recovery by stimulating brain plasticity and working in tandem with physical therapy to restore lost functions. Finally, we explore the future of BCIs, particularly the emerging use of multimodal techniques that combine brain-computer interfacing with other advanced technologies, such as virtual reality (VR), robotics, and machine learning. These hybrid approaches show promise in enhancing the accuracy and effectiveness of BCIs in both assistive and rehabilitative settings.

2 Historical Development of BCIs

The concept of BCIs has evolved significantly over the past several decades, shaped by advances in neuroscience, biomedical engineering, and computing technology. While the term "BCI" is relatively modern, the scientific principles that underline this technology have deep historical roots in the study of the brain's electrical activity and the ongoing quest to understand how neural processes govern behavior and interaction.

2.1 Early Foundations: The Discovery of Electrical Brain Activity

The origins of BCIs date back to the eighteenth century when an Italian scientist Luigi Galvani and his wife showed that the muscles of dead frogs' legs could be twitched when struck by an electrical spark [7]. This discovery provides the foundation for understanding how electricity can govern motion and how the brain and body communicate electrically. Years later, Richard Caton (1842–1926) discovered that the electrical activity generated from the brain can be measured using galvanometer and he measured the electrical activity of the rabbit and monkey's brain [8]. Subsequently, the first human EEG was recorded by a German physiologist, Hans Berger, in 1924 and now, we are marking the centenary of the first human EEG recording [9]. It was the first time when electrical activity was recorded, named as brain scripts, made the way for the advancements that led to development of BCIs. In the trials with Berger, the measurements of electrical activity of the human brain depicted rhythmic oscillations, which were later known as the alpha waves as well as the beta waves associated with varying states of consciousness. This discovery was revolutionary as it considered EEG to record the brain's electrical activity and later use to understand the functioning of the brain. In the subsequent years, investigators began to establish a relationship between EEG and other mental states and motor abilities, a line of conceptual evolution that would inevitably give rise to BCIs.

2.2 The Emergence of Neurophysiology and Control Mechanisms

In mid twentieth century, neurophysiology and the concept of neuronal circuits became basic to our understanding of how the nervous system coordinates the body. Scientists realized that signs captured by an EEG and other techniques were not only side effects of brain function; they were the ways of deciphering how neurons signal to produce motion, feeling, and thought. As early as the 1960s the principles of brain-computer interface started to emerge as researchers such as Jose Delgado attempted direct methods of brain stimulation. The community was allured and also concerned when Delgado successfully used electrodes to control animal behavior [10].

2.3 Early BCI Research and the Role of Cybernetics

The field of cybernetics was developed during 1970–1980 as an interdisciplinary field which deals with the control and communication of the various systems, including artificial and living systems. When dealing with the Communication between the living and artificial machine/system, the BCI played an important role, mainly when investigators were trying to interface external devices using brain signals in animals. In the 1970s, Dr. Jacques J. Vidal from the university of California, Los Angeles (UCLA) used signal processors to translate brain signals into direct control of computers. He developed the first system allowing direct Communication between the brain and a machine and coined the term "Brain-Computer Interface" [11]. The Vidal's work has also shown that the EEG signal can be used to speak to a computer thus a step to human BCIs. He was able to prove that brain signals could be digitized, and in effect, the brain could communicate with a computer without coaxing the body into speaking. This important study demonstrated the potential that BCIs had potential for helping individuals with disabilities to regain or enhance volitional control over movement. It also became evident for the practical application of BCIs in clinical practice as opposed to what was evidently more of a theoretical orientation in previous years.

2.4 The Shift to Clinical Applications in the Late 20th Century

The technological advances in computing and signal processing accelerated the development of BCIs. Researchers began to apply BCIs in medical contexts, focusing on patients with severe disabilities such as ALS, spinal cord injuries, and locked-in syndrome. One of the key challenges that arose during this period was improving the accuracy of signal acquisition and interpretation. Early systems struggled with signal noise and interference, limiting their real-world applications.

In the 1980s and 1990s, advancements in computing as well as advancements in signal processing contributed more towards the development of BCIs. Scientists started to use BCIs in therapeutic practice and practice in the treatment of severe patients, for example, those with ALS, spinal cord injuries or locked-in syndrome. One of the major issues that emerged in earlier studies was improving the accuracy of signal acquisition and interpretation of the signals which was limiting the real-world applications. In addition to the non-invasive BCI, at the same time there were some attempts also made towards more invasive BCI's where electrodes are inserted deep into the brain to record brain activity. In 1998, Philip Kennedy made a ground breaking finding in which a paralyzed man could operate a computer cursor through the utilization of implantable electrodes laid in the motor cortex [12]. This was one of the first cases of invasive BCIs in humans and showed that implanted electrodes and devices may provide accurate and stable control over external environment. Moreover, the work from scientists such as Dr. Niels Birbaumer at the University of Tübingen in Germany have contributed to the creation of non-invasive application of BCI for patients with locked-in syndrome that completely conscious persons retain no ability to move or gesture. Specifically, Birbaumer's group was able to show that patients could use EEG-based BCIs in order to select letters on a screen, which would offer a way to communicate for people who could not otherwise interact with the world around them [13].

2.5 *The Rise of Neuroprosthetics and Modern BCIs*

BCI has grown rapidly in the twenty-first century, both invasive and noninvasive BCIs, particularly with the development of neuroprosthetics. These devices, often used to assist individuals with damage to the spinal cord, nerves, or muscles and are able to generate motor movements. These devices directly connect the nervous system to artificial limbs, which helps paralyzed people to control prosthetics with their thoughts. One of the most notable examples is the BrainGate (Cyberkinetics, Inc.), which is a neuromotor prostheses system and has been a launched as an assistive technology for the individual with paralysis [14]. The BrainGate system uses neurosurgically placed microelectrode arrays that capture signals in the motor cortex and transmit them to external equipment, such as robotic limbs or cursors. These developments have allowed for better interpretation of the neural signals and thus better practical application of the BCIs in the service of people with motor disabilities. For example, EEG based BCI are now used for different purposes from controlling wheelchairs and Robotic Devices to Communication for ALS patients. This advancement is well illustrated by recent studies showing how machine learning approaches can improve signal processing hence improving the accuracy of motor signals or communication pointers produced by the users [15]. Such innovations have made it possible to develop more advanced, more friendly and easy to use assistive technologies thus enhancing the quality of life of people with severe motor disabilities.

3 Understanding BCIs: Concepts and Working Mechanism

The basic concept of operation of BCIs involves recording and interpreting brain signals, mainly related to movement or cognitive functions, and then using these as signals to control different other systems outside the body such as a computer, prosthetic limbs or communication gadgets. Here, we describe how BCIs function and selectively review its components of signal acquisition, decoding, and feedback. Understanding these mechanisms will shed light on both the technical challenges and the remarkable potential that BCIs hold for various applications in assistive technology and rehabilitation.

3.1 Brain Signals and Neural Communication

The basis of BCI lies in the brain's capacity to generate electrical activity and display hemodynamic responses. These signals are directly or indirectly produced by neurons. All neurons have the ability to generate electrical signals, known as action potentials, which are responsible for transmitting signals within the neuron. Neurons connect with other neurons chemically, forming synapses. At pre- and post-synaptic sites, due to the differential release of chemicals, changes occur in the local field potential, which generate pre- and post-synaptic potentials, primarily measured through EEG. Since functional neurons require more oxygen to operate, changes in the local oxygen and deoxygenation levels occur, which can be captured using fMRI and fNIRS. Neuronal units function as part of a network, forming neural pathways that enable motor activities, sensing, thinking, and feeling. BCIs operate by recording such brain signals from areas in the brain. For example, when a person imagines moving their hand, the neural circuits in the motor cortex produce some unique firing patterns. While the patterns of neural activity are being recorded and decoded, the BCI is able to deliver the intention to move in terms of an external action in a device such as moving a cursor on a display or controlling a robotic limb.

3.2 Types of Brain Signals Used in BCIs

There is the possibility of using different types of brain signals in BCI systems with varying pros and cons [11]. The most commonly used brain signals in BCIs include:

- **Magnetoencephalography (MEG)**: MEG records the magnetic fields that are generated by the neurons' activity. Temporal and spatial resolution is also high which allows you to accurately map the activity within the brain. However, MEG systems are large, costly and not ideal for routine BCI use due to the reasons described above.
- **Electroencephalography (EEG)**: Among all the non-invasive procedures of recording the brain signals, EEG is commonly employed. This method involves use of electrodes on the scalp to record electrical activity produced by large numbers of neurons. EEG signals are best for use in BCIs because they are non-invasively acquired and having a good temporal resolution. Further, they are easy to record and give real-time results. However, EEG signals can only be measured at the surface of the scalp and so they provide less detail about the activity arising from particular regions of the brain. Despite this limitation, EEG has received a lot of attention because it is safe, easy to use, and readily available.
- **Functional Near-Infrared Spectroscopy (fNIRS)**: fNIRS is a non-invasive technique that assesses oxygen supply of the brain. When neurons are activated, blood flow to the region increases and this can be measured using infrared light [16]. As compared to EEG; fNIRS, has a higher spatial resolution compared to EEG, but has a slow temporal response because blood flow changes are slower than electrical signals; generally, fNIRS is combined with other BCI systems to improve signal analysis.
- **Electrocorticography (ECoG)**: ECoG records electrical activity in real-time and involves the use of electrodes that are placed beneath the skull. This invasive method is superior to EEG in terms of spatial resolution and signal quality because it records electrical activity directly from the brain cortex. ECoG is somewhat invasive compared to EEG and fMRI but it is employed clinically for studying brain activity before operations, as it offers a possibility to localize the activity with high accuracy.
- **Local Field Potential (LFP)**: Field potentials are electrical signals recorded at the extracellular space of the brain tissue. These signals capture the electric activities generated by the collective activity of groups of neurons and are typically recorded using microelectrodes implanted in or near the brain tissue. FP recordings provide high spatial and temporal resolution, making them ideal for studying localized neural circuits and precise motor commands (Table 1).

3.3 The BCI Working Process: From Signal Acquisition to Feedback

The functioning of a BCI system can be broken down into three main stages: signal acquisition, signal processing (including feature extraction and classification), and the generation of feedback [11]. Each stage plays a crucial role in transforming raw brain signals into meaningful actions that the user can control (Fig. 1).

Table 1 Brain signal in brain computer interface

Technique	Description	Spatial resolution	Temporal resolution	Invasiveness	Applications and limitations
MEG	Records magnetic fields generated by neuronal activity	High	High	Non-invasive	Accurate brain activity mapping: impractical for routine use due to large size and high cost
EEG	Records electrical activity from the scalp using electrodes	Low	High	Non-invasive	Easy to use, real-time results, and widely available; limited detail on specific brain regions due to surface-level recording
fNIRS	Measures brain oxygen supply using infrared light to detect blood flow changes during neural activity	Higher than EEG	Slow	Non-invasive	Less sensitivity to movement; limited by slow response due to blood flow dynamics
ECoG	Measures electrical activity using electrodes placed beneath the skull	Very high	High	Moderately invasive	Superior signal quality; used to measure localize brain activity
Local field potential (FP)	Captures electrical activity from groups of neurons using implanted microelectrodes	Very high	Very high	Highly invasive	Ideal for studying localized neural circuits and fine motor control

3.3.1 Signal Acquisition

The first step found in any BCI system is signal acquisition. This concerns obtaining the brain signals by the correct procedure depending on the type of BCI, as well as the signals which will be recorded, i.e. EEG, fNIRS etc. In the case of non-invasive BCIs, electrodes are fixed on the scalp while in invasive BCIs they are placed deep in the brain. They detect electrical activity or monitor other physiological changes like flow. Once these signals are captured, they are transmitted to a computer for further processing. When the signal is acquired, they might contain various kinds of noise depending upon the nature of the signals, the source of the noise can be different. Brain signals are often weak and can be affected by various factors, such as muscle movement, external electrical devices, and other sources of noise. One of the key challenges at this stage is minimizing noise and interference. Signal pre-processing

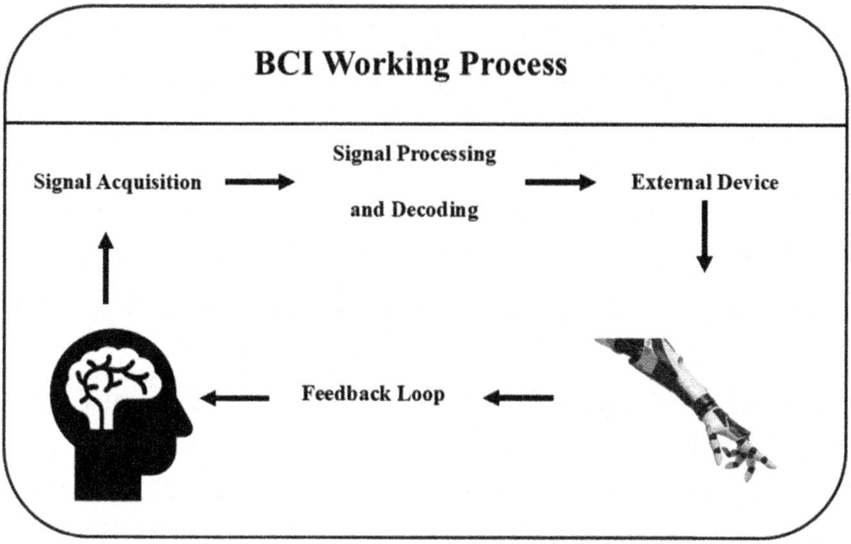

Fig. 1 Three stages of brain computer interfaces for assistive and rehabilitation applications: (1) signal acquisition (2) signal processing (3) feedback loop

techniques, such as filtering, are employed to clean the data and improve the quality of the recorded signals.

3.3.2 Signal Processing and Decoding

After the brain signals are acquired, they must be processed and decoded to extract meaningful information [17]. This step involves several sub-processes:

Feature Extraction: The raw signals hold a lot of information which can be relevant or irrelevant to the brain activity. In fact, most of the data are not specifically relevant to the problem at hand. Feature extraction can be understood as the act of separating the relevant signal segment in order to obtain a representation of the user's intent. For example, in motor-imagery based BCI, the system may filter selected frequency bands of EEG signals that are related to mental imaging of motion.

Classification and Prediction: After the extraction of relevant features, what follows next is the categorization of the brain signal with reference to the user's intention. In this step, the signals which correspond to different actions are distinguished with the help of the features by the means of machine learning algorithms. However, it is important to consider that the BCIs features are non-stationary and rapidly vary over time. This results in a reduction of signal quality, which can make it more challenging to predict the exact meaning of the state with high accuracy [18]. Therefore, the use of adaptive machine learning models and personalized-based models is crucial to address these challenges. These approaches dynamically adjust to the user's unique and evolving neural patterns, ensuring that the system remains effective and intuitive.

Translation: After classification, the system translates the brain signals into concrete actions. For example, if a BCI system detects that a user is imagining moving their right hand, it may translate that signal into the movement of a robotic arm or cursor in a specific direction [17].

Recent advancements in AI modelling have shown great promise in improving signal processing and decoding for BCIs [19]. One of the biggest challenges in the field is losing signals because of the low data quality, but generative AI offers a solution by synthesizing signals based on existing data, effectively augmenting the available datasets [20, 21]. AI is also making strides in enhancing signal resolution, helping to extract more precise and reliable information. Additionally, transfer learning has emerged as a powerful tool for addressing inter-subject variability, enabling models to adapt more effectively across different individuals.

3.3.3 Feedback Loop

The feedback stage of a BCI system is the last stage of the procedure of this system. Additionally, feedback in the BCI system is also important for the user. It can be near or far from the user, depending on the application and it can include forms of visual, auditory or even haptic feedback. For instance, if a user is using a BCI to control a robotic arm, the user may get the visual feedback when the arm is moving along with the brain signal or he may get the sensory feedback if the robotic arm touched something [17]. The feedback loop is a key part of BCI systems because it enables the user to change the strategy in response to the system's response. This simply implies that over time, users can develop the skills to gain better control over the system by changing how it encodes the brain signals. This learning process is known as 'co-adaptation', by which the user and the BCI system learn to perform optimally.

4 Types of BCIs Techniques

The BCI techniques can be broadly categorized into two types based on how the brain signals are acquired: (1) invasive and (2) non-invasive BCIs. Although these two approaches have the common goal of facilitating Communication between the brain and external devices, they achieve this through very different mechanisms and have their own advantages and disadvantages. This section describes both invasive and non-invasive BCI techniques and explores their strengths, weaknesses, and potential applications (Table 2).

Table 2 Type of brain computer interface

BCI type	Description	Advantages	Limitations
Invasive BCIs	Involves the surgical implantation of electrodes directly into the brain for high-precision signal recording	High spatial resolution: direct access to neural signals ability to perform complex tasks, such as controlling robotic limbs	Surgical risks (infection, bleeding); limited device longevity due to immune response; ethical and regulatory concerns
Non-invasive BCIs	Use external devices (e.g., scalp electrodes) to detect brain signals without requiring surgery	Safer and more accessible; lower risk of complications; easier and faster development for broad applications	Lower signal quality due to noise and tissue interference; limited control capabilities; training requirements for noise handling and signal interpretation

4.1 Invasive BCIs

Invasive BCIs involve the surgical implantation of electrodes or other devices directly into the brain. This method allows for high-precision recording of neural activity, as the electrodes are placed in close proximity to the neurons themselves [22]. The main advantage of invasive BCIs is their ability to capture detailed, high-resolution signals that provide precise control over external devices. The invasive BCIs can be performed by placing the electrodes on the surface of the brain, just beneath the skull but above the cerebral cortex or by implanting microelectrode arrays directly into the cerebral cortex, known as electrocorticography (ECoG) and Intracortical BCIs respectively [23]. Since the electrode is placed beneath the scalp in ECoG, it offers a middle ground between intracortical BCIs and non-invasive EEG systems. It provides better signal quality than non-invasive BCIs without the same level of risk associated with intracortical implants. EcoG-BCIs are used in both research and clinical applications, particularly for mapping brain activity before surgeries. In intracortical BCIs, microelectrode implants directly into the cerebral cortex allowing for extremely precise recording at the activity of individual neurons [24].

4.1.1 Advantages of Invasive BCIs

High Spatial Resolution: Invasive BCIs are able to deliver spatial specificity even with the ability to record from one or a few neurons [24]. This makes it possible to decode the simplicity of the signal in the brain and control the movement of devices outside the body in the first place.

Direct Access to Neural Signals: As invasive BCIs implant electrodes into targeted regions of the brain without affecting the skull and other tissues, signal quality is of high quality as it avoids interference through noise and other contamination from the outside. This makes them perfect for applications like operations that need real time control like operation of robotic limbs.

Potential for Complex Tasks: The dense stream of information that an invasive BCI provides to the user includes allowing one to manipulate outside equipment. For instance, intracortical BCIs have been successfully applied to control dual degree robotic limbs schematically, where patients can play out diverse operations such as grasping, lifting objects and performing detailed work with nice precision [15, 25].

4.1.2 Limitations of Invasive BCIs

Surgical Risks: The invasive BCIs need electrodes or sensors to be inserted in the brain, which can be done via surgical procedural. This procedure has many risks associated with it such as infection, bleeding and injuries to the rest of the brain matter. 'When used as permanent devices in the body for a long term, the electrodes might cause scarring of tissues around them,' or 'Long term recording causes the quality of the recorded signals to degrade round the electrodes.'

Limited Longevity: Longevity of the implanted devices is one of the vital challenges posed by invasive BCIs. The immune system can build up resistance around the electric electrodes, which over a period of time, begins to corrode [26]. Some efforts are being made to create matte and coatings that are biocompatible to improve the lifespan of these devices [27].

Ethical and Regulatory Issues: Concerns about invasive BCIs include Consent, Privacy and presence of harm over a long period of time. Legal control is important in order to guarantee that invasive BCI technologies are secure for people and that potential drawbacks are excluded.

4.2 Non-invasive BCIs

Non-invasive BCIs do not require surgery and instead use external devices, such as electrodes placed on the scalp, to detect brain signals. These systems are more widely used and accessible due to their non-intrusive nature. However, they generally provide lower signal quality, and less precision compared to invasive BCIs. The EEG, and fNIRS are the most commonly used techniques for the non-invasive. The EEG records brain signals through electrodes placed on the scalp. The fNIRS measure changes in blood oxygenation levels in the brain by passing the infrared light through the scalp and detecting how much it is absorbed by the blood.

4.2.1 Advantages of Non-invasive BCIs

Safety and Accessibility: Due to the non-invasive nature of these techniques are safer and do not requires any surgical intervention therefore can be used by the broader population. This also make EEG to be used for the experimental and clinical used for the temporary basis.

Lower Risk of Complications: Since the non-invasive BCIs do not require surgery and therefore the user is not at risk of developing complications such as infection or a damaged brain tissue. This may well make them a more attractive proposition for many patients and researchers out there.

Rapid Prototyping and Development: Noninvasive BCIs are also easier to create, debug, and modify to ones liking than invasive systems. This has resulted in a broader application area and faster technological advancement in application, including neurofeedback, rehabilitation and brain-computer gaming.

4.2.2 Challenges and Limitations of Non-invasive BCIs

Lower Signal Quality: The major drawback of non-invasive BCIs is that the recorded signals in the scalp are smaller and contaminated with noises in comparison with those recorded directly from the neurons. The skull and other tissues attenuate the electric fields, making the recorded signals weak, low on resolution and noisy. Hence, non-invasive BCIs are generally characterized by lower spatial resolution and lower accuracy of the decoding of brain activity.

Limited Control Capabilities: Since the quality of the non-invasive signals are not good compared to the invasive, which leads to coarser control over external devices than with invasive systems.

Training Requirements: As non-invasive BCIs signals contains more noise than the invasive signals, a proper skill is requiring to deal with various types of the noise for their detection and removal. Some of the non-invasive BCIs demand a set up procedure that trains the users on how to produce the desired brain signals that control the system. This can be tiresome and often leads to experiencing unreliable performance, especially with the user with low cognitive or motor skills.

4.3 Choosing Between Invasive and Non-invasive BCIs

The decision to use an invasive or non-invasive BCI depends on several factors, including the specific application, the user's medical condition, and the level of control required. Invasive BCIs are generally preferred in cases where high precision and direct brain control are necessary, such as for neuroprosthetics or restoring movement in individuals with paralysis. However, due to the surgical risks involved, invasive BCIs are typically only considered for patients with severe disabilities or in clinical research settings. Non-invasive BCIs, on the other hand, are more practical for everyday use and are suitable for a wider range of applications. They are commonly used in assistive technologies for Communication, rehabilitation, and neurofeedback training. Non-invasive BCIs are also increasingly being explored in consumer applications, such as gaming and virtual reality, where safety and ease of use are paramount.

5 BCIs for Assistive Applications

BCIs have emerged as a transformative technology for individuals with severe motor impairments or other physical disabilities. By establishing a direct communication channel between the brain and external devices, assistive BCIs offer new avenues for interacting with the environment, enabling users to perform tasks that were previously impossible due to their disabilities. In this section, we explore the key applications of BCIs in assistive technology, particularly focusing on communication systems, mobility aids, and control of prosthetic limbs.

5.1 The Role of BCIs in Assistive Technology

Assistive BCI has emerged as a novel technology to improve the quality of life of people with lock-in syndrome and other severe motor disability conditions [28]. These conditions often result in loss of capabilities to communicate, entertain, move or even perform day-to-day activities. Assistive BCIs create new possibilities to provide a solution to this problem by bypassing the damaged or non-functional motor pathway. Instead, control signals can be obtained directly from the brain and applied to external devices. The primary concept of assistive BCIs is to independently command devices using only one's mind. For, instance, patients with Amyotrophic Lateral Sclerosis (ALS), who cannot talk or move anymore can use BCI to control the communication aid or a computer which allow them to relay their thoughts and communicate [1, 23]. Likewise, via implanted devices, patients with spinal cord injuries can use BCIs to control wheelchairs or robotic replacements gaining back some measure of normality in life.

5.2 BCI-Based Communication Systems

The greatest usefulness of assistive BCIs is in the realm of Communication. Patients with ALS or LIS who have lost voluntary control of muscles can particularly benefit from a BCIs which allows them to communicate directly through brain signal. These are specifically important for patients in the late-stage ALS when even the muscles used in speech production are compromised by the disease.

5.2.1 EEG-Based Communication

The most common approach in applying assistive BCI is by using EEG. EEG records electrical signals in the brain through electrodes connected to the scalp. EEG-based communication systems assist users in choosing letters/words displayed on the computer screen just by focusing their attention on specific items. These systems rely on event-related potentials (ERPs) or Steady-State Visual Evoked Potentials (SSVEPs) to monitor the user's intention.

Event-Related Potentials (ERP): ERPs are the brain responses elicited by specific stimuli which may be a light flash or slight change in environment. For BCI communication system, its interface displays a set of letters arranged in a grid and each letter is flashed on the screen in a sequence. When the target or desired letter is highlighted, the brain generates a unique ERP which the EEG System picks and proceeds to select the letter. A common ERP used in assistive BCI is P300 brain wave. P300 typically is a positive deflection in the EEG signal that is elicited approximately 300 ms after the presentation of the target stimulus [29, 30].

Steady-State Visual Evoked Potentials (SSVEPs): SSVEPs are a steady state visually evoked potentials which are elicited when individual is attending to a flickering visual stimulus at a specific frequency. Thus, in a SSVEP based communication system, every letter or symbol on the screen visually flickers at a specific frequency. If the user focuses on a specific letter, he or she elicits an SSVEP at the corresponding frequency which is recognizable by the chosen BCI to determine the letter [31].

These systems have been used in providing basic communication for those people with severe motor disability, and these people are able to build, sentences, messages or control other digital goods. Yet, they generally need to be first trained and, thereafter, fine-tuned for improving their performance and wise depends on signal quality, comprehending capacity of the user, etc.

5.2.2 fNIRS-Based BCI Communication

Besides the EEG-based systems, research has been conducted to find novel methods to improve the BCI communication. For instance, some of the systems employ fNIRS to quantify the changes in blood oxygenation level in the brain when the user is asked to pay attention to specific letters. However, fNIRS have lower temporal resolution than EEG but has high spatial resolution. The EEG, on other hand, has low spatial resolution but high temporal resolution. The limitations of both modalities lead to new multi-modal system, known as the hybrid EEG-fNIRS BCI approach which significantly improves the system in terms of accuracy and stability [16].

5.2.3 Other Approaches to BCI Communication

Another emerging approach is utilizing direct intracortical BCIs with the help of microelectrodes, placed directly into the cortex to record activity of individually cortical neurons. While this approach offers higher speed and accuracy, these are invasive solutions and require surgery to implant microelectrodes and are still prototypes.

5.3 *BCIs for Mobility and Wheelchair Control*

Another important application of assistive BCIs is to help with mobility impairments, especially among people with spinal cord injuries, muscular dystrophy, or related problems. BCIs can also be integrated with powered mobility and wheelchair systems, allowing a user to drive the wheelchair by using brain signals. Using BCIs that are based on EEG signals, patients can command the movement of wheelchairs or air conditioning by just thinking about it, giving a degree of freedom to severely disabled people [32]. This technology provides opportunities for personal autonomy for people who otherwise would require assistance for their mobility and other needs.

5.3.1 EEG-Based Wheelchair Control

Similar to communication systems, EEG is the most used technique in BCI controlled wheelchairs today. EEG based system uses motor imagery where user imagines a movement such as imagining moving their hand or foot. Such motor imagery generates a corresponding brain activity in the motor cortex. The BCI system records this activity and coverts the signal into mental command such as move forward, turning left or stop and transmit these commands to the wheelchair to execute the desired movement [33]. For the BCI-controlled wheelchair movement the subject has to be trained on how to use his brains to issue specific commands in order to execute certain movement. For instance, imagining a hand movement to move wheelchair forward and imagining the foot motion to turn the chair. A better precision and accuracy can be achieved through practice, although performance may vary based on factors like fatigue, concentration, and environmental noise [34].

5.3.2 Hybrid Approaches to Wheelchair Control

BCI systems that incorporate EEG information with other sorts of sensors like eye trackers or EMG provide better control over the mobility devices. For instance, the eye-tracking technology can be utilized to identify the direction of the user's gaze for direction control; and EEG signal for acceleration or braking [35]. The use of this multimodal approach decreases the user's cognitive load and increases the accuracy and efficacy of the system, especially in demanding or rapidly changing conditions. Another promising technology is the integrating BCI systems with the autonomous navigation for a mobile robot system [36]. The integrated system dynamically adjusts the fusion of the BCI for gross-motor control signals such as 'go forward' or 'turn left', while wheelchair itself constructs environmental maps including path planning and obstacle detection. This hybrid approach enables a dynamic and user-friendly approach to reduce cognitive load on user with significant mobility and cognitive impairments.

5.4 Control of Prosthetic Limbs with BCIs

One of the remarkable applications of assistive BCI is its ability to control prosthetic limbs. The prosthetic limb is an artificial limb that replaces an amputated or injured limb and is traditionally controlled by mechanical switch or muscle contraction. These techniques of control can be limiting and require a significant effort by the user. BCI brings the possibility of commanding an artificial limb using their thoughts. The BCI system analyzes the user's neural impulses and convert them into commands for the prosthetic limb [37].

5.4.1 Invasive BCIs for Neuroprosthetics

Invasive BCIs are those that involve the implantation of the electrodes directly in the brain eliminating the need of manual control and providing the greatest level of control of prosthetic limbs. BCI technology is based on the finding that movement-related activity can be recorded directly from the motor cortex after paralysis or other impairments affecting the spinal cord, nerves or muscles. Due to the interface of the BCI with the motor cortex, such systems can provide alternate "neural pathway" can record the activity in individual neurons and thereby reproducing those signals in the form of highly detailed movement commands for the prosthetic limb. For instance, the intracortical BCIs have created an opportunity for people with tetraplegia to control robotic arms enabling them to pick objects, grasp and manipulate them [25, 38]. These systems functions by mapping specific neural activity in the motor cortex to specific movements of the prosthetic limb. Once the users get familiarized with the system, they improve precision as well as smoothness during the movements. However, invasive BCI prosthetics also come with several challenges like surgical risks, device maintenance and stability [39].

5.4.2 Non-invasive BCIs for Prosthetic Control

Non-invasive BCIs like those using EEG are also effective in controlling prosthetic limbs but provide less precision as compared to invasive systems. EEG-based control of prosthetics is usually achieved using motor imagery, in which the user's imagines specific movements to produce required brain potentials for controlling the prosthetic. Although these systems are relatively safer and present fewer challenges than invasive BCIs, the performance of these systems is relatively slow and less accurate compared to invasive BCIs. Researchers are currently trying to enhance non-invasive BCIs for prosthetic control by working with hybrid BCIs which integrate EEG with EMG signals from residual muscles or sensory feedback from the prosthetic. These approaches are intended to improve the user's ability to control prosthetics with better efficiency [40].

6 BCIs for Rehabilitation

Besides helping those with disability, BCIs are increasingly finding applications in rehabilitation especially in people with neurological injury due to stroke, TBI or Spinal cord Injury. By utilizing neuroplasticity, or the brain's capacity to reorganize itself after an injury, BCIs play a transformational role in the field of neurorehabilitation. Rehabilitation BCIs are developed to stimulate the brain's circuits to encourage the rewiring of the brain and restore the lost functionality. In this section, the applications of BCIs are discussed with specific reference to motor rehabilitation following stroke, neuroplasticity and the combination with conventional rehabilitation interventions.

6.1 Motor Rehabilitation After Stroke

Stroke is one of the leading causes of motor disability worldwide. Conventional forms of rehabilitation approach are to use physical therapy that encourages the patient to perform certain movements with the affected limbs to help restore the motor function. However, these are partially effective, and recovery can be slow or incomplete. BCIs present a new and highly effective system of stroke rehabilitation, which works to stimulate the motor cortex of the human brain to help restore motor coordination. In BCI-based rehabilitation, patients use their motor imagery to control the external apparatus, such as a robotic arm or exoskeleton [41]. The brain signals are filtered and processed to selectively extract the target brain activity, and the external device is later controlled accordingly. The BCI rehabilitation systems also feature real-time feedback providing users with information about their performance. A sensory, auditory of visual feedback about the performance allows the patients to adapt their brain activity patterns and, in such a way, that "retrain" their brain to perform the intended motor function.

There are studies which show that BCI-driven motor rehabilitation can translate into a significant increase in upper limb function in stroke patients, even years after the initial injury [42]. In addition, BCIs are used in combination with spinal cord injury

(SCI) to allow people suffering from immobility and loss of motor control use the BCI to bypass damaged spinal circuits and directly trigger movements based on these intentions by controlling a prosthetic device. These interventions may improve motor outcome through enhanced neuroplasticity [43]. Ramos-Murguialday and team [42] demonstrated that chronic stroke patients improved significantly in motor function after the use of a BCI controlled robotic orthosis compared to those who received conventional therapy. This observation thus establishes the theoretical possibility of brain-computer interface recovering function even decades after the initiating lesion.

6.1.1 EEG-Based Motor Rehabilitation

The most of the BCIs based on EEG are applied in stroke rehabilitation to restore the motor functions of the affected limbs. Recent research literature have highlighted EEG-based BCIs as one of the most promising technique for the rehabilitation of stroke survivors [44]. Most of these systems operate based on motor imagery in which the patient requires to imagine the movement of the affected limb. The BCI system then offers immediate feedback to the patient via signals, be it in visual form such as showing a virtual limb moving on the screen or physically presenting cues such as activating assistive robotics to move the functional body limb. This feedback loop assists the patient to re-engage neural networks responsible for motor motivation and helps in boosting connectivity between actual and intended motor activity. Eventually with continued practice, the process can restore the damaged motor pathways and improve a patient's ability to control their limbs. For example, Zhan et al. used functional electrical stimulation (FES) an electrical stimulation method for stimulating muscles using electrical currents, causing them to contract and produce movement [45]. They then coupled FES with BCI, when BCI detected the motor intention (e.g., imagining hand movement), it activated the FES device which in turn elicits hand movement. This pairing of upper brain control and movement rehabilitation training resulted in clinically significant improvements in motor function of patients with chronic stroke.

6.1.2 Benefits and Challenges of BCI-Based Motor Rehabilitation

BCI-based motor rehabilitation offers several advantages over traditional rehabilitation methods. BCI-based motor rehabilitation offers enhanced engagement by providing a realtime feedback enabling patients to be able to visualize the outcome of their work immediately. This as a positive reinforcer and will keep the patients motivated to continue with their therapy [46]. BCIs also allows precise targeting of affected motor pathways making rehabilitation process more effective. As BCIs capture the motor intentions using mental imagery they can train the brain to re-establish connection between the brain and muscles even when the neural connection is compromised [47]. Another advantage of BCI-based neurorehabilitation is the ability to implement rehabilitation even when the patient is unable to perform voluntary movements during early phases of recovery. The early intervention is crucial for stroke recovery, as this window is often associated with better long-term outcomes.

However, there are a number of challenges as well, which are associated with BCI systems, which include:

Training Requirements: In order to use the BCI systems the patients are required to undergo training, particularly those requiring motor imagery. This can be very challenging to some people, especially those with cognitive disabilities or severe neurological injuries.

Variability in Outcomes: The extent to which patients with stroke benefit from BCI based rehabilitation also varies from patient to patient. The effects of the therapy depend on where and to what extent the brain has been damaged, cognitive abilities of the patient as well as motivation to go through the therapy.

Cost and Accessibility: BCI systems, especially, coupled with robotic or electrical stimulation devices can be expensive and may not even be readily available. Therefore, there is need for more research to develop cost-effective solutions which can be implemented at a wider scale.

6.2 Neuroplasticity and BCI-Assisted Recovery

Neuroplasticity is a key concept in BCI based rehabilitation, defined as the brain's ability to reorganize itself by forming new connections. Neuroplasticity plays a crucial role in recovering from any kind of brain injury. If some part of the brain is injured, other, healthy parts of the brain can be rewired to handle the functions of the injured or damaged brain areas. BCIs enable neuroplasticity by providing consistent feedback that re-enforces specific pathways in the brain. For instance, as discussed in the previous sections when a stroke patient tries to imagine moving the affected limb and BCI provides feedback in real time inform of a virtual movement on the computer screen, the brain starts to forge new connections between the intention of the motor movement and actual movement. With practice a period of time will lead to functional gains even if conventional therapy response has been poor.

Hebbian Learning in BCI-Based Rehabilitation

BCI-based rehabilitation often relies on the principle of Hebbian learning, which is reflected in the concept "cells that fire together, wire together". This principle suggests that the repeated activation of neural circuits strengthens the connection between neurons, making them more likely to fire together. So, during rehabilitation, this means that when patients repeatedly attempt to engage motor cortex through mental imagery using BCI training, they strengthen the neural circuits involved in motor control. This repeated activity reinforces neural pathways associated with the intended movement, leading to improved movement and promoting recovery. BCI systems also accelerates this process of recovery by providing real-time feedback, ensuring that the correct circuits are activated during each training session.

6.3 BCIs and Cognitive Rehabilitation

Besides motor rehabilitation, new protocols in BCIs are being utilized for cognitive rehabilitation. Cognitive rehabilitation aims to improve cognitive skills such as problem solving, attention, and memory that often impaired in individuals after traumatic brain injuries (TBI), neurodegenerative diseases, or other conditions that affect cognitive function.

6.3.1 Cognitive Training with BCIs

BCIs can be utilized to study the brain's activity during certain cognitive tasks and can also offer real time feedback to help improve their performance. For example, EEG based BCIs can detect the attention of the patient declining during an exercise and provide feedback to the patient to sustain focused attention. Gradually, through reinforcing neural circuits associated with attention to improve cognitive function by strengthening the circuits representing attention and other cognitive processes. A widely practiced method of cognitive rehabilitation with BCIs is neurofeedback training, through which patients receive realtime feedback of their brain activity, while performing a cognitive task. The feedback help them to learn to regulate their brain activity and help to improve their cognitive functions [48]. Another important area, in which BCIs are currently under research, is memory rehabilitation. For instance, systems based on the EEG can be used to monitor patients' brain responses while performing various memory based cognitive tasks and feedback enhances their memory storage and retrieval functions [49].

6.3.2 Applications in Neurodegenerative Diseases

BCI-based cognitive rehabilitation holds a lot of potential for people with neurodegenerative disorders including Alzheimer's or Parkinson's sickness. These neurodegenerative conditions frequently lead to a gradual worsening of a patient's cognitive abilities, and in many cases the application of traditional forms of rehabilitation have proven to only be marginally effective in slowing the cognitive decline. BCIs offer a new approach by providing targeted, real-time feedback that can help preserve cognitive function for longer periods [50].

6.4 *Integration of BCIs with Traditional Rehabilitation Techniques*

BCIs are most effective when combined with other conventional forms of rehabilitation including physical therapy, occupational therapy, and cognitive behavioral therapy. This combined approach allows for a more comprehensive rehabilitation process, addressing both the physical and cognitive aspects of recovery. For instance, in motor rehabilitation after stroke, the use of BCIs can complement along physical therapy to improve the results of therapy. It allows patients perform some movements with the help of BCI—controlled robotic device, meanwhile engaging in conventional exercises to strengthen muscles and enhance coordination. This approach has been found more effective than either of the approaches used alone [51]. Likewise in cognitive rehabilitation, the BCIs can be employed to support cognition behavioral therapy by providing real time feedback of brain activity during therapy sessions. This enables the patients to learn to regulate their brain activity and improve cognitive functions effectively [52].

7 Multimodal Approaches in BCIs for Assistance and Rehabilitation

A multimodal BCI or hybrid BCI combines a BCI system with another system resulting in multiple input sources such as brain signals, muscle activity, eye tracking or any other bio-signal. This type of BCI offers improved precision, accuracy and reliability along with offering a larger number of action commands compared to single-mode BCIs. In this section, we discuss how Multimodal techniques improve the applicability of BCI in performance in assistive and rehabilitation technologies.

7.1 *The Rationale for Multimodal BCIs*

Each modality used in BCIs has its unique advantages and limitations, therefore using multimodal BCIs can complement various techniques to each other's. For example, a single modality based BCIs systems such as only EEG-based systems can pose problems of spatial localization of the signals and is very susceptible to the motions noise which leads to the variability in users' performance. The EEG signals also have the problem for its interference by muscle activity, movement and also other sources of the environment decrease the reliability of the system. Moreover, it is ill suited for decoding in terms of its electrical activity at high precision through noninvasive methods. These challenges are resolved in Multimodal BCIs because they use other signals that can help augment the brain signals giving better control of external devices. The EEG fused with EMG signals of muscles, or with gaze tracking as an additional or replacement input modality for Communication or motor control. Enabling different information sources, these systems can have better accuracy, less workload for the user, and improve overall user experience [53, 54].

7.2 *Combining EEG with EMG*

One of the most common multimodal approaches in BCIs is the integration of EEG with EMG, which measures electrical activity in the muscles. This combination is particularly useful in rehabilitation and assistive applications where residual muscle function is present. For prosthetic limb control, EEG might be employed to discern the user motor purpose, whereas EMG signals from the remaining muscles in the limb could also prove useful in determining movement patterns. For instance, in partial limb amputees, muscles of the residual limb can produce electrical signals associated with movement which can be measured by EMG. It is useful in rehabilitation and assistive applications where residual muscle function is present. For example, recently Jacob et al., 2019 proposed an Artificial Muscle Intelligence with Deep Learning (AMIDL) system that uses EEG signals and EMG control system to assist post-stroke patients in regaining control over paralyzed limbs [55]. By reducing the burden of exoskeletons and integrating gesture recognition for Communication, AMIDL enhances rehabilitation and quality of life for patients. Another important application of EEG-EMG integration is in motor rehabilitation after stroke or spinal cord injury. Using EEG signals, the BCI can detect the user's intention to move (via EEG), and EMG signals can trigger functional electrical stimulation (FES) to initiate actual movement [56]. This combined approach helps re-engage the neural and muscular pathways involved in motor control, promoting recovery and neuroplasticity.

7.3 *Combining EEG with Eye Tracking*

Eye tracking is another modality that is frequently combined with EEG in multimodal BCIs. Eye tracking evaluates the position of focus and is an easy and accurate method of controlling gadgets for disabled individuals with no fine voluntary motor control. Eye tracking is used in combination with EEG in the BCI-based communication systems to improve the selection of the letters or words. For instance, whereas an ordinary user can write a word on the keyboard, this user utilizes eye scanning to type a word by identifying the vicinity of the preferred letter on the keyboard. The EEG then confirms the user's intention by measuring P300 in response to stimuli associated with attention or via movement imagery. By combining eye tracking technologies with EEG, the rate of Communication is improved as well as reliability because the system detects and corrects for errors that originate from each input source [57, 58].

In mobility applications, such as BCI-controlled wheelchairs, eye tracking can be combined with EEG to improve navigation [59]. The user might use their gaze to indicate the desired direction (e.g., looking left to turn the wheelchair), while EEG signals provide confirmation or additional control (e.g., imagining a movement to move forward). This hybrid approach reduces the cognitive load on the user, as they do not need to rely solely on motor imagery, which can be mentally exhausting. Moreover, an important application of this multimodal approach is in working with people with ALS, locked-in syndrome, or other diseases that affect motor control but allow eye movement. For interaction, the chosen approach of eye tracking looks natural and intuitive, and the addition of EEG as an extra layer enhances the results and increases the system's efficiency [60].

7.4 Combining EEG with fNIRS

The integration of EEG with fNIRS can offer more extensive electrical as well as hemodynamic data about the brain to individual during cognitive and motor rehabilitation period where EEG offer a good temporal resolution but fNIRS offer a better spatial resolution as well as less susceptibility for the motion [61]. This dual-modality approach gives more accurate and long feedback to the patient and thus gives better location of the neural pathway to motor control. Hence, tracking both forms of signals can help clinician better understand the patient's brain activity to improve their rehabilitation strategy [62].

7.5 Multimodal BCIs and Virtual Reality Integration

Multimodal BCIs themselves are an exciting area of research, but one of the most interesting recent developments in this field is the possibility of interfacing them with virtual reality. VR allows one to have a high, engaging level of interaction which makes it suitable for assistive as well as rehabilitative applications, which implies that integrating BCIs with other input modalities is best done in VR.

In motor rehabilitation, VR environment can duplicate a real-life task and enable the patient to practice some movements. As a result, it is possible to combine BCIs with VR, for example, enhancing the patient's movement of virtual figures or even robotic models, using their brain signals as inputs; this offers a great deal of motivation and activity to practice different motor movements [48]. For instance, stroke patients can move his/her hand in a VR environment using the BCI to perform some simple tasks such as grasping or reaching. The VR system has both visual and tactile aspects and thus has the ability to strengthen the existing neural link between thought and action, which can stimulate plastic change and enhance motor function. The combination of EEG, EMG, and VR allows for more natural and intuitive control of the virtual environment, as users can generate both brain and muscle signals to perform complex tasks. This multimodal approach enhances the realism of the VR experience, making the rehabilitation process more effective and enjoyable for patients.

In addition, combining VR in BCIs can also be helpful for the cognitive rehabilitation, VR environments can simulate real-life scenarios that challenge memory, attention, and problem-solving skills. By integrating EEG and other modalities such as eye tracking or heart rate monitoring, VR-based BCI systems can provide real-time feedback on the patient's cognitive performance, helping them improve their skills over time. For example, Gomez-Pilar et al. developed a neurofeedback training tool for BCIs centered around motor imagery. Following NFT sessions, significant improvements were observed in three cognitive functions: visual perception, articulate expression, and immediate memory. Salisbury et al., 2016 demonstrated the potential for incorporating advanced technologies during the subacute phase of neurorehabilitation. Participants tolerated these modalities well, and they could be seamlessly integrated into both inpatient and outpatient rehabilitation programs without causing disruptions to their treatment schedules [63].

8 Roadmap and Outlook for BCI Research in Assistive and Rehabilitation Technology

The landscape of BCI in assistive and rehabilitation technology is rapidly evolving, with many significant advances expected over the coming years (Fig. 2). BCIs interprets neural activity and enables direct Communication between brain and external devices offering groundbreaking possibilities for individuals with motor and cognitive disabilities. However, for BCIs to become even more accessible and efficient, a number of technical and practical issues need to be resolved. Here are the key developments anticipated in future.

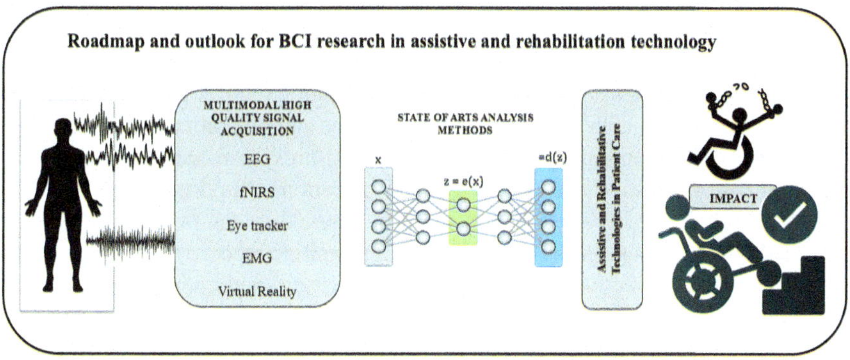

Fig. 2 Research roadmap for BCI in assistive and rehabilitation technologies

New advances are expected to be made regarding signal acquisition, enhancing the machine learning algorithms in the system, and the integration of multiple modalities. The immediate research directions focus on enhancing the precision and the validity of the non-invasive BCIs. The primary goal involves improving signal acquisition and processing methods mainly in EEG based systems. Interference and noise from extraneous factors such as muscle movement, eye blinks or other environmental conditions can be reduced with advances in electrode materials, sensor technology, and signal processing algorithms. These advancements are crucial for translating BCIs into more dependable and effective in real world environment applications. Furthermore, machine learning and artificial intelligence (AI) are expected to get significantly advanced which will enhance the accuracy and efficiency of decoding brain signals and make brain-computer interfaces better. Better real-time algorithms capable of adapting as per individual users will enhance BCI performance, making these systems faster and more accurate (Fig. 2). However, developing algorithms that can process brain signals in real-time remains a significant challenge, especially given the complexity and variability of neural activity. Furthermore, AI and machine learning rely heavily on large datasets to work effectively. Unfortunately, one of the biggest challenges in this field is the lack of sufficient data. While sharing data could help address this, it comes with its own set of problems, such as variability in datasets and concerns about privacy. The advancements in techniques like transfer learning and generative AI modeling can offer new solutions. For example, Wei and colleagues recently introduced a Sandwich meta-framework that addresses privacy issues in EEG data sharing [64]. Approaches like this not only help protect sensitive information but also provide a roadmap for models to work across different datasets. As these methods continue to evolve, they hold great promises for making BCIs more effective and widely applicable. Another immediate goal is the incorporation of BCIs with wearable and wireless systems and devices. The access to BCIs will improve as they become portable, less invasive, easier to adjust and install. Wireless Communication will enhance the applicability of BCIs in both home and clinical settings, hence enabling patients to interact with the technology even outside clinical settings. However, challenges remain in ensuring stable wireless connections and creating energy-efficient systems that do not require frequent recharging.

Over the subsequent years, research will probably pay attention to enhancing multimodal BCI systems. Integrated with technologies such as EEG, EMG, or eye tracking, will offer higher performance for assistive and rehabilitation technologies. Multimodal BCIs combining multiple inputs will provide more robust and accurate control for assistive and rehabilitation applications. Multimodal BCIs will be particularly effective in rehabilitation settings, where they can provide comprehensive feedback to patients working on motor or cognitive recovery. The same period is also expected to see expansion of BCIs with virtual and augmented reality for rehabilitation experiences that are more immersive and interactive. But it is very important to note that the use of multimodal interfaces might make the system more complex, could pose a barrier to widespread adoption of these technologies. The priority should be to simplify the user experience and reducing cognitive load. BCIs will also find more applications in cognitive rehabilitation, particularly for patients with Traumatic Brain Injuries or neurodenerative conditions such as Alzheimer's or Parkinson's disease. Cognitive rehabilitation poses unique challenges, as brain networks involved in cognition are more complex and less understood than those involved in motor function. BCIs can assess cognitive functions such as attention, memory, cognitive load or problem solving along with providing real-time feedback. Advances in technology using tools like fNIRS will mean that interventions can be aimed at specific areas of the brain, leading to more effective interventions.

In the long term, BCIs are likely to become more individualized and integrated into more general assistive and rehabilitative care. The key goal will be to integrate the most advanced technologies for more personalized approaches to cater individual demands and needs. The future BCIs will integrate stimuli-specific or subject-specific components by leveraging advancements in AI and neuroplasticity, offering highly tailored interventions that optimize performance and recovery outcomes. These personalized systems will bring improved rehabilitation outcomes for patients with a variety of neurological conditions. Moreover, newer possibilities will open up for with the convergence of other emerging fields of technology like nanotechnology, brain stimulation, and neurogenesis, to name a few. Such convergence will bring the possibility to address next even severe forms of neurological disorders.

At the moment BCIs is still in its infancy, these interfaces will likely transition from being experimental technologies to mainstream tools used in both clinical and everyday settings. But to realize this vision we need to address the current issues of cost, efficiency, and accessibility. Accessibility is also crucial, as the cost and complexity of current systems limit their availability in low-resource settings. More collaborative efforts between academia, healthcare providers, and industry are needed to create cost-effective, scalable solutions. As we progress with BCI technologies, the ethical frameworks surrounding BCI use, particularly regarding privacy, data security, and the autonomy of users, must be prioritized. It is crucial to ensure the privacy and security of neural data, especially for vulnerable populations such as those with cognitive impairments is essential to fostering trust; promoting responsible development and deployment of BCI technologies. Despite the challenges the future of BCIs specifically in assistive and rehabilitation technologies is promising and further research will lead to opportunities to life-altering technologies for those with motor disorders, learning disabilities and neurological disorders.

9 Conclusion

BCIs are one of the most significant innovations in the health care industry, leading to an increased level of functionality for patients care for assistive and rehabilitation purposes. BCIs can help paralyzed patients use their thoughts to speak and control lights or other devices. Recent years have seen developments in multimodal BCIs employing brain signals together with other signals such as EMG signals, eye tracking, and virtual reality to enhance its applicability in assistive technology and rehabilitation. The advances in BCI reflect an optimistic perception and hope of a better quality of life to people suffering with severe motor and cognitive disabilities. Looking ahead, we can expect more sophisticated, accessible, and personalized systems that can meet the diverse needs of patients with motor impairments, cognitive disabilities, or even speech disorders. Further research will pave the way for BCIs from being the experimental technologies to mainstream tool extending beyond traditional rehabilitation to encompass care. BCIs have the potential to revolutionize rehabilitation practices, improving the quality of life of millions living with disabilities to regain autonomy and control over their lives.

References

1. Wolpaw, J.R., Birbaumer, N., Heetderks, W.J., McFarland, D.J., Peckham, P.H., Schalk, G., Donchin, E., Quatrano, L.A., Robinson, C.J., Vaughan, T.M.: Brain-computer interface technology: a review of the first international meeting. IEEE Trans. Rehabil. Eng. **8**, 164–173 (2000). https://doi.org/10.1109/tre.2000.847807
2. Colucci, A., Vermehren, M., Cavallo, A., Angerhöfer, C., Peekhaus, N., Zollo, L., Kim, W.-S., Paik, N.-J., Soekadar, S.R.: Brain-computer interface-controlled exoskeletons in clinical neurorehabilitation: ready or not? Neurorehabil. Neural Repair **36**, 747–756 (2022). https://doi.org/10.1177/15459683221138751
3. Velasco-Álvarez, F., Fernández-Rodríguez, Á., Ron-Angevin, R.: Brain-computer interface (BCI)-generated speech to control domotic devices. Neurocomputing **509**, 121–136 (2022). https://doi.org/10.1016/j.neucom.2022.08.068
4. Chaudhary, U., Vlachos, I., Zimmermann, J.B., Espinosa, A., Tonin, A., Jaramillo-Gonzalez, A., Khalili-Ardali, M., Topka, H., Lehmberg, J., Friehs, G.M., Woodtli, A., Donoghue, J.P., Birbaumer, N.: Spelling interface using intracortical signals in a completely locked-in patient enabled via auditory neurofeedback training. Nat. Commun. **13**, 1236 (2022). https://doi.org/10.1038/s41467-022-28859-8
5. Simeral, J.D., Kim, S.-P., Black, M.J., Donoghue, J.P., Hochberg, L.R.: Neural control of cursor trajectory and click by a human with tetraplegia 1000 days after implant of an intracortical microelectrode array. J. Neural Eng. **8**, 025027 (2011). https://doi.org/10.1088/1741-2560/8/2/025027
6. Zotey, V., Andhale, A., Shegekar, T., Juganavar, A.: Adaptive neuroplasticity in brain injury recovery: strategies and insights. Cureus **15**, e45873. https://doi.org/10.7759/cureus.45873
7. Piccolino, M.: Luigi Galvani and animal electricity: two centuries after the foundation of electrophysiology. Trends Neurosci. **20**, 443–448 (1997). https://doi.org/10.1016/s0166-2236(97)01101-6
8. Ormerod, W.: Richard Caton (1842–1926): pioneer electrophysiologist and cardiologist. J. Med. Biogr. **14**, 30–35 (2006). https://doi.org/10.1258/j.jmb.2006.04-22
9. Tudor, M., Tudor, L., Tudor, K.I.: Hans Berger (1873–1941)—the history of electroencephalography. Acta Med. Croatica **59**, 307–313 (2005)
10. Jonge, L.D., Petrykiv, S., Fennema, J., Michielsen, P., Arts, M.: A tribute to Jose M.R. Delgado (1915–2011): the pioneer of electric brain-stimulation. Eur. Psychiatr. **41**, S768–S768 (2017). https://doi.org/10.1016/j.eurpsy.2017.01.1447
11. Kawala-Sterniuk, A., Browarska, N., Al-Bakri, A., Pelc, M., Zygarlicki, J., Sidikova, M., Martinek, R., Gorzelanczyk, E.J.: Summary of over fifty years with brain-computer interfaces—a review. Brain Sci. **11**, 43 (2021). https://doi.org/10.3390/brainsci11010043
12. Kennedy, P.R., Bakay, R.A.: Restoration of neural output from a paralyzed patient by a direct brain connection. NeuroReport **9**, 1707–1711 (1998). https://doi.org/10.1097/00001756-199806010-00007

13. Birbaumer, N.: Breaking the silence: brain-computer interfaces (BCI) for communication and motor control. Psychophysiology **43**, 517–532 (2006). https://doi.org/10.1111/j.1469-8986.2006.00456.x
14. Hochberg, L.R., Serruya, M.D., Friehs, G.M., Mukand, J.A., Saleh, M., Caplan, A.H., Branner, A., Chen, D., Penn, R.D., Donoghue, J.P.: Neuronal ensemble control of prosthetic devices by a human with tetraplegia. Nature **442**, 164–171 (2006). https://doi.org/10.1038/nature04970
15. Bockbrader, M.A., Francisco, G., Lee, R., Olson, J., Solinsky, R., Boninger, M.L.: Brain computer interfaces in rehabilitation medicine. PMR **10**, S233–S243 (2018). https://doi.org/10.1016/j.pmrj.2018.05.028
16. Naseer, N., Hong, K.-S.: fNIRS-based brain-computer interfaces: a review. Front. Hum. Neurosci. **9** (2015). https://doi.org/10.3389/fnhum.2015.00003
17. Iturrate, I., Chavarriaga, R., Millán, J.D.R.: General principles of machine learning for brain-computer interfacing. Handb. Clin. Neurol. **168**, 311–328 (2020). https://doi.org/10.1016/B978-0-444-63934-9.00023-8
18. Lotte, F., Congedo, M., Lécuyer, A., Fabrice, L., Arnaldi, B.: A review of classification algorithms for EEG-based brain-computer interfaces. J. Neural Eng. **4** (2007). https://doi.org/10.1088/1741-2560/4/2/R01
19. Eldawlatly, S.: On the role of generative artificial intelligence in the development of brain-computer interfaces. BMC Biomed. Eng. **6**, 4 (2024). https://doi.org/10.1186/s42490-024-00080-2
20. Zhang, R., Zeng, Y., Tong, L., Shu, J., Lu, R., Yang, K., Li, Z., Yan, B.: ERP-WGAN: a data augmentation method for EEG single-trial detection. J. Neurosci. Methods **376**, 109621 (2022). https://doi.org/10.1016/j.jneumeth.2022.109621
21. Kunanbayev, K., Abibullaev, B., Zollanvari, A.: Data augmentation for P300-based brain-computer interfaces using generative adversarial networks. In: 2021 9th International Winter Conference on Brain-Computer Interface (BCI), pp. 1–7 (2021)
22. Klaes, C.: Chapter 28—Invasive brain-computer interfaces and neural recordings from humans. In: Manahan-Vaughan, D. (ed.) Handbook of Behavioral Neuroscience, pp. 527–539. Elsevier (2018)
23. Shih, J.J., Krusienski, D.J., Wolpaw, J.R.: Brain-computer interfaces in medicine. Mayo Clin. Proc. **87**, 268–279 (2012). https://doi.org/10.1016/j.mayocp.2011.12.008
24. Brandman, D.M., Cash, S.S., Hochberg, L.R.: Review: human intracortical recording and neural decoding for brain-computer interfaces. IEEE Trans. Neural Syst. Rehabil. Eng. **25**, 1687–1696 (2017). https://doi.org/10.1109/TNSRE.2017.2677443
25. Bockbrader, M., Annetta, N., Friedenberg, D., Schwemmer, M., Skomrock, N., Colachis, S., Zhang, M., Bouton, C., Rezai, A., Sharma, G., Mysiw, W.J.: Clinically significant gains in skillful grasp coordination by an individual with tetraplegia using an implanted brain-computer interface with forearm transcutaneous muscle stimulation. Arch. Phys. Med. Rehabil. **100**, 1201–1217 (2019). https://doi.org/10.1016/j.apmr.2018.07.445
26. Gulino, M., Kim, D., Pané, S., Santos, S.D., Pêgo, A.P.: Tissue response to neural implants: the use of model systems toward new design solutions of implantable microelectrodes. Front. Neurosci. **13**, 689 (2019). https://doi.org/10.3389/fnins.2019.00689
27. Green, R.A., Lovell, N.H., Wallace, G.G., Poole-Warren, L.A.: Conducting polymers for neural interfaces: challenges in developing an effective long-term implant. Biomaterials **29**, 3393–3399 (2008). https://doi.org/10.1016/j.biomaterials.2008.04.047
28. Millán, J.R., Rupp, R., Müller-Putz, G.R., Murray-Smith, R., Giugliemma, C., Tangermann, M., Vidaurre, C., Cincotti, F., Kübler, A., Leeb, R., Neuper, C., Müller, K.-R., Mattia, D.: Combining brain–computer interfaces and assistive technologies: state-of-the-art and challenges. Front. Neurosci. **4**, 161 (2010). https://doi.org/10.3389/fnins.2010.00161
29. Arvaneh, M., Robertson, I.H., Ward, T.E.: A P300-based brain-computer interface for improving attention. Front. Hum. Neurosci. **12** (2019). https://doi.org/10.3389/fnhum.2018.00524
30. Mak, J.N., Wolpaw, J.R.: Clinical applications of brain-computer interfaces: current state and future prospects. IEEE Rev. Biomed. Eng. **2**, 187–199 (2009). https://doi.org/10.1109/RBME.2009.2035356

31. Reitelbach, C., Oyibo, K.: Optimal stimulus properties for steady-state visually evoked potential brain-computer interfaces: a scoping review. Multimodal Technol. Interact. **8**, 6 (2024). https://doi.org/10.3390/mti8020006
32. Cruz, A., Pires, G., Lopes, A., Carona, C., Nunes, U.: A self-paced BCI with a collaborative controller for highly reliable wheelchair driving: experimental tests with physically disabled individuals. IEEE Trans. Human-Mach. Syst. **51**, 109–119 (2020). https://doi.org/10.1109/THMS.2020.3047597
33. Birbaumer, N., Cohen, L.G.: Brain–computer interfaces: communication and restoration of movement in paralysis. J. Physiol. **579**, 621–636 (2007). https://doi.org/10.1113/jphysiol.2006.125633
34. Galán, F., Nuttin, M., Lew, E., Ferrez, P.W., Vanacker, G., Philips, J., Millán, J.D.R.: A brain-actuated wheelchair: asynchronous and non-invasive brain-computer interfaces for continuous control of robots. Clin. Neurophysiol. **119**, 2159–2169 (2008). https://doi.org/10.1016/j.clinph.2008.06.001
35. Ben Taher, F., Ben Amor, N., Jallouli, M.: A multimodal wheelchair control system based on EEG signals and eye tracking fusion. In: 2015 International Symposium on Innovations in Intelligent SysTems and Applications (INISTA), pp. 1–8 (2015)
36. Xu, B., Liu, D., Xue, M., Miao, M., Hu, C., Song, A.: Continuous shared control of a mobile robot with brain-computer interface and autonomous navigation for daily assistance. Comput. Struct. Biotechnol. J. **22**, 3–16 (2023). https://doi.org/10.1016/j.csbj.2023.07.033
37. Vilela, M., Hochberg, L.R.: Applications of brain-computer interfaces to the control of robotic and prosthetic arms. Handb. Clin. Neurol. **168**, 87–99 (2020). https://doi.org/10.1016/B978-0-444-63934-9.00008-1
38. Cajigas, I., Davis, K.C., Meschede-Krasa, B., Prins, N.W., Gallo, S., Naeem, J.A., Palermo, A., Wilson, A., Guerra, S., Parks, B.A., Zimmerman, L., Gant, K., Levi, A.D., Dietrich, W.D., Fisher, L., Vanni, S., Tauber, J.M., Garwood, I.C., Abel, J.H., Brown, E.N., Ivan, M.E., Prasad, A., Jagid, J.: Implantable brain-computer interface for neuroprosthetic-enabled volitional hand grasp restoration in spinal cord injury. Brain Commun. **3**, fcab248 (2021). https://doi.org/10.1093/braincomms/fcab248
39. Rapeaux, A.B., Constandinou, T.G.: Implantable brain machine interfaces: first-in-human studies, technology challenges and trends. Curr. Opin. Biotechnol. **72**, 102–111 (2021). https://doi.org/10.1016/j.copbio.2021.10.001
40. Nsugbe, E., Samuel, O., Asogbon, M., Li, P.: Phantom motion intent decoding for transhumeral prosthesis control with fused neuromuscular and brain wave signals. IET Cyber-Syst. Robot. (2021). https://doi.org/10.1049/csy2.12009
41. Cantillo-Negrete, J., Carino-Escobar, R.I., Ortega-Robles, E., Arias-Carrión, O.: A comprehensive guide to BCI-based stroke neurorehabilitation interventions. MethodsX **11**, 102452 (2023). https://doi.org/10.1016/j.mex.2023.102452
42. Ramos-Murguialday, A., Broetz, D., Rea, M., Läer, L., Yilmaz, O., Brasil, F.L., Liberati, G., Curado, M.R., Garcia-Cossio, E., Vyziotis, A., Cho, W., Agostini, M., Soares, E., Soekadar, S., Caria, A., Cohen, L.G., Birbaumer, N.: Brain-machine interface in chronic stroke rehabilitation: a controlled study. Ann. Neurol. **74**, 100–108 (2013). https://doi.org/10.1002/ana.23879
43. Ang, K.K., Guan, C., Phua, K.S., Wang, C., Zhou, L., Tang, K.Y., Joseph, G.J.E., Kuah, C.W.K., Chua, K.S.G.: Brain-computer interface-based robotic end effector system for wrist and hand rehabilitation: results of a three-armed randomized controlled trial for chronic stroke. Front. Neuroeng. **7** (2014). https://doi.org/10.3389/fneng.2014.00030
44. Simon, C., Bolton, D.A.E., Kennedy, N.C., Soekadar, S.R., Ruddy, K.L.: Challenges and opportunities for the future of brain-computer interface in neurorehabilitation. Front. Neurosci. **15** (2021). https://doi.org/10.3389/fnins.2021.699428
45. Zhan, G., Chen, S., Ji, Y., Xu, Y., Song, Z., Wang, J., Niu, L., Bin, J., Kang, X., Jia, J.: EEG-based brain network analysis of chronic stroke patients after BCI rehabilitation training. Front. Hum. Neurosci. **16**, 909610 (2022). https://doi.org/10.3389/fnhum.2022.909610
46. Wang, J., Wang, W., Hou, Z.-G.: Toward improving engagement in neural rehabilitation: attention enhancement based on brain–computer interface and audiovisual feedback. IEEE Trans. Cogn. Dev. Syst. 1–1 (2020). https://doi.org/10.1109/TCDS.2019.2959055

47. Baniqued, P.D.E., Stanyer, E.C., Awais, M., Alazmani, A., Jackson, A.E., Mon-Williams, M.A., Mushtaq, F., Holt, R.J.: Brain–computer interface robotics for hand rehabilitation after stroke: a systematic review. J. Neuroeng. Rehabil. **18**, 15 (2021). https://doi.org/10.1186/s12984-021-00820-8
48. Yang, S., Li, R., Li, H., Xu, K., Shi, Y., Wang, Q., Yang, T., Sun, X.: Exploring the use of brain-computer interfaces in stroke neurorehabilitation. Biomed. Res. Int. **2021**, 9967348 (2021). https://doi.org/10.1155/2021/9967348
49. Burke, J.F., Merkow, M.B., Jacobs, J., Kahana, M.J., Zaghloul, K.A.: Brain computer interface to enhance episodic memory in human participants. Front. Hum. Neurosci. **8** (2015). https://doi.org/10.3389/fnhum.2014.01055
50. Tayebi, H., Azadnajafabad, S., Maroufi, S.F., Pour-Rashidi, A., Khorasanizadeh, M., Faramarzi, S., Slavin, K.V.: Applications of brain-computer interfaces in neurodegenerative diseases. Neurosurg. Rev. **46**, 131 (2023). https://doi.org/10.1007/s10143-023-02038-9
51. Mane, R., Wu, Z., Wang, D.: Poststroke motor, cognitive and speech rehabilitation with brain-computer interface: a perspective review. Stroke Vasc. Neurol. **7**, 541–549 (2022). https://doi.org/10.1136/svn-2022-001506
52. Radončić, A., Hadžić, S., Laković, J.: Brain-Computer Interface Through the Prism of Modern Age (2024)
53. D'Mello, S., Kory, J.: Consistent but modest: a meta-analysis on unimodal and multimodal affect detection accuracies from 30 studies. In: ICMI'12—Proceedings of the ACM International Conference on Multimodal Interaction (2012). https://doi.org/10.1145/2388676.2388686
54. Vortmann, L.-M., Ceh, S., Putze, F.: Multimodal EEG and eye tracking feature fusion approaches for attention classification in hybrid BCIs. Front. Comput. Sci. **4** (2022). https://doi.org/10.3389/fcomp.2022.780580
55. Jacob, S., Menon, V., Al-Turjman, F., Vinoj, P.G., Mostarda, L.: Artificial muscle intelligence system with deep learning for post-stroke assistance and rehabilitation. IEEE Access 1–1 (2019). https://doi.org/10.1109/ACCESS.2019.2941491
56. Sarhan, S.M., Al-Faiz, M.Z., Takhakh, A.M.: A review on EMG/EEG based control scheme of upper limb rehabilitation robots for stroke patients. Heliyon **9**, e18308 (2023). https://doi.org/10.1016/j.heliyon.2023.e18308
57. Brunner, P., Joshi, S., Briskin, S., Wolpaw, J.R., Bischof, H., Schalk, G.: Does the "P300" speller depend on eye gaze? J. Neural Eng. **7**, 056013 (2010). https://doi.org/10.1088/1741-2560/7/5/056013
58. Kalika, D., Collins, L., Caves, K., Throckmorton, C.: Fusion of P300 and eye-tracker data for spelling using BCI2000. J. Neural Eng. **14**, 056010 (2017). https://doi.org/10.1088/1741-2552/aa776b
59. Ghasemi, S.: Empowering Mobility: Brain-Computer Interface for Enhancing Wheelchair Control for Individuals with Physical Disabilities (2024)
60. Ngo, T.D., Kieu, H.D., Nguyen, M.H., Nguyen, T.H.-A., Can, V.M., Nguyen, B.H., Le, T.H.: An EEG and eye-tracking dataset of ALS patients and healthy people during eye-tracking-based spelling system usage. Sci. Data **11**, 664 (2024). https://doi.org/10.1038/s41597-024-03501-y
61. Li, R., Yang, D., Fang, F., Hong, K.-S., Reiss, A.L., Zhang, Y.: Concurrent fNIRS and EEG for brain function investigation: a systematic, methodology-focused review. Sensors (Basel) **22**, 5865 (2022). https://doi.org/10.3390/s22155865
62. Maher, A., Mian Qaisar, S., Salankar, N., Jiang, F., Tadeusiewicz, R., Pławiak, P., Abd El-Latif, A., Hammad, M.: Hybrid EEG-fNIRS brain-computer interface based on the non-linear features extraction and stacking ensemble learning. J. Appl. Biomed. **43** (2023). https://doi.org/10.1016/j.bbe.2023.05.001
63. Salisbury, D.B., Dahdah, M., Driver, S., Parsons, T.D., Richter, K.M.: Virtual reality and brain computer interface in neurorehabilitation. Proc. (Bayl. Univ. Med. Cent.) **29**, 124–127 (2016). https://doi.org/10.1080/08998280.2016.11929386

64. Wei, X., Narayan, J., Faisal, A.: The Sandwich Meta-Framework for Architecture Agnostic Deep Privacy-Preserving Transfer Learning for Non-invasive Brainwave Decoding. Presented at the April 10 (2024)

Assessing Performance and Mechanics of the PUCP IArm Upper Limb Prosthesis

Estefany Macedo, Enzo Romero, Victoria E. Abarca, and Dante A. Elias

Abstract Access to upper limb prostheses in developing countries, including Peru, is constrained. Models with features like increased individual finger strength and greater degrees of freedom tend to be costly, compelling individuals to opt for less advanced passive prostheses or aesthetic options. This creates a significant necessity for affordable prosthetics that do not compromise essential functionality and allow users to perform daily activities effectively. An example of this development is the PUCP IArm prosthesis, a laboratory model designed and developed at the "Pontificia Universidad Católica del Perú" (PUCP, for its initials in Spanish). This functional and accessible prosthesis is manufactured using additive manufacturing technologies. This study aims to characterize the mechatronic properties and evaluate the functionality of the hand component of the PUCP IArm prosthesis. Evaluations included videogrammetric assessments, finger force measurements, distal tensile tests, static downward bending tests, and response latency measurements between gesture detection and actuation. The assessed prosthesis weighs 230 g and has a maximum load capacity of 1.83 N in pinch gestures and 19.05 N in cylindrical grip. Compared with myoelectric activation prosthesis models developed using digital fabrication techniques, the results obtained by the PUCP IArm prosthesis are promising and stand out.

Keywords Upper limb prosthesis · Myoelectric control · Mechatronic characterization

E. Macedo · E. Romero · V. E. Abarca (✉) · D. A. Elias
Pontificia Universidad Católica del Perú, Lima, Peru
e-mail: victoria.abarca@pucp.edu.pe

E. Macedo
e-mail: estefany.macedom@pucp.edu.pe

E. Romero
e-mail: enzo.romero@pucp.edu.pe

D. A. Elias
e-mail: delias@pucp.pe

J. Narayan and H. Gritli (eds.), *Design and Control of Rehabilitation Robots*, Studies in Systems, Decision and Control 585, https://doi.org/10.1007/978-3-031-86977-8_2

1 Introduction

The global prevalence of people with amputations is a significant public health concern, with the World Health Organization (WHO) estimating that approximately 1.5 million individuals undergo such procedures annually; however, access to prosthetic devices remains limited in developing countries, where only 5–15% of those in need can avail themselves of these assistive technologies [1]. This disparity underscores the pressing need to address accessibility issues, particularly in regions such as Peru, where a substantial portion of the population-approximately 485,211 individuals-faces challenges related to mobility or limb impairment; yet only a fraction, 11.4%, receive requisite physical therapies or treatments [2, 3]. The multifaceted nature of this accessibility gap requires diverse strategies and solutions, with a primary focus on cost reduction, expedited manufacturing processes, and ensuring that prosthetic functionalities meet minimum standards [1, 4, 5].

Hand prosthetics are crucial in restoring functional capabilities for individuals with limb loss, particularly in facilitating essential grasping movements for everyday activities. These movements encompass various grips, including powerful and precision grips, each essential for tasks requiring different degrees of force and accuracy [6]. Central to effective grip performance is the role of the thumb, which facilitates stability and precision, crucial for tasks such as lateral pinching [7]. Wrist functionality contributes to overall stability and functionality, but its incorporation into prosthetic designs remains limited due to inherent complexities [8].

An updated systematic review of prosthetic and robotic hands [8] identifies shared mechanical features, particularly the use of motors as actuators, with most devices utilizing 5–6 motors and an overall power consumption below 20 W. The motors include stepper, servo, Alternating Current (AC), Direct Current (DC), and Brushless Direct Current (BLDC) types. Advancements in materials and manufacturing techniques offer promising avenues for enhancing prosthetic design and functionality. While traditional metals remain prevalent in prosthetic frames, emerging alternatives such as 3D printing, compliant silicone, carbon fiber, and 3D printed steel monocoques present opportunities for improved performance and affordability [9].

In the academic domain, electronic activation prosthetic models demonstrate the capacity to exert finger pressure forces of up to 15 N through specific mechanisms [7, 8]. Nonetheless, attaining such levels of force necessitates the incorporation of particular technical attributes, including weight considerations (850 g), alongside utilizing specialized materials in the construction of the mechanical framework, such as aluminum alloy supplemented with a carbon fiber coating.

In addition to the comparative study between prosthesis characterization methods, it is pertinent to briefly focus on the developments presented by academia in the immediate years regarding the design and implementation of upper limb myoelectric prostheses manufactured using digital fabrication technologies and additive manufacturing.

Among the notable works is the one presented by Sharma and colleagues in 2024 [10]. This hand prosthesis is manufactured using Acrylonitrile Butadiene

Styrene (ABS) filaments and features a Force-Sensitive Resistor (FSR) force sensor; its activation is generated by proportional control of electromyographic (EMG) readings. Although the author does not report quantitative force analyses, the prosthesis was tested by volunteers without amputation in daily life tasks. While it can provide force feedback thanks to the force sensor, it has limited grip patterns due to the shape of the hand prosthesis and a lack of reporting on the functional characterization of the prosthesis.

On the other hand, in 2024, Ved colleagues [11] presented the design, development, and implementation of a low-cost 3D-printed hand prosthesis. This prosthesis, made with Polylactic Acid (PLA) and Thermoplastic Polyurethane (TPU) filaments for flexible joints, also features EMG signal control that is then actuated by servomotors within the hand. The tests carried out included a finite element analysis to evaluate the mechanical strength of the hand, where it was found that the digits could withstand loads of up to 50 N in flexion without deforming. Although the model stands out for its design to replicate the natural movement of the fingers, the lack of quantitative information about reaction times and latency between electromyographic reading and prosthetic activation limits the complete technical description of the developed prototype.

Thirdly, in 2023, Nibe and colleagues [12] presented a hand prosthesis prototype made of PLA filaments for structural parts and elastic bands to provide flexibility in finger movement. The prosthesis has a servomotor for gesture activation. Control is performed by an Atmel microcontroller that processes the EMG signal reading. This study evaluated the hand prosthesis's ability to grip objects and hold weighing up to 2.0 kg. The authors indicate that the response time obtained due to noise in EMG signal processing can reach 0.7 s; however, this time corresponds only to that which elapses between the sensor readings and their arrival at the microcontroller. The authors did not calculate the time between recording the movement intention and the time it took for the servomotors to carry it out.

Finally, Solano Rico and colleagues [13] also presented in 2023 a prototype of a myoelectric prosthesis designed and developed by 3D printing. This model, made of PLA filaments, incorporates servomotors in the palm for actuation. The prosthetic design was carried out using Autodesk Fusion 360. The myoelectric control focuses on pinch and cylindrical grips. Although no characterization tests were carried out on the prototype, a structural analysis was performed to evaluate load displacements. Thus, a load limit result of up to 1.5 kg was obtained. On the other hand, even though the model adds an EMG filtering stage to improve detection accuracy, the study does not report data on latency in the tests carried out with users.

From the studies reviewed, a technological need to be resolved has been identified: To carry out mechatronic characterization tests on 3D-printed myoelectric prostheses after the structural analysis stage by software that allows knowing the load capacity, as well as the response time between the muscle gesture intention stage and the actuator activation stage.

It is noteworthy to emphasize the significance of mechanical characterization procedures in upper limb prostheses, as they contribute to comprehending the kinetic and strength capabilities inherent to such prosthetic devices. This holds particular relevance in accessible prosthetic advancements utilizing digital manufacturing technologies. Casings and internal mechanisms, typically fabricated from PLA, Polyethylene Terephthalate Glycol (PETG), and ABS polymers, are subject to varied orientations and layer heights, with the number of perimeters unspecified during lamination. Consequently, a single prosthesis manufactured through 3D printing with differing lamination configurations may exhibit distinct static and dynamic mechanical properties [14, 15].

Within the literature, characterization predominantly focuses on evaluating the performance of prosthetic devices. Recent contributions include the work by Ivani et al. [16], who studied tactile perception in four upper limb prostheses through vibratory trials from the prosthesis to the user's skin. Their findings indicated that less complex prosthetic hands demonstrated an 83% accuracy rate in contact discrimination, surpassing more intricate prosthetic models with increased degrees of freedom. Furthermore, characterization is a valuable tool for comparing two prosthetic devices, mainly when one exhibits mechanical, electronic, or control innovations. For instance, [17] conducted a quantitative comparison between the "mHand" model prosthesis and its predecessor, the "Delft Self-Grasping Hand". Their analysis revealed that mHand exhibited superior functionality scores, reduced interaction time with the contralateral hand during bilateral activities, and a broader range of occupational tasks achievable compared to the previous model.

This study focuses on characterizing only the hand of the PUCP IArm transradial prosthesis, as it is the component responsible for executing active movements. The prosthesis integrates a gesture recognition system based on artificial intelligence algorithms using EMG signals [18]. Characterization involves dynamic, functional grip, and kinetic tests. Furthermore, the study emphasizes the importance of developing and implementing inclusive technologies that address populations' specific needs and constraints in resource-limited settings, promoting equitable access to essential healthcare resources.

The present study is structured as follows. Section 2 delineates the technical specifications and general characteristics of the PUCP IArm prosthesis. Additionally, it elucidates the methodologies employed in the characterization tests. Section 3 presents the results of each test applied. Section 4 provides a comprehensive discussion of the findings and establishes a comparison with the state-of-the-art studied. Finally, conclusions are presented in Sect. 5.

2 The PUCP IArm Prosthetic Hand

The prosthetic hand from PUCP IArm is distinguished by its comprehensive design comprising five articulated digits with rubber tips, mimicking the structure of a human hand, including the palm, back, and wrist. The digits are constructed in two segments: the proximal phalanx and a combination of the medial and distal phalanx, as meticulously illustrated in Fig. 1. In the palm, motors responsible for the finger movements endow the prosthesis with precision and agility.

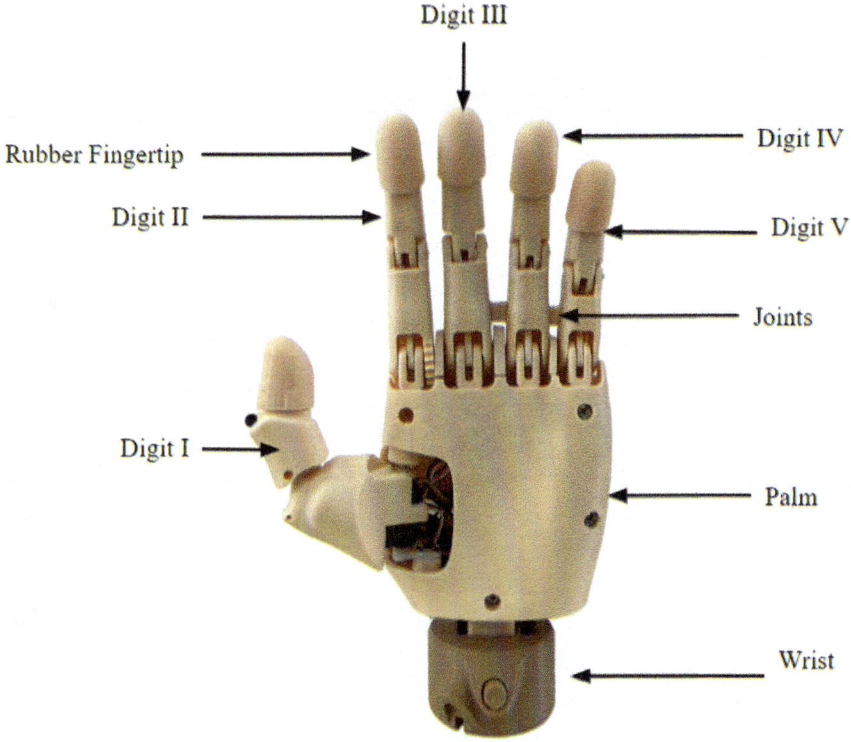

Fig. 1 Components of the PUCP IArm prosthetic hand

The mechanical components of the prosthesis are fabricated using PLA through 3D printing, while the fingertips are made from SILIKAMOLDS Platinum RTV-1510 silicone rubber. This choice of materials replicates the appearance of nails and fingerprints and ensures optimal durability and functionality. Additionally, the prosthesis is available in two skin tones, light and dark, and comes in two sizes and medium to accommodate different users' needs.

The prosthetic hand described in Table 1 incorporates a design utilizing three servomotors for digit articulation, each individually assigned to optimize movement and precision. Digit I is capable of performing abduction and adduction in an active state with the aid of the ES09MD II servomotor. Although the flexion and extension movements of Digit I are in a passive state, the use of this servomotor suggests that it can be activated for other movements. On the other hand, Digit II and the group of Digits III, IV, and V present active states for flexion and extension, each powered by their dedicated servomotor, B13DLM V2, which allows for coordinated movement.

As shown in Fig. 2a, the servomotors are securely housed within the palm of the prosthesis, ensuring protection and optimal performance. Additionally, Fig 2b–d illustrate the movement mechanism driven by each servomotor. Specifically, the B13DLM V2 [19] servomotor, used for both Digit I and Digit V, operates within a

Table 1 Joint kinematics

Prosthetic hand	Movement	State	Servomotors
Digit I	Abduccion–adducción	Active	ES09MD II
Digit I	Flexion–extension	Passive	–
Digit II	Flexion–extension	Active	B13DLM V2
Digit III, IV and V	Flexion–extension	Active	B13DLM V2

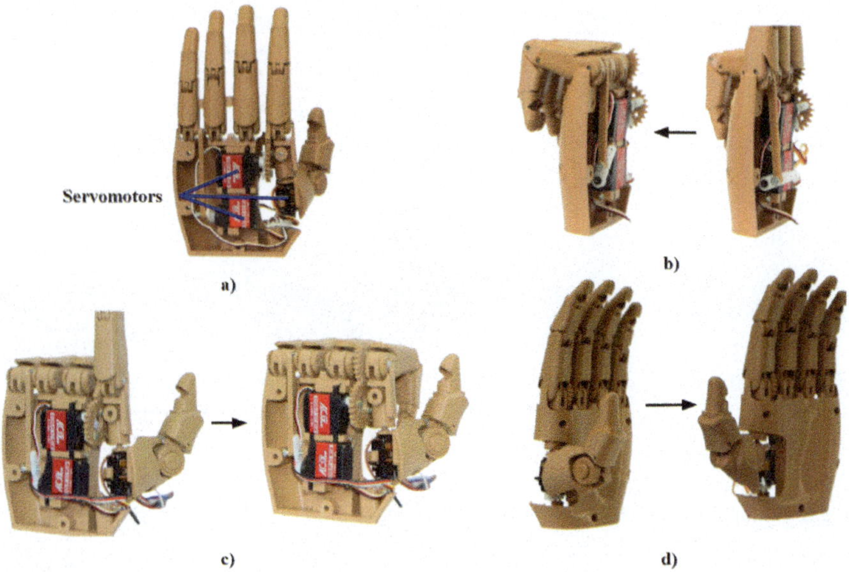

Fig. 2 Positioning of servomotors in the PUCP IArm prosthetic hand: **a** placement of servomotors in the palm; **b** range of motion mechanism for digits III, IV, and V via four-bar linkage; **c** flexion and extension mechanism of digit II via gear system; **d** adduction and abduction mechanism of digit I

voltage range of 4.8–8.4 V and offers a stall torque of 3.0–4.5 kg-cm with a speed of 0.10 s/60°. In contrast, for Digit II, the ES09MD II servomotor [20] is employed, operating at a voltage of 4.8–6 V, providing a stall torque of 2.3–2.6 kg-cm and a speed of 0.08 s/60°. These specifications ensure that each digit of the prosthesis can move with the precision and strength necessary to perform everyday tasks.

The prosthesis incorporates an algorithm that detects three forearm muscle movements: flexion, relaxation, and extension. The combination of these movements allows the prosthesis to execute four types of gestures: Pinch, Cylindrical grasp, C-grasp, and *Keyboard* gestures, the latter being defined by the manufacturers as a movement where only Digit II is extended, allowing the user to type on a computer. Flexion and extension do not need to be maintained to execute the gestures.

The PUCP IArm prosthesis employs a myoelectric control system that interprets EMG signals from the user's residual limb, allowing for intuitive and responsive movement. This control system relies firstly on our development called SEMGUI software. SEMGUI adapts a Graphical User Interface for organized data collection, capturing, and labeling EMG signals associated with movements such as flexion, extension, and rest. The dataset generated through SEMGUI is segmented into temporal windows (50, 100, 150, and 200 data points) and enriched with a feature vector containing nine characteristics per channel. This structured dataset is the foundation for training the control model, enabling accurate gesture recognition.

To refine signal interpretation, the control system incorporates a stacked model architecture. This dual-layer structure consists of a general movement detector and a gesture classifier. The first model detects movement initiation, filtering out unintended signals, while the second model identifies the specific gesture (flexion or extension). This approach reduces false positives and enhances control precision, ensuring the prosthesis responds accurately to user intent. Together, the use of SEMGUI software for dataset generation and the stacked classification system provide a responsive myoelectric control experience.

3 Evaluation of Prosthetic Finger Performance

The prosthetic finger performance evaluation focuses on the mechanical and functional characterization of the PUCP IArm prosthetic hand. This analysis includes several tests illustrated in Fig. 3: videogrammetry tests to assess finger movement, static downward bending tests to measure structural resistance, distal tensile tests for durability, and finger strength tests to evaluate force capabilities during various gestures. Additionally, response latency measurements between user intent and prosthetic actuation are conducted.

3.1 Videogrammetry Test

A videogrammetry test of the finger movements of a hand prosthesis in the sagittal and transversal planes, a controlled recording environment, is first established by placing high-contrast markers on essential joints of the prosthesis. Using a high-definition capable camera (Sony FDR-AX700), the movements are captured from angles corresponding to the sagittal and transversal planes, ensuring adequate lighting and the markers remain visible. The videos are then imported into *Kinovea* [21], a software specifically designed to analyze sports movements. Within *Kinovea*, spatial calibration is carried out using a reference object to establish the scale. The software's tracking tools are then employed to mark and track the points of interest throughout the video sequence, enabling a detailed analysis of the trajectories and dynamics of the prosthetic movements.

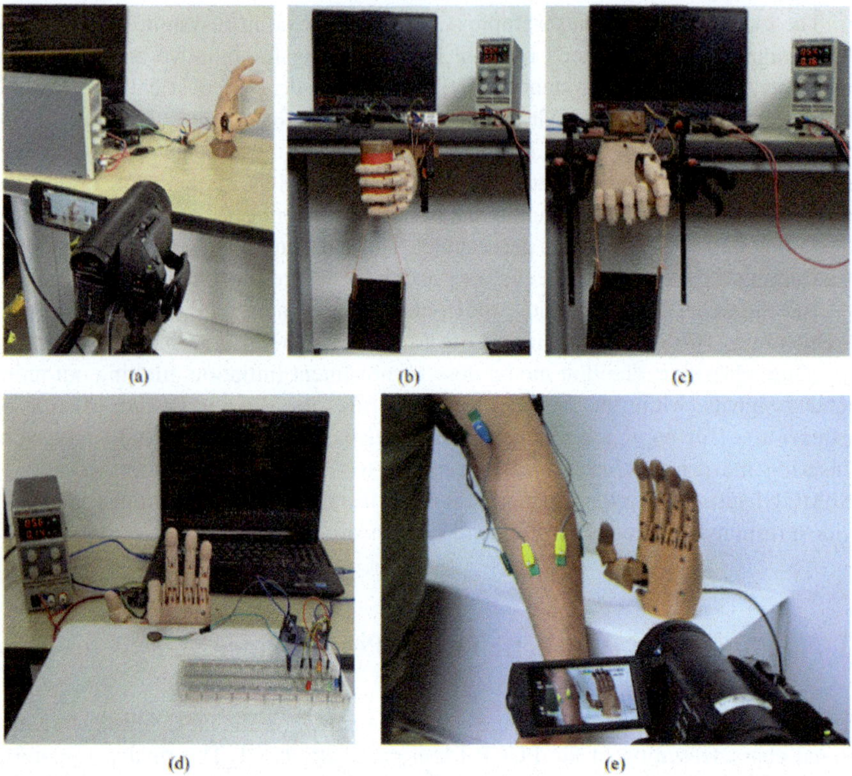

Fig. 3 Test: **a** videogrammetry, **b** static downward bending test, **c** distal tensile test, **d** prosthetic finger strength, **e** response latency measurements

During the evaluation, each finger underwent its maximum range of active motion, with five repetitions performed for each movement: extension and flexion for Digits II (in the sagittal anterior plane), III, IV, V (in the sagittal posterior plane), and Digit I, abduction, and adduction (in the transverse plane) for Digit II, as illustrated in Fig. 3a. The tip and the proximal joint of the Digit II and IV were analyzed, while only the remnants of the tip were analyzed.

3.2 Static Downward Bending Test

Mount the test sample in the test equipment by the test configuration illustrated in Fig. 3b. Apply the test force at the load application point to the lower (distal) end attachment or the particular grip device for the terminal device with the load line parallel to the long axis of the upper arm segment of the test sample as illustrated in Fig. 3b. Increase the test force to the value Fmin = 10 N. Measure and record the initial position of the load application point on the load line. Increase the test force smoothly at a rate of between 1 and 10 N/s. Continuously record the corresponding displacement of the load application point along the load line from its initial position. Continue the test until the test force has reached the test load level specified by the manufacturer/submitter or failure occurs. Record the maximum test force value reached. If failure occurs, record this together with the mode of failure. Record the final displacement of the load application point from its initial position, measured either with the test force F = Fmin applied or immediately before failure if this occurs [22].

3.3 Distal Tensile Test

Mount the test sample onto the test equipment according to the test configuration shown in Fig. 3c. Apply the test force at the load application point at the lower (distal) end attachment or the specialized grip device designed for the terminal device. Ensure the load line is aligned or parallel to the test sample's long axis. Increment the test force to a minimum value of Fmin = 10 N. Record the initial position of the lower (distal) load application point, P (or the designated reference point) along the load line. Gradually increase the test force at a rate of 1–10 N/s, and continuously monitor and record the displacement of the load application point, P (or the designated reference point) along the load line from its initial position. Proceed with the test until the test force F attains the load level specified by the manufacturer or until failure. Document the peak test force value achieved. In the event of failure, note the occurrence and failure mode. Finally, measure and document the end displacement of the load application point from its initial position, with the test force F = Fmin applied or just before failure, should it occur [22].

3.4 Prosthetic Finger Strength Test

This test comprises two parts: determining the maximum force an individual digit can exert, as depicted in Fig. 3d, and assessing the force exerted during specific gestures: pressure exerted by Digit II when extended and pinched.

The force exerted by Digit II on Digit I during the pinch gesture was examined under two conditions: one without any intervening object and another with an intervening object present. Two types of objects, rigid and soft, were utilized to discern potential variations in applied force. Specifically, a rectangular prism measuring 29 mm × 18 mm × 12.5 mm was made of PLA, and a sponge was employed. Conversely, to address the pressure exerted by Digit II when extended, the sensor was positioned on a keyboard to press the gesture involving the extended Digit II to replicate an occupational scenario.

Before starting the measurements, the MF01 force sensor was calibrated. This sensor operates as a variable resistor, changing its resistance based on the degree of deformation. The calibration was performed using a set of standard weights ranging from 20 to 200 g to establish a relationship between the applied force (in Newtons) and the sensor's output (in Bits), as expressed in Eq. (1). The resulting calibration equation captures the sensor's exponential response, characteristic of its design:

$$\text{Force} = 0.0643 \cdot e^{0.0042 \cdot \text{Bits}} \tag{1}$$

The sensor height was adjusted to align with the starting point of the digit's range of motion to measure the maximum force exerted by each digit. This ensured a progressive increase in applied force during flexion (Digits II, III, IV, and V) or abduction and adduction (Digit I). The force sensor was centrally placed between the digits for the gesture tests to optimize force application.

Custom grips securely held the prosthesis in position during trials, with movements controlled via a microcontroller to minimize errors. The MF01 force sensor was positioned between the prosthetic finger's tip and the object, recording forces during flexion and specific gestures. Digits I, II, III, IV, and V measurements were captured using an Arduino UNO, with the calibration equation applied to convert sensor output from Bits to Newtons.

3.5 Response Latency Measurements

The test aims to determine the latency time between the user's movement intention and the prosthesis response using the Sony FDR-AX700, which allows for recording up to 960 fps. The recordings were taken from a frontal plane, capturing both the prosthesis and the user in the same frame, with the camera positioned at a distance of 1.05 m, as illustrated in Fig. 3e. During the trials, the participant's arm was required to be completely relaxed before initiating any movement to avoid detecting unintended muscle activity. The participant was a 35-year-old Peruvian male with both upper limbs present. The study followed the guidelines approved by the ethics committee (approval code 79-2024-CEICVyT/PUCP), and informed consent to participate and to publish was obtained from the individual participant.

In each gesture performed, the first digit to move is either Digit I (Pinch) or Digit II (Keyboard, Cylindrical grasp, and C-grasp). Therefore, ten measurements were taken, where the first digit to move was Digit I, and ten measurements were taken where Digit II was the first to move, using the corresponding digit as the reference point for latency measurement. In addition to latency time, the execution time of each servomotor mechanism was also measured for Digits I, II, III, IV, and V. These measurements were taken with a resolution of 10 ms. The videos were then imported into *Kinovea* for analysis. *Latency time* was defined as the interval from detecting the user's initial slight movement to the first displacement of the digits.

4 Results of the Evaluation of Prosthetic Finger Performance

4.1 Videogrammetry Test

The trajectories of Digit I and Digits II, III, IV, and V are depicted in Fig. 4, while the ranges of motion are presented in Table 2.

The obtained values are consistent with the manufacturer's proposal and a human hand's average range of motion (0°–90° in the proximal joint) [23]. However, it was observed that there are not many intermediate angles between the maximum extension and flexion values of Digits III, IV, and V. This can be explained by the internal mechanism of the prosthesis, which employs a servomotor for the movement of these three digits through a linkage. This differs from Digits I and II, which receive direct motion transmission from a servo motor.

4.2 Static Downward Bending Test and Distal Tensile Test

During this test, the C-grip and cylindrical grip gestures were analyzed, following the guidelines of the ISO 22523:2006 standard "Prostheses and orthoses of external limbs: requirements and test methods" [11]. A grip was designed to correctly position the hand along with a basket, to which 100 g were progressively added until a significant change was observed. The results indicate that the prosthesis can withstand forces of 29.70 N in the C grip and 19.05 N in the cylindrical grip.

4.3 Prosthetic Finger Strength Test

During the initial trial, where maximum values are recorded, Digit I exhibited the highest force, measuring 2.30 N in extension and 2.78 N in flexion. Additionally, Digit II showed a force of 2.02 N, while Digits III, IV, and V displayed forces of 0.71 N each.

The results of the second part of the trial are depicted in Fig. 5 and Table 3, where a progressive increase in force can be observed as the pressure of the gesture increases. Furthermore, it is noted that the force reaches a maximum plateau it can exert.

4.4 Response Latency Measurements

The response latency measurements evaluate the time delay between the user's muscle signal intention and the corresponding movement of the prosthetic hand. The test results for latency times and the execution duration of the complete motion trajectory for each mechanism are presented in Table 4.

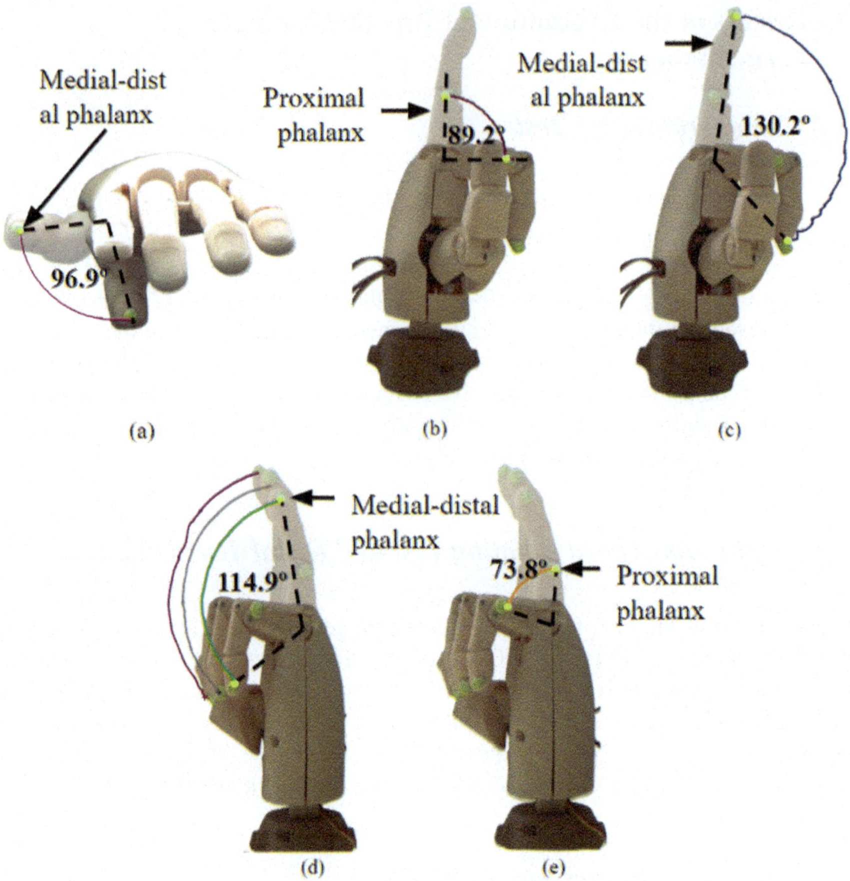

Fig. 4 Range of motion: **a** medial-distal phalanx of Digit I, **b** proximal phalanx of Digit II, **c** medial-distal phalanx of Digit II, **d** medial-distal phalanx of Digits III, IV, and V, **e** medial-distal phalanx of Digits III, IV and V

5 Discussion

5.1 Kinematic Analysis and Movement Limitations

The kinematic analysis results indicate that the movement of digits III, IV, and V in the prosthesis does not allow for gradual displacement, which limits its usability. This limitation arises because movement is delivered through a connecting rod rather than a direct actuator. Despite this, the range of movement aligns closely with the manufacturer's proposed specifications.

Table 2 Results of kinematic Analysis

Digit	Joint	Range of motion (°)
Digit I—thumb	Distal	96.9 ± 0.30
Digit II—index finger	Proximal	89.32 ± 0.36
Digit II—index finger	Distal	130.12 ± 0.519
Digit III—middle finger	Distal	106.07 ± 0.60
Digit IV—ring finger	Distal	110.12 ± 0.63
Digit V—little finger	Proximal	73.85 ± 0.28
Digit V—little finger	Distal	$114, 93 \pm 0.67$

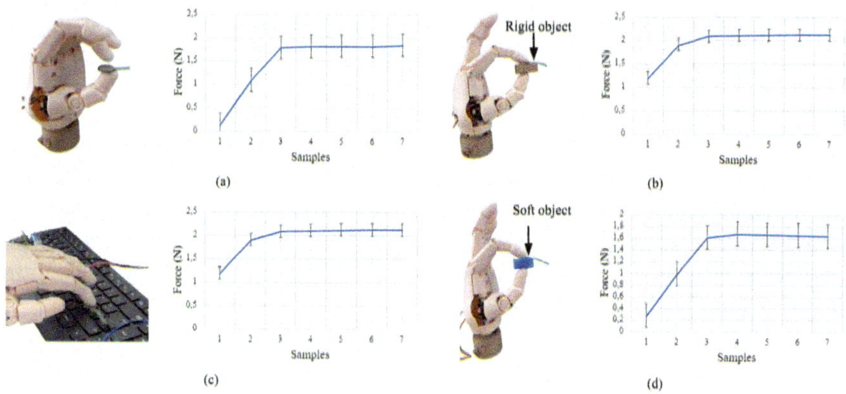

Fig. 5 Forces in gestures: **a** pinch without object, **b** pinch with a rigid object, **c** extended digit II, **d** pinch with a soft object

Table 3 Maximum forces in gestures

Gestures	Maximum forces (N)
Pinch without an object	1.84
Pinch with a rigid object	2.12
Pressure exerted by extended Digit II	1.96
Pinch with a soft object	1.67

5.2 *Grip Performance and Force Transmission*

During the cylindrical grip test, the prosthesis showed signs of instability due to the wrist connection, causing the cylindrical object to slide. In contrast, in the C-grip, digits III, IV, and V were the first to give way, which is expected given that these digits lack direct force transmission compared to digits I and II. The grip force provided by the prosthesis digits competes with similar prostheses, such as [10–13], considering

Table 4 Results of latency analysis

Digit	Latency time (s)	Execution time (s)
Digit I—thumb	1.89 ± 1.36	3.62 ± 0.096
Digit II—index finger	3.24 ± 1.08	3.14 ± 0.072
Digits II, IV and V	–	1.6 ± 0.91

factors such grip force, grip types, DoF, weight material, execution time, Number of EMG channels and latency time as seen at Table 5.

5.3 Force Application and Stability

The prosthesis's capacity for progressive force application in gestures is limited. While the servomotor enables movement, the force stabilizes at a maximum level that mirrors each finger's inherent strength. These experiments were conducted using a passive wrist interface to determine force values with the prosthetic hand. The interface's design facilitated hand positioning adjustments, effectively transmitting the maximum force load to the interface.

5.4 Response Latency and Execution Times

Response latency analysis revealed a delayed reaction compared to similar models. Recent myoelectric prostheses manufactured through digital methods rarely report this data, as digital prosthetic designs are still emerging. For comparison, in [12], a latency value of 0.7 s is provided, but this was measured solely from the EMG sensor signal's arrival at the microcontroller, unlike our study's complete measurement using videogrammetry. Factors such as noise during testing and the use of an intact limb with preserved muscle mass-yielding higher EMG signal amplitude-may have influenced latency results. Consequently, the latency likely reflects algorithm performance rather than the fully integrated system's response.

Execution times for each digit's movement align with the expected range, with Digit I showing the longest duration, while digits III, IV, and V exhibit shorter response times.

Table 5 Comparison of prosthetic hands with different characteristics

Prostheses, author	Grip force (N)	Grip types	DoF	Weight (g)	Material	Execution time (s)	EMG channels	Latency time (s)
Transradial, "PUCP IArm"	29.70	Cylindrical, C and pinch	4	230	PLA	3.62	6	1.89–3.24
Transradial, Sharma et al. [10]	–	Cylindrical	2	350	PLA	0.350	1	–
Transhumeral, Nibe et al. [12]	23.54	Cylindrical	2	–	PLA	0.200	1	–
Transradial, Rico et al. [13]	Up to 14	Cylindrical and pinch	3	285.5	PLA	–	2	0.7
Transradial, Borges da Silva et al. [24]	–	Cylindrical	1	420	ABS	–	1	–
Transradial, Naik et al. [25]	–	Cylindrical, C, fist, claw grip and pinch	6	–	PLA	–	3	–
Robotic hand, Zheng et al. [26]	Up to 5.88 in grasping force	Cylindrical, C and pinch	3	370	Resin and aluminum alloy	–	1	–
Transradial, Calado et al. [27]	Up to 10 in pinch grip	Fist, hook, tripod A, tripod B and pinch	3–4	600	PLA structure and TPU elements	0.3	1	0.5–0.7
Transradial, Koprnický et al. [28]	Up to 8 in various grips	–	4	450	PLA structure and TPU elements	0.250	1	0.8
Transradial, Abarca et al. [5]	3.48–5.06	Tip, tripod, power, spherical and extension	8	950	ABS	164.57/s	–	–

5.5 Durability and Material Performance

An essential consideration in the design of 3D-printed prostheses is the durability of the materials under repeated loading, mainly when using polylactic acid (PLA) as the primary material. While PLA offers notable advantages in terms of cost, accessibility, and compatibility with consumer-grade 3D printing technology [29], it is inherently limited by its mechanical resilience, especially under cyclic loading and environmental stressors [30]. The choice of PLA for this prototype was guided by its accessibility, ease of fabrication, and suitability for rapid prototyping in low-resource settings. This material enables the development of an affordable and functionally effective prosthesis, which aligns with our objectives of accessibility and rapid deployment in regions with limited manufacturing infrastructure.

All prosthetic components, manufactured from PLA polymer using 3D printing, demonstrated durability during testing. No signs of breakage or displacement were due to heat-induced deformation from the servomotors inside the palm. The addition of silicone on the digits enhanced friction during object manipulation, particularly benefiting cylindrical grip performance compared to tests conducted with bare 3D-printed digits.

Despite the advantages provided by PLA, we acknowledge its limitations, particularly its susceptibility to mechanical degradation when exposed to prolonged or high-frequency stress and its vulnerability to environmental factors such as humidity and temperature fluctuations. Initial testing of the PUCP IArm prosthesis has shown satisfactory performance in controlled laboratory conditions; however, extended durability tests, including cyclic loading assessments, are ongoing to evaluate the material's suitability for sustained real-world application. These tests aim to quantify the long-term mechanical stability of PLA in a prosthetic context, thus addressing a critical need for data on how the material behaves under continuous use. The insights from these evaluations will guide the selection of materials for subsequent iterations, especially as we move toward models intended for long-term use by amputees.

In response to these material challenges, alternative polymers such as PETG and TPU are currently under consideration, as they offer superior durability and flexibility compared to PLA. Furthermore, multi-material 3D printing presents a promising avenue for future development, allowing for the integration of rigid structural components with softer, more flexible elements to enhance user comfort and adaptability to repetitive movements [31, 32]. For instance, carbon fiber-reinforced polymers (CFRPs) have been extensively studied for their superior mechanical properties. Research indicates that CFRPs exhibit higher tensile strength and modulus, making them suitable for prosthetic load-bearing applications [33]. Similarly, glass fiber-reinforced composites have demonstrated enhanced strength and toughness, contributing to the durability of prosthetic devices [34].

5.6 *Future Developments in Control Model*

The present article centers on the mechatronic characterization of the PUCP IArm, focusing on its mechanical performance rather than control model optimization. It has been considered to explore in future developments advanced machine learning techniques for EMG signal classification, such as the Convolutional Recurrent Neural Networks (CRNNs) presented in studies such us [35]. CRNNs, which combine CNNs for spatial feature extraction and LSTMs for capturing temporal patterns, have demonstrated superior EMG classification accuracy, outperforming traditional models like Support Vector Machines (SVMs) by effectively capturing the temporal dynamics of EMG data.

Similarly, the hybrid CNN-SVM architecture, which combines CNN for feature extraction with SVM for classification, has shown substantial success in sEMG classification, particularly for hand and wrist movements in rehabilitation contexts [36]. Additionally, a broader overview of machine learning applications in EMG-based control emphasizes the role of feature extraction and classification models in enhancing real-time responsiveness and accuracy in rehabilitation systems [37].

For future work, integrating CRNN-based classification and hybrid models could enhance the responsiveness and adaptability of the PUCP IArm, enabling it to handle complex EMG patterns in real-time. Although this study focuses on mechatronic characterization, these techniques offer promising pathways for further refining control capabilities in the PUCP IArm.

6 Conclusions

The PUCP IArm prosthesis demonstrates the potential of additive manufacturing to develop prosthetic accessibility in resource-limited settings. By using accessible materials such as PLA and silicone rubber, this low-cost prosthesis achieves functional capabilities, including a range of motion and grip strength comparable to more expensive models. The maximum force of 29.70 N in the C grip and 19.05 N in the cylindrical grip highlights its capability to support daily tasks, challenging the assumption that affordability must compromise functionality. This manufacturing approach reduces costs and enables rapid prototyping and customization, addressing a significant global need for affordable, accessible prosthetic solutions.

However, limitations remain in the fine motor control capabilities of the PUCP IArm, particularly in the intermediate positioning of specific digits, which restricts its versatility for tasks requiring precision. Future research will focus on refining control algorithms and exploring alternative actuation mechanisms to enhance movement granularity. This study provides a comprehensive foundation for ongoing improvements in low-cost prosthetic design, offering valuable insights and a structured evaluation framework that can guide the development of sophisticated, affordable prostheses suited to users' needs in developing regions.

Competing Interests This work is supported by The National Program for Scientific Research and Advanced Studies Prociencia-CONCYTEC project number PE501079967-2022. The authors have no conflicts of interest to declare that they are relevant to the content of this chapter.

Ethics Approval This study was performed in line with the principles of the Declaration of Helsinki. Approval was granted by the Ethics Committee of "Pontificia Universidad Católica del Perú" University with approval code 79-2024-CEICVyT/PUCP.

References

1. Liao, C., et al.: Product Narrative: Protheses (2020). https://at2030.org/pn-prostheses/. Last accessed 26 Feb 2024
2. Nacional, Instituto, de Estadística e Informática (INEI): Perfil Sociodemográfico de la población con discapacidad, 2017. INEI, Perú (2017)
3. Nacional, Instituto, de Estadística e Informática (INEI): En el Perú 1 millón 575 mil personas presentan algún tipo de discapacidad. INEI, Perú (2017)
4. Segura, D., Romero, E., Abarca, V.E., Elias, D.A.: Upper limb prostheses by the level of amputation: a systematic review. Prosthesis **6**(2), 277–300 (2024). https://doi.org/10.3390/prosthesis6020022
5. Abarca, V.E., Flores, K.M., Elías, D.: The octa hand: an affordable multi-grasping 3D-printed robotic prosthesis for transradial amputees. In: Proceedings of the 5th International Conference on Control, Automation and Robotics (ICCAR), pp. 92–97. IEEE (2019). https://doi.org/10.1109/ICCAR.2019.8813417
6. Napier, J.R.: The prehensile movements of the human hand. J. Bone Joint Surg. Br. **38-B**(4), 902–913 (1956). https://doi.org/10.1302/0301-620x.38b4.902.f
7. Vertongen, J., Kamper, D.G., Smit, G., Vallery, H.: Mechanical aspects of robot hands, active hand orthoses, and prostheses: a comparative review. IEEE-ASME Trans. Mechatron. **26**(2), 955–965 (2021). https://doi.org/10.1109/tmech.2020.3014182
8. Cutkosky, M.R.: On grasp choice, grasp models, and the design of hands for manufacturing tasks. IEEE Trans. Robot. Autom. **5**(3), 269–279 (1989). https://doi.org/10.1109/70.34763
9. Brack, R., Amalu, E.H.: A review of technology, materials and R&D challenges of upper limb prosthesis for improved user suitability. J. Orthop. **23**, 88–96 (2021). https://doi.org/10.1016/j.jor.2020.12.009
10. Sharma, N., Jadhav, A., Prakash, A., Sharma, S.: Development of low-cost myoelectric prosthetic hand with force feedback. In: 2024 9th International Conference on Integrated Circuits, Design, and Verification (ICDV), pp. 197–200 (2024). https://doi.org/10.1109/ICDV61346.2024.10616604
11. Vedi, H., Bhaiswar, V., Walke, P.: Design, analysis and development of low-cost 3D printed prosthetic hand. AIP Conf. Proc. 3139 (2024). https://doi.org/10.1063/5.0225224
12. Nibe, N., Barpute, N., Pawar, Y., Sutar, R., Wakchaure, K.: Design and development of low-cost myoelectric prosthetic arm for upper limb amputees. In: 2023 4th International Conference on Computation, Automation And Knowledge Management (ICCAKM), pp. 1–6 (2023). https://doi.org/10.1109/ICCAKM58659.2023.10449544
13. Rico, C., Sierra, P., Hermosillo, P.: Ergonomic Design of a Bionic Upper-limb Prosthesis Controlled by Electromyographic Signals (2024). https://doi.org/10.24054/rcta.v1i43.2825
14. Saldarriaga, A.M., Romero, E., Abarca, V.E., Elias, D.A.: A parametric design approach for affordable customized 3D socket for transradial upper limb prostheses. In: 2024 10th International Conference on Control, Decision and Information Technologies (CoDIT), pp. 745–750, Vallette, Malta (2024). https://doi.org/10.1109/CoDIT62066.2024.10708382

15. Saldarriaga, A.M., Abarca, V.E., Elias, D.A.: Design and implementation of a physical simulator for hand and wrist joint range of motion validation for rehabilitation devices. In: 2023 IEEE CHILEAN Conference on Electrical, Electronics Engineering, Information and Communication Technologies (CHILECON), pp. 1–6, Valdivia, Chile (2023). https://doi.org/10.1109/CHILECON60335.2023.10418754
16. Ivani, A.S., Catalano, M.G., Grioli, G., Bianchi, M., Visell, Y., Bicchi, A.: Tactile Perception in Upper Limb Prostheses: Mechanical Characterization, Human Experiments, and Computational Findings (2024). arXiv:2402.12989. https://doi.org/10.1109/TOH.2024.3436827
17. Krinis, S.L., Chadwell, A., Kenney, L., Smit, G.: A quantitative comparison between the mHand adapt passive adjustable hand prosthesis and its predecessor, the delft self-grasping hand. PLoS ONE **19**(3), e0300469 (2024). https://doi.org/10.1371/journal.pone.0300469
18. Cifuentes-Cuadros, A.A., Romero, E., Caballa, S., Vega-Centeno, D., Elias, D.A.: The LIBRA NeuroLimb: hybrid real-time control and mechatronic design for affordable prosthetics in developing regions. Sensors **24**(1), 70–70 (2023). https://doi.org/10.3390/s24010070
19. Huizhou AGFRC Electronic Technology Co., Ltd.: AGFRC 4.5KG Upgraded High Voltage High Torque 17g Micro Digital Brushed Servo(B13DLM V2). https://www.agf-rc.com/servo-B13DLM-p1669120.html. Last accessed 26 Mar 2024
20. Emax USA: EMX-SV-0287 ES09MD II (Dual-bearing) Specific Swash Servo for 450 Helicopters. https://emax-usa.com/products/emx-sv-0287-es09md. Last accessed 26 Mar 2024
21. Kinovea: Kinovea. https://www.kinovea.org. Last accessed 26 Mar 2024
22. ISO 22523:2006: External Limb Prostheses and External Orthoses—Requirements and Test Methods. International Organization for Standardization (2006). http://www.iso.org. Last accessed 26 Mar 2024
23. Taboadela, C.H.: Parte III: Goniometría de los miembros superiores. In: Goniometría: Una herramienta para la evaluación de las incapacidades laborales, pp. 70–75. Asociart S.A. ART, Buenos Aires (2007). ISBN: 978-987-9274-0-40
24. da Silva, B.B., Porsani, R.N., Hellmeister, L.A.V., Medola, F.O., Paschoarelli, L.C.: Design and development of a myoelectric upper limb prosthesis with 3D printing: a low-cost alternative. In: Di Nicolantonio, M., Rossi, E., Alexander, T. (eds.) Advances in Additive Manufacturing, Modeling Systems and 3D Prototyping. AHFE 2019. Advances in Intelligent Systems and Computing, vol. 975. Springer, Cham (2020). https://doi.org/10.1007/978-3-030-20216-330
25. Naik, I., Raut, D., Patel, T., Sutar, R., Kasambe, P.: 3D printed myoelectric prosthetic arm with feedback. In: 2022 International Conference for Advancement in Technology (ICONAT), pp. 1–5. IEEE (2022). https://doi.org/10.1109/ICONAT53423.2022.9725943
26. Zheng, Y., Li, X., Tian, L., Li, G.: Design of a low-cost and humanoid myoelectric prosthetic hand driven by a single actuator to realize basic hand functions. In: 2018 IEEE International Conference on Cyborg and Bionic Systems (CBS), pp. 603–606. IEEE (2018). https://doi.org/10.1109/CBS.2018.8612255
27. Calado, A., Soares, F., Matos, D.: A review on commercially available anthropomorphic myoelectric prosthetic hands, pattern-recognition-based microcontrollers and sEMG sensors used for prosthetic control. In: 2019 IEEE International Conference on Autonomous Robot Systems and Competitions (ICARSC), pp. 1–6, Porto, Portugal (2019). https://doi.org/10.1109/ICARSC.2019.8733629
28. Koprnický, J., Najman, P., Šafka, J.: 3D printed bionic prosthetic hands. In: 2017 IEEE International Workshop of Electronics, Control, Measurement, Signals and their Application to Mechatronics (ECMSM), pp. 1–6, Donostia, Spain (2017). https://doi.org/10.1109/ECMSM.2017.7945898
29. Raj, S.A., Muthukumaran, E., Jayakrishna, K.: A case study of 3D printed PLA and its mechanical properties. Mater. Today: Proc. **5**(5), 11219–11226. Elsevier (2018). https://doi.org/10.1016/j.matpr.2018.01.146
30. Forster, A.M.: Materials Testing Standards for Additive Manufacturing of Polymer Materials: State of the Art and Standards Applicability. US Department of Commerce, National Institute of Standards and Technology (2015). http://dx.doi.org/10.6028/NIST.IR.8059

31. Baron, J., Hazubski, S., Otte, A.: 3D multi-material printing of an anthropomorphic, personalized replacement hand for use in neuroprosthetics using 3D scanning and computer-aided design: first proof-of-technical-concept study. Prosthesis **2**(4), 362–370 (2020). https://doi.org/10.3390/prosthesis2040030
32. Bijadi, S., de Bruijn, E., Tempelman, E.Y., Oberdorf, J.: Application of multi-material 3D printing for improved functionality and modularity of open source low-cost prosthetics: a case study. In: Frontiers in Biomedical Devices, vol. 40672, p. V001T10A003. American Society of Mechanical Engineers (2017). https://doi.org/10.1115/BIOMED2017-3590
33. Kim, S., Yalla, S., Shetty, S., Rosenblatt, N.J.: 3D printed transtibial prosthetic sockets: a systematic review. PLOS ONE **17**(10), e0275161 (2022). https://doi.org/10.1371/journal.pone.0275161
34. Ramlee, M.H., Ammarullah, M.I., Mohd Sukri, N.S., Faidzul Hassan, N.S., Baharuddin, M.H., Abdul Kadir, M.R.: Investigation on three-dimensional printed prosthetics leg sockets coated with different reinforcement materials: analysis on mechanical strength and microstructural. Sci. Rep. **14**(1), 6842 (2024). https://doi.org/10.1038/s41598-023-53999-3
35. Briouza, S., Gritli, H., Khraief, N., Belghith, S.: CRNN-based classification of EMG signals for the rehabilitation of the human arm. In: Campos-Cantón, E., Huerta-Cuellar, G., Zambrano-Serrano, E., Tlelo-Cuautle, E. (eds.) Complex Systems and Their Applications. EDIESCA 2023. Springer, Cham (2024). https://doi.org/10.1007/978-3-031-51224-75
36. Briouza, S., Gritli, H., Khraief, N., Belghith, S., Singh, D.: Classification of sEMG biomedical signals for upper-limb and hand rehabilitation using a hybrid CNN-SVM architecture. In: 2022 International Conference on Data Analytics for Business and Industry (ICDABI), pp. 467–471. IEEE (2022). https://doi.org/10.1109/ICDABI56818.2022.10041524
37. Briouza, S., Gritli, H., Khraief, N., Belghith, S., Singh, D.: A brief overview on machine learning in rehabilitation of the human arm via an exoskeleton robot. In: 2021 International Conference on Data Analytics for Business and Industry (ICDABI), pp. 129–134. IEEE (2021). https://doi.org/10.1109/ICDABI53623.2021.9655865

Generative Design of Wearable Assistive Devices: Strategies and Selection

Sanchit Jhunjhunwala and Jyotindra Narayan

Abstract Generative methods have evolved over the previous decade to become reliable for hardware design. Given their characteristics, these methods have found notable use in designing wearable assistive devices (WADs) such as orthoses, prostheses, and exoskeletons. Challenges in designing such devices may be characterized as lightweighting, actuator load reduction, adjustability, and responsiveness. In this work, we look at the strategies employed by these methods and their features—and generalize a framework for the design of assistive wearables. An overview of the evolution of the design of WADs is included, along with an exploration of the advent of generative design, to contextualize the two together. Works concerning generative methods to design wearable assistive hardware are examined for their objectives, design methods, and generative strategies. Presented thereafter is a selection philosophy—tabled as a mapping of strategy alternatives and methods to objectives. The same has been exemplified in the context of a case—a short-term injury necessitating the use of an ankle–foot orthotic (AFO) device.

Keywords Generative methods · Wearable assistive devices · Orthoses · Exoskeletons · Actuator load reduction · Adjustability · Responsiveness · Ankle–foot orthotic device

S. Jhunjhunwala (✉)
Translead Medtech, Indian Institute of Science, Bengaluru, Karnataka, India
e-mail: sanchit.jhun@alumni.iitg.ac.in

J. Narayan
Department of Mechanical Engineering, Indian Institute of Technology Patna, Patna, Bihar, India

Department of Computing, Imperial College London, London, UK

J. Narayan and H. Gritli (eds.), *Design and Control of Rehabilitation Robots*, Studies in Systems, Decision and Control 585, https://doi.org/10.1007/978-3-031-86977-8_3

1 Introduction

Mechanical design of orthotic devices is motivated by three essential requirements: maximal assistance to the target biomechanical members; compatibility with user specifications—geometric, kinematic, and material; and minimal hindrance to non-affected biomechanical functions that may or may not be adjacent to the targeted site. Techniques for the development, design, and fabrication of orthoses have therefore evolved given these prerequisites, with the additional constraint of cost-effectiveness in the case of devices intended for deployment en-masse. Development of orthotic concepts requires a multidisciplinary view of the specific biomechanical system(s), and corresponding medicinal/therapeutic procedures for various possible anomalies arising from disorder/disease/injury. Their design, however, specifically with manufacturability as an objective, specifies dimensional parameters and their variability.

Compatibility with user specifications is crucial to the performance of wearable assistive devices (WADs) [1]. Devices, both active and passive, are required to be adaptable and adjustable to the user [2–4], therefore implying a variability of geometric dimensions associated with functions and features requiring adjustability/adaptability.

Further, a requirement for optimality arises from the fact that design objectives may be incompatible with each other. For example, customization solves user-specification compatibility but increases fabrication complexity and therefore costs—given the non-scalability of the manufacturing process arising from the non-uniformity of design dimensions. Generative methods enable the location of optimality basins of involved parameters for such multivariate problems.

Given their generality, generative design techniques offer the promise of consolidating such a basket of approaches into a cohesive yet general policy for the design of wearable assistive devices. While clubbed under the term *generative design*, the strategies constituting this policy vary in approach and implementation [5]. Proposed in this work is a generalized framework for the generative design of wearable assistive devices. While such devices extend beyond orthoses, prostheses, and exoskeletons, into wearable [6] audio, visual, and cognitive aids, this chapter focuses on the mechanical design capabilities of generative methods. The following sections examine existing practices to address these challenges—manual and algorithmic—while listing objectives and their sequence of priority. A selection philosophy is tabulated to allow contextual evaluation of strategy alternatives. Finally, a procedure for generative design of wearable assistive devices is proposed and exemplified in the case of an ankle–foot orthotic (AFO) used for recovery after a short-term injury.

1.1 Evolution of Design of Wearable Assistive Devices and Motivation

Design and methodologies thereof evolve with usage and production techniques. Functional wearables such as prosthetics and orthotics have evolved in both functionality and manufacturability over the previous century. While early shifts in their design involved switching to durable materials; wars [7] and epidemics [8] incentivized the development of mass-manufactured designs. The introduction of materials such as plastics and fibers enabled lightweight and comfort enhancement. For instance, Velcro straps allowed replacement of leather-based fastenings thereby increasing breathability and ease of adjustment. Improved fabrication techniques allowed for cost-efficient customizability of the devices. Processes such as investment casting, vacuum molding, and then 3D printing were incorporated by developers once each technology reached a point of maturity. With the advent of each of these, costs per component dropped for lower production numbers [9].

While fabrication techniques allowed for the materialization of the designs, improvements were made on fronts for acquiring necessary design dimensions from the users themselves. This constituted the skill-intensive aspect of the problem, with customization requiring trained prosthetics to make custom-built adapters, while also needing the extended presence of the patients themselves during the procedures to acquire the fit specifications (fiber casts, etc.). Technologies such as 3D scanning allowed for improved accuracy and reduced data acquisition time, with direct implications on operational costs.

Modularization allows for the segmentation of the devices into regions requiring customization and those that could do without it, thus allowing for mass customization. The custom modules themselves can either be built-to-spec from parameters acquired either from the user, or from statistical medians established by appropriate literature or survey.

Assistive devices can be characterized as functional structures that operate in parallel with human anatomical members, while also being attached to them, to augment, replace, or strengthen a set of motion characteristics—such as load-bearing capacity, range of motion, etc. Among classes of wearable assistive devices, exoskeletons—specifically upper and lower limb exoskeletons—have seen tremendous interdisciplinary effort for development in the last two decades. Advances in exoskeleton design are largely driven by the impetus for the improvement of the efficacy of physical human–robot interaction (pHRI), by addressing challenges of kinematic compatibility and control adaptability [10]. The interaction is largely affected by

(a) kinematic design for specific biological features and systems
(b) associated compliance/stiffness factors
(c) selected control method.

Given the complexity and variability of human anatomical structures, models for the same cannot be uniform [11]. This necessitates adaptability to parameters of anatomical members such as geometric dimensions, weight, center of mass, etc.

Further, the selection of design parameters for the wearable depends on the targeted functionality.

2 Overview of Generative Design Strategies

Evolving from computer-aided design (CAD) during the early 2000s, algorithmic and generative methods for component design focused heavily on optimization for function and scalability, mass customization, and their ability to generate complexity and novelty. There are three sequential aspects to a generative design problem: seeding, exploration, and evaluation. Seeding of solutions refers to the selection of an initial point in the design space where the optimization journey starts. In the present context, human input can be relied upon for the definition of input models. Exploration of the design space further consists of two aspects: iteration and branching. While iteration refers to the magnitude of steps taken to vary parameters through successive rounds of design generation, branching refers to the directionality of these steps. Directions of subsequent steps are linked with outcomes of evaluation upon specified objectives. In multi-objective problems, such directionality can be non-singular. Such locations in the design space, referred to hereby as nodes, can serve as junctions for design decisions, thus requiring manual evaluation. Resulting iterations upon geometry can be subtractive, as in topology optimization; additive, as in constructive solid geometry-based parametric growth models; or adaptive, as in machine learning-based optimization. While the term *generative design* is increasingly being used interchangeably with *artificial intelligence (AI) based design techniques*, methods using machine learning and other modalities of statistical intelligence form only a subset of the umbrella of generative methods—alongside algorithmic methods which do not use statistical inference or decision modalities to score or rank across design alternatives (Fig. 1).

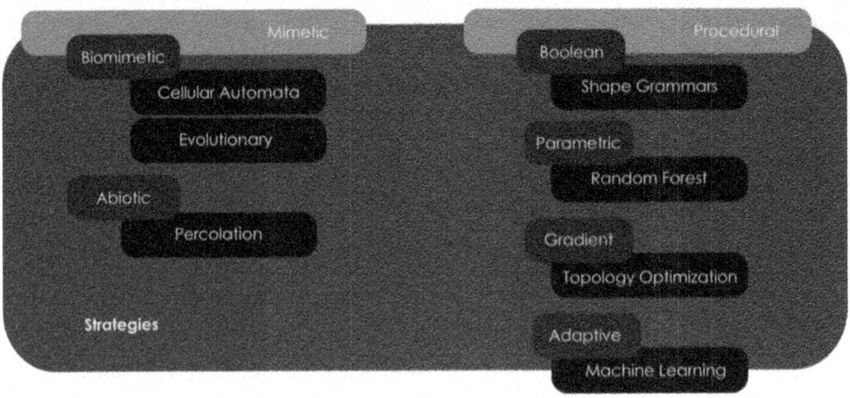

Fig. 1 Optimization strategies for generative design

Broadly, these methods can be characterized into two categories—mimetic, where optimization pathways are modeled after natural phenomenon: both biological (biomimicry) and otherwise (abiotic); and procedural, where logic structures are defined programmatically to guide the iterative steps towards a convergent optimality. Among these are four notable strategies—the first three of which were outlined in a seminal piece of work by McCormack et al. [12]:

1. Self-organization and self-assembly: Inspired by behaviors studied in chemistry, these methods involve the modeling of simple elements that attach based on specified laws of attraction and repulsion. The challenge here is the enormous computational load implied by such an emergent system. A simplified version involves loosely coupled elements called agents which perceive their surroundings, including other agents. These agents are also referred to as cellular automata.
2. Evolutionary systems: Used in animation and graphics applications, this generative technique begins with a parameterized model of the required outcome, with a general range for the variation of parameters—wide enough to produce all possible outcomes of the desired component geometry. A random set of designs is initiated with randomized parameters, referred to as a seed population. Manual selection based on functional and aesthetic considerations is then used to select a few alternatives, which are then *bred* with each other to produce intermediate options that derive parametric qualities from the selected *parents*. A function may be defined to iteratively cull sections of the population.
3. Generative grammar: Shape grammar-based techniques specify a mapping of a string of characters with geometric features. These could be feature shapes, or sets of operations required to produce specific shapes—such as in computer-aided manufacturing—depending on the fabrication method that has been planned for production. An initial string of characters along with a set of production rules is specified. In discrete time steps, these rules define how the characters are replaced with others. This process iterates until the location of a string that completely specifies the design of a component—either as shapes or as instructions for production.
4. Topology Optimization: Based on iterative elimination of material from zones of low stress (under a specified/adapting threshold), followed by recalculation of stresses for the modified geometry, this is one of the most popular contemporary methods for optimization of structures to maximize performance while minimizing weight.

3 Methodology

While each distinct generative method requires a specific procedural implementation of computational tools and inputs, a general workflow for the generative design of WADs has been proposed in Fig. 2.

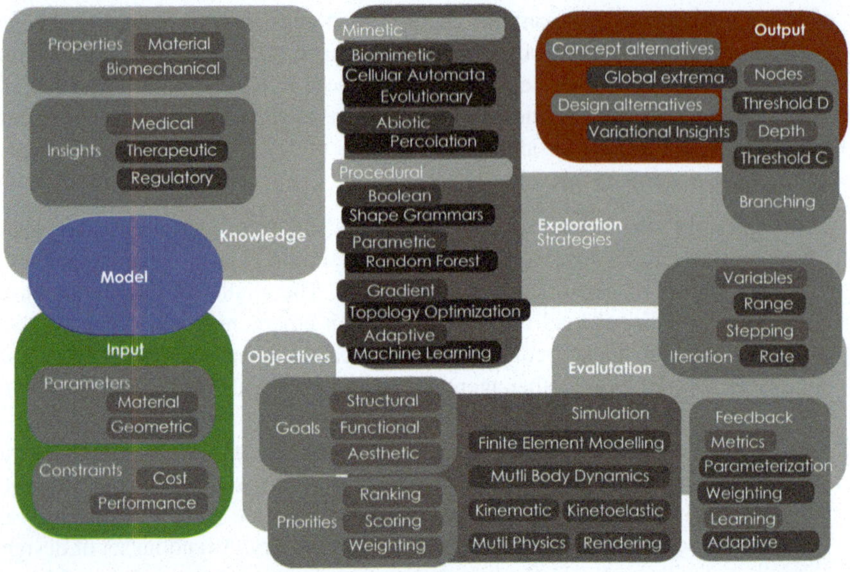

Fig. 2 Process workflow of a generative design exercise

An input model is provided by the human designer. Depending on the desired output, conceptual abstractions for the same can be arrived at using information from available insights into the target limbs and joints, such as their biomechanical properties—joint trajectories, kinematic dimensions, elasticities, viscosities, etc. Insights into the therapeutic framework required to protect, assist, or treat the region are crucial to setting exploration objectives.

Input Specification: The process begins with the definition of parameters—values of which the algorithm iterates through, and constraints that specify the boundaries within which these iterative explorations occur. Necessary parameters associated with geometric features, material properties, and constraints such as performance indicators and cost are defined. Geometric parameters are associated with dimensions of features of interest. These could be features on components associated with the fit, kinematics, or structural rigidity. While material properties can be set to constant in the case of design refinement, exploration exercises benefit from the specification of desired material properties such as strength, biocompatibility, elasticity, manufacturability, cost, etc. Optimization approaches can be tasked to tackle material and geometry objectives simultaneously [13]. Cost implications are critical in generative design. The model input must establish cost limits for materials, manufacturing processes, and assembly, to ensure the viability of production. Metrics such as load-bearing capacity, thermal resistance, reliability, customizability, etc. are defined to measure the performance of exploration outcomes and guide the algorithm toward desired functional requirements.

Objectives: Desired function requirements are defined as goals for the generative exercise. These may be structural objectives such as strength, weight, rigidity, flexibility, etc.; functional objectives such as kinematic degrees of freedom, travel paths, kineto-elastic requirements such as energy conversion and storage in elastic members, spring-back, etc.; or ergonomic objectives such as comfort, aesthetic objectives such as appearance or affordance, etc.

Prioritization of defined objectives is important as multi-objective problems need to balance competing goals. This may be specified by the human designer as a ranking or may be arrived at using mathematical or semantic approaches [14].

Exploration: Upon the input of a seed model, a specified algorithm generates variations of the model by varying parameters within specified ranges defined by the constraints. Strategies for such exploration, as described in a previous section, are selected based on objectives such as diversity of solutions and constraints such as computational power.

Evaluation and Optimization: Generated design alternatives are analyzed using context-appropriate simulation tools: for structural objectives, finite element methods are employed; for interaction and kinematic objectives, motion analysis frameworks such as multi-body dynamics, kineto-elastic analysis, etc. are employed; for aesthetic objectives, rendering methods such as raytracing are employed. The analyses yield specific metrics for evaluation and feedback, which are then used to guide subsequent design space exploration. Notably, adaptive algorithms incorporate machine learning techniques to learn from previous evaluations. By building a history of iteration-performance behavior, relationships can be established between parameters and their effects on performance mapped to specific objectives. These relationships can help reduce the extensive branching of the solution tree thus reducing computational costs. Further, insights provided by such relationships can help designers identify design aspects such as features and segments with relatively greater impact on design outcomes, which enables focused redefinition of parameter spaces and design objectives. This is crucial to the efficacy and efficiency of the next step, as these metrics help isolate the most viable design alternatives.

Design Selection: The algorithm is allowed to terminate either upon completion of a prescribed range of solution alternatives or upon meeting the computational budget. Upon completion of this exercise, the designer must select from the presented concept alternatives and design refinements, based on metrics provided by the evaluation framework. Techniques such as Multi-Criteria Decision Analysis (MCDA) [15] can help with quantitative decision-making. This structured approach helps in making informed decisions about which designs best meet the project goals. Collaborative evaluation with stakeholders can provide necessary interdisciplinary insights.

3.1 Design Objectives for Wearable Assistive Devices

The selection of an optimization strategy and the prioritization of objectives is dependent largely on the context of the design exercise. These contexts can be classified into two categories. In terms of the diversity of output design alternatives, generative methods can be used either to generate concepts for the mechanical arrangement of the entire device or its subassemblies or to present refinements on specific aspects of the mechanical arrangement towards a specified performance metric—referred to as design refinement. These contexts can be overlapped to build a design approach.

Concept generation requires a knowledge bank of medical insight into the therapeutic objective of the assistive device, and mechanical specification of a kinematic framework associated with the targeted biomechanical function, e.g. degrees of freedom at the targeted human body joint. It is a process of mechanism exploration, with the option to constrain the types of kinematic joints based on available fabrication and assembly processes.

Design refinement, however, is initiated with singular objectives that may be coupled with or competed against each other. This section lists some of these objectives.

3.1.1 Lightweighting

All wearable devices add to the mass of the human-device system. The larger the mass, the greater the hindrance to the no-targeted functions of the user. The field of structural optimization, a precursor—and now elemental—to generative design, was birthed [16] by the need to reduce the material content on members of trusses and frames. The majority of work done in the generative design of assistive robotics, especially exoskeletons, in the last 5 years is driven by the objectives of lightweighting and ergonomics. Chen et al. [17] presented an exoskeleton for spinal rehabilitation and correction in patients with scoliosis designed using generative methods to optimize for weight, wearability, and appearance. Verification of this design yielded similar results as regular orthotics, with 40–60% of their hindering influence—implying its usability for non-surgical treatment and auxiliary rehabilitation after surgical treatment. Optimization of static structures requires defined boundary conditions and calculation stress flows arising from specified force loads at fixed points. The objective functions are broadly aimed towards minimizing deformation and displacements. However, compliant mechanisms allow for flexible structures to deform within limits of elasticity. Loads may be described either as force conditions or as displacement load conditions defined at points that are required to move along desired path lines.

3.1.2 Actuator Elimination

While the primary intent of lightweighting exercises is to reduce the load borne by the user, the approach can also be tasked to eliminate or reduce the requirement of actuation within the framework of the device. Exoskeleton actuators are expensive and heavy. Significant effort is required towards the design of structure and control towards the mere elimination of the effect of their weight, such as in transparency modes available in industrial and military exoskeletons [18]. The objective function in such a structural optimization exercise would differ from that of lightweighting. Rather than simply incentivizing the reduction of overall mass, members need to be designed for reduced moments of inertia around rotational centers—thereby further constraining mass distribution for optimal placement of the center of mass (COM) [19].

3.1.3 Passive Assistance

Alternatively, algorithms can be tasked with the exploration of kineto-elastic arrangements to supplement the energy required at a joint. Movements of the human body joints are generally cyclic, i.e., limbs return to home positions upon completion of activities to attain a minimum energy requirement for gait stabilization [20]. This implies that joint angle variables return to initial conditions, perhaps after a hysteretic pathway. Segmented appropriately, inertial gait phases such as midstance [21] can be used to store energy into a compliant member, while other phases restore the energy towards load-bearing or propelling portions of these movements. Cyclicities can be ranked in order of continuity and differentiability between the external ends of the periodic boundaries of joint angle motion. The physical implication of such an order of cyclicity manifests as desirable elasticities and viscosities of corresponding compliant members used to store and release energy. These insights can be derived from relative displacement gradients of joint displacement vectors. The same may be extended therapeutically, such that the designs of compliant members allow for adjustable stiffness and viscosity. Controlled adjustment of mechanical feedback parameters can allow adjustment of assistance or resistance provided by the orthoses [22].

3.1.4 Customizability

Adherence to user parameters is critical for the safety and efficacy of assistive devices [23–25]. However, tailoring products to consumer specifications raises costs of production, per customized component. While fabrication methods such as additive manufacturing enable lower costs for low production quantities when compared to scalable manufacturing processes such as molding and casting, they require the collection of data from the customer before processing [26, 27], resulting in skill-intensive labor requirements and long waiting times. While there are products that promise to reduce the interval between patient measurement and subsequent delivery [28], the user specification process requires their physical presence with skilled personnel. Over-the-counter devices require on-the-go adjustments. The challenge is therefore to find design space optimalities between the competing objectives of product-tailoring and repeatable production-friendly features. Modularization is a popular method to address this issue [29]. The underlying strategy is to identify segments of the device that require customizability to user-specific geometric fitment and kinematic compatibility characteristics. The segments with the highest reported variance across user demographics are then produced using low-production volume processes such as 3D printing, while components with minimal variance are mass-produced with scalable processes such as injection molding. Algorithmically, solutions to mass customizability may be explored either as a generation of concepts—to identify segments requiring customization and propose alternative mechanical arrangements for fitment; or in the refinement of components in a customizability parameter space—to prescribe ranges for design dimensions at such locations. The knowledge bank required for such an exercise is

(a) statistical variance data of biomechanical parameters associated with fitment, and
(b) sensitivities associated with kinematic parameters corresponding to the target human body joint(s).

3.2 *Multi-objective Optimization*

When dealing with a multitude of objectives in a design problem, defining a schema for prioritization across objectives is important to resolve conflicts in competing parameters that may exclusively benefit separate objectives. Multi-objective optimization (MOO) methods allow simultaneous addressing of conflicting design goals. Metrics such as cost, performance, and reliability are inversely related to each other, which often implies the existence of a set of design alternatives where these metrics cannot be improved without worsening the others. A Pareto front can help represent such a set. Further, these objectives can be collapsed into a smaller set of objectives, or even a single objective—using dimensionality reduction or scalarization methods such as weighted sum methods where objective functions are added to each other after multiplication with individual weights.

4 Selection of Strategy

While a general framework for the generative design of wearable assistive devices has been outlined in previous sections, strategy selection is objective dependent. Optimization methods vary in efficacy for different types of design objectives [30]. Further, objectives require a gradation in their relative priorities. Quantitative definition of the same can be done using methods based on ranking and scoring. Qualitatively complex and semantically described objectives such as ergonomics, aesthetics, and affordance, and their relative priorities can be graded using techniques such as fuzzy inference (Table 1) [31].

Table 1 Design objectives, required analyses, and corresponding strategies for optimization

Objectives			Analysis	Strategy	
Structural	Lightweighting		Finite element analysis	Design refinement	Topology optimization
	Flexibility		Finite element analysis	Design refinement	Topology optimization
Functional	Actuator elimination		Multi-body dynamics	Design refinement	Topology optimization
	Passive assistance		Kineto-elastic analysis	Concept generation	Shape grammars
	Customizability	Fit	Manufacturability analysis	Concept generation	Parametric search
		Motion	Kinematic analysis	Concept generation	Adaptive search
	Usability		Ergonomic analysis	Design refinement	Percolation
Aesthetic			Visual rendering	Concept generation	Biomimicry

4.1 *Example*

The human ankle joint forms the kinetic linkage between the foot and the lower leg, allowing the body to interact with the ground. Despite high stability and laid bearing capacity, it is susceptible to injury. For instance, sudden and excessive supination—also referred to as twisting of the ankle—which may arise during high activity periods such as sporting, can render lateral ligaments sprained, leading to deviations from normal in characteristics such as joint motion, kinetics, and muscle activation [32]. Thus, an ankle–foot orthotic (AFO) device may be prescribed for the patient's use during the period of recovery. In such a case, the desired outcomes of a generative design exercise would be:

(a) mapping of geometric parameters to their functions/features, to identify critical parameters requiring adjustability
(b) the ranges of values within which said parameters would need to be allowed to vary—derived from medical literature, adjustment then prescribed by the patient's caregiver, and adjustment carried out by the patient; and
(c) concept generation for mechanical arrangements to facilitate the same.

The input model consists of a kinematic arrangement specifying the underlying joint biomechanics such as degrees of freedom and characteristic segment lengths; joint angle characteristics such as range of motion and elasticities of flexible members; therapeutic insights such as safe attachment locations for the orthosis and required kineto-elastic constraints. These inputs can be structured as a set of parameters and constraints. Multi-modal models can further allow for the input to be in the form of 2D or 3D schematics.

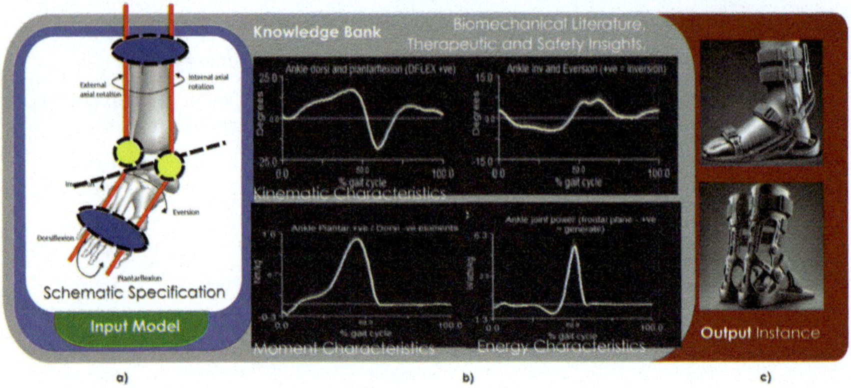

Fig. 3 I/O diagram for an exemplified visualization of the presented generative design framework with **a** Input model specified as a kinematic schematic, **b** Biomechanical characteristics of the ankle joint [33], **c** Expected generative output instance

The objectives in such a case are those of structural integrity, kinematic compatibility, adjustability at the attachment location, and adjustable stiffnesses for kinetoelastic constraining of the orthotic's motion in accordance with the prescribed treatment. Additional objectives of breathability of the geometry and/or material, aesthetic appeal, ease of wearing and removal, etc. can be included (Fig. 3).

5 Conclusion

Presented hereby is a general framework for the generative design of wearable assistive devices, specifically orthoses. Strategies for optimization have been listed and outlined, along with a workflow for concept exploration and design refinement. While a parameterized design space is required for initiating generative iterations, a mapping of design objectives to optimization strategies alongside a scoring of priorities is required to arrive at a solution set, within the constrained design space. Design objectives can be structural, functional, or aesthetic. The workflow inputs and specifications are exemplified using the case of an ankle–foot orthosis designed to address a sporting injury.

The generative design of wearable assistive devices presents several opportunities. It enables enhanced user experience [34] and efficacy, given lightweighting and customizability. It facilitates customization en-masse by enabling efficient production of tailored solutions. It integrates seamlessly smart manufacturing processes—including data-driven customization [35]—by allowing the leveraging of additive manufacturing technologies [36]. The inherent optimization of material usage and production processing promotes sustainability in these solutions for accessibility, by minimizing waste during production and energy consumption during the operation of the designed devices. It also enables avenues for novel characteristics in assistive paradigms, such as responsiveness [37], which requires kineto-elastic behaviors of mechanisms to be programmable envelopes rather than parameterized curves.

References

1. Scherb, D., Wartzack, S., Miehling, J.: Modelling the interaction between wearable assistive devices and digital human models—a systematic review. Front. Bioeng. Biotechnol. **10**, 1044275 (2023)
2. Tachiquin, R., Velázquez, R., Del-Valle-Soto, C., Gutiérrez, C.A., Carrasco, M., De Fazio, R., et al.: Wearable urban mobility assistive device for visually impaired pedestrians using a smartphone and a tactile-foot interface. Sensors **21**(16), 5274 (2021)
3. Covi, E., Donati, E., Liang, X., Kappel, D., Heidari, H., Payvand, M., Wang, W.: Adaptive extreme edge computing for wearable devices. Front. Neurosci. **15**, 611300 (2021)
4. Huhn, S., Axt, M., Gunga, H.C., Maggioni, M.A., Munga, S., Obor, D., et al.: The impact of wearable technologies in health research: scoping review. JMIR mHealth uHealth **10**(1), e34384 (2022)
5. Krish, S.: A practical generative design method. Comput. Aided Des. **43**(1), 88–100 (2011)
6. Dilibal, C., Davis, B.L., Chakraborty, C.: Generative design methodology for internet of medical things (IoMT)-based wearable biomedical devices. In: 2021 3rd International Congress on Human-Computer Interaction, Optimization and Robotic Applications (HORA), pp. 1–4. IEEE (2021)
7. Dougherty, P.J., DeMaio, M.: Major general Norman T. Kirk and amputee care during world war II. Clin. Orthop. Relat. Res. ® **472**(10), 3107–3113 (2014)
8. Murkute, U.: Historical perspectives of critical care in India and worldwide. Indian J. Hist. Sci. 1–9 (2024)
9. Jhunjhunwala, S., Kapil, S.: Rapid manufacturing of biomedical devices: Process alternatives, selection and planning. In: Advanced Micro- and Nano-manufacturing Technologies: Applications in Biochemical and Biomedical Engineering, pp. 77–104. Springer Singapore, Singapore (2021)
10. Gull, M.A., Bai, S., Bak, T.: A review on design of upper limb exoskeletons. Robotics **9**(1), 16 (2020)
11. Jarrassé, N., Proietti, T., Crocher, V., Robertson, J., Sahbani, A., Morel, G., Roby-Brami, A.: Robotic exoskeletons: a perspective for the rehabilitation of arm coordination in stroke patients. Front. Hum. Neurosci. **8**, 947 (2014)
12. McCormack, J., Dorin, A., Innocent, T.: Generative Design: A Paradigm for Design Research. (2004)
13. Ananthasuresh, G.K., Ashby, M.F.: Workshop: Optimal Design. (2003)
14. Gunantara, N.: A review of multi-objective optimization: methods and its applications. Cogent Eng. **5**(1), 1502242 (2018)
15. Taherdoost, H., Madanchian, M.: Multi-criteria decision making (MCDM) methods and concepts. Encyclopedia **3**(1), 77–87 (2023)
16. Bendsoe, M.P., Lund, E., Olhoff, N., Sigmund, O.: Topology optimization-broadening the areas of application. Control Cybern. **34**(1), 7–35 (2005)
17. Chen, L., Li, Y., Han, L., Yuan, L., Sun, Y., Tang, X., et al.: Generative design and simulation of the exoskeleton in patients with scoliosis. Elderly Health Serv. Remote Health Monit. 15–24 (2020)
18. Andrade, R.M., Sapienza, S., Bonato, P.: Development of a "transparent operation mode" for a lower-limb exoskeleton designed for children with cerebral palsy. In: 2019 IEEE 16th International Conference on Rehabilitation Robotics (ICORR), pp. 512–517. IEEE (2019)

19. Narayan, J., Jhunjhunwala, S., Gupta, M., Dwivedy, S.K.: Backpropagation neural network based design of a novel sit-to-stand exoskeleton at seat-off position for paraplegic children. In: 2020 6th International Conference on Control, Automation and Robotics (ICCAR), pp. 546–552. IEEE (2020)
20. Alexander, R.M.: Optimization and gaits in the locomotion of vertebrates. Physiol. Rev. **69**(4), 1199–1227 (1989)
21. Quraishi, H.A., Shepherd, M.K., McManus, L., Harlaar, J., Plettenburg, D.H., Rouse, E.J.: A passive mechanism for decoupling energy storage and return in ankle–foot prostheses: a case study in recycling collision energy. Wear. Technol. **2**, e9 (2021)
22. Stoeffler, C., Kumar, S., Peters, H., Brüls, O., Müller, A., Kirchner, F.: Conceptual design of a variable stiffness mechanism in a humanoid ankle using parallel redundant actuation. In: 2018 IEEE-RAS 18th International Conference on Humanoid Robots (Humanoids), pp. 462–468. IEEE (2018)
23. Poggensee, K.L., Collins, S.H.: How adaptation, training, and customization contribute to benefits from exoskeleton assistance. Sci. Robot. **6**(58), eabf1078 (2021)
24. Narayan, J., Abbas, M., Dwivedy, S.K.: Design and validation of a pediatric gait assistance exoskeleton system with fast non-singular terminal sliding mode controller. Med. Eng. Phys. **123**, 104080 (2024)
25. Reina-Bueno, M., Vázquez-Bautista, M.D.C., Pérez-García, S., Rosende-Bautista, C., Sáez-Díaz, A., Munuera-Martínez, P.V.: Effectiveness of custom-made foot orthoses in patients with rheumatoid arthritis: a randomized controlled trial. Clin. Rehabil. **33**(4), 661–669 (2019)
26. Dombroski, C.E., Balsdon, M.E., Froats, A.: The use of a low cost 3D scanning and printing tool in the manufacture of custom-made foot orthoses: a preliminary study. BMC Res. Notes **7**, 1–4 (2014)
27. Schrank, E.S., Stanhope, S.J.: Dimensional accuracy of ankle-foot orthoses constructed by rapid customization and manufacturing framework. J. Rehabil. Res. Dev. **48**(1) (2011)
28. Cre-aid Labs: https://www.creaidlabs.com/why-mithril (n.d.). Accessed 9 October 2024
29. Ramírez-Rios, L.Y., Camargo-Wilson, C., Olguín-Tiznado, J.E., López-Barreras, J.A., Inzunza-González, E., García-Alcaraz, J.L.: Design of a modular plantar orthosis system through the application of TRIZ methodology tools. Appl. Sci. **11**(5), 2051 (2021)
30. Selvarajan, S.: A comprehensive study on modern optimization techniques for engineering applications. Artif. Intell. Rev. **57**(8), 194 (2024)
31. Lee, S.S.Y.: Using fuzzy-set qualitative comparative analysis. Epidemiol. Health **36** (2014)
32. Li, Y., Ko, J., Zhang, S., Brown, C.N., Simpson, K.J.: Biomechanics of ankle giving way: a case report of accidental ankle giving way during the drop landing test. J. Sport Health Sci. **8**(5), 494–502 (2019)
33. Brockett, C.L., Chapman, G.J.: Biomechanics of the ankle. Orthop. Trauma **30**(3), 232–238 (2016)
34. Bonello, M., Farrugia, P., Buhagiar, N.: Investigating children's experiences for smart wearables: implications on design for additive manufacturing. In: International Design Engineering Technical Conferences and Computers and Information in Engineering Conference, vol. 87332, p. V005T05A012. American Society of Mechanical Engineers (2023)
35. Jiang, Z., Wen, H., Han, F., Tang, Y., Xiong, Y.: Data-driven generative design for mass customization: a case study. Adv. Eng. Inform. **54**, 101786 (2022)
36. Bendoly, E., Chandrasekaran, A., Lima, M.D.R.F., Handfield, R., Khajavi, S.H., Roscoe, S.: The role of generative design and additive manufacturing capabilities in developing human–AI symbiosis: evidence from multiple case studies. Decis. Sci. (2023)
37. Jhunjhunwala, S., Kumar, R.M., KL, C., Patil, S., Shaikh, A., Ananthasuresh, G.K.: A customizable and responsive mechanism for a chair to assist during all phases of stand-to-sit and sit-to-stand maneuvers. In: International Design Engineering Technical Conferences and Computers and Information in Engineering Conference, vol. 87363, p. V008T08A042. American Society of Mechanical Engineers (2023)

Comparative Evaluation of Analytical Techniques for Estimating Ground Reaction Force in Human Walking

Sekar Anup Chander, Siddharth Garg, Ravjeet Singh, Vhatkar Dattatraya Shivling, and Ashish Singla

Abstract Estimating Ground Reaction Force (GRF) is important for deriving and understanding human gait dynamics and for the effective design of assistive devices. Accurate force estimation can lead to enhanced control strategies, promoting natural and efficient movement assistance. In this simulation study, various analytical techniques for GRF estimation are explored using an open standard gait dataset. The study began with the Smooth Transition Approach (STA), followed by the inverted pendulum model for estimating GRF during the double stance phase. The study compared multiple force estimation methods across different gait phases, ultimately exploring and improving a single model capable of predicting GRF throughout the entire gait cycle. The results indicated that one method provided high accuracy by using different equations for each stance phase, while another method offered consistent performance across the entire phase using a unified model. The methods were compared both visually and quantitatively using RMSE values to assess their performance in estimating GRF. The findings offer valuable insights into the selection and optimization of analytical models for GRF estimation, with potential applications not only in assistive device design but also in the development of physics-informed learning models.

Keywords Ground reaction force · Gait analysis · Estimation model · Biomechanics

---

S. A. Chander · V. D. Shivling
CSIR-Central Scientific Instruments Organisation, Sector-30C, Chandigarh, India

Academy of Scientific and Innovative Research (AcSIR), Ghaziabad, Uttar Pradesh, India

S. Garg
Netaji Subhas University of Technology, Delhi, India

R. Singh · A. Singla (✉)
Thapar Institute of Engineering and Technology, Patiala, Punjab, India
e-mail: ashish.singla@thapar.edu

J. Narayan and H. Gritli (eds.), *Design and Control of Rehabilitation Robots*, Studies in Systems, Decision and Control 585, https://doi.org/10.1007/978-3-031-86977-8_4

1 Introduction

Estimating ground reaction forces (GRFs) during human locomotion is important for a wide range of applications in biomechanics, sports science, rehabilitation, and towards design of prosthetics or implants. GRF data is also essential for diagnosing gait abnormalities, conducting ergonomic assessments, and understanding user intent while using an exoskeleton. In addition to human applications, it is also used for real-time balance control and dynamic stability of bipedal robots and humanoids during locomotion. Recent studies indicate the use of GRF values within virtual reality-based environments [1]. GRFs are usually measured using devices such as force plates, and pressure-sensitive mats, which are equipped with an array of sensors to capture the distribution of forces over the entire contact area. These devices provide highly accurate measurements of the vertical, anterior-posterior, and medial-lateral force components. However, they have limitations such as high cost, limited portability, and operational complexity due to large size. To address these challenges, researchers have attempted to estimate GRFs without relying on these devices. The estimation techniques can be broadly categorized as inverted-pendulum-based techniques, multibody segment techniques, and data-driven or neural-network based techniques.

Several techniques have been developed to estimate GRFs using an inverted pendulum model of human gait. For instance, Buczek et al. [2] utilized the telescopic motion of an inverted pendulum to predict ground reaction forces, accounting for muscle activity during normal gait. The work focusses on the roles of joint powers and gravitational force during body support and forward propulsion of human gait. Similarly, a bipedal model incorporating a roller-foot to replicate human walking dynamics was also reported to describe the smooth transition between the single-support and double-support phases of walking [3]. The study also includes the approach for estimating the vertical and horizontal GRFs. The simple spring-based inverted pendulum models were then appended with a damper component also, to examine the oscillatory behavior of center-of-mass (CoM) during the single support phase [4]. The damper in the model was used to simulate how the muscles and tendons dissipate energy to avoid excessive oscillations and unstable gait. The effect of the damper on the relative size of GRF peaks was observed [5]. A similar spring-loaded inverted pendulum model was proposed by Kim et al. [6], which utilizes the leg length as a variable parameter to predict the GRF. The study analyses leg stiffness at different speeds, thereby improving the contribution of spring-like behavior of lower limb during gait dynamics. The 'M-shaped' trend was replicated using the spring-like behavior model. This modification was further refined to match the joint torque profile with the ankle torque profile [7]. The conventional single-body representation of human body was replaced by a multi-body approach. This shift allowed for the application of traditional mechanical engineering techniques of multi-body systems to analyze human gait.

Considering the techniques based on the multibody segment method, joint kinematics are also employed to predict the GRF values [8–10]. In some of these approaches, the human body is modelled as a system composed of different interconnected segments, and the dynamics of each segment are combined to calculate the overall force, which indirectly represents the GRFs. The effect of the segment masses on the vertical GRF is presented in the work by Komaris et al. [11]. Ripic et al. presented their work regarding prediction of GRF and moments using RGB-markerless kinematics and musculoskeletal modeling [12]. Similar technique was employed by Eltoukhy et al. to predict GRF using a Kinect-driven musculoskeletal gait analysis model [13]. The paper by Shourijeh [14] describes the foot-ground contact model which can be used for GRF estimation. Different types of contact modeling elements like Kelvin-Voigt, linear volumetric, and hyper volumetric were explored.

While initial GRF estimation methods were based on models like the inverted pendulum and multi-body based models, recent advancements in data-driven techniques have enabled the use of regression models and neural networks for more accurate and robust predictions. Some of the recent developments in the use of neural networks for the estimation of GRF use long short-term memory (LSTM) networks [15, 16]. In the same work, the authors have demonstrated predicting GRF using a markerless motion capture system. The LSTM networks were also trained to predict Parkinson's disease from the GRF data [17]. Whole-body kinematics was used to train bilateral long short-term memory (BiLSTM) networks [18]. In all of the above techniques discussed, the kinematics is derived using reflective markers, accelerometer sensors or inertial measurement unit sensors (IMUs) placed on specific segments of the body. The IMUs along with pressure insoles were used to develop a regression model that uses high-dimensional approximation for estimating vertical ground reaction force [19]. Uniaxial load cells to collect GRF and the same data was later used to train LSTM network in the work by Kim et al. [20].

In addition to the use of IMUs, other techniques involving depth sensors, ultrasound sensors, and vision-based sensors are also used for estimating GRF. The use of 2D video data in prediction of 3D GRF was demonstrated in the paper by Morris et al. [21]. In the research about the use of NN, several studies highlight the advantages of incorporating physical models to enhance the prediction accuracy of neural network predictions. One such approach is the Physics Informed Neural Networks (PINN) approach, which integrates the governing physical laws, into the NN's loss function [22]. The process of implementation of PINN becomes easier when the equation of the system is available for the entire stance phase. It eliminates the need for approximations or segmenting the stance phase into smaller intervals with different governing equations, and performing the training process. An example using the concept of PINN and its implementation in MATLAB software is presented in the work by Alqezweeni et al. [23]. Considering this in mind, the authors of this chapter have focused on different analytical techniques, and also about which of those can be used for implementation of PINN. The analytical techniques are implemented and analysed on a standard and published gait dataset. The estimated GRF and the reference GRF from the dataset are compared through visual plots and also by calculating the root mean square error (RMSE) values. Comparing various estimation techniques is also essential to identify the accurate and adaptable method, depending on the application. This chapter aims to provide insights into the strengths and limitations of each approach. The novelty of the present work includes comprehensive comparison of GRF estimation techniques. The work not only compares existing techniques but also enhances their accuracy by modifying existing formulations. The study explores both sectional models (e.g., for double and single stance phase) and unified models for GRF estimation, highlighting the differences in formulation. As mentioned previously, the study identifies analytical techniques that can be integrated into physics-informed learning models, bridging the gap between analytical and data-driven approaches. The book chapter is structured as follows. In Sect. 2, the dataset used for the study, along with the data processing steps is detailed. Section 3 describes the five different methods used to estimate GRF. The results after simulation are presented and discussed in Sect. 4. Section 5 concludes the findings of the research work presented in this book chapter.

2 Dataset Used and the Initial Processing Required for the Study

In order to provide a robust benchmarking against different GRF estimation models used, a standard dataset is required. Therefore, the dataset is sourced from the study by Lencioni et al. [24]. It includes data from 50 healthy individuals of both genders across age groups—young, adult and elderly. The participants' ages range from 6 to 72 years, with heights between 117 to 185 cm, and body masses ranging from 18 to 110 kg. All participants reported no known locomotor disorders or health issues that could affect their performance. Data collection was conducted in the Laboratory for Movement Analysis (LAM) at the Don Carlo Gnocchi Foundation, Milan, Italy, using state-of-the art equipment, including a nine-camera motion capture system, force platforms, and a wireless eight-channel EMG (Electromyography) recording system. The dataset provides comprehensive biomechanical information, including the 3D coordinates of external markers, joint angles of the lower limbs in three anatomical planes, GRF and torque values, centre of pressure (CoP), centre of mass (CoM), and surface electromyography (EMG) data. For each subject, approximately 10 walking trials are available, resulting in a total of around 500 trials. In addition to the data of level walking, the trials also include heel walking, toe walking, step ascending and step descending data. In the present work, the data corresponding to level walking are used for analysis. Figure 1 represents the data of vertical GRF of a subject walking on level ground at different speeds.

The measured GRF values from the dataset are used as a reference for comparison with the estimated GRF. The raw dataset was initially processed using MATLAB software. The analysis and comparison with different estimation techniques followed a structured pipeline, beginning with data loading and preprocessing, followed by filtering, interpolation and comparison. The GRF frequency is retrieved from the dataset, along with the time corresponding to the initiation of force plate data. The

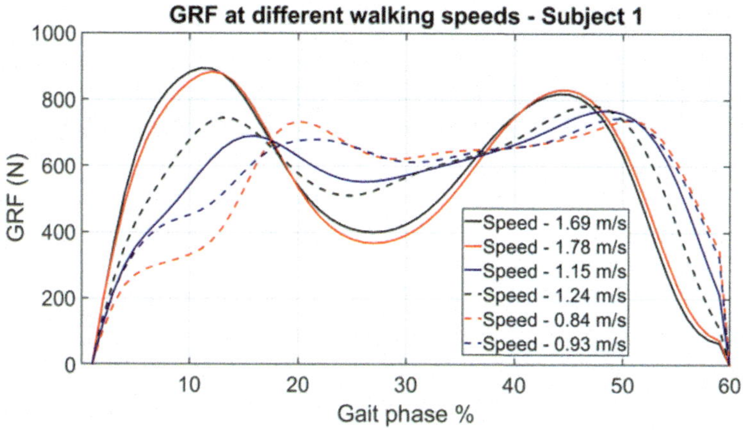

Fig. 1 GRF of subject 1 at different walking speeds

cycle time for each gait cycle is calculated based on the subject's cadence. To reduce the noise in the GRF data, a low-pass Butterworth filter is applied with a cut-off frequency of around 10 Hz, and a fourth-order filter is applied to prevent phase distortion. The suitable cut-off frequencies for different types of gait data are presented in the paper by Bogert et al. [25]. Since the GRF data is available during the stance phase (about 60% of the gait cycle), the filtered GRF is interpolated using cubic spline interpolation. This method provides smooth estimates of GRF at 61 evenly spaced points across the stance phase. The GRF is assumed to be zero, for the remaining 40% of the gait cycle (during the swing phase). Thus, the GRF data is finally ready for comparison for the entire gait cycle.

To ensure a representative analysis, subjects 1, 11, 21, and 31 were randomly selected for the present study. Their respective GRF values are used as reference values for comparison with the estimated GRF values. The height, mass, age and gender of these subjects are detailed in Table 1.

The GRF acting during the stance phase is indicated using Fig. 2. The overall GRF vector represents the cumulative force exerted by the ground in response to the gait. The two components: vertical GRF ($\boldsymbol{F}_v$), and the horizontal GRF ($\boldsymbol{F}_h$) are also indicated in the above figure. The orientation and magnitude of these vectors change dynamically with respect to the phase of gait. The actual values of $\boldsymbol{F}_v$, and $\boldsymbol{F}_h$ are

Table 1 Data of the subjects used in this study

S. no.	Subject	Age	Gender	Height (cm)	Mass (kg)
1	01	70	Male	185	70
2	11	69	Male	187	69
3	21	55	Female	154	55
4	31	53	Female	166	53

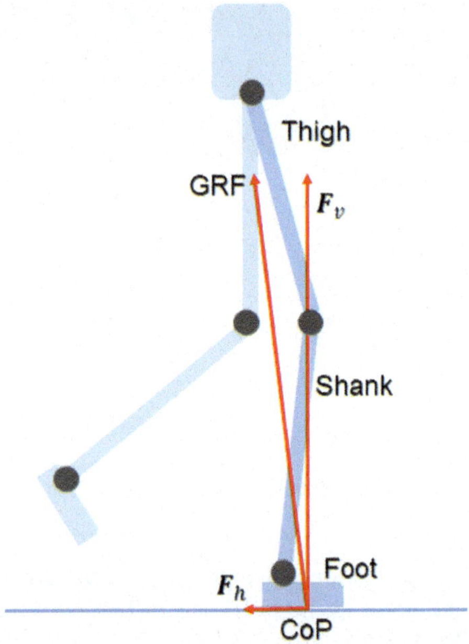

Fig. 2 GRF on the stance foot

provided in the dataset. However, for the estimation of these values, the segment (thigh, shank, and foot) masses along with the segment lengths are required. These mass and length values are approximated as percentages of the total body mass and height, respectively [26], and are computed separately for each of the subjects selected in this study. The next section details the estimation techniques used.

3 Estimation Techniques

In this section the different estimation techniques are described. The formulation for each of these techniques used for simulation is also detailed.

3.1 Method 1: Smooth Transition Assumption (STA) Technique

This technique for estimating ground reaction forces (GRF) involves the use of a three-dimensional (3D) whole-body multi-segment model, using kinematic data instead of force plate data. The kinematic data is captured through a 3D motion analysis system. The study by Ren et al. [9] combines the translational equations of motion of different body segments under consideration to estimate the GRF, as the

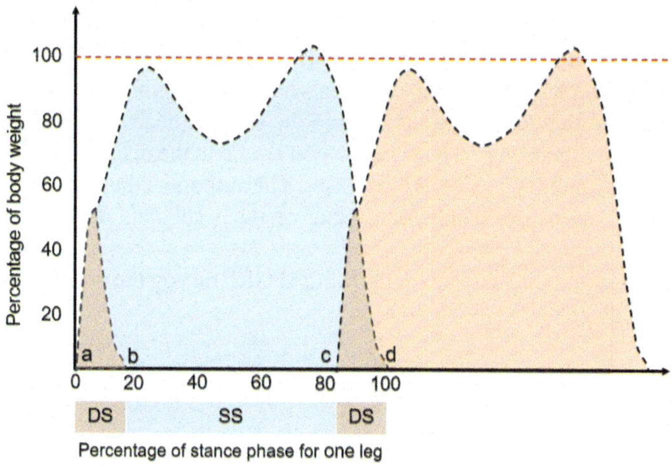

Fig. 3 Stance phase of human gait during level-walking

only external force acting on the body during the stance phase is ground reaction force. The human body was divided into 13 rigid segments comprising of head, torso, upper arms, forearms, pelvis thighs, shanks and feet.

The technique allows for a portable and versatile approach, making it suitable for applications where use of force plates may not be feasible. However, the GRF acting on each leg needs to be determined for the double-support section of the stance phase. Figure 3 is used to explain the different types of stance phases. The section between 'a' and 'd' indicates the complete stance phase. The sections 'a–b' and 'c–d' indicate the double-support section, whereas section 'b–c' represents the single support section of the stance phase. The sections are used to accurately model the GRF dynamics within each section. The equation for computing GRF during the single support section is given by (1).

$$\boldsymbol{F}_{left} + \boldsymbol{F}_{right} = \sum_{j=1}^{n} \left[m_j \left(\ddot{\boldsymbol{r}}_j - \boldsymbol{g} \right) \right] \quad (1)$$

where

$\boldsymbol{F}_{left}, \boldsymbol{F}_{right}$	is the GRF on left and right foot respectively
m_j	is the mass of the jth body segment
$\ddot{\boldsymbol{r}}_j$	is the translation acceleration of the centre-of-mass of the jth body segment
$\boldsymbol{g}$	is the gravitational acceleration vector

For the double-support phase, (2) is used to compute the GRF. The equation is a part of the smooth-transition approach which uses simple analytical functions derived from empirical data. The technique highlights how the GRF and moments on the trailing foot gradually diminish to zero as the foot lifts off. The transition functions are developed by combining the analytical insights with experimental data, refined through trial-and-error process. T_{ds} represents half of the double-support duration and is used to ensure that the equations capture the smooth decay of forces as the foot transitions during this phase. The negative power of e in the decay function ensures a reduction in the function's value as t increases. The authors of the technique presented have validated the model using force plate of three subjects walking at different speeds.

The equation of interest to derive the vertical GRF during the double-stance phase is given by (2).

$$\boldsymbol{F}_v = F_{y0} e^{-\left\{\frac{t}{T_{ds}}\right\}^3} \tag{2}$$

where,

F_{y0} is the GRF value at the beginning of the single-support section
t is the time instance of the gait
T_{ds} is half of the double-support section time

In this chapter, for simulation, the entire gait cycle is divided into sections as previously discussed. For the single stance phase, the Anterior Superior Iliac Spine (ASIS) marker of the right leg is used to compute the acceleration. The actual name of the marker used is labelled as 'YRxAsis' in the dataset, which is one of the 3D coordinates of the ASIS position. It is used as a simplified representation of the CoM's vertical acceleration. Although alternate techniques exist for approximating the CoM motion, this current study adopts the ASIS marker approach. The velocity (v_y) and acceleration (a_y) are further computed from this marker using the forward-difference method. The effective acceleration of the subject is given by adding the hip acceleration to the acceleration due to gravity. The sections along with the Eqs. (3–5) used are summarized in Table 2.

In the initial phase, the force is influenced by the mass and the vertical acceleration, but also decreases based on the exponential function and a time constant. It is the difference between the total GRF and the GRF value of the foot which is undergoing 'toe-off'. As the foot transitions from being fully grounded to 'toe-off' state, the

Table 2 Equations to compute vertical GRF using STA

S. no.	Percentage of stance phase	Equation used	
1	0–11	$F_v = mass * a_y - F_{y0} * \left(e^{-\left\{\frac{t}{T_{ds}}\right\}^3}\right)$	(3)
2	12–49	$F_v = mass * (g + a_y)$	(4)
3	50–60	$F_v = \text{mass} * a_y * \left(e^{-\left\{\frac{t}{T_{ds}}\right\}^3}\right)$	(5)

Table 3 Equations to compute vertical GRF using improved technique

S. no.	Percentage of stance phase	Equation used	
1	0–11	$F_v = a * \log(b(x - 11)))$	(6)
2	12–49	$F_v = mass * (g + a_y)$	(7)
3	50–60	$F_v = F_{v,12-49} - F_{v,0-11}$	(8)

GRF decreases. Equation (3) effectively computes the net vertical force. The initial technique relied on exponential terms to compute vertical GRFs. A new technique, using logarithmic terms, was introduced to improve the precision of the estimated GRF values.

3.2 *Method 2: Improvements to the STA Technique*

Upon initial simulation, a visual comparison was made between the reference vertical GRF and the estimated GRF values, using equations as shown in Table 2. For the double-support phases, it was noticed that the resultant graph is concave upwards for the sections 'a–b'. This deviated from the desired output curve, which is a concave downwards curve. Therefore, the equations from the previous technique were refined, leading to the updated Table 3. The results of each technique are discussed in the 'Results and Discussion' section.

Equation (6) uses a logarithmic function, wherein parameters 'a' and 'b' are constants which are determined based on the empirical data. The logarithmic nature of the equation allows to achieve the desired curve shape. Equation (7) remains unchanged from the previous technique. The 'fminsearch' function of MATLAB is used to estimate the parameters of (6). Equation (8) can be simply described as the difference of Eqs. (7) and (6). These equations are expected to yield a more accurate representation of the vertical GRF. The efficacy of both the original and improved techniques is evaluated and presented in the 'Results and Discussion' section, providing a comprehensive analysis of their performance. Method 1 and 2 divide the stance phase into multiple sections and estimate GRF for each section. The subsequent techniques, however, treat the entire stance phase as a single unit for GRF estimation.

3.3 *Method 3: Inverted-Pendulum Model of Human Gait*

Another approach to predict GRF is through the inverted pendulum model, as discussed in the work by Buczek et al. [2]. The paper evaluates the performance of an inverted pendulum model applied to normal human gait. The model in the paper serves as a valuable tool for analyzing human gait by simplifying the complex movement of gait. The authors hypothesized that an inverted pendulum model, without joint power contributions, would be inadequate for predicting the GRF accurately. The length of the pendulum was defined between the centre of pressure (CoP) and the centre of mass (CoM) of the body. The model allows for a variable pendulum length, instead of a fixed one, making it more representative compared to earlier inverted pendulum models. In the same paper, the equations of motion were derived in both Cartesian and polar coordinates to describe the vertical and horizontal ground reaction forces at the pivot point (CoP). As the focus of this book chapter is on vertical ground reaction force ($\boldsymbol{F}_v$), it is expressed as shown in (9).

$$\boldsymbol{F}_v = m\left[\left(\ddot{l} - l\dot{\boldsymbol{\theta}}^2\right)\sin\boldsymbol{\theta} + (l\boldsymbol{\alpha} + 2\dot{l}\dot{\boldsymbol{\theta}})\cos\boldsymbol{\theta}\right] + mg \tag{9}$$

where,

- m is the mass of the body
- l is the length of the pendulum i.e., the distance between CoP and CoM, which is computed using the co-ordinates of CoP and CoM
- θ is the angle of inclination between the pendulum and the horizontal axis
- $\boldsymbol{\alpha}$ is the angular acceleration calculated using θ
- g is the acceleration due to gravity

The hip and knee joint angles are extracted from the dataset. From these joint angles, the segment angles are computed as described in the book by Winter [27]. From these segment angles, the hip joint position is calculated for each time step. The CoM position relative to the hip joint is computed, as this information is not provided in the dataset. With all these parameters, the GRF is finally computed using (9).

3.4 Method 4: Spring-Loaded Inverted Pendulum (SLIP) Model

The previous approach presented by Buczek et al. [2] discusses the human gait with an inverted pendulum where the CoM moves like a pendulum-like motion. On the other hand, the current SLIP model by Kim et al. [6] introduces stiffness as a parameter that modulates gait dynamics and GRFs. By considering the lower limb as a spring, the model estimates the GRF based on the deflection of the limb. This deflection is translated into vertical and Anterior-Posterior (AP) forces. The process involves calculating the leg length and angle at different points of the stance phase. Later, the spring stiffness is determined based on the body weight and leg deflection. Finally, the GRF is estimated using (10).

$$F_v = k\Delta l\cos\theta \tag{10}$$

where,

k is the spring constant, which describes how stiff the leg
Δl is the change in the length of the pendulum w.r.t rest length
θ is the angle between the leg and the vertical axis

The change in length depends on the contact point '$x_{contact}$'. The contact point is computed using (11).

$$(x_{heelstrike} - x_{contact})^2 + y_{heelstrike}{}^2 = (x_{toeoff} - x_{contact})^2 + y_{toeoff}{}^2 = l_{rest}{}^2 \quad (11)$$

where,

$(x_{heelstrike}, y_{heelstrike})$ are the coordinates of the CoM at the moment of heel-strike
(x_{toeoff}, y_{toeoff}) are the coordinates of the CoM at the moment of toe-off
$x_{contact}$ is the x-coordinate of the contact point, and y-coordinate is zero
l_{rest} is the length of the leg while normal resting

Using the contact point, the varying CoM position (x_{CoM}, y_{CoM}), the length of the leg during and the angle for vertical axis is calculated using (12), (13).

$$l(t) = \sqrt{(x(t) - x_{contact})^2 + y(t)^2} \quad (12)$$

$$\theta(t) = \tan^{-1}\left(\frac{x(t) - x_{contact}}{y(t)}\right) \quad (13)$$

It is assumed that the average GRF is approximately 95% times the weight of the body during the single support phase. The spring stiffness as specified in the model is calculated using (14).

$$k\Delta l_{average} = 0.95\text{mg} \quad (14)$$

where m, and g are values of subject mass and gravitational constant respectively. After computing these values, the vertical GRF is computed using (10), discussed above.

3.5 *Method 5: Improved Spring-Loaded Inverted Pendulum (SLIP) Model*

During the process of human walking, the CoP and hip joint both change positions with time. While the change in position of the hip joint is defined in Kim et al. [6], the same is not considered for the CoP which could be a possible reason for the low accuracy using that method. But in Mauersberger et al. [28], the change in positions

Table 4 Values of 'k' used for different subjects

S. no.	Subject no.	'k'
1	01	28,008
2	11	26,827
3	21	22,232
4	31	25,736

of the hip joint as well as the CoP, both have been accounted for when calculating the change in length of the pendulum. This also meant that the value of 'k' needs to be defined accurately and separately for each subject. The stiffness parameter, 'k' is estimated values using trial-and-error approach. However, as stiffness is function of walking speed, it can be calculated as a function of walking speed, which is not considered in the scope of current work. The improved version of Method 4 is labelled as Method 5 for comparison in the next section. The values of 'k' used for the different subjects is tabulated in Table 4.

The 'k' values are obtained by finding the best value between the range of values from 20,000 to 30,000 by coding in MATLAB software. An alternate approach can be using the relation between 'k' and parameters like walking speed as presented in [4]. All the techniques discussed so far are simulated and the results are presented in the following section.

4 Results and Discussion

In this section, the performance of five distinct methods for estimating GRF is presented and analysed. Each method offers a unique approach, and the results obtained are compared to highlight their respective strengths and limitations. Figure 5, and Fig. 6 indicate estimated GRF values using Method 1 and Method 2 respectively. Through visual comparison, it was observed that the values estimated using Method 1 align more closely with the actual or reference GRF during the single support phase (i.e., 12–49%). However, this alignment is less accurate during the double-support phase. This is because the GRF during the single-support phase belongs to force acting only on one leg, which is calculated by (4) or (7). In contrast, Method 2 shows improved agreement with the actual GRF values, especially for subject-11 and subject-31. The estimation can be further enhanced by experimenting with different curve-fitting equations. The accuracy can be improved by experimenting with different curve-fitting equations, refining the approach for different datasets, and using alternative optimization techniques to determine equation parameters (Fig. 4).

The simulation was later performed using the equations of Method 3 and Method 5 as shown in Fig. 7 and Fig. 9 respectively. The estimated GRF values did not match the actual values in the case of Method 3. Though it did not match in the case of Method 4 as shown in Fig. 8, the estimated GRF values showed some similarity to

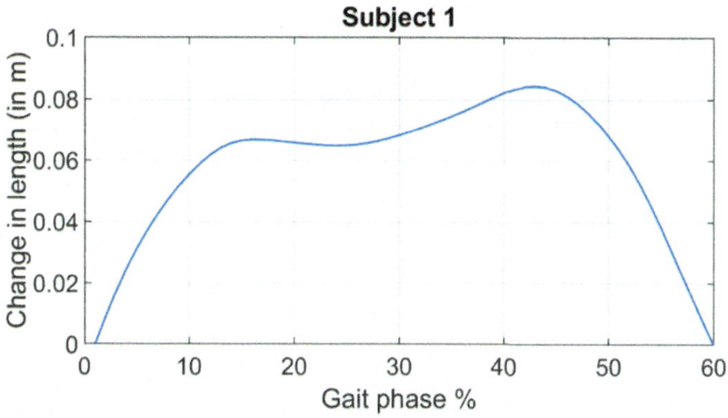

Fig. 4 Change in length versus Gait phase % for subject 1

the 'M' shape of 'change in length of pendulum', evident from the similarity between 'change in length of the pendulum' and 'the vertical GRF' during the stance phase as in Fig. 4. This motivated the authors to further explore on the spring-loaded inverted pendulum model of human gait. While working with the model from Kim et al. [6], it was understood that the value of 'k' governs the magnitude of GRF.

The work done by Mauersberger et al. [28] was explored and it was found that this work employed a rather different approach to calculate the value of 'change in length of pendulum'. The 'change in length of the pendulum' is similar to the straight-line distance between CoP and the hip joint in an actual human leg. So far, the estimated values of GRF were visually compared with the reference values. Box-plots were also plotted for understanding the distribution of output error for each subject across the five different estimation techniques, as shown in Fig. 10. Method 5 has the lowest median error for subjects 1, 11 and 21. The plots suggest that Method 5 could be a reliable method overall. To compare these techniques quantitatively, RMSE values were calculated to provide a more objective assessment of each method's performance, tabulated in Table 5. The phase wise RMSE values are calculated and presented in Tables 6, 7 and 8.

From the above tables, it can be observed that for subject 1, Method 5 shows the lowest RMSE (65.80) considering all the stance phases, indicating a more accurate estimation compared to other methods. Similarly, Method 5 provides the lowest error for subject 11 (58.67) and subject 21 (69.39), demonstrating its consistency in accuracy. Method 2 uses different equations for different sections of the stance phase. However, if one has to utilize a single model, Method 5 can be employed which shows lower RMSE values across all subjects mostly. The error using Method 5 can be further reduced by studying and utilizing the relation between the value of 'k' and the physical parameters of the subject and gait parameter like speed. Therefore, the techniques evaluated in this study provide valuable insights into estimating ground reaction forces during human gait. These techniques can further be optimized for even

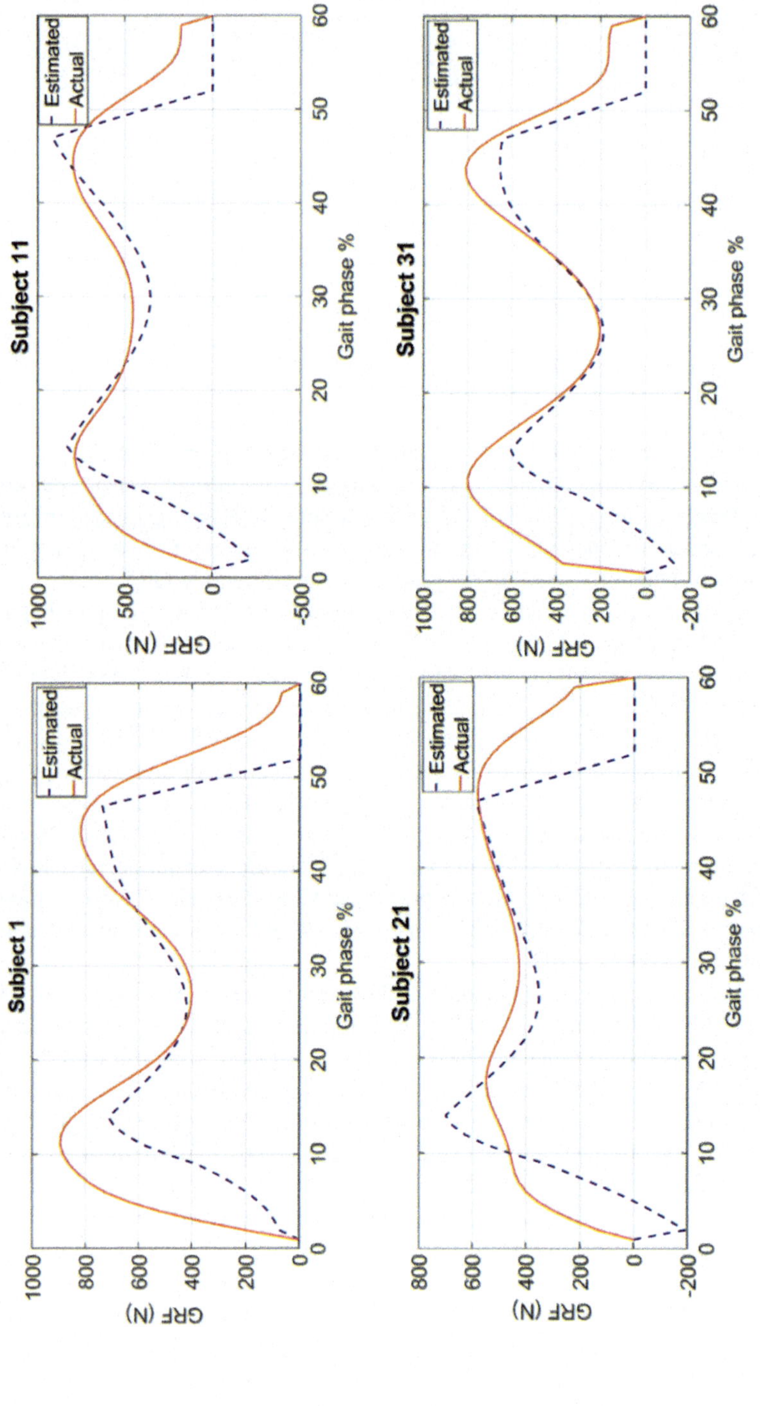

Fig. 5 Comparison of GRF values estimated using Method 1

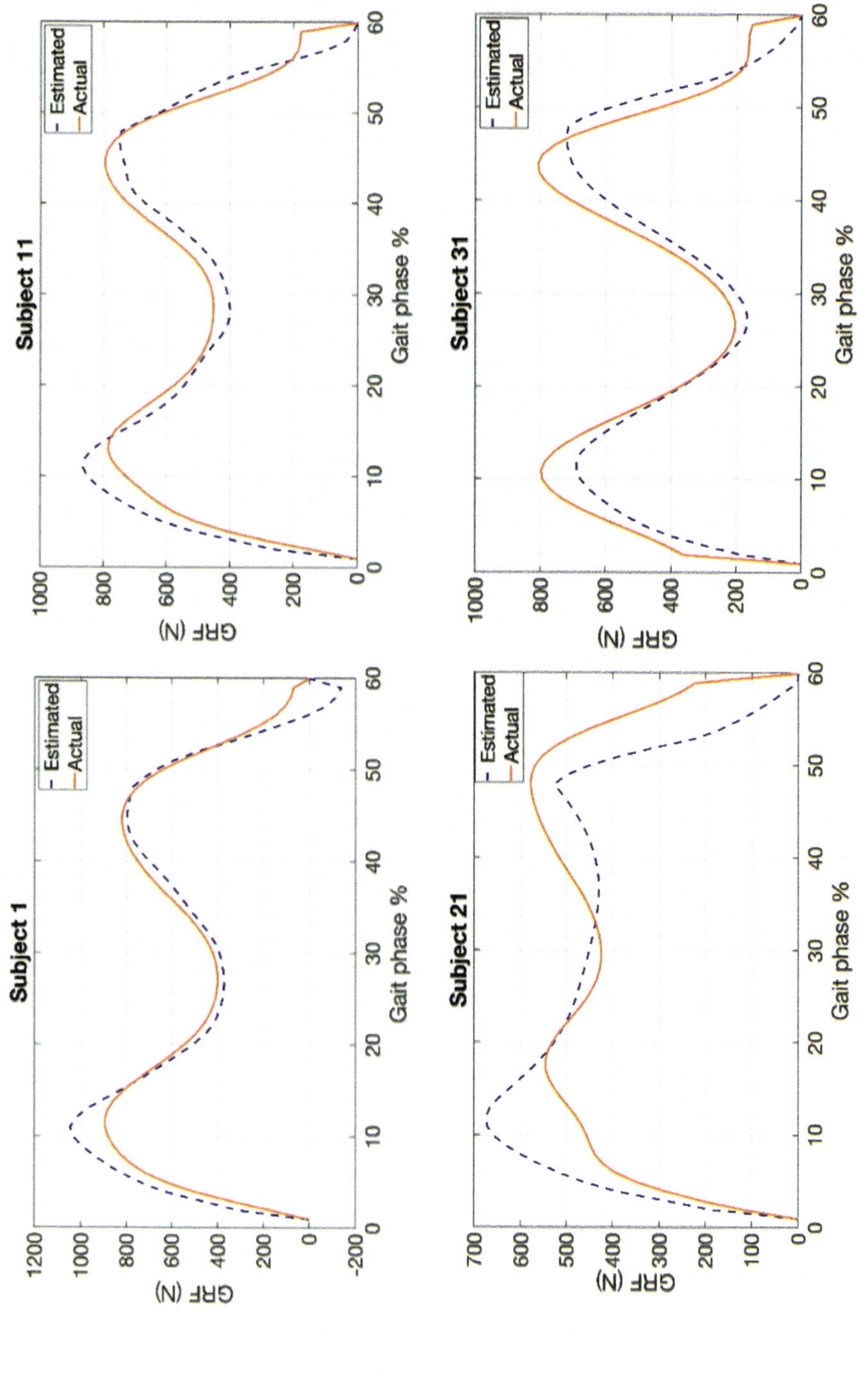

Fig. 6 Comparison of GRF values estimated using Method 2

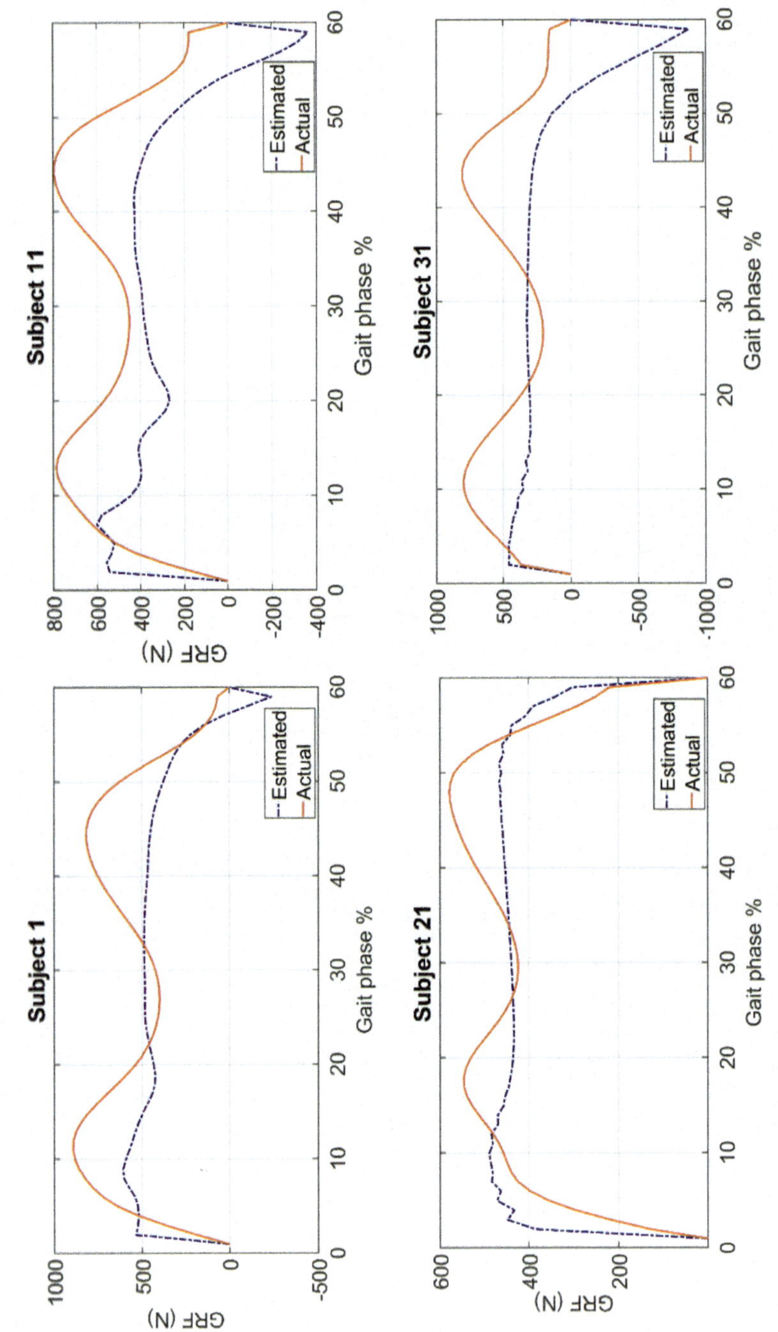

Fig. 7 Comparison of GRF values estimated using Method 3

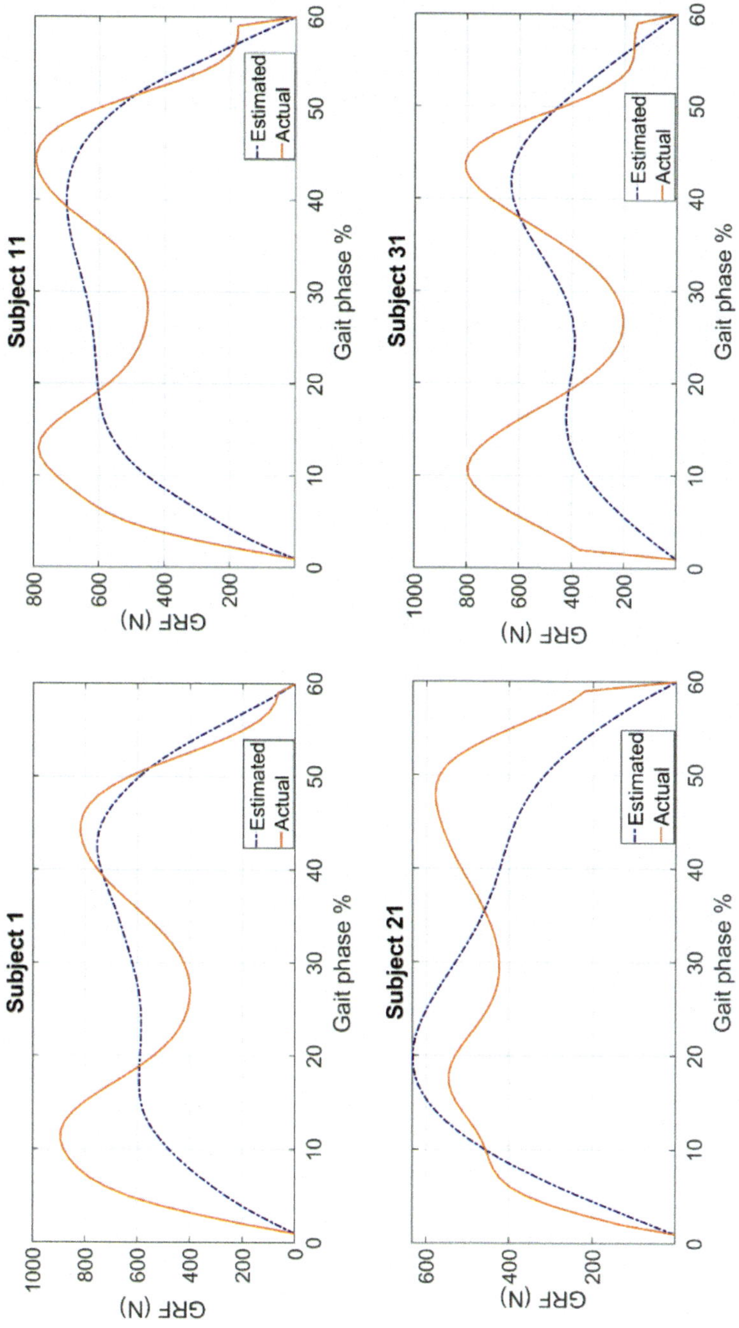

Fig. 8 Comparison of GRF values estimated using Method 4

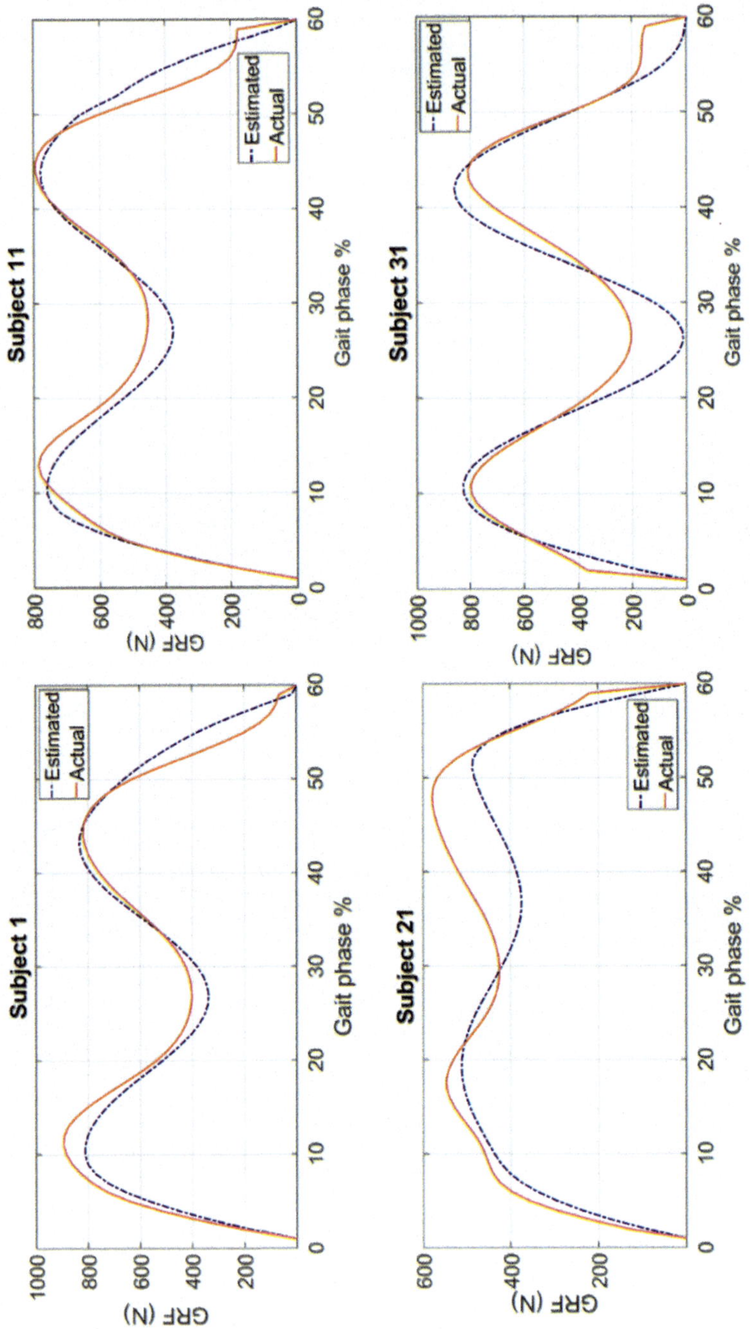

Fig. 9 Comparison of GRF values estimated using Method 5

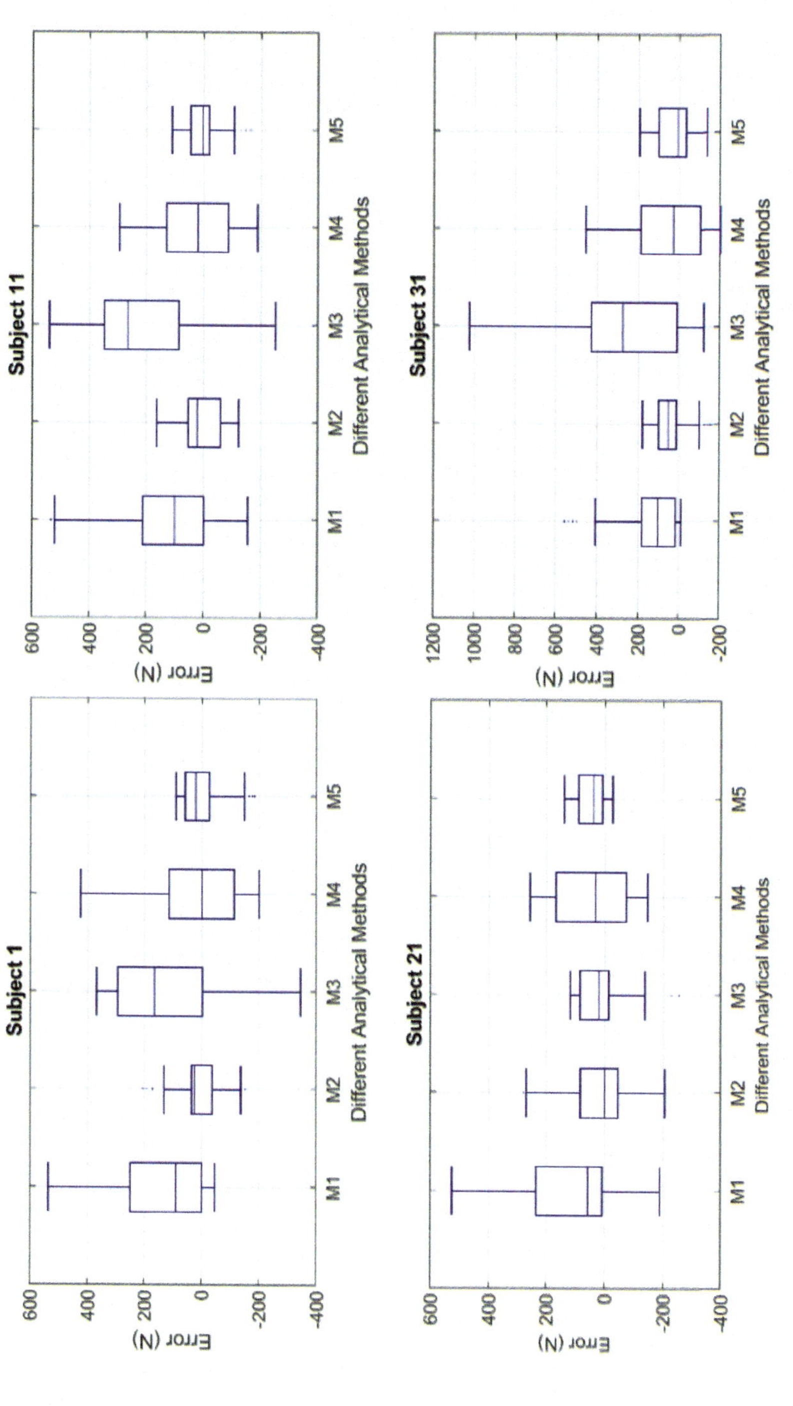

Fig. 10 Box-plot of vertical GRF error for different methods

Table 5 RMSE values of estimated GRF for different methods

S. no.	Subject	Method 1	Method 2	Method 3	Method 4	Method 5
1	1	219.99	80.16	215.68	193.68	65.80
2	11	213.58	70.11	278.16	154.79	58.67
3	21	202.12	125.31	80.66	136.42	69.39
4	31	230.10	82.91	362.77	215.93	92.70

Table 6 RMSE values of estimated GRF between **0–11%** phase for different methods

S. no.	Subject	Method 1	Method 2	Method 3	Method 4	Method 5
1	1	405.74	125.76	220.09	346.07	59.75
2	11	400.10	107.74	206.14	245.89	30.64
3	21	255.99	140.08	121.24	82.90	41.53
4	31	479.20	109.17	265.29	385.72	73.53

Table 7 RMSE values of estimated GRF between **12–49%** phase for different methods

S. no.	Subject	Method 1	Method 2	Method 3	Method 4	Method 5
1	1	92.60	42.35	230.11	150.32	43.64
2	11	75.94	46.94	273.97	139.25	42.52
3	21	82.04	69.69	67.62	122.04	79.20
4	31	86.71	67.68	301.66	170.51	98.78

Table 8 RMSE values of estimated GRF between **50–60%** phase for different methods.

S. no.	Subject	Method 1	Method 2	Method 3	Method 4	Method 5
1	1	264.10	114.19	149.59	73.77	116.04
2	11	262.33	87.14	346.78	56.92	107.65
3	21	366.14	221.94	70.69	207.82	53.59
4	31	182.19	98.77	577.13	71.43	88.13

greater precision, making them useful for both clinical and research applications in gait analysis.

5 Conclusion

In conclusion, this chapter has explored various techniques for estimating GRF. Through simulation studies and visual comparisons, it was observed that different techniques resulted in varying degrees of alignment with actual GRF values, particularly during the single-support and double-support phases of gait. Method 5 emerged relatively as the most accurate technique across multiple subjects, considering the low RMSE values. This indicates that adopting this method could enhance GRF predictions significantly, especially when coupled with more understanding of relationships between spring-constant and gait parameters. The integration of machine learning techniques, may also enhance the prediction accuracy by capturing complex patterns in diverse datasets. Additionally, the techniques can also be tested for various gait scenarios, including stair climbing, ramp walking, or on uneven surfaces. Overall, the techniques evaluated in this study not only provide valuable insights into estimating GRF during human gait but also pave the way for future advancements in the field of biomechanical research.

Acknowledgment The authors of this chapter are grateful to the Science and Engineering Research Board, Government of India for funding the project through grant CRG/2022/005578.

References

1. Kantha, P., et al.: A novel balance training approach: biomechanical study of virtual reality-based skateboarding. Front. Bioeng. Biotechnol. **11**, 1136368 (2023)
2. Buczek, F.L., et al.: Performance of an inverted pendulum model directly applied to normal human gait. Clin. Biomech. **21**(3), 288–296 (2006)
3. Whittington, B.R., Thelen, D.G.: A Simple Mass-Spring Model with Roller Feet Can Induce the Ground Reactions Observed in Human Walking (2009)
4. Kim, S., Park, S.: Leg stiffness increases with speed to modulate gait frequency and propulsion energy. J. Biomech. **44**(7), 1253–1258 (2011)
5. Li, T., Li, Q., Liu, T.: An actuated dissipative spring-mass walking model: predicting human-like ground reaction forces and the effects of model parameters. J. Biomech. **90**, 58–64 (2019)
6. Kim, B., Lim, H., Park, S.: Spring-loaded inverted pendulum modeling improves neural network estimation of ground reaction forces. J. Biomech. **113**, 110069 (2020)
7. Lim, H., Park, H., Park, S.: Compliant walking model with a curvy foot reflecting the position of ankle on reproducing the ankle torque profile. J. Mech. Sci. Technol. **29**, 2307–2311 (2015)
8. Jung, Y., et al.: Ground reaction force estimation using an insole-type pressure mat and joint kinematics during walking. J. Biomech. **47**(11), 2693–2699 (2014)
9. Ren, L., Jones, R.K., Howard, D.: Whole body inverse dynamics over a complete gait cycle based only on measured kinematics. J. Biomech. **41**(12), 2750–2759 (2008)
10. Shahabpoor, E., Pavic, A.: Estimation of vertical walking ground reaction force in real-life environments using single IMU sensor. J. Biomech. **79**, 181–190 (2018)
11. Komaris, D.-S., et al.: Effects of segment masses and cut-off frequencies on the estimation of vertical ground reaction forces in running. J. Biomech. **99**, 109552 (2020)
12. Ripic, Z., et al.: Prediction of gait kinetics using Markerless-driven musculoskeletal modeling. J. Biomech. **157**, 111712 (2023)

13. Eltoukhy, M., et al.: Prediction of ground reaction forces for Parkinson's disease patients using a Kinect-driven musculoskeletal gait analysis model. Med. Eng. Phys. **50**, 75–82 (2017)
14. Shourijeh, M.S., McPhee, J.: Foot–ground contact modeling within human gait simulations: from Kelvin-Voigt to hyper-volumetric models. Multibody Syst. Dyn. **35**(4), 393–407 (2015)
15. Lichtwark, G.A., et al.: Markerless motion capture provides accurate predictions of ground reaction forces across a range of movement tasks. J. Biomech. **166**, 112051 (2024)
16. Vonstad, E.K., et al.: Performance of machine learning models in estimation of ground reaction forces during balance exergaming. J. NeuroEng. Rehabil. **19**(1), 18 (2022)
17. Shalin, G., et al.: Prediction and detection of freezing of gait in Parkinson's disease from plantar pressure data using long short-term memory neural-networks. J. Neuroeng. Rehabil. **18**, 1–15 (2021)
18. Liu, D., et al.: Deep learning based ground reaction force estimation for stair walking using kinematic data. Measurement **198**, 111344 (2022)
19. Weidensager, L., et al.: Estimating vertical ground reaction forces from plantar pressure using interpretable high-dimensional approximation. Sports Eng. **27**(1), 3 (2024)
20. Kim, J., et al.: A deep learning model for 3D ground reaction force estimation using shoes with three uniaxial load cells. Sensors **23**(7), 3428 (2023)
21. Morris, C., et al.: Predicting 3D ground reaction force from 2D video via neural networks in sidestepping tasks. ISBS Proc. Arch. **39**(1), 300 (2021)
22. Raissi, M., Perdikaris, P., Karniadakis, G.E.: Physics-informed neural networks: a deep learning framework for solving forward and inverse problems involving nonlinear partial differential equations. J. Comput. Phys. **378**, 686–707 (2019)
23. Alqezweeni, M.M., Gorbachenko, V.I., Karmanova, Z.A.: Physics-informed neural networks and their implementation in MATLAB. In: 2022 8th International Conference on Contemporary Information Technology and Mathematics (ICCITM). IEEE (2022)
24. Lencioni, T., et al.: Human kinematic, kinetic and EMG data during different walking and stair ascending and descending tasks. Sci. Data **6**(1), 309 (2019)
25. Van den Bogert, A., De Koning, J.: On optimal filtering for inverse dynamics analysis. In: Proceedings of the IXth Biennial Conference of the Canadian Society for Biomechanics. Simon Fraser University Vancouver (1996)
26. Drillis, R., Contini, R., Bluestein, M.: Body segment parameters. Artif. Limbs **8**(1), 44–66 (1964)
27. Winter, D.A.: Biomechanics and Motor Control of Human Movement. Wiley (2009)
28. Mauersberger, M., et al.: Predicting ground reaction forces of human gait using a simple bipedal spring-mass model. R. Soc. Open Sci. **9**(7), 211582 (2022)

A Comprehensive Comparative Analysis of Gait Phases Using Machine Learning: Binary and Ternary Classifications with Dual Training Methods and Scaling Techniques

Amal Mekni, Jyotindra Narayan, and Hassène Gritli

Abstract Human walking is a fundamental activity, and understanding its intricacies is crucial for diagnosing and treating various gait abnormalities and musculoskeletal disorders. This chapter explores the classification of gait phases using advanced machine learning techniques, dividing these phases into binary and three subphases. The study draws on data from 100 individuals sourced from an open-source platform and applies two distinct training methodologies. The first method employs stratified random sampling, with 80% of the data from each subphase used for training and 20% for testing. The second method involves training on data from 80% of participants and testing on the remaining 20%. Before applying various machine learning algorithms, the dataset was evaluated using different scaling and preprocessing methods, including Min-Max Scaling (MMS), Standard Scaling (SS), and Principal Component Analysis (PCA). After proper scaling, we implemented and assessed the performance of several machine learning models, such as k-Nearest Neighbors (k-NN), Logistic Regression (LR), Decision Tree (DT), Random Forest (RF), Support Vector Machine (SVM), Naive Bayes (Gaussian, Bernoulli, Multinomial) (NB), Linear Discriminant Analysis (LDA), and Quadratic Discriminant Analysis (QDA). Each model was evaluated based on multiple metrics, including cross-validation score, Mean Squared Error (MSE), Root Mean Squared Error (RMSE), accuracy, and R^2

A. Mekni · H. Gritli (✉)
Laboratory of Robotics, Informatics and Complex Systems (RISC Lab—LR16ES07), National Engineering School of Tunis, University of Tunis El Manar, Tunis, Tunisia
e-mail: grhass@yahoo.fr

A. Mekni
e-mail: amal.mekni@istic.ucar.tn

J. Narayan
Department of Mechanical Engineering, Indian Institute of Technology Patna, Patna, India

Department of Computing, Imperial College London, London, UK

H. Gritli
Higher Institute of Information and Communication Technologies, University of Carthage, Hammam Chatt, Ben Arous, Tunisia

J. Narayan and H. Gritli (eds.), *Design and Control of Rehabilitation Robots*, Studies in Systems, Decision and Control 585, https://doi.org/10.1007/978-3-031-86977-8_5

score. In binary classification, RF achieved the highest accuracy of 97.33% and R^2 score of 88.91%, with k-NN closely following with 96.98% accuracy and a cross-validation score of 0.9409. For ternary classification, RF showed the best results with an accuracy of 96.58% and MSE of 0.0743, while k-NN reached 96.04% accuracy.

Keywords Gait phase classification · Machine learning · Random forest · Principal component analysis · k-nearest neighbors · Support vector machine · Logistic regression · Min-max scaling · Standard scaling

1 Introduction

Walking, a fundamental human activity, often taken for granted, involves the coordinated movement of various muscles, joints, and nerves. It is a complex process crucial for clinicians and researchers to understand, enabling the diagnosis and treatment of diverse gait abnormalities and musculoskeletal disorders [1–5]. Gait, the coordinated movement of the legs during walking, varies across individuals based on factors such as age and health. It comprises mainly two distinct phases (see Fig. 1): the stance phase (0–60%) and the swing phase (60–100%). The stance phase includes initial contact (0–2%): the moment when the foot first touches the ground, loading response (2–10%): the period after initial contact until the opposite foot leaves the ground, mid stance (10–30%): the body weight is aligned over the supporting foot, terminal stance (30–50%): the heel of the supporting foot begins to lift off the ground, and pre-swing (50–60%): the period just before the foot leaves the ground, preparing for the swing phase. The swing phase includes the initial swing (60–73%): for which the foot leaves the ground, and the leg begins to move forward, mid swing (73–87%): the leg is moving forward, with the foot clearing the ground, and terminal swing (87–100%): the leg is extending forward, preparing for initial contact with the ground again [1]. Recognizing different gait stages is essential for patient assessment and diagnosis, as well as in the automation of leg exoskeletons for individuals with movement difficulties [2, 4, 6–10].

Recent studies have explored using advanced algorithms in machine and deep learning models for gait analysis. For instance, Fricke et al. [11] achieved high accuracy in classifying normal and abnormal gait patterns using convolutional neural networks (CNN), with SVM and KNN performing less effectively. Krutaraniyom et al. [12] focused on classifying gait cycle phases, achieving high accuracy using a multilayer perceptron (MLP) classifier. This has implications for detecting periodic phases in both healthy individuals and those with Parkinson's disease. Bauman and Brandon [13] developed a machine learning algorithm to detect activity and gait phase in real-world scenarios, focusing on lower limb exoskeleton control. The study achieves high accuracy in activity-specific models, with potential applications in motion assistive devices. Park et al. [14] recently found that muscle synergy features improved gait phase classification accuracy when used with SVM, DT, k-NN, and neural network (NN) algorithms.

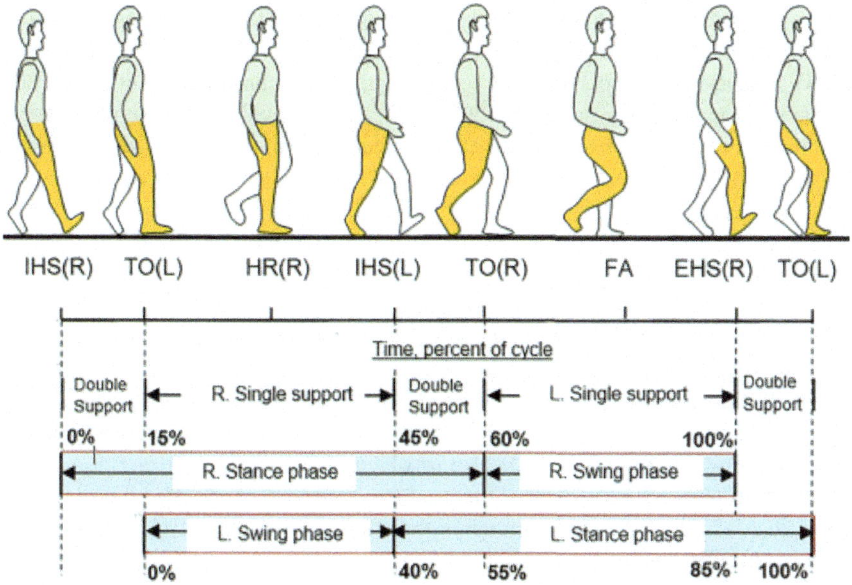

Fig. 1 Different gait phases of human walking [1]

Semwal et al. [4] described the algorithms employed for forward propagation in Long short-term memory (LSTM) and Gated Recurrent Unit (GRU) models, highlighting the crucial role of weight matrices, input vectors, and hidden states in accurately predicting gait trajectories based on input data and temporal dependencies. Furthermore, Lan et al. [15] applied LSTM, ResNet, and InceptionTime algorithms for deep learning analysis of 3D gait data, achieving high diagnostic accuracy in distinguishing between healthy and pathological gait patterns, identifying various pathologies, and determining the stroke onset time in children with gait disorders. Additionally, researchers have developed innovative models, such as a lightweight CNN-LSTM model designed by Moura-Coelho et al. [16] for classifying footstep phases across different terrains, and spatiotemporal deep learning programs like ST-DeepGait, introduced by Konz et al. [17], which use multi-layer recurrent neural networks (RNNs) to improve step detection accuracy under various walking conditions. LSTM models have shown strong performance in gait estimation, even with noisy data [18].

Recent advancements in gait analysis include noteworthy comparisons with classifiers like XGBoost and RF, underscoring the efficacy of SVM in predicting human activities [19]. In another work by Moura-Coelho et al. [16], an innovative lightweight CNN-LSTM model is introduced for diverse footstep phase classifications in various terrains. Additionally, spatiotemporal DL programs like ST-DeepGait, introduced by Konz et al. [17], utilized multi-layer RNNs to enhance step detection accuracy under varying walking conditions, with LSTM demonstrating commendable performance in gait estimation, especially in noisy data scenarios [18].

Two different training strategies are applied. The first approach randomly allocated 80% of the dataset for training and the remaining 20% for testing, ensuring each gait phase was proportionately represented. The second strategy allocated 80% of the participants for training and 20% for testing, providing diverse phase representations and consistent evaluation of the model performance across different users. Furthermore, in the study of gait phase classification, we will compare various levels of granularity, ranging from binary sub-phases to three sub-phases. The binary phases consist of the stance phase, which spans from 0 to 59% of the gait cycle (from foot contact to foot lift-off), and the swing phase, which spans from 60 to 100% of the gait cycle (from foot lift-off to foot contact). For a more detailed classification, we also considered the three phases: the stance phase, the swing phase, and the double support phase (when both feet are in contact with the ground), in order to increase the precision further.

The structure of the rest of this chapter is as follows: Sect. 2 covers the methodology, including details about the dataset involved in this work, the description of the different ML algorithms as well as the used hyperparameters. Section 3 provides the numerical results of classification using the employed ML models. Finally, Sect. 5 presents the conclusion and future works.

2 Methodology

2.1 Dataset Involved

In this study, we utilized a dataset comprising 100 healthy adults aged between 21 and 79 years (mean ± SD: 41.9 ± 15.3), with 51 males and 49 females. The dataset includes demographic details such as age, gender, height, and weight, providing valuable insights into the variation of gait patterns across different age groups and genders. Additionally, it features extensive biomechanical, kinematic, and kinetic data collected under various walking conditions. To ensure participant anonymity and maintain data integrity, each individual was assigned a unique identifier. For more information about the dataset, please refer to the work by Bahadori et al. [20, 21], accessible at https://data.mendeley.com/datasets/wwnvw28n2m/1. Our research concentrated on specific lower-body movements, including left and right hip flexion/extension (Fig. 2), abduction/adduction (Fig. 3), intra/extra rotation (Fig. 4), knee flexion/extension (Fig. 5), and ankle dorsiflexion/plantarflexion (Fig. 6), totaling ten movements. Figure below depicts the movements of the left and right limbs throughout the gait cycle.

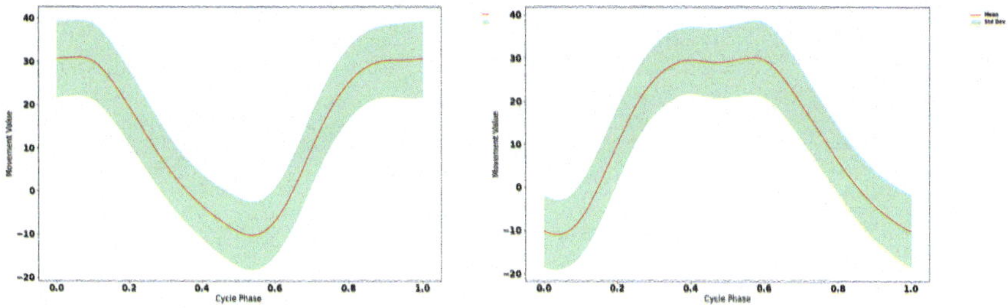

(a) Curve 1: Hip Flexion/Extension (Left) (b) Curve 2: Hip Flexion/Extension (Right)

Fig. 2 Left and right hip flexion/extension

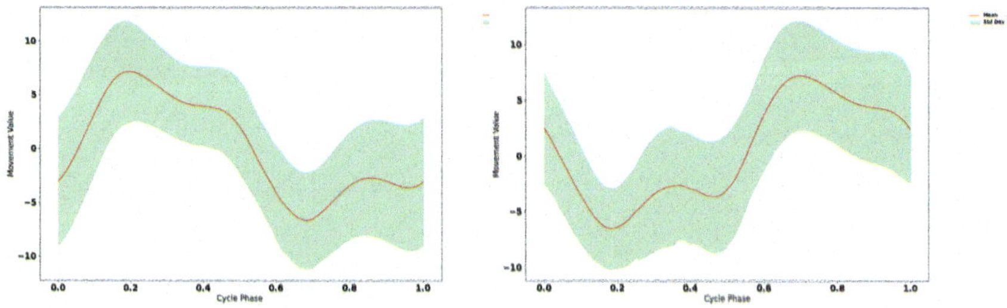

(a) Curve 3: Hip Abduction/Adduction (Left) (b) Curve 4: Hip Abduction/Adduction (Right)

Fig. 3 Hip abduction/adduction (left and right)

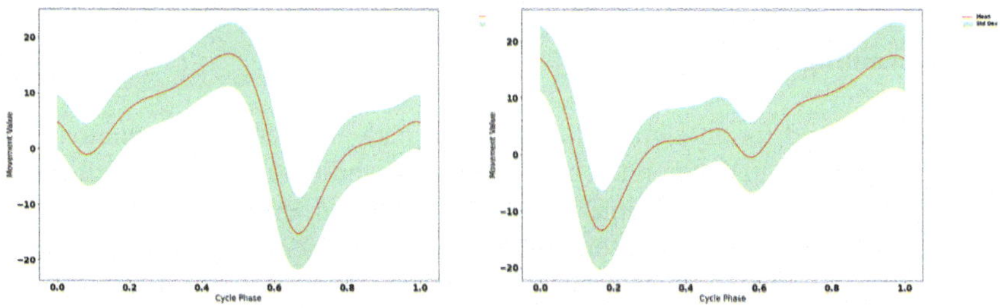

(a) Curve 5: Ankle Flexion/Extension (Left) (b) Curve 6: Ankle Flexion/Extension (Right)

Fig. 4 Ankle flexion/extension (left and right)

2.2 *Evaluation Metrics*

The assessment criteria are critical in evaluating the efficiency of ML models and helping in the choice of the most suitable model for a specific problem.

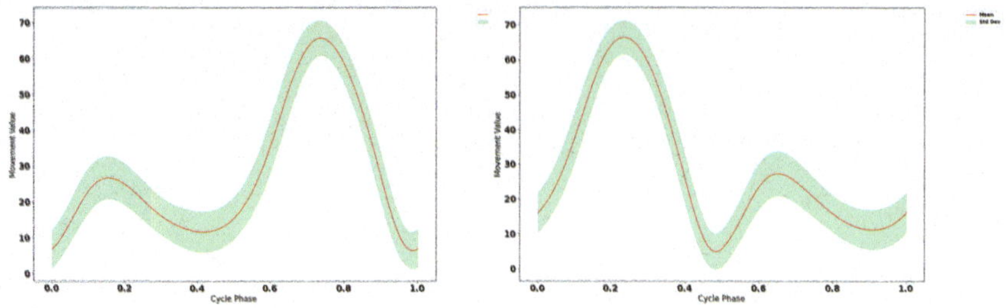

(a) Curve 7: Knee Flexion/Extension (Left) (b) Curve 8: Knee Flexion/Extension (Right)

Fig. 5 Knee flexion/extension (left and right)

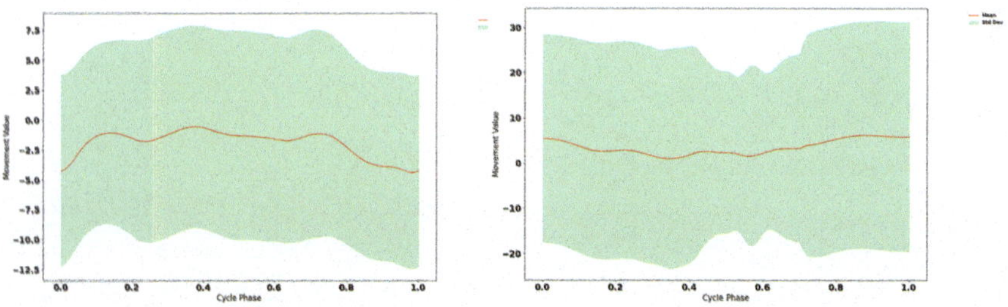

(a) Curve 9: Hip Rotation (Left) (b) Curve 10: Hip Rotation (Right)

Fig. 6 Hip rotation (left and right)

2.2.1 Training Time

The time required for an algorithm to train on data and generate predictions is a crucial evaluative parameter, reflecting the model's ability to process real-time data efficiently. A shorter training time is generally preferred, enabling faster and more efficient processing, which is essential for real-time applications. Training and prediction times can vary based on data size, complexity, the algorithm chosen, and the available hardware and software resources.

2.2.2 Cross-validation (CV) Score

The performance and generalization ability of ML models are evaluated using this approach. It involves partitioning the dataset into multiple subsets (folds) and then training and testing the model using different combinations of these folds. In k-fold cross-validation, the dataset is split into k subsets, and the model is trained on $k-1$ subsets while being tested on the remaining one. This process is reiterated k times, and the outcomes from each iteration are averaged to gauge the overall performance of the model [22–24].

2.2.3 Mean Squared Error (MSE)

The error between the predicted and actual values in regression, known as MSE, is measured to evaluate the model's performance. This involves calculating the average of the squared variances between the actual and predicted values:

$$\text{MSE} = \frac{1}{n}\sum_{i=1}^{n}(y_i - \hat{y}_i)^2 \quad (1)$$

In the given formula (1), n represents the total number of data points. y_i stands for the actual value of the i-th data point, while $\hat{y}_i$ represents its predicted value. The summation symbol, $\sum$, indicates the summation over all the data points as mentioned in the Refs. [22–24].

2.2.4 Root Mean Squared Error (RMSE)

The RMSE assesses the difference between expected and observed values, providing a more accurate measure of prediction inaccuracies than the MSE. The RMSE formula is:

$$\text{RMSE} = \sqrt{\text{MSE}} \quad (2)$$

In the given Eq. (2), the symbol n denotes the total count of observations. y_i denotes the observed value, and $\hat{y}_i$ The predicted value is represented by it.

2.2.5 R^2 Score

The R^2 score quantifies the fit between the model and the data, and is defined via the following expression:

$$R^2 = 1 - \frac{SS_{\text{res}}}{SS_{\text{tot}}} \quad (3)$$

The total variance of the observed values is represented by the sum of squared residuals (SS_{res}), while SS_{tot} denotes the total variance of the observed values. An R^2 score of 1 indicates a perfect fit, while a score of 0 suggests that the model does not explain any variability. Additionally, a negative score signifies that the model performs worse than using the mean prediction [22–24].

2.2.6 Accuracy

The metric of accuracy determines the percentage of precisely identified data points compared to the whole number of data points, and it is commonly utilized in classification assignments. The calculation of accuracy is as follows [25]:

$$\text{Accuracy} = \frac{\text{TP} + \text{TN}}{\text{TP} + \text{TN} + \text{FP} + \text{FN}} \tag{4}$$

The count of True Positives is represented by TP, the count of False Positives is denoted by FP, TN stands for True Negatives, and FN indicates False Negatives.

2.2.7 Precision

The model's precision measures how accurately it identifies true positive instances out of all the instances predicted as positive [22–24]:

$$\text{Precision} = \frac{\text{TP}}{\text{TP} + \text{FP}} \tag{5}$$

2.2.8 Recall

The effectiveness of the ML model in correctly classifying positive cases among all actual positive instances is computed via the Recall function, which is defined as follows:

$$\text{Recall} = \frac{\text{TP}}{\text{TP} + \text{FN}} \tag{6}$$

2.2.9 F1-Score

The F1-score evaluates the model's performance by combining precision and recall in a balanced manner:

$$\text{F1-score} = 2 \times \frac{\text{Precision} \times \text{Recall}}{\text{Precision} + \text{Recall}} \tag{7}$$

The F1-score varies between 0 and 1, with greater values indicating enhanced performance and a more optimal balance between precision and recall [26].

2.3 Scaling Techniques

Before applying ML algorithms, preprocessing the data with scaling techniques is essential. These methods standardize or normalize the features, the goal is to make sure that every feature has an equal impact on the model and enhances its overall performance. The following are the three scaling techniques that are frequently utilized: MMS, SS, and PCA.

2.3.1 Min-Max Scaling (MMS)

MMS is a normalization technique that adjusts features by rescaling them within a defined range, typically from 0 to 1. This technique is especially valuable when features have varying units or scales, as it standardizes all features to the same range, preventing those with larger scales from disproportionately influencing the model [27, 28].

$$X_{\text{scaled}} = \frac{X - X_{\text{min}}}{X_{\text{max}} - X_{\text{min}}} \tag{8}$$

In the given expression (8), the initial feature value is denoted by X, the minimum value of the feature is represented as X_{min}, and the maximum value of the feature is denoted by X_{max}.

2.3.2 Standard Scaling (SS)

The SS approach normalizes the features, which are firstly adjusted by centering them at 0, and then are scaled to have a standard deviation of 1. It works especially well when the dataset is normally distributed, guaranteeing that each feature has an equal impact on the model [28].

$$X_{\text{scaled}} = \frac{X - \mu}{\sigma} \tag{9}$$

In the feature value Eq. (9), the value of X is denoted by the feature, the mean of the feature is represented by μ, and the standard deviation of the feature is indicated by σ.

2.3.3 Principal Component Analysis (PCA)

PCA is not purely a scaling technique but a dimensionality reduction method. It converts the dataset. The given data is transformed into a new collection of perpendicular components, which are called principal components. These components can identify the highest amount of variance that can be present in the data. By simplifying the dataset and retaining most of its essential information, PCA enhances computational efficiency and model performance, particularly when dealing with numerous correlated features [29].

$$X_{\text{scaled}} = XW \tag{10}$$

The original data matrix is denoted by X, the matrix of eigenvectors corresponding to the principal components is denoted by W, and the transformed data matrix consisting of the principal components is denoted by X_{scaled}.

By utilizing these previous scaling methods, we can guarantee that the data is properly prepared, which is a crucial part of enhancing the efficiency of ML algorithms.

2.4 Machine Learning Techniques

In classification models, training data is used to develop an ML model that predicts or classifies the classes of new samples [25, 30, 31]. To achieve this, we first applied scaling methods to the dataset features before preparing both training and testing datasets. We proceeded to use a range of classification algorithms [32], which include k-NN, LR, DT, RF, SVM, NB, LDA, and QDA, to classify the different walking subphases.

2.4.1 *k*-Nearest Neighbors (*k*-NN)

The k-NN algorithm categorizes a new data point The most common can be determined by the class among its closest neighbors in the feature space, guaranteeing accurate classification:

$$\text{Prediction} = \text{majority class among } k \text{ nearest neighbors} \tag{11}$$

The predicted class is denoted by "Prediction" in this Eq. (11), and k specifies the number of nearest neighbors utilized for classification.

2.4.2 Logistic Regression (LR)

LR is a method for binary classification that involves using one or more predictor variables to predict the probability of a binary outcome.

$$y = \frac{e^{(b_0+b_1x)}}{1+e^{(b_0+b_1x)}} \tag{12}$$

In this previous expression (12), the input value is denoted by x, the value predicted is denoted by y, the bias or intercept term is indicated by b_0, and the coefficient associated with the input x is denoted by b_1. The regularization strength (C) is utilized to avoid overfitting by implementing a log-loss function derived from maximizing the log-likelihood function of the equation.

2.4.3 Decision Trees (DTs)

DTs are characterized by their hierarchical structures, where nodes represent features and branches embody potential values, playing a crucial role in categorization. They efficiently order instances based on their feature values, thereby aiding the classification process. For example:

$$\begin{cases} \text{if (Feature} \leq \text{ Threshold) : Go to left branch} \\ \text{else: Go to right branch} \end{cases} \tag{13}$$

In this context, the Feature signifies a specific attribute of the data, and the Threshold refers to a value used to guide classification at each node. The total number of nodes is referred to as the maximum depth.

2.4.4 Random Forest (RF)

RF is an ensemble learning method that generates multiple DTs and aggregates their outputs to reduce the risk of overfitting. By randomly selecting subsets of data and features for each tree, RF generates robust predictions and enhances feature importance:

$$\text{Prediction} = \frac{1}{N}\sum_{i=1}^{N} \text{DecisionTree}_i(\text{Input}) \tag{14}$$

In this Eq. (14), Prediction denotes the final classification outcome, the number (N) of DTs is known as estimators, and DecisionTree_i represents the output of the i-th DT.

2.4.5 Support Vector Machine (SVM)

The SVM is based on the idea of "margin", which is the area on both sides of a hyperplane that divides two sets of data. The decision boundary is represented as:

$$w^T x + c = 0 \tag{15}$$

where w denotes the weight vector of feature coefficients, x denotes the input data vector, and c is the bias hyperparameter that affects the positioning of the decision boundary.

2.4.6 Naive Bayes (NB)

NB is efficient in memory usage and predictive accuracy due to its minimal storage requirements for prior and conditional probabilities:

$$P(z \mid x_1, \ldots, x_n) \propto P(z) \prod_{i=1}^{n} P(x_i \mid z) \tag{16}$$

where z represents the class variable, x_i refers to the features, and $P(\cdot)$ and $P(\cdot \mid \cdot)$ denote prior and conditional probabilities, respectively. Variance smoothing ($\text{Var}_{\text{smoothing}}$) enhances the distribution's variance, accommodating distant samples from the mean.

2.4.7 Linear Discriminant Analysis (LDA)

LDA involves identifying a linear combination of features to distinguish between two or more classes in data classification. The separability of classes is enhanced by maximizing the ratio of between-class variance to within-class variance.

$$\text{LDA} = \arg\max \frac{|\mu_1 - \mu_2|^2}{\sigma^2} \tag{17}$$

The means of the classes are denoted as μ_1 and μ_2, and the pooled variance is represented by σ^2.

2.4.8 Quadratic Discriminant Analysis (QDA)

QDA extends LDA by assigning a unique covariance matrix to each class, it captures more intricate relationships between the features:

$$P(x \mid z) = \frac{1}{\sqrt{(2\pi)^k |\Sigma_z|}} e^{-\frac{1}{2}(x-\mu_z)^T \Sigma_z^{-1} (x-\mu_z)} \tag{18}$$

where μ_z is the mean vector for class z, and Σ_z is the covariance matrix.

2.5 *Performance Metrics*

Performance in classification tasks is primarily evaluated using **accuracy**, which measures the proportion of correctly classified instances among all instances, A confusion matrix provides a detailed analysis of classification results by comparing predicted and actual classes in a tabular format. Its components-True Positives (TP), True Negatives (TN), False Positives (FP), and False Negatives (FN)-enable the calculation of various performance metrics, including accuracy, F1-score, recall, and precision [25].

2.6 Dataset Preparation Methods for Training

Highlighting the significance of training and testing datasets is crucial before applying classification algorithms. Careful consideration is given to data segmentation before model implementation, ensuring clarity and accurate model performance through effective data-splitting strategies [8–10].

2.6.1 First Method

The training and testing sets were created from the dataset to ensure that each phase was equally represented. Specifically, the training set received 80% of each participant's data through random assignment, while the remaining 20% was set aside for testing purposes. This method aims to provide diverse instances for comprehensive exposure, enabling a robust evaluation of classification capabilities across all phases.

2.6.2 Second Method

Participant-based data partitioning entails allocating 80% of participants for training and setting aside 20% for testing. This approach yields a mixed output covering both the stance and swing phases. By strategically distributing participants among gait phases, we aim for a heterogeneous representation, ensuring consistent assessment of model performance across various users.

3 Scaling-Based Binary Classification

3.1 Comparative Analysis of MMS, SS, and PCA Performance

As shown in Tables 1 and 2, for the first training method, RF consistently outperforms other algorithms, showing the highest accuracy 97.33%, lowest MSE 0.0267, and best R^2 score 88.91%. Meanwhile, k-NN follows closely behind, particularly in terms of cross-validation score 0.9409 and accuracy 96.98%. Both RF and k-NN show exceptional performance in minimizing error and maximizing accuracy.

In the second training method, k-NN performs strongly, with a high cross-validation score of 0.9480 and an accuracy of 94.38%. However, QDA demonstrates competitive results in terms of RMSE 0.2439 and R^2 score 75.35%, indicating its ability to maintain high performance across both methods.

Under SS, as shown in Tables 3 and 4, under SS for the first method, RF again exhibits the best results, with an accuracy of 96.69% and an impressive R^2 score of 86.26%. *k*-NN also shows notable performance with a high cross-validation score of 0.9423 and an accuracy of 94.27%. These results suggest that RF and *k*-NN maintain strong performance across different training and scaling methods.

In the second method, RF achieves the highest cross-validation score 0.9300, and accuracy of 93.74%, reinforcing its reliability. However, *k*-NN continues to demonstrate solid performance with a cross-validation score of 0.9400 and an accuracy of 94.41%, indicating its adaptability across scaling methods.

When applying PCA, as seen in Tables 5 and 6, When applying PCA, RF shows the highest accuracy of 96.82% and lowest RMSE of 0.1800 in the first method, demonstrating its robustness in reducing dimensionality without compromising performance. *k*-NN also performs well, with a cross-validation score of 0.9494 and accuracy of 95.29%, indicating its strong capacity to handle dimensionality reduction.

Table 1 Min-max scaling for each algorithm (first training method)

Algorithm	Cross-validation score	MSE	RMSE	Accuracy (%)	R^2 score (%)
k-NN	**0.9409**	**0.0302**	**0.174**	**96.98**	**87.48**
LR	0.9334	0.0703	0.2651	92.97	70.85
DT	0.9011	0.0470	0.2170	95.30	80.50
RF	**0.9454**	**0.0267**	**0.1635**	**97.33**	**88.91**
SVM	0.9578	0.0396	0.1990	96.04	83.58
NB	0.9184	0.0876	0.2960	91.24	63.66
LDA	0.9228	0.0807	0.2841	91.93	66.54
QDA	0.9489	0.0574	0.2396	94.26	76.19

Table 2 Min-max scaling for each algorithm (second training method)

Algorithm	Cross-validation score	MSE	RMSE	Accuracy (%)	R^2 score (%)
k-NN	**0.9480**	**0.0562**	**0.2371**	**94.38**	**76.69**
LR	0.8708	0.1144	0.3146	88.56	52.58
DT	0.9000	0.0900	0.2900	91.23	63.63
RF	0.9300	0.0600	0.2500	93.71	73.92
SVM	0.8700	0.1100	0.3100	89.40	56.04
NB	0.8200	0.1800	0.4000	82.29	26.60
LDA	0.9197	0.0823	0.2868	91.77	65.90
QDA	**0.9459**	**0.0595**	**0.2439**	**94.05**	**75.35**

Table 3 Standard scaling for each algorithm (first training method)

Algorithm	Cross-validation score	MSE	RMSE	Accuracy (%)	R^2 score (%)
k-NN	**0.9423**	**0.0573**	**0.2386**	**94.27**	**76.25**
LR	0.9324	0.0695	0.2634	93.05	71.19
DT	0.9000	0.0600	0.2500	93.61	73.52
RF	**0.9400**	**0.0300**	**0.1800**	**96.69**	**86.26**
SVM	0.9300	0.0700	0.2600	93.31	72.25
NB	0.8400	0.1600	0.3700	84.35	61.56
LDA	0.9228	0.0807	0.2841	91.93	66.54
QDA	0.9348	0.0733	0.2707	92.67	69.62

Table 4 Standard scaling for each algorithm (second training method)

Algorithm	Cross-validation score	MSE	RMSE	Accuracy (%)	R^2 score (%)
k-NN	**0.9400**	**0.0559**	**0.2364**	**94.41**	**76.81**
LR	0.9286	0.0636	0.2519	93.64	73.64
DT	0.9000	0.0900	0.3000	91.14	63.27
RF	**0.9300**	**0.0600**	**0.2500**	**93.74**	**74.03**
SVM	0.9300	0.0600	0.2500	93.86	74.53
NB	0.8400	0.1500	0.3600	85.08	62.23
LDA	0.9197	0.0823	0.2868	91.77	65.90
QDA	0.9344	0.0758	0.2754	92.42	68.57

Table 5 PCA for each algorithm (first training method)

Algorithm	Cross-validation score	MSE	RMSE	Accuracy (%)	R^2 score (%)
k-NN	**0.9494**	**0.0471**	**0.2167**	**95.29**	**80.47**
LR	0.9354	0.0664	0.2577	93.36	72.46
DT	0.9200	0.0500	0.2200	94.75	78.24
RF	**0.9400**	**0.0300**	**0.1800**	**96.82**	**86.80**
SVM	0.9400	0.0600	0.2500	93.52	73.14
NB	0.8000	0.2000	0.4100	80.50	66.54
LDA	0.9228	0.0807	0.2841	91.93	66.54
QDA	0.9271	0.0762	0.2761	92.38	68.39

Table 6 PCA for each algorithm (second training method)

Algorithm	Cross-validation score	MSE	RMSE	Accuracy (%)	R^2 score (%)
k-NN	**0.9490**	**0.0515**	**0.2270**	**94.85**	**78.64**
LR	0.9305	0.0600	0.2450	94.00	75.11
DT	0.9200	0.0700	0.2600	92.94	70.73
RF	0.9400	0.0600	0.2500	93.92	74.82
SVM	**0.9300**	**0.0600**	**0.2400**	**94.09**	**75.49**
NB	0.8000	0.2000	0.4200	79.76	58.66
LDA	0.9197	0.0823	0.2868	91.77	65.90
QDA	0.9284	0.0818	0.2859	91.82	66.11

For the second method, k-NN leads with a cross-validation score of 0.9490 and an accuracy of 94.85%. SVM, however, achieves competitive performance with an accuracy of 94.09% and R^2 score of 75.49%, highlighting its efficiency in the PCA scaling method.

3.2 Performance Metrics

Figures 19, 20, 21, 22, 23, 24, 25 and 26 show the confusion matrices obtained using the various ML algorithms adopted for the two training methods. We considered only the MMS technique for scaling. The numbers from 0 to 1 represent the class labels (Figs. 7, 8, 9, 10, 11, 12, 13, 14, 15, 16, 17 and 18).

In the first training method, the k-NN algorithm achieved a maximum correct classification rate of 98.05% for Class 0, with a misclassification rate of 1.95% where instances were incorrectly labeled as Class 1. The LR model performed best for Classes 1, correctly classifying 93.50% of the instances, though 6.50% were misclassified as Classes 0. DT achieved a maximum classification rate of 95.08% for

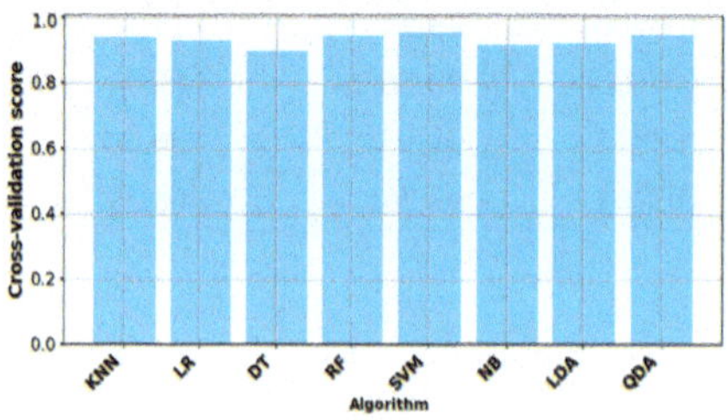

(a) First Method

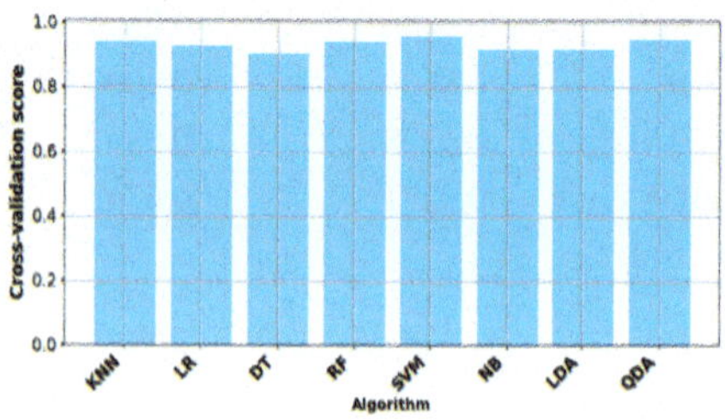

(b) Second Method

Fig. 7 Cross-validation scores for each algorithm (min-max scaling with two training methods)

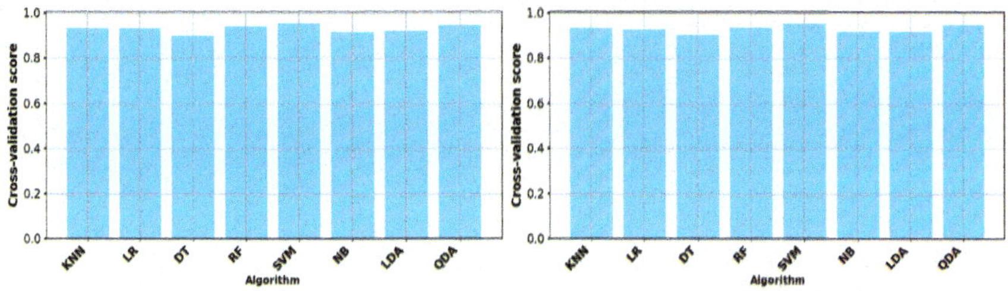

(a) First Method (b) Second Method

Fig. 8 Cross-validation scores for each algorithm (standard scaling with two training methods)

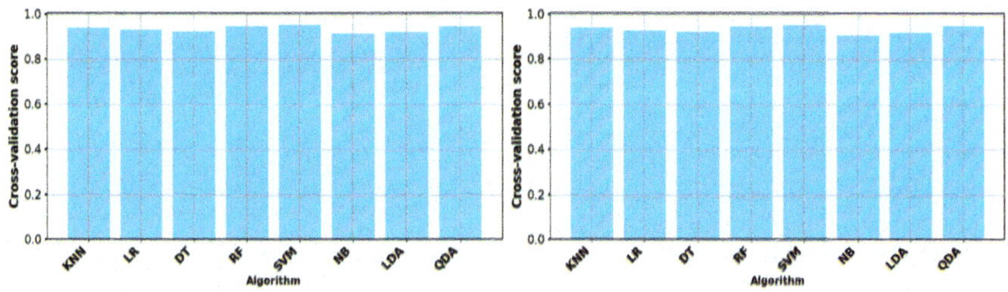

(a) First Method (b) Second Method

Fig. 9 Cross-validation scores for each algorithm (PCA with two training methods)

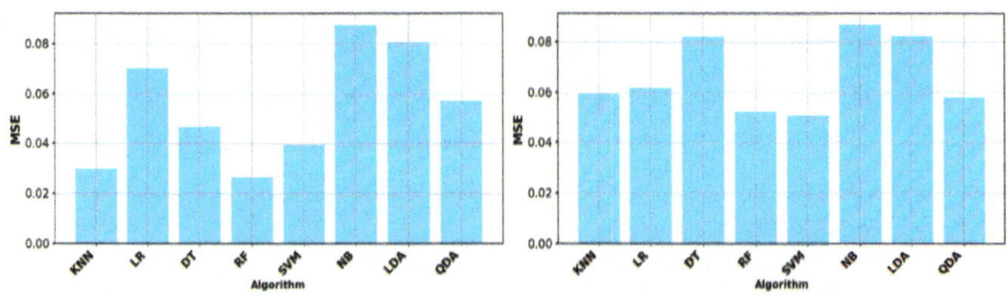

(a) First Method (b) Second Method

Fig. 10 MSE for each algorithm (min-max scaling with two training methods)

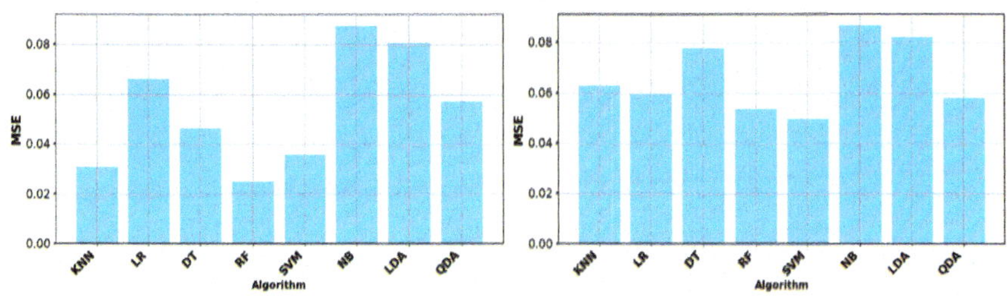

(a) First Method (b) Second Method

Fig. 11 MSE for each algorithm (standard scaling with two training methods)

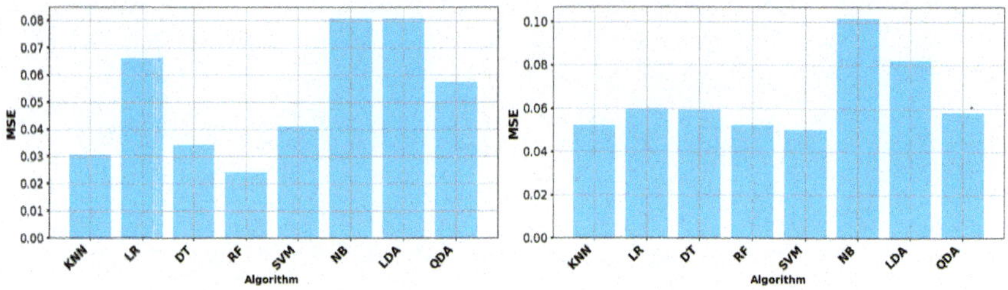

(a) First Method (b) Second Method

Fig. 12 MSE for each algorithm (PCA with two training methods)

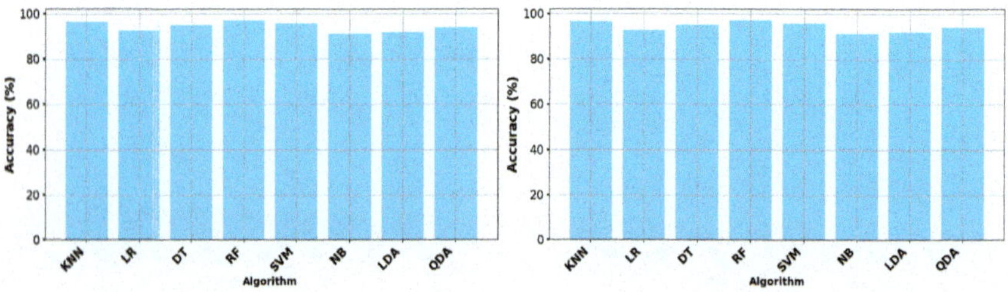

(a) First Method (b) Second Method

Fig. 13 Accuracy (%) for each algorithm (min-max scaling with two training methods)

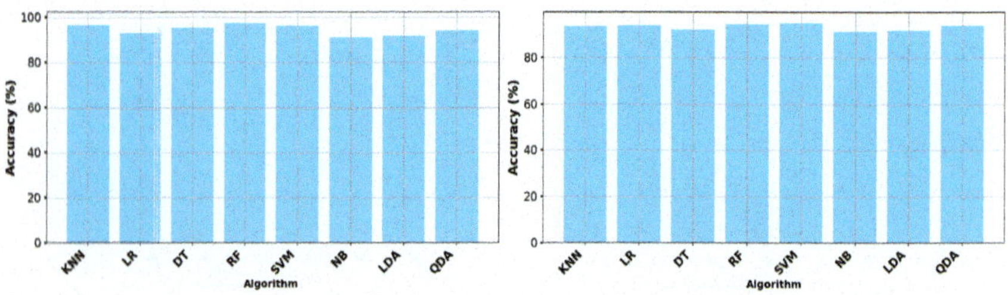

(a) First Method (b) Second Method

Fig. 14 Accuracy (%) for each algorithm (standard scaling with two training methods)

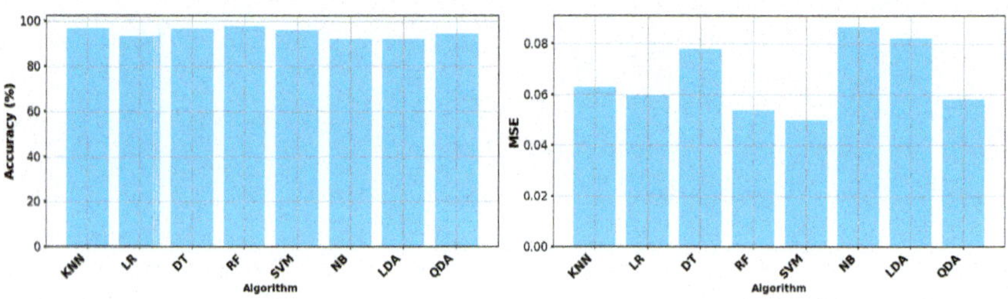

(a) First Method (b) Second Method

Fig. 15 Accuracy (%) for each algorithm (PCA with two training methods)

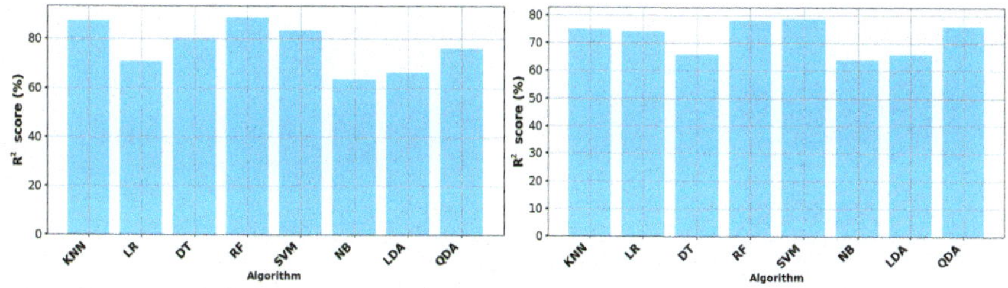

(a) First Method

(b) Second Method

Fig. 16 R^2 score for each algorithm (min-max scaling with two training methods)

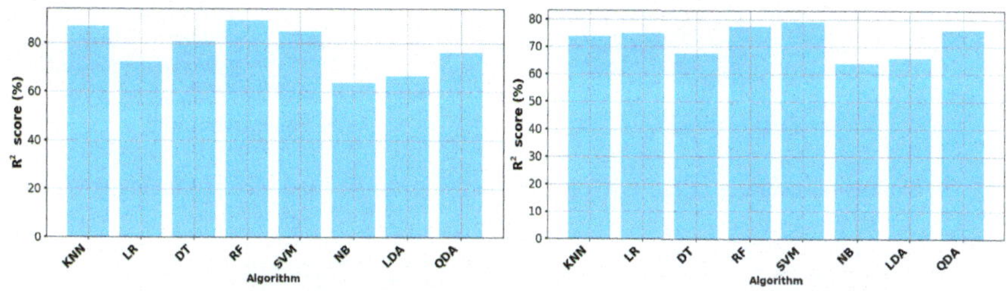

(a) First Method

(b) Second Method

Fig. 17 R^2 score for each algorithm (standard scaling with two training methods)

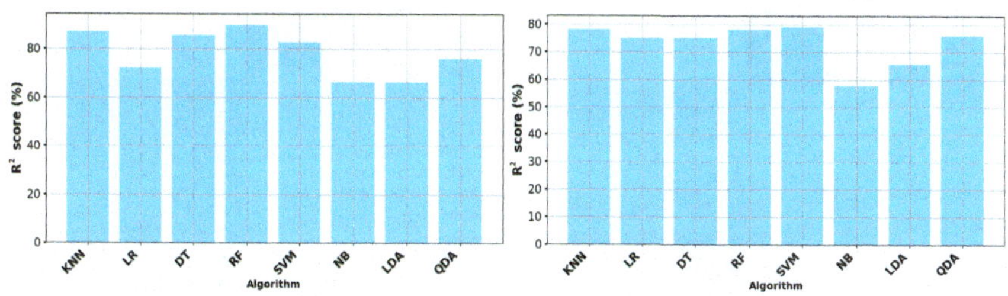

(a) First Method

(b) Second Method

Fig. 18 R^2 score for each algorithm (PCA with two training methods)

Class 1, with a 4.92% misclassification rate for Class 0. The RF algorithm demonstrated strong performance, achieving the highest classification rate of 97.17% for Class 1, and a 2.83% misclassification rate for Class 0. The SVM algorithm showed a maximum accuracy of 93.83% for Class 1, but 6.17% of the instances were misclassified as Class 0. NB reached a correct classification rate of 90.37% for Class 0, though 9.63% were misclassified as Class 1. The LDA model performed best for Class 0, with a correct classification rate of 92.44%, while 7.56% of the instances were misclassified as Class 1. The QDA model achieved a maximum correct classification rate of 96.65% for Class 0, though 3.05% of the instances were misclassified as Class 1.

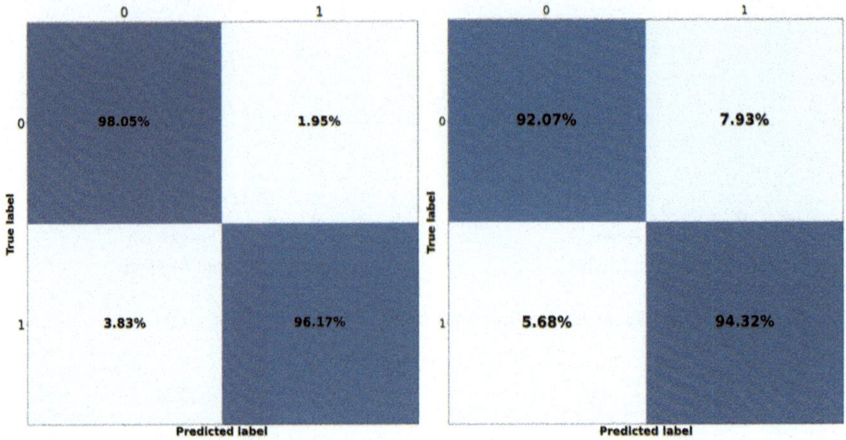

(a) First method (b) Second method

Fig. 19 Confusion matrices of the *k*-NN algorithm using the first and second training methods

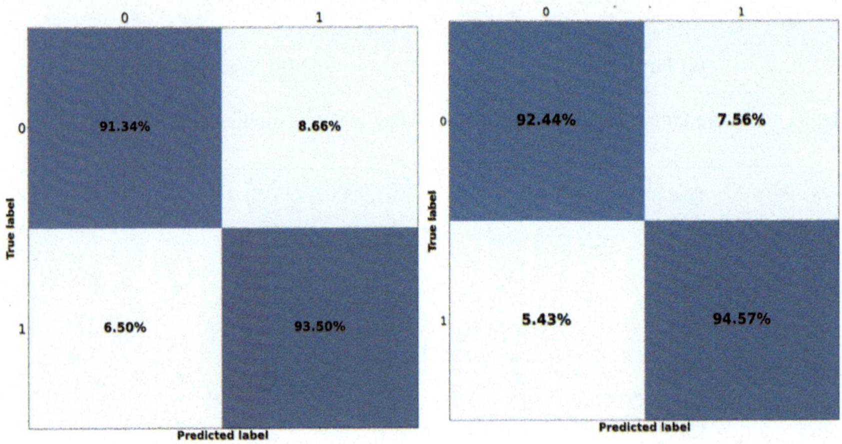

(a) First method (b) Second method

Fig. 20 Confusion matrices of the LR classification algorithm using the first and second training methods

In the second training method, the *k*-NN algorithm achieved a maximum correct classification rate of 94.32% for Class 1, with a misclassification rate of 5.68% where instances were incorrectly labeled as Class 0. The LR model performed best for Class 1, correctly classifying 94.57% of the instances, though 5.43% were misclassified as Class 0. DT achieved a maximum classification rate of 94.07% for Class 1, with a 5.93% misclassification rate for Class 0. The RF algorithm demonstrated strong performance, achieving the highest classification rate of 94.74% for Class 1, and a 5.26% misclassification rate for Class 0. The SVM algorithm showed a correct classification of 94.07%

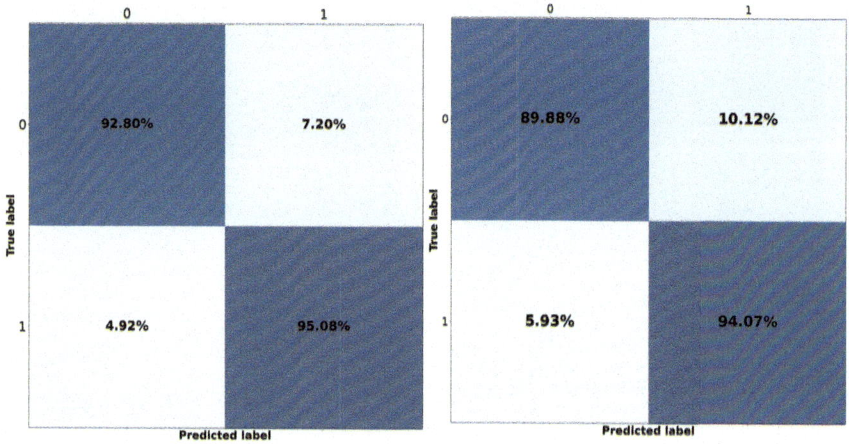

(a) First method (b) Second method

Fig. 21 Confusion matrices of the DT classifier using the first and second training methods

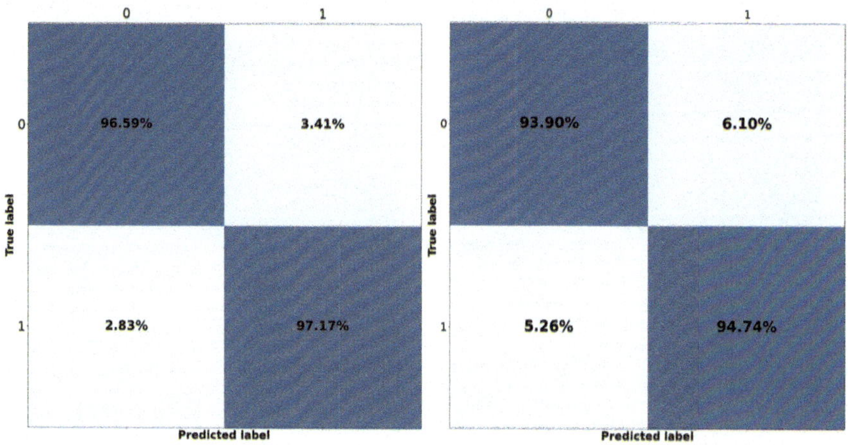

(a) First method (b) Second method

Fig. 22 Confusion matrices obtained using the RF classification algorithm and for the first and second training methods

for Class 1, but 5.93% of the instances were misclassified as Class 0. NB reached a correct classification rate of 91.34% for Class 0, though 8.66% were misclassified as Class 1. The LDA model performed best for Class 0, with a correct classification rate of 90.65%, while 9.35% of the instances were misclassified as Class 0. The QDA model achieved a maximum correct classification rate of 96.59% for Class 0, though 3.41% of the instances were misclassified as Class 1.

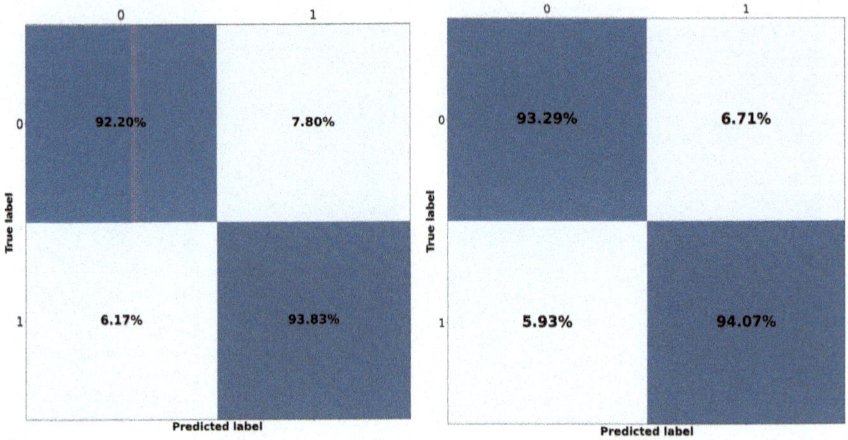

(a) First method (b) Second method

Fig. 23 Confusion matrices obtained via the SVM classifier and using the first and second training methods

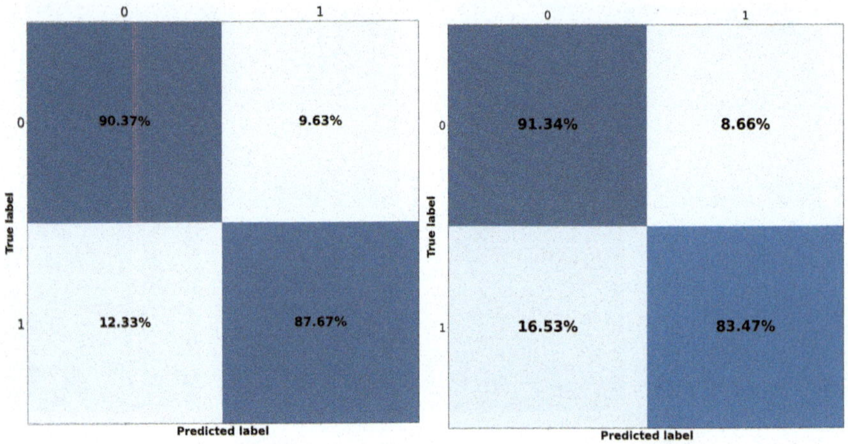

(a) First method (b) Second method

Fig. 24 Confusion matrices obtained with the NB classifier and by using the first and second training methods

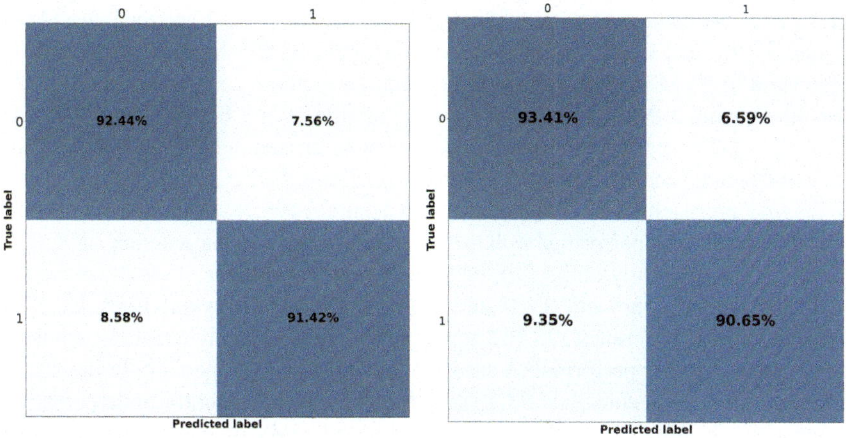

(a) First method (b) Second method

Fig. 25 Confusion matrices obtained using the LDA classification algorithm and for the first and second training methods

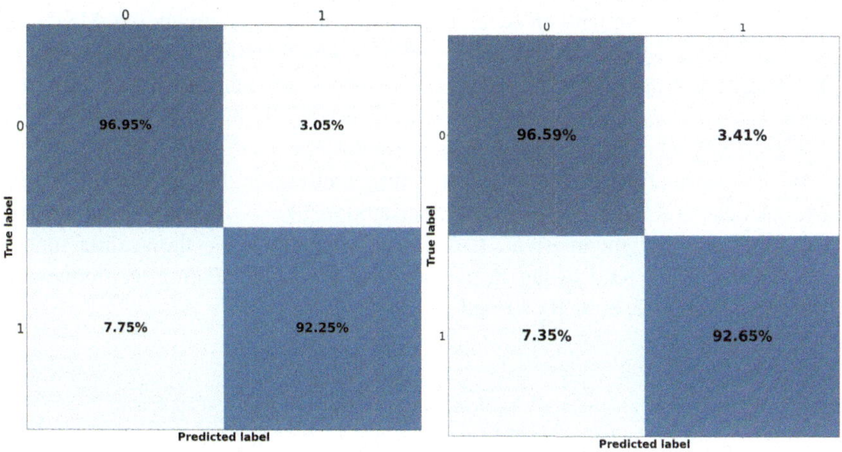

(a) First method (b) Second method

Fig. 26 Confusion matrices obtained using the QDA classifier and by adopting the first and second training methods

4 Scaling-Based Ternary Classification

4.1 Comparative Analysis of MMS, SS, and PCA Performance

In the first approach under MMS, The RF consistently achieved high accuracy, cross-validation scores, and R^2 values across both approaches. Specifically, RF achieved the best results in the first approach, with an accuracy of 96.58% and the lowest Mean Squared Error (MSE) of 0.0743. In the second approach, RF slightly underperformed compared to SVM but still remained competitive with an accuracy of 93.16%. On the other hand, k-NN demonstrated strong results in the first approach (96.04% accuracy), but its performance declined slightly in the second approach (92.37%).

In the first approach under SS, RF again emerged as the top performer, particularly in the first approach where it had the highest accuracy (96.78%) and the lowest MSE (0.0663). k-NN also performed well in the first approach, with an accuracy of 96.09%, although it was slightly outperformed by RF in overall metrics.

In the second approach, SVM and LR both showed competitive results. SVM achieved the best accuracy (93.56%) and cross-validation score (0.9423), while LR demonstrated stable performance across both approaches, especially in the second approach with an accuracy of 93.46%. NB and LDA consistently underperformed relative to other models, with NB achieving an accuracy of 90.83% in the second approach and LDA showing a drop in accuracy (91.28%) under Standard Scaling.

Using PCA for dimensionality reduction, RF once again stood out, particularly in the first approach with an accuracy of 97.18% and the lowest MSE of 0.0594. RF maintained its dominance across both approaches, with SVM being a close second in the second approach with an accuracy of 93.21% and a relatively low MSE.

While k-NN continued to show strong results with a 96.09% accuracy in the first approach, its performance dropped in the second approach. DT improved in this scaling method compared to Standard Scaling, especially in the first approach, but still lagged behind other models such as RF and SVM.

Overall, the analysis across different scaling methods and approaches highlights the robustness of RF and SVM, while other models like k-NN and LR demonstrate good, albeit slightly inconsistent, performance depending on the scaling method used. NB and LDA tend to lag in performance across all scaling techniques and approaches (Tables 7, 8, 9, 10, 11 and 12).

Table 7 Performance of various ML models using min-max scaling (first approach)

Algorithms	Cross validation score	MSE	RMSE	Accuracy (%)	R^2 score (%)
k-NN	**0.9261**	**0.0782**	**0.2797**	**96.04**	**88.68**
LR	0.9386	0.1074	0.3278	94.01	84.46
DT	0.8993	0.0955	0.3091	95.20	86.18
RF	**0.9374**	**0.0743**	**0.2725**	**96.58**	**89.26**
SVM	0.9428	0.1074	0.3278	94.60	84.46
NB	0.9025	0.1525	0.3905	90.40	77.94
LDA	0.9197	0.1327	0.3642	91.93	80.80
QDA	0.9272	0.1248	0.3532	93.17	81.95

Table 8 Performance of various ML models using min-max scaling (second approach)

Algorithms	Cross validation score	MSE	RMSE	Accuracy (%)	R^2 score (%)
k-NN	0.9237	0.1298	0.3603	92.37	81.23
LR	**0.9396**	**0.1204**	**0.3470**	**93.31**	**82.59**
DT	0.8858	0.2052	0.4529	88.11	70.34
RF	0.9340	0.1204	0.3470	93.16	82.59
SVM	**0.9438**	**0.1169**	**0.3419**	**93.36**	**83.09**
NB	0.8999	0.1348	0.3671	90.83	80.52
LDA	0.9198	0.1511	0.3888	91.28	78.15
QDA	0.9291	0.1343	0.3665	92.37	80.59

Table 9 Performance of various ML models using standard scaling (first approach)

Algorithms	Cross validation score	MSE	RMSE	Accuracy (%)	R^2 score (%)
k-NN	**0.9146**	**0.0748**	**0.2734**	**96.09**	**89.18**
LR	0.9401	0.1035	0.3217	94.26	85.03
DT	0.9001	0.0871	0.2952	95.45	87.39
RF	**0.9362**	**0.0663**	**0.2576**	**96.78**	**90.40**
SVM	0.9407	0.0985	0.3139	95.05	85.75
NB	0.9025	0.1525	0.3905	90.40	77.94
LDA	0.9197	0.1327	0.3642	91.93	80.80
QDA	0.9272	0.1248	0.3532	93.17	81.95

4.2 *Performance Metrics*

In the analysis of different ML classifiers using two distinct training methods with the MMS metrics as cited in Figs. 39, 40, 41, 42, 43, 44, 45 and 46, the confusion matrices highlight variations in correct classifications across models (Figs. 27, 28, 29, 30, 31, 32, 33, 34, 35, 36, 37 and 38).

For K-NN, the first method correctly classifies 97.07% of class 0, with a misclassification rate of 2.07% as class 2. In the second method, k-NN achieves a correct classification rate of 94.50% for class 2, with a misclassification rate of 3.50% for class 1.

LR correctly classifies 99.15% of class 0 in the first method, with a misclassification rate of 0.85% as class 2. In the second method, it also correctly classifies 99.39% of class 0, with a misclassification rate of 0.61% as class 2.

Table 10 Performance of various ML models using standard scaling (second approach)

Algorithms	Cross-validation score	MSE	RMSE	Accuracy (%)	R^2 score (%)
k-NN	0.9166	0.1477	0.3843	91.48	78.65
LR	**0.9405**	**0.1174**	**0.3427**	**93.46**	**83.02**
DT	0.8849	0.1987	0.4458	88.60	71.28
RF	0.9312	0.1179	0.3434	93.11	82.95
SVM	**0.9423**	**0.1179**	**0.3434**	**93.56**	**82.95**
NB	0.8999	0.1348	0.3671	90.83	80.52
LDA	0.9198	0.1511	0.3888	91.28	78.15
QDA	0.9291	0.1343	0.3665	92.37	80.59

Table 11 Performance of various ML models using PCA (first approach)

Algorithms	Cross validation score	MSE	RMSE	Accuracy (%)	R^2 score (%)
k-NN	**0.9269**	**0.0762**	**0.2761**	**96.09**	**88.97**
LR	0.9402	0.1030	0.3209	94.31	85.10
DT	0.9088	0.1124	0.3352	94.70	83.74
RF	**0.9308**	**0.0594**	**0.2437**	**97.18**	**91.40**
SVM	0.9416	0.1059	0.3255	94.46	84.67
NB	0.9182	0.1525	0.3905	92.03	77.94
LDA	0.9197	0.1327	0.3642	91.93	80.80
QDA	0.9272	0.1248	0.3532	93.17	81.95

Table 12 Performance of various ML models using PCA (second approach)

Algorithms	Cross validation score	MSE	RMSE	Accuracy (%)	R^2 score (%)
k-NN	0.9237	0.1293	0.3596	93.01	81.30
LR	**0.9402**	**0.1219**	**0.3491**	**93.31**	**82.38**
DT	0.9077	0.1368	0.3698	92.72	80.23
RF	**0.9355**	**0.1110**	**0.3332**	**93.36**	**83.95**
SVM	0.9418	0.1199	0.3463	93.21	82.66
NB	0.9239	0.1492	0.3862	92.22	78.44
LDA	0.9198	0.1511	0.3888	91.28	80.80
QDA	0.9291	0.1343	0.3665	92.37	80.59

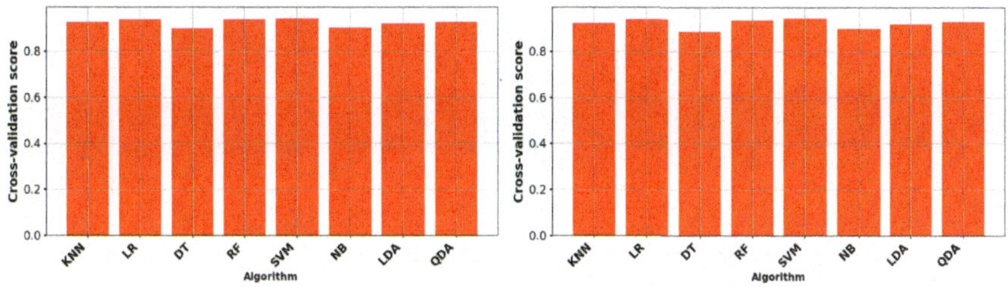

(a) First Method (b) Second Method

Fig. 27 Cross-validation scores for each algorithm (min-max scaling with two training methods)

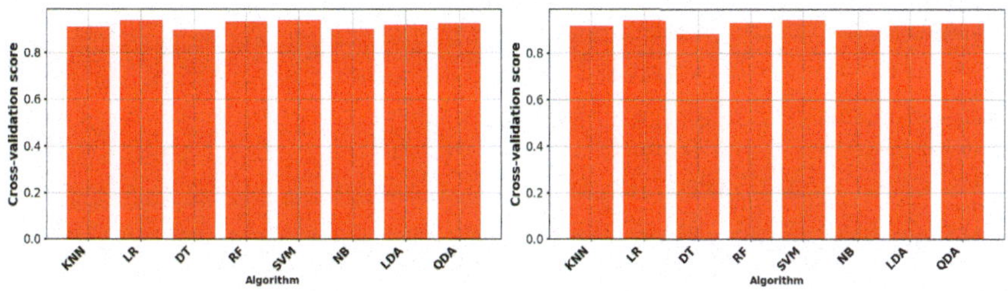

(a) First Method (b) Second Method

Fig. 28 Cross-validation scores for each algorithm (standard scaling with two training methods)

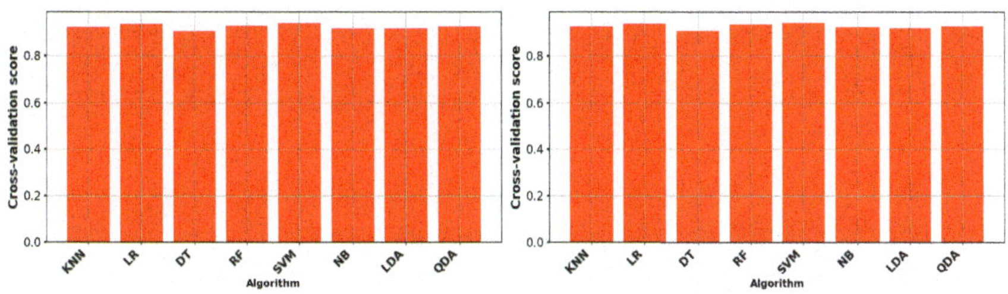

(a) First Method (b) Second Method

Fig. 29 Cross-validation scores for each algorithm (PCA with two training methods)

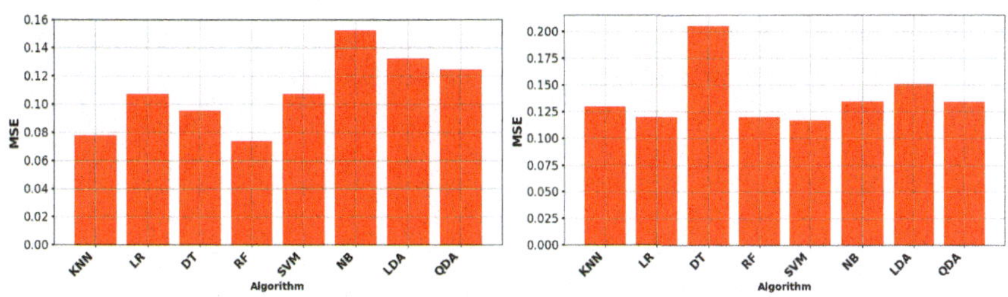

(a) First Method (b) Second Method

Fig. 30 MSE for each algorithm (min-max scaling with two training methods)

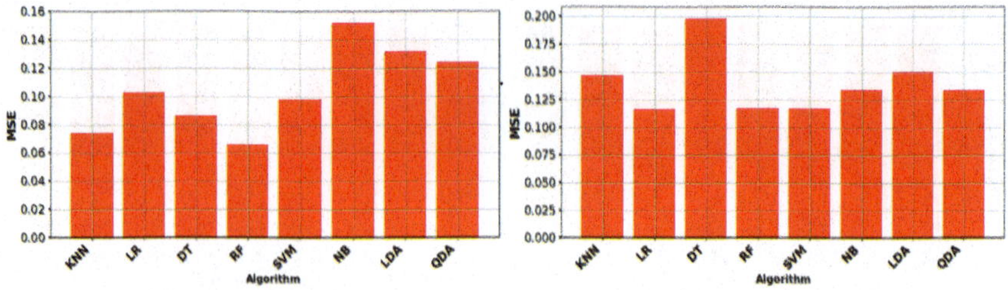

(a) First Method (b) Second Method

Fig. 31 MSE for each algorithm (standard scaling with two training methods)

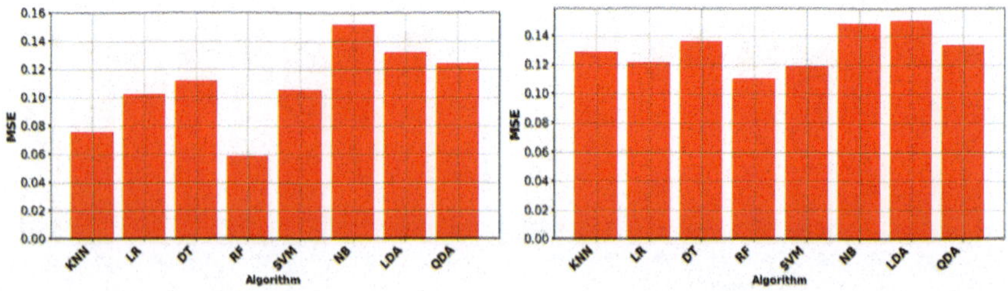

(a) First Method (b) Second Method

Fig. 32 MSE for each algorithm (PCA with two training methods)

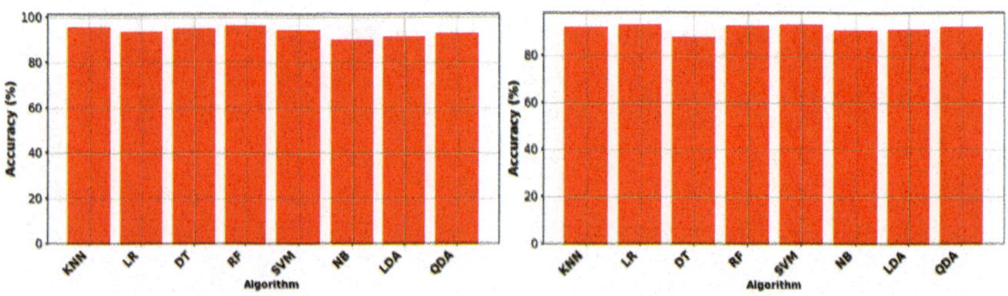

(a) First Method (b) Second Method

Fig. 33 Accuracy (%) for each algorithm (min-max scaling with two training methods)

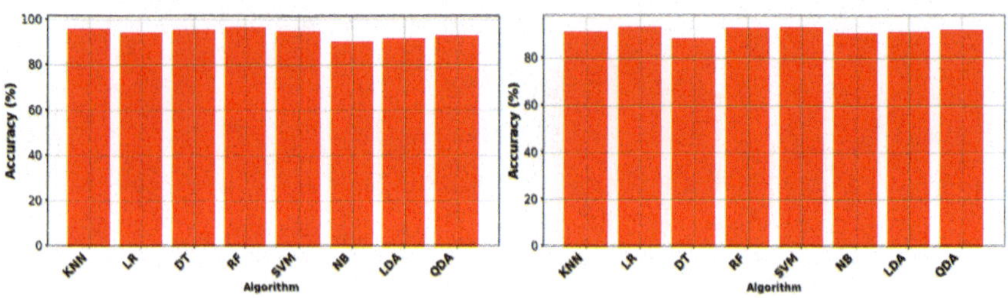

(a) First Method (b) Second Method

Fig. 34 Accuracy (%) for each algorithm (standard scaling with two training methods)

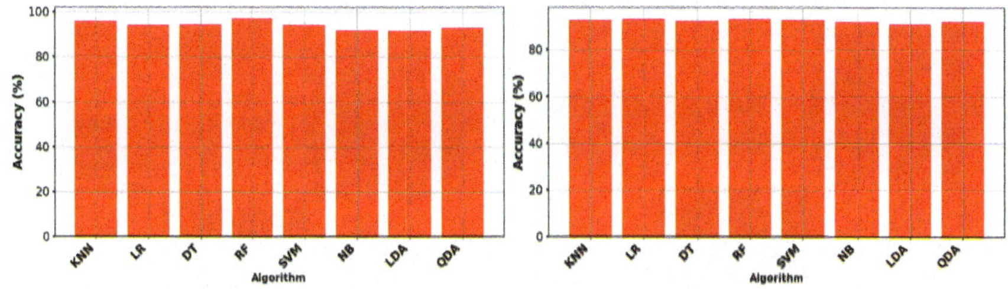

(a) First Method (b) Second Method

Fig. 35 Accuracy (%) for each algorithm (PCA with two training methods)

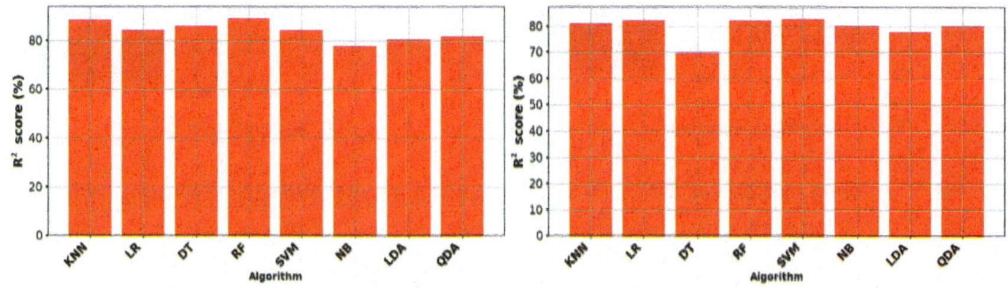

(a) First Method (b) Second Method

Fig. 36 R^2 score for each algorithm (min-max scaling with two training methods)

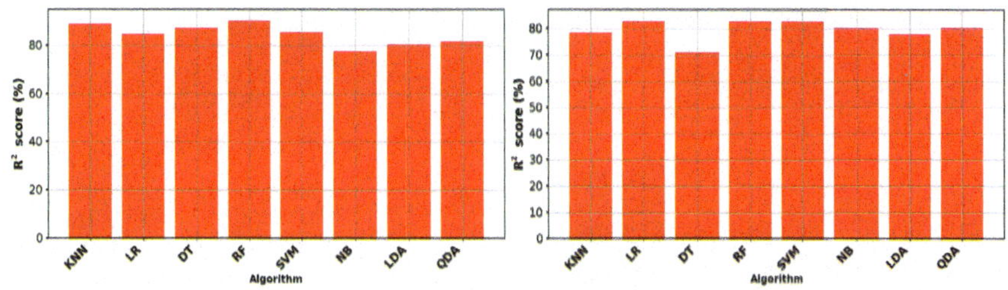

(a) First Method (b) Second Method

Fig. 37 R^2 score for each algorithm (standard scaling with two training methods)

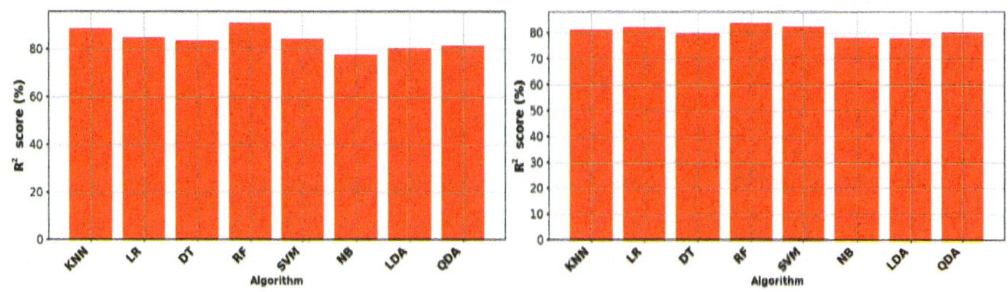

(a) First Method (b) Second Method

Fig. 38 R^2 score for each algorithm (PCA with two training methods)

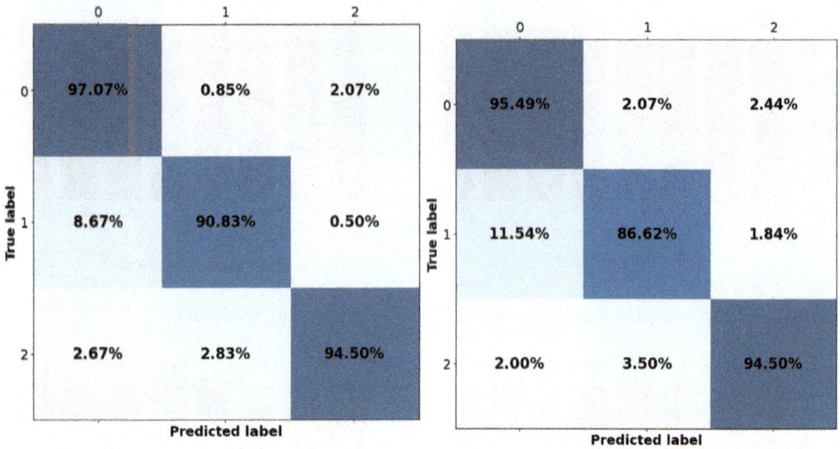

(a) First method (b) Second method

Fig. 39 Confusion matrices of the k-NN algorithm using the first and second training methods

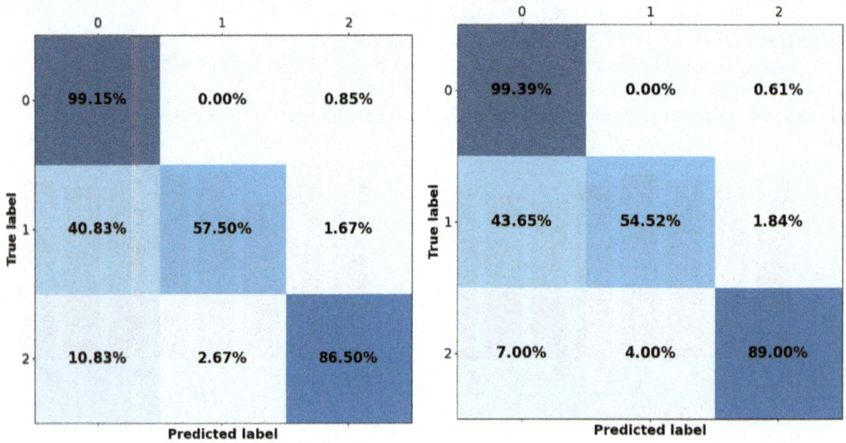

(a) First method (b) Second method

Fig. 40 Confusion matrices of the LR classification algorithm using the first and second training methods

DT correctly classifies 96.50% of class 2 in the first method, with a misclassification rate of 2.33% as class 1. In the second method, DT achieves a correct classification rate of 93.67%, with a misclassification rate of 5.67%.

RF achieves a correct classification rate of 96.50% for class 2 in the first method, with a misclassification rate of 2.33%. In the second method, it incorrectly classifies 4% as class 1, with a correct classification rate of 95%.

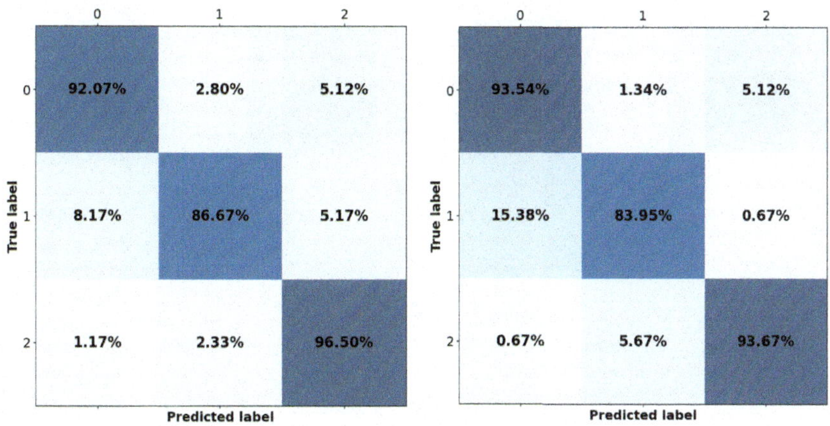

(a) First method (b) Second method

Fig. 41 Confusion matrices of the DT classifier using the first and second training methods

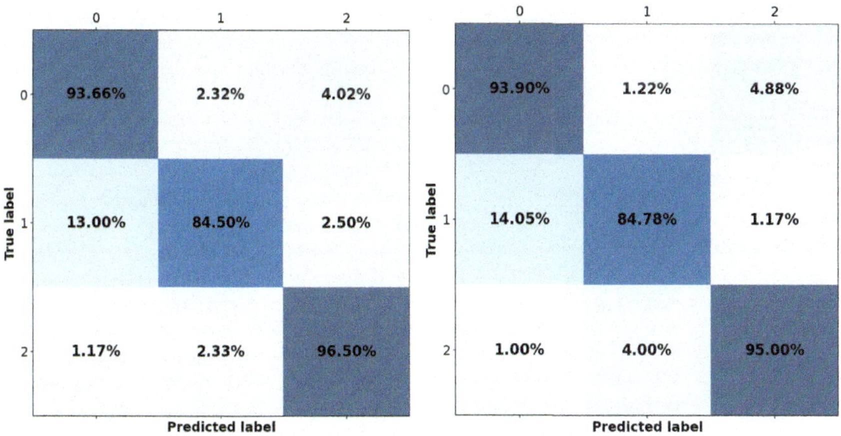

(a) First method (b) Second method

Fig. 42 Confusion matrices obtained using the RF classification algorithm for the first and second training methods

SVM correctly classifies 94.37% as class 0 in the first method, with a misclassification rate of 4.63% as class 2. In the second method, SVM achieves a correct classification rate of 94.15% as class 1, with a misclassification rate of 5.85%.

NB correctly classifies 94.83% as class 2 in the first method, with a misclassification rate of 4% as class 1. In the second method, it achieves a correct classification rate of 94.50% as class 2, with a misclassification rate of 4.17%.

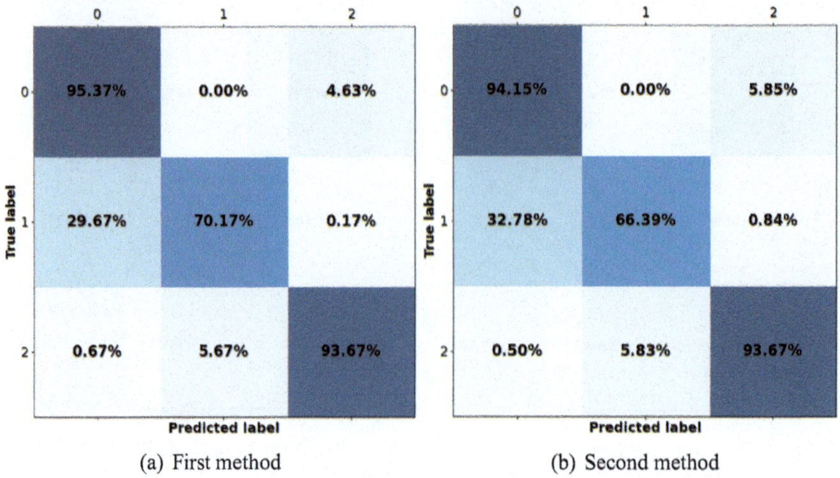

Fig. 43 Confusion matrices obtained using the SVM classifier for the first and second training methods

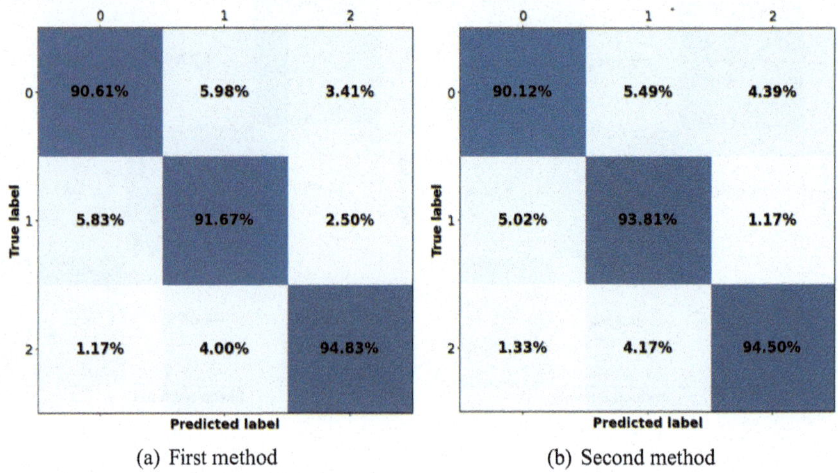

Fig. 44 Confusion matrices obtained using the NB classifier for the first and second training methods

LDA correctly classifies 93.29% as class 0 in the first method, with a misclassification rate of 3.54% as class 2. In the second method, it achieves a correct classification rate of 93.41% as class 0, with a misclassification rate of 5.12%.

QDA correctly classifies 99.51% as class 0 in the first method, with a misclassification rate of 0.49% as class 2. In the second method, it achieves a correct classification rate of 99.63% as class 0, with a misclassification rate of 0.37%.

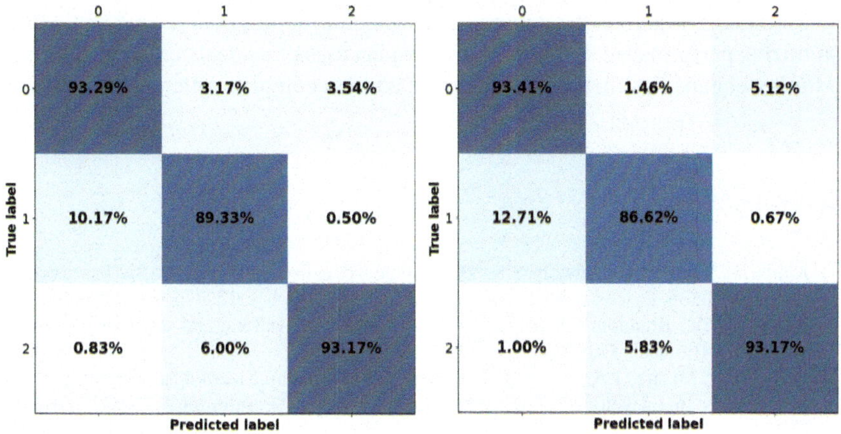

(a) First method (b) Second method

Fig. 45 Confusion matrices obtained using the LDA classification algorithm for the first and second training methods

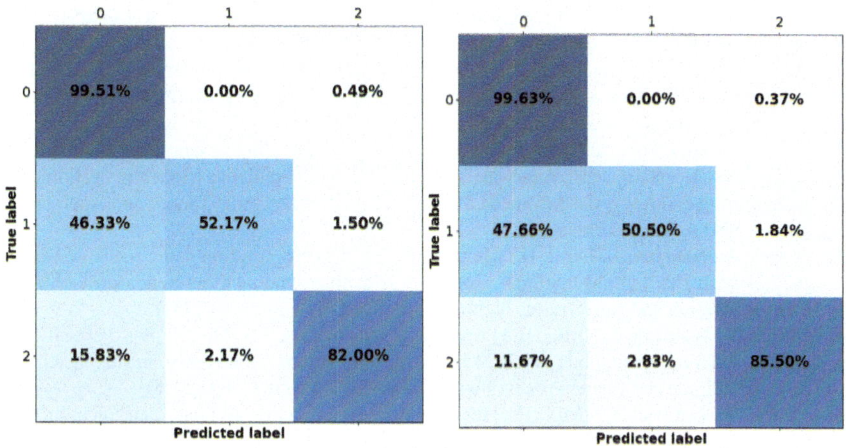

(a) First method (b) Second method

Fig. 46 Confusion matrices obtained using the QDA classifier for the first and second training methods

5 Conclusion and Future Works

In this study, we explored various machine learning models, including RF, k-NN, LR, DT, SVM, NB, LDA, and QDA, across different scaling methods (MMS, SS, PCA). The models' performances were evaluated using key metrics such as accuracy, CV score, and MSE. The results demonstrated that scaling techniques significantly impact model accuracy, especially for models like SVM and k-NN.

Our findings highlight the importance of selecting appropriate scaling methods for optimizing performance. In future work, we plan to explore deep learning techniques to further enhance prediction accuracy and address complex patterns in the dataset.

References

1. Kalita, B., Narayan, J., Dwivedy, S.K.: Development of active lower limb robotic-based orthosis and exoskeleton devices: a systematic review. Int. J. Soc. Robot. **13**, 775–793 (2021)
2. Bauman, V.V., Brandon, S.C.E.: Gait phase detection in walking and stairs using machine learning. J. Biomech. Eng. **144**(12), 121,007 (2022)
3. Bhoir, A.A., Mishra, T.A., Narayan, J., Dwivedy, S.K.: Machine learning algorithms in human gait analysis. In: Encyclopedia of Data Science and Machine Learning, pp. 922–937. IGI Global (2023)
4. Semwal, V.B., Jain, R., Maheshwari, P., Khatwani, S.: Gait reference trajectory generation at different walking speeds using LSTM and CNN. Multimed. Tools Appl. **82**(21), 33401–33419 (2023)
5. Narayan, J., Gritli, H., Dwivedy, S.K.: Lower limb joint torque estimation via Bayesian regularized backpropagation neural networks. In: 2024 IEEE International Conference on Interdisciplinary Approaches in Technology and Management for Social Innovation (IATMSI), vol. 2, pp. 1–6. IEEE (2024)
6. Albuquerque, P., Verlekar, T.T., Correia, P.L., Soares, L.D.: A spatiotemporal deep learning approach for automatic pathological gait classification. Sensors **21**(18) (2021)
7. Narayan, J., Jhunjhunwala, S., Mishra, S., Dwivedy, S.K.: 5—a comparative performance analysis of backpropagation training optimizers to estimate clinical gait mechanics. In: Roy, S., Goyal, L.M., Balas, V.E., Agarwal, B., Mittal, M. (eds.) Predictive Modeling in Biomedical Data Mining and Analysis, Advanced Studies in Complex Systems: Theory and Applications, pp. 83–104. Academic Press (2022)
8. Mekni, A., Narayan, J., Gritli, H.: Multi-class gait phase recognition using machine learning models with two training methods. In: 2024 15th International Conference on Computing Communication and Networking Technologies (ICCCNT), pp. 1–6 (2024)
9. Mekni, A., Narayan, J., Gritli, H.: Leveraging machine learning for gait phase classification with varied training methods. In: 2024 IEEE 7th International Conference on Advanced Technologies, Signal and Image Processing (ATSIP), vol. 1, pp. 582–587 (2024)
10. Mekni, A., Narayan, J., Gritli, H.: Binary and ternary human gait phase classification using machine learning algorithms. In: 2024 International Conference on Innovation and Intelligence for Informatics, Computing, and Technologies (3ICT), pp. 72–79. University of Bahrain, Bahrain (2024)
11. Fricke, C., Alizadeh, J., Zakhary, N., Woost, T.B., Bogdan, M., Classen, J.: Evaluation of three machine learning algorithms for the automatic classification of EMG patterns in gait disorders. Front. Neurol. **12**, 666,458 (2021)
12. Krutaraniyom, S., Sengchuai, K., Booranawong, A., Jaruenpunyasak, J.: Pilot study on gait classification using machine learning. In: 2022 International Electrical Engineering Congress (iEECON), pp. 1–4. IEEE (2022)
13. Bauman, V.V., Brandon, S.C.: Gait phase detection in walking and stairs using machine learning. J. Biomech. Eng. **144**(12), 121,007 (2022)
14. Park, H., Han, S., Sung, J., Hwang, S., Youn, I., Kim, S.J.: Classification of gait phases based on a machine learning approach using muscle synergy. Front. Human Neurosci. **17**, 1201,935 (2023)

15. Lan, Z., Lempereur, M., Gueret, G., Houx, L., Cacioppo, M., Pons, C., Mensah, J., Rémy-Néris, O., Aïssa-El-Bey, A., Rousseau, F., et al.: Towards a diagnostic tool for neurological gait disorders in childhood combining 3d gait kinematics and deep learning. Comput. Biol. Med. 108095 (2024)
16. Moura Coelho, R., Gouveia, J., Botto, M.A., Krebs, H.I., Martins, J.: Real-time walking gait terrain classification from foot-mounted inertial measurement unit using convolutional long short-term memory neural network. Expert Syst. Appl. **203**, 117,306 (2022)
17. Konz, L., Hill, A., Banaei-Kashani, F.: ST-DeepGait: a spatiotemporal deep learning model for human gait recognition. Sensors **22**(20) (2022)
18. Sarshar, M., Polturi, S., Schega, L.: Gait phase estimation by using LSTM in IMU-based gait analysis-proof of concept. Sensors **21**(17), 5749 (2021)
19. Halim, A., Abdellatif, A., Awad, M.I., Atia, M.R.: Prediction of human gait activities using wearable sensors. Proc. Inst. Mech. Eng. Part H: J. Eng. Med. **235**(6), 676–687 (2021)
20. Bahadori, S., Williams, J., Wainwright, T.: Lower limb kinematic, kinetic and spatial-temporal gait data for healthy adults using a self-paced treadmill (2020). https://data.mendeley.com/datasets/wwnvw28n2m/1. https://doi.org/10.17632/wwnvw28n2m.1. Mendeley Data, V1
21. Bahadori, S., Williams, J.M., Wainwright, T.W.: Lower limb kinematic, kinetic and spatial-temporal gait data for healthy adults using a self-paced treadmill. Data Brief **34**, 106,613 (2021)
22. Hossin, M., Sulaiman, M.N.: A review on evaluation metrics for data classification evaluations. Int. J. Data Mining Knowl. Manag. Process **5**, 01–11 (2015)
23. Shaha, R., Talukder, D., Iqbal, M.A., Haque, M.M.: TOS: a relative metric approach for model selection in machine learning solutions. In: 2021 IEEE International Conference on Robotics, Automation, Artificial-Intelligence and Internet-of-Things (RAAICON), pp. 26–31 (2021)
24. Abaker, M., Dafaalla, H., Eisa, T.A.E., Abdelgader, H., Mohammed, A., Burhanur, M., Hasabelrsoul, A., Alfakey, M.I., Morsi, M.A.: Deep learning- and IoT-based framework for rock-fall early warning. Appl. Sci. **13**, 9978 (2023)
25. Tharwat, A.: Classification assessment methods. Appl. Comput. Inform. **17**(1), 168–192 (2021)
26. Hicks, S.A., Strümke, I., Thambawita, V., Hammou, M., Riegler, M.A., Halvorsen, P., Parasa, S.: On evaluation metrics for medical applications of artificial intelligence. Sci. Rep. **12**(1), 5979 (2022)
27. de Amorim, L.B., Cavalcanti, G.D., Cruz, R.M.: The choice of scaling technique matters for classification performance. Appl. Soft Comput. **133**, 109,924 (2023)
28. Ahsan, M.M., Mahmud, M.A.P., Saha, P.K., Gupta, K.D., Siddique, Z.: Effect of data scaling methods on machine learning algorithms and model performance. Technologies **9**(3) (2021)
29. Greenacre, M., Groenen, P.J.F., Hastie, T., D'Enza, A.I., Markos, A., Tuzhilina, E.: Principal component analysis. Nat. Rev. Methods Primers **2**(1), 100 (2022)
30. Pratap, S., Narayan, J., Hatta, Y., Ito, K., Hazarika, S.M.: From tactile signals to grasp classification: exploring patterns with machine learning. In: 2024 IEEE International Conference on Interdisciplinary Approaches in Technology and Management for Social Innovation (IATMSI), vol. 2, pp. 1–6 (2024)
31. Vujovic, Z.D.: Classification model evaluation metrics. Int. J. Adv. Comput. Sci. Appl. **12**(6) (2021)
32. Mahesh, B.: Machine learning algorithms—a review. Int. J. Sci. Res. **9**(1), 381–386 (2020)

Solving Inverse Kinematics of a 2-DoF Upper Limb Exoskeleton Robot Using a Hybrid Approach Combining Metaheuristic Algorithms and Artificial Neural Networks

Rania Bouzid, Hassène Gritli, and Jyotindra Narayan

Abstract Dealing with the complexities of Inverse Kinematics is essential for effectively controlling exoskeletons in complex environments. Traditional analytical methods frequently encounter difficulties due to the non-linearity of these problems. Our approach begins with using Particle Swarm Optimization (PSO), Genetic Algorithm (GA), BAT Algorithm (BAT), and Grey Wolf Optimizer (GWO) to address the inverse kinematics model for two-degree-of-freedom (2-DoF) upper Limb Exoskeletons. These metaheuristic solutions are then enhanced by integrating ANNs, utilizing both random step-size and sinusoidal-signal datasets to refine the results. This hybrid method optimizes the ANN hyperparameters, such as the number of hidden layers and activation functions. Our methodology introduces unique hyperparameter optimization techniques for training the ANNs, effectively leveraging the strengths of both metaheuristic algorithms and ANNs. Experimental results demonstrate the obtained lowest error rates, alongside the best hyperparameters for the ANN architecture. These findings suggest the scalability of our hybrid model to more complex robotic systems.

Keywords Inverse kinematics · Exoskeletons · Metaheuristic algorithms · Artificial neural networks · Hyperparameter optimization

R. Bouzid
Polytechnic School of Tunisia, University of Carthage, La Marsa, Tunisia
e-mail: rania.bouzid@istic.ucar.tn

R. Bouzid · H. Gritli (✉)
Laboratory of Robotics, Informatics and Complex Systems (RISC Lab—LR16ES07), National Engineering School of Tunis, University of Tunis El Manar, Tunis, Tunisia
e-mail: grhass@yahoo.fr

H. Gritli
Higher Institute of Information and Communication Technologies, University of Carthage, Hammam Chatt, Ben Arous, Tunisia

J. Narayan
Department of Mechanical Engineering, Indian Institute of Technology Patna, Patna, India

Department of Computing, Imperial College London, London, UK

J. Narayan and H. Gritli (eds.), *Design and Control of Rehabilitation Robots*, Studies in Systems, Decision and Control 585, https://doi.org/10.1007/978-3-031-86977-8_6

1 Introduction

The development of advanced robotics systems, particularly those related to human-assistive technologies, has seen substantial growth in recent years [1–8]. Among these, upper limb and low limb exoskeletons have emerged as a crucial area of research, driven by the need to aid individuals with motor impairments, enhance human capabilities, and provide rehabilitation support [8–14]. Upper limb exoskeletons are wearable technologies designed to augment or restore the movements of the human arm by actuating joints in a way that emulates natural limb motion [15–17]. The design and control of such exoskeletons present significant challenges, particularly in tackling the inverse kinematics challenge [11–14, 18], which involves determining the joint parameters that produce a target position and orientation of the end-effector [19]. Upper limb exoskeletons play a pivotal role in rehabilitation engineering [15, 20], offering the potential to restore arm movements in patients suffering from conditions such as stroke, spinal cord injuries, or neurodegenerative diseases [21, 22]. These devices typically consist of articulated joints replicating the rotations and translations of the shoulder, elbow, and wrist controlled by motors or actuators [23, 24]. The control of these joints is highly complex due to the need to accommodate the anatomical and biomechanical intricacies of the human arm, such as joint limits, muscle dynamics, and redundancy in movement [25, 26]. Moreover, upper limb exoskeletons must ensure both accuracy in following desired trajectories and safety in human-robot interaction [27, 28].

The inverse kinematics problem is central to achieving these goals, as it directly affects the system's ability to execute precise and natural movements, concerned with finding the angular positions of the joints that attain a specific placement and direction of the robot's end-effector, which in this case is the terminal part of the exoskeleton attached to the user's limb [18]. For robots with multiple motion capabilities, the inverse kinematics problem is inherently nonlinear and often involves multiple solutions or, in some cases, no solution at all [29].

Traditional analytical methods, such as geometric or algebraic approaches, may be effective for simple, low-DoF robots but tend to fail or become intractable for more complex systems, such as a human-exoskeleton interface, where the kinematic equations can be highly nonlinear and coupled [30–36]. The complexity increases when considering the need for real-time solutions in a dynamic environment, where delays or inaccuracies in solving the inverse kinematics problem can lead to poor performance or even unsafe operation of the exoskeleton [37, 38].

To address the challenges posed by traditional methods, researchers have turned to metaheuristic algorithms, which are nature-inspired optimization methods, to solve the inverse kinematics problem [39–42]. Metaheuristic algorithms, such as Genetic Algorithms (GA) inspired by the process of natural selection, have been applied to evolve solutions over generations, producing robust inverse kinematics solutions for complex robotic systems [43], Particle Swarm Optimization (PSO) which is inspired by the social behavior of birds flocking, has been effectively used to solve inverse kinematics problems by iteratively adjusting the joint angles to converge on an optimal solution [44–46], and Differential Evolution (DE) [47, 48], Dragonfly Algorithm (DA) [49], BAT Algorithm (BAT) [50] and Grey Wolf Optimizer (GWO) [51] etc., are particularly suited for complex, multi-modal search spaces where the objective function is difficult to express analytically. These algorithms operate by iteratively exploring the solution space to find the optimal set of joint angles that reduce the gap between the expected and realized end-effector positions. One of the significant advantages of metaheuristic algorithms is their flexibility in handling a wide range of objective functions and constraints, making them highly adaptable to various robotic systems [52].

Artificial Neural Networks (ANNs) have gained prominence in robotics due to their ability to model complex, nonlinear input-output relationships [53]. ANNs are particularly powerful when applied to problems where the relationship between the input (e.g., end-effector position) and the output (e.g., joint angles) is difficult to model explicitly [54], as is the case with the forward and inverse kinematics in exoskeletons and robotic manipulators [55–58]. An ANN can be trained on a dataset of known input-output pairs, learning the underlying mapping through a process of weight adjustment using algorithms such as backpropagation. Once trained, the ANN can quickly produce accurate joint angle predictions for new, unseen end-effector positions, making it an attractive option for real-time inverse kinematics solutions [59].

The generalization capabilities of ANNs, combined with their efficiency in handling high-dimensional data, make them suitable for applications in robotics where fast and accurate decision-making is critical [55–58, 60–64]. While metaheuristic algorithms are effective for finding optimal solutions to complex problems, they can be computationally intensive, especially for real-time use cases. On the other hand, ANNs, once trained, can provide rapid predictions but may suffer from issues such as overfitting or local minima if not properly optimized. A hybrid approach that combines the strengths of both methods offers a promising solution. In this approach, a metaheuristic algorithm can be used to optimize the parameters of the ANN, such as the weights and biases, during the training phase [65]. This ensures that the ANN is fine-tuned to produce the most accurate inverse kinematics solutions possible. Alternatively, the metaheuristic algorithm can be used to search the solution space initially, providing a set of potential solutions that the ANN can refine and generalize across the solution space [66]. This hybrid strategy not only enhances the accuracy and robustness of the inverse kinematics solutions but also reduces the computational burden by leveraging the rapid inference capabilities of ANNs in real-time scenarios.

The specific focus of this chapter is on a 2-DoF upper limb exoskeleton, which simplifies the inverse kinematics problem to a more manageable scope while still offering meaningful insights and applications. The exoskeleton is designed to assist with shoulder and elbow movements, crucial for performing daily tasks such as reaching, lifting, and manipulating objects. Despite its lower complexity compared to full-arm exoskeletons, solving the inverse kinematics problem for a 2-DoF system remains challenging due to the need for precise joint control and the potential for multiple or no valid solutions. By applying a hybrid approach of metaheuristic algorithms and ANNs, this chapter aims to demonstrate a method that not only solves the inverse kinematics problem with high accuracy but also identifies the optimal hyperparameters for the ANN architecture.

Several studies have explored the application of metaheuristic techniques and ANNs for solving inverse kinematics problems, particularly in the context of robotic manipulators and exoskeletons [67–70]. For example, authors in [71] utilized PSO-DNN and Grey Wolf Optimization (GWO)-DNN to optimize the inverse kinematics solutions for a 6-DoF robotic arm, achieving significant improvements in computational efficiency and solution performance compared with traditional techniques. Similarly, authors in [72] integrated GA with ANNs to address the inverse kinematics challenge for a redundant robot, demonstrating the effectiveness of hybrid approaches in handling complex kinematic configurations. In the domain of exoskeletons, the work in [73] proposed an all-encompassing metaheuristic approach for designing a 5-DoF medical robot for upper limb rehabilitation, including topological optimization, kinematic modeling, human-robot interaction, control strategies, and experimental validation. The approach integrates engineering design perspectives and Industry 4.0 pillars, achieving a 40% weight reduction, a versatile interface, and biomechanically viable models, with average Root Mean Squared Error (RMSE) of 0.08 and 0.11 rad in therapeutic routine evaluations. The research work in [74] proposed a modified PSO (m-PSO) and its multi-objective variant (MO-m-PSO) for solving inverse kinematics of a 5-DoF robotic arm, using fitness-based particle adaptation to improve optimization. The approach outperforms classical methods and mitigates inverse kinematics challenges like singularities, ensuring accurate position and orientation solutions. In their research, Slim et al. [50] proposed IK-BA, a modified Bat Algorithm for solving robotic arm inverse kinematics, incorporating an elongation control mechanism to minimize joint variation. They demonstrate its effectiveness through comparative analysis and real-world applications, including geometric path planning.

This chapter contributes to the field of robotics by presenting a novel hybrid approach for solving the inverse kinematics problem in a 2-DoF upper limb exoskeleton. The proposed method integrates PSO, GA, BAT, and GWO metaheuristic algorithms, with an ANN trained on two datasets of joint angles and end-effector positions. The hybrid approach is evaluated in terms of accuracy, computational efficiency, and robustness, with results demonstrating its superiority over traditional methods and standalone ANN or metaheuristic approaches.

The remainder of the chapter is structured as follows: Sect. 2 provides a detailed description of the 2-DoF upper limb exoskeleton model and the mathematical formulation of the inverse kinematics problem. Section 3 discusses the implementation of the hybrid approach, including the design of the ANN and the parameter settings for metaheuristic algorithms. Section 4 presents the experimental results and performance analysis, followed by a discussion of the implications for real-time applications. Finally, Sect. 5 concludes our work with a summary of the findings and suggestions for future research directions.

2 Adopted 2-DoF Upper Limb Exoskeleton Model and Mathematical Formulation of the Inverse Kinematics Problem

2.1 On the 2-DoF Upper Limb Exoskeleton

The 2-DoF upper limb exoskeleton presented in this study is designed to assist with movements at the shoulder and elbow joints, enabling a range of arm motions crucial for performing daily tasks such as reaching, grasping, and lifting. The exoskeleton is composed of articulated segments that correspond to the human upper arm and forearm, connected by two rotational joints (Fig. 1).

2.2 Kinematic Model of the 2-DoF Exoskeleton Robot

The kinematic model of the 2-DoF upper limb exoskeleton provides a mathematical representation of the correlation between joint angles and the end-effector's location. This model is essential for both forward and inverse kinematics calculations.

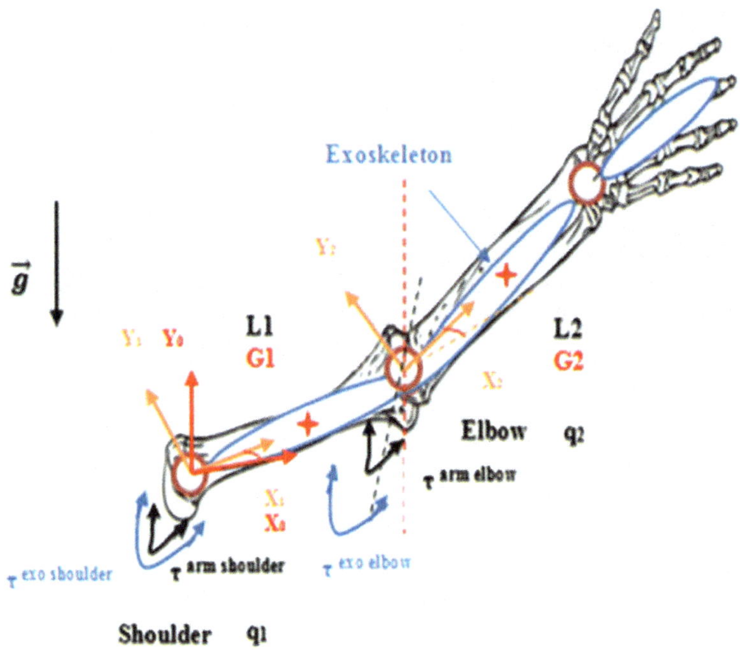

Fig. 1 Representation of the adopted 2-DoF exoskeleton robot model for the rehabilitation of the human lower limb [17]

2.2.1 Forward Kinematics

In forward kinematics, the end-effector's placement and angular configuration are computed based on known joint angles. For the 2-DoF exoskeleton, the forward kinematics can be expressed as:

$$x = L_1 \cos(\theta_1) + L_2 \cos(\theta_1 + \theta_2), \tag{1}$$

$$y = L_1 \sin(\theta_1) + L_2 \sin(\theta_1 + \theta_2), \tag{2}$$

where

- θ_1 is the angle of the shoulder joint,
- θ_2 is the angle of the elbow joint,
- L_1 is the length of the upper arm (first segment),
- L_2 is the length of the forearm (second segment),
- x and y represent the coordinates of the end-effector in a 2D plane.

These equations describe the position of the end-effector (hand) in the plane based on the known angles of the shoulder and elbow joints.

2.2.2 Inverse Kinematics

To solve for θ_1 and θ_2 given x and y, we proceed with the following steps:

1. **Step 1**: Compute the distance to the end-effector
   First, compute the distance r from the origin (shoulder joint) to the end-effector:

$$r = \sqrt{x^2 + y^2} \tag{3}$$

2. **Step 2**: Reachability check
   For the end-effector position to be reachable, the following condition must be satisfied:

$$r \leq L_1 + L_2 \tag{4}$$

   If $r > L_1 + L_2$, the position is outside the exoskeleton's workspace and cannot be reached.
3. **Step 3**: Solve for the elbow angle θ_2
   Using the law of cosines, the elbow angle θ_2 can be determined as:

$$\cos(\theta_2) = \frac{r^2 - L_1^2 - L_2^2}{2L_1L_2} \tag{5}$$

   Thus, we obtain:

$$\theta_2 = \arccos\left(\frac{r^2 - L_1^2 - L_2^2}{2L_1L_2}\right) \tag{6}$$

4. **Step 4**: Solve for the shoulder angle θ_1
   To find θ_1, we first define the angle α as:

$$\alpha = \arctan 2(y, x) \tag{7}$$

   Next, compute the auxiliary angle β:

$$\beta = \arctan 2\left(\frac{L_2 \sin(\theta_2)}{L_1 + L_2 \cos(\theta_2)}\right) \tag{8}$$

   Finally, the shoulder angle θ_1 is given by:

$$\theta_1 = \alpha - \beta \tag{9}$$

The joint angles that solve the inverse kinematics problem are then defined as follows:

$$\theta_1 = \arctan 2(y, x) - \arctan 2\left(\frac{L_2 \sin(\theta_2)}{L_1 + L_2 \cos(\theta_2)}\right) \tag{10}$$

$$\theta_2 = \arccos\left(\frac{r^2 - L_1^2 - L_2^2}{2L_1L_2}\right) \tag{11}$$

These angles provide the configuration of the exoskeleton's joints to reach the desired end-effector position (x, y).

The inverse kinematics problem for the 2-DoF exoskeleton can yield multiple solutions due to the non-linear nature of the equations. Specifically, for a given end-effector position, there may be two valid sets of joint angles (θ_1, θ_2) that satisfy the kinematic equations, corresponding to different configurations of the arm (elbow up or elbow down). This phenomenon is known as redundancy, and selecting the appropriate solution is crucial for ensuring natural and safe movements.

Singularities occur when the determinant of the Jacobian matrix, which relates the joint velocities to the end-effector velocities, approaches zero. At a singularity, the system loses one or more degrees of freedom, making it impossible to achieve certain end-effector velocities regardless of the joint velocities. For the 2-DoF exoskeleton, singularities typically occur when the arm is fully extended or fully folded, resulting in a loss of control over certain directions of motion. These singular configurations must be carefully managed to avoid instability and ensure smooth operation of the exoskeleton.

2.3 Objective Function Formulation for Solving the Inverse Kinematics Problem Using ANN

Optimizing inverse kinematics models for manipulator robots involves defining an objective function and Fitness evaluation.

Given

- $\mathbf{h}$: Vector of hyperparameters
- $\text{ANN}_{\mathbf{h}}$: ANN model configured with hyperparameters $\mathbf{h}$
- $\mathbf{X}_{\text{train}}$: Training input data
- $\mathbf{Y}_{\text{train}}$: Training target data
- $\mathbf{X}_{\text{val}}$: Validation input data
- $\mathbf{Y}_{\text{val}}$: Validation target data
- $\hat{\mathbf{Y}}_{\text{val}}$: Predicted output matrix for validation data
- $f(\mathbf{h})$: Objective function to be minimized (Mean Squared Error (MSE))

Objective Function
The objective function to minimize is the MSE on the validation set:

$$f(\mathbf{h}) = \frac{1}{N} \sum_{i=1}^{N} \left(\text{ANN}_{\mathbf{h}}(\mathbf{X}_{\text{val}})_i - \mathbf{Y}_{\text{val},i} \right)^2 \tag{12}$$

2.4 General Process of Metaheuristic Algorithms

- **Initialization**
  The algorithm begins by initializing a set of potential solutions. This could be done randomly or based on some heuristic. The solutions could be a population of individuals, particles, or a single candidate, depending on the algorithm.
- **Evaluation**
  Each solution is evaluated using an objective function (or fitness function), which defines the quality of the solution with respect to the problem being solved. This step involves calculating how well each solution meets the criteria set by the problem.
- **Exploration and Exploitation**
  The algorithm explores the solution space by applying search operators. These operators modify current solutions to explore new possibilities. Common operators include mutation, crossover, local search, or random perturbations. The goal is to balance:

  - **Exploration**: Searching new areas in the solution space.
  - **Exploitation**: Improving existing solutions.

- **Selection or Acceptance**
  The algorithm decides which solutions to keep for the next iteration. This can be based on a solution's fitness, or sometimes less optimal solutions are kept to avoid being stuck in local optima. In population-based algorithms, a selection process chooses the best-performing individuals for reproduction or further exploration.
- **Termination**
  The algorithm stops when a predefined termination condition is met, such as:

  - A specified maximum of iterations or generations.
  - A time limit.
  - No significant improvement in the best solution over a fixed number of iterations.

- **Output**
  The best solution found throughout the search process is returned as the final output.

3 Proposed Methodology and Algorithms Used

We used the random step size and sinusoidal signal-based datasets to simulate diverse scenarios and test the robustness of our hybrid approach under different input conditions. The random step-size dataset introduces variability and unpredictability, reflecting real-world noise and irregularities, while the sinusoidal signal-based dataset mimics periodic movements typical in human-like motion. Both datasets were generated by us to ensure full control over parameters and consistency with the problem specifications, allowing for a thorough evaluation of the proposed method's performance.

3.1 Proposed Methodology

Table 1 presents the key parameters used to generate the datasets for the 2-DoF exoskeleton, which includes the lengths of the first and second links and the total number of samples used for data generation, both for the sinusoidal signal dataset and random step size dataset.

Table 1 Parameters used for generating sinusoidal and random datasets for a 2-DoF exoskeleton

Parameter	Value	Description
L_1	0.39 m	Length of the first link
L_2	0.156 m	Length of the second link
Number of samples	10,000	Total number of samples for dataset generation

3.1.1 Random Step-Size Dataset

For generating the random dataset, joint angles were sampled randomly within the range of $-\pi$ to π. This was achieved by multiplying random values, generated uniformly between 0 and 1, by 2π and then subtracting π to shift the range. The formula used for the random dataset generation is:

$$\theta_i = 2\pi \times \text{rand}(\text{size}, 2) - \pi \tag{13}$$

Here, `rand(size, 2)` produces random values between 0 and 1 for two joint angles, and the transformation ensures that the resulting joint angles θ_1 and θ_2 lie between $-\pi$ and π. This method was employed to generate a dataset of 10,000 samples, covering a wide range of possible joint configurations for the 2-DoF exoskeleton as presented in Fig. 2.

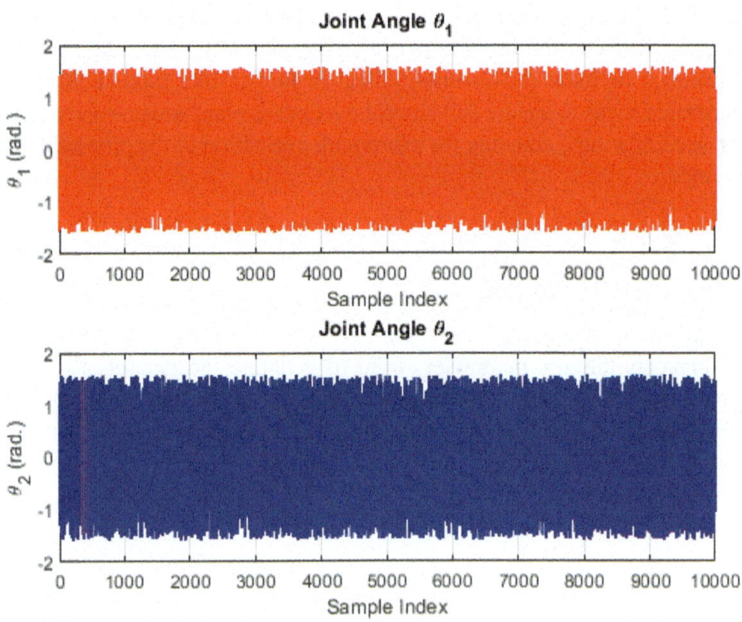

Fig. 2 Representation of different joint variables for the generated Random dataset

Table 2 Parameters used for generating sinusoidal signal dataset for a 2-DoF exoskeleton

Parameter	Value	Description
Amplitude ($\text{amp}_1, \text{amp}_2$)	[1, 2]	Amplitudes for the two DoFs
Frequency ($\text{freq}_1, \text{freq}_2$)	[0.5 Hz, 1.0 Hz]	Frequencies for the two DoFs
Phase ($\text{phase}_1, \text{phase}_2$)	[0, $\pi/4$]	Phases for the two DoFs

3.1.2 Sinusoidal Signal-Based Dataset

Table 2 presents the key parameters involved in creating the dataset. These parameters include the amplitude, frequency, and phase for the two DoFs of the exoskeleton. Each parameter helps define the nature of the sinusoidal signal for generating movement data for the exoskeleton's joints.

The joint angles for the sinusoidal signal dataset were generated using the following general equation:

$$\theta_i(t) = \text{amp}(i) \times \sin(2\pi \times \text{freq}(i) \times t + \text{phase}(i)), \tag{14}$$

where amp(i) is the amplitude, freq(i) is the frequency, and phase(i) is the phase for each degree of freedom (DoF), $i = 1, 2$. For this dataset, the amplitudes were set as $[1, 2]$, the frequencies as $[0.5, 1.0]$ Hz, and the phases as $[0, \pi/4]$.

Figure 3 visually depicts the joint angles or variables corresponding to the sinusoidal signal dataset for a 2-DoF exoskeleton. The graph likely shows the variations in joint positions over time, illustrating how the joint angles fluctuate based on the sinusoidal parameters.

3.2 *Used Metaheuristic Algorithms*

3.2.1 PSO Metaheuristic Technique

PSO is an optimization technique that draws inspiration from the social behavior of birds and fish in groups. In PSO, a group of particles (potential solutions) moves around the search space to find the optimal solution. Each particle modifies its position based on its previous experience and the experience of its neighbors. Key components of PSO include particle position and velocity, personal best (pBest), and global best (gBest). The velocity and position of each particle are updated using specific equations involving inertia weight, cognitive weight, social weight, and random factors.

$$\mathbf{x}_i = q_{\min} + (q_{\max} - q_{\min}) \cdot r \tag{15}$$

$$\mathbf{v}_i = r' - 0.5 \tag{16}$$

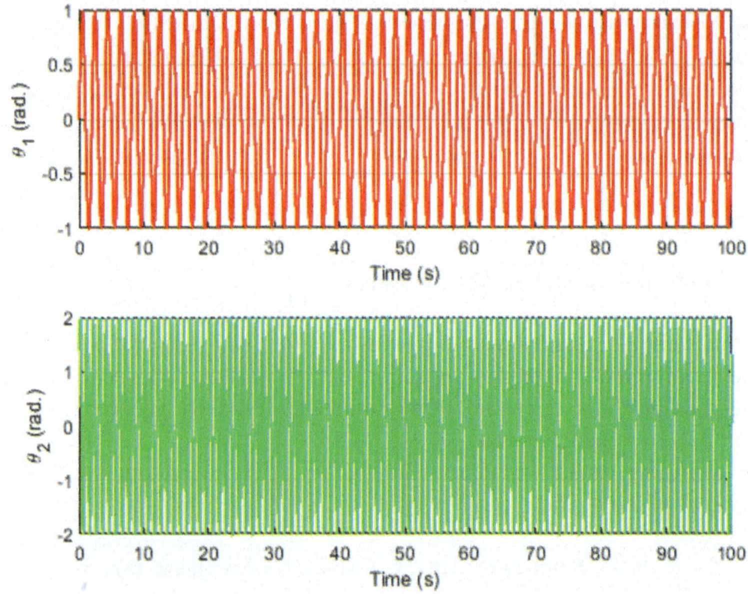

Fig. 3 Representation of different joint variables for the generated sinusoidal signal dataset

where

- $\mathbf{x}_i$: Initial position of particle i.
- $q_{\min}$: Minimum bound for the joint angles.
- $q_{\max}$: Maximum bound for the joint angles.
- r: Random vector generated uniformly in the range [0, 1] (size 2 for 2-DoF).
- $\mathbf{v}_i$: Initial velocity of particle i.
- r': Random vector generated uniformly in the range [0, 1] (size 2).

The velocity of each particle is updated using the following formula:

$$\mathbf{v}_i = w \cdot \mathbf{v}_i + c_1 \cdot r_1 \cdot (\mathbf{p}_i^\star - \mathbf{x}_i) + c_2 \cdot r_2 \cdot (\mathbf{p}_g - \mathbf{x}_i) \tag{17}$$

For solving inverse kinematics, PSO optimizes the joint parameters (angles) to reduce the discrepancy between the target and actual end-effector position calculated using the kinematic equations. For a 2-DoF exoskeleton robot with arm lengths $L_1 = 1$, and $L_2 = 1$, the PSO parameters might include 100 particles, 200 iterations, an inertia weight of 0.5, and cognitive and social weights of 2.0 each as presented in Table 3.

Table 3 PSO parameters

Parameter	Value
Number of particles (`numParticles`)	100
Number of iterations (`numIterations`)	200
Inertia weight (`w`)	0.5
Cognitive weight (`c1`)	2.0
Social weight (`c2`)	2.0

Table 4 GA parameters

Parameter	Value
Population size (`populationSize`)	100
Number of generations (`numGenerations`)	200
Crossover fraction (`crossoverFraction`)	0.8

3.2.2 GA Metaheuristic Technique

GA is a search heuristic inspired by the process of natural selection. It works with a population of candidate solutions, using genetic operators such as selection, crossover, and mutation to iteratively improve the solutions. The basic steps in GA include initializing a population, evaluating fitness, selecting the fittest individuals for reproduction, combining parts of two parents to create offspring, introducing random changes to maintain diversity, and forming a new population to repeat the process until convergence. For solving inverse kinematics, GA also optimizes the joint parameters to minimize the gap between the desired and achieved end-effector positions. For a 2-DoF exoskeleton robot, GA parameters might include a population size of 100, 200 generations, and a crossover fraction of 0.8 as shown in Table 4.

For a 2-DoF exoskeleton robot, the forward kinematics equations are represented in Sect. 2. To solve the inverse kinematics problem, we aim to find the joint angles θ_1 and θ_2 such that the end-effector reaches a desired position (x_d, y_d). Both PSO and GA can be used to minimize the error function defined as:

$$\text{Error} = X - X_d. \tag{18}$$

Joint limits: Represented as vectors of minimum and maximum joint angles:

$$\mathbf{q}_{\text{min}} = \begin{bmatrix} -\frac{3\pi}{4} \\ -\pi \end{bmatrix}, \qquad \mathbf{q}_{\text{max}} = \begin{bmatrix} \pi \\ \frac{3\pi}{4} \end{bmatrix} \tag{19}$$

Initialize Population: The function initializePopulation creates the initial population based on the joint limits:

$$\mathbf{P}_0 = \text{initializePopulation}(N, \mathbf{q}_{\text{min}}, \mathbf{q}_{\text{max}}) \tag{20}$$

Select Parents: Parents are selected based on fitness.

$$\text{Parents} = \text{selectParents}(\mathbf{P}, \text{Fitness}) \tag{21}$$

Crossover: A new population is created using the crossover operator:

$$\mathbf{P}' = \text{crossover}(\text{Parents}) \tag{22}$$

Mutation: The population undergoes mutation, with the mutation rate and joint limits applied:

$$\mathbf{P}_{\text{new}} = \text{mutate}(\mathbf{P}', m, \mathbf{q}_{\text{min}}, \mathbf{q}_{\text{max}}) \tag{23}$$

Update Population: The population is updated with the new generation:

$$\mathbf{P} = \mathbf{P}_{\text{new}} \tag{24}$$

In the PSO approach, particles representing joint angles are initialized randomly, their positions are evaluated based on the error, and their velocities and positions are updated iteratively until the error is minimized or the maximum number of iterations is reached. However, in the GA approach, an initial population of chromosomes representing joint angles is generated, evaluated based on the error, selected for reproduction, combined through crossover, mutated to maintain diversity, and iterated through generations until the error is minimized or the maximum number of generations is reached. Thus, both PSO and GA provide robust methods for optimizing the joint parameters in 2-DoF manipulators to achieve precise forward kinematics and inverse kinematics solutions.

3.2.3 GWO Metaheuristic Technique

Grey Wolf Optimizer (GWO) is a metaheuristic algorithm modeled after the social structure and hunting strategies of grey wolves. In GWO, the wolves are divided into four categories: alpha (α), beta (β), delta (δ), and omega (ω). The hunting process is mainly guided by the first three categories, where the omega wolves follow the guidance provided by α, β, and δ.

In GWO, the position of each wolf is updated iteratively by mimicking the hunting mechanism, which includes tracking, encircling, and attacking the prey. The positions of the wolves in the population are updated based on their distances from the alpha, beta, and delta wolves, which are assumed to have the best, second-best, and third-best fitness values, respectively. The wolves are initialized randomly within the specified joint limits:

$$\mathbf{W}(i) = \mathbf{q}_{\min} + (\mathbf{q}_{\max} - \mathbf{q}_{\min}) \cdot \text{rand}(\text{numWolves}, 2) \tag{25}$$

The GWO uses three main components, represented by coefficients, to direct the search process: A_1, A_2, and A_3, which are calculated as follows:

$$A_k = 2a \cdot \text{rand}(1, 2) - a, \quad k = 1, 2, 3 \tag{26}$$

where a is a coefficient that decreases linearly from 2 to 0 over the iterations, encouraging exploration at the beginning and exploitation as the algorithm converges.

The new position of a wolf is determined by the average of its movement towards the three leading wolves (alpha, beta, and delta). The position updates are given by:

$$\mathbf{X}_1 = \mathbf{W}_\alpha - A_1 \cdot |2\mathbf{W}_\alpha - \mathbf{W}(i)| \tag{27}$$

$$\mathbf{X}_2 = \mathbf{W}_\beta - A_2 \cdot |2\mathbf{W}_\beta - \mathbf{W}(i)| \tag{28}$$

$$\mathbf{X}_3 = \mathbf{W}_\delta - A_3 \cdot |2\mathbf{W}_\delta - \mathbf{W}(i)| \tag{29}$$

The final position of the wolf i is updated as the average of the calculated positions:

$$\mathbf{W}(i) = \frac{\mathbf{X}_1 + \mathbf{X}_2 + \mathbf{X}_3}{3} \tag{30}$$

To ensure that the wolves' positions remain within the joint limits, the updated positions are clipped to be within the bounds:

$$\mathbf{W}(i) = \max(\min(\mathbf{W}(i), \mathbf{q}_{\max}), \mathbf{q}_{\min}) \tag{31}$$

This iterative process continues for a predefined number of iterations, refining the search towards the optimal solution.

The key parameters for the GWO algorithm used in this study are summarized in Table 5.

Table 5 GWO algorithm parameters

Parameter	Value
Number of wolves	100
Number of iterations	200
Coefficient a	Linearly decreases from 2 to 0

Table 6 BAT parameters for solving inverse kinematics of a 2-DoF exoskeleton

Parameter	Description	Value
`numBats`	Population size	100
`maxIterations`	Maximum number of iterations	50
`Qmin`	Minimum frequency	0
`Qmax`	Maximum frequency	2
`A`	Loudness	0.5
`r`	Pulse rate	0.5

3.2.4 BAT Metaheuristic Technique

The Bat Algorithm (BAT) is an optimization method inspired by the echolocation abilities of bats. It leverages the principles of sound pulse emission and frequency modulation to explore and exploit the search space efficiently. For the inverse kinematics in 2-DoF exoskeletons, BAT optimizes the joint angles to minimize the positional error between the desired and actual end-effector positions. Each bat in the algorithm represents a potential solution, adjusting its position in the search space based on frequency, pulse rate, and loudness. The frequency range $[Q_{min}, Q_{max}]$ controls the step size for position updates, where lower frequencies encourage global exploration and higher frequencies promote local refinement. The loudness (A) decreases over time to enhance convergence as the solution nears the optimum, while the pulse rate (r) increases to ensure more precise movements. The search space for this problem includes two dimensions: the hidden layer size and the activation function for an ANN, with defined lower and upper bounds. This approach efficiently handles the non-linear and complex nature of inverse kinematics problems, providing accurate joint configurations while minimizing computational costs. Table 6 defines the algorithm's parameters.

4 Results of Training with Different Algorithms and Datasets

This section presents the results of using various optimization algorithms (PSO, GA, GWO, and BAT) for solving inverse kinematics problems. It discusses the performance of each algorithm, focusing on the best fitness achieved, joint angles, and position accuracy, while providing visual representations of the error convergence over iterations, illustrating the efficiency of each algorithm in minimizing error and reaching optimal solutions.

Table 7 PSO algorithm results for inverse kinematics

Best solution found	Value
Best fitness	5.77005×10^{-11}
Convergence reached	At iteration 84
Best desired position X_d	−1.3088, −1.0974
Best joint angles	0.32583, 1.0943
X Position found	−1.3088, −1.0974

4.1 *Results of Training with PSO Algorithm*

The results of the PSO algorithm, as presented in the Table 7, highlight the effectiveness and precision of the optimization process. The Best Fitness achieved is 5.77005×10^{-11}, indicating a highly optimal solution with minimal error or deviation from the desired objective. The algorithm successfully Convergence Reached at iteration 84, demonstrating its efficiency in identifying the best solution within a relatively small number of iterations. The Best Desired Position X_d found is (−1.3088, −1.0974), which matches exactly with the X Position Found at (−1.3088, −1.0974), showcasing the algorithm's precision in reaching the target position. Additionally, the Best Joint Angles identified for this optimal position are 0.32583 and 1.0943, which are critical parameters for achieving the desired configuration in applications like robotics or mechanical systems. This alignment between desired and achieved positions, along with optimal joint angles, underscores the PSO algorithm's capability to solve complex optimization problems efficiently.

Figure 4 titled "Error versus Iterations (Convergence at Iteration 97)," visually represents the error convergence process for PSO during the training of an ANN for an inverse kinematics problem. the blue line tracks the error reduction over 200 iterations of the PSO algorithm. The error begins at approximately 0.07 and drops sharply within the first 10 iterations, indicating a rapid improvement in the solution quality. As the iterations progress, the error continues to decrease, albeit at a slower rate, eventually converging to a minimal value close to zero. Convergence is achieved by the 97th iteration, as indicated by the red circle, suggesting that the PSO algorithm has effectively minimized the error.

4.2 *Results of Training with GA Algorithm*

Table 8 displays specific outcomes: the best desired position coordinates $X_d = (-1.3088, -1.0974)$, corresponding joint angles, which is different with the X Position Found at (−1.1848, −1.0422)the closest achieved position (0.18766, 1.3234), and the best fitness value of 0.037822.

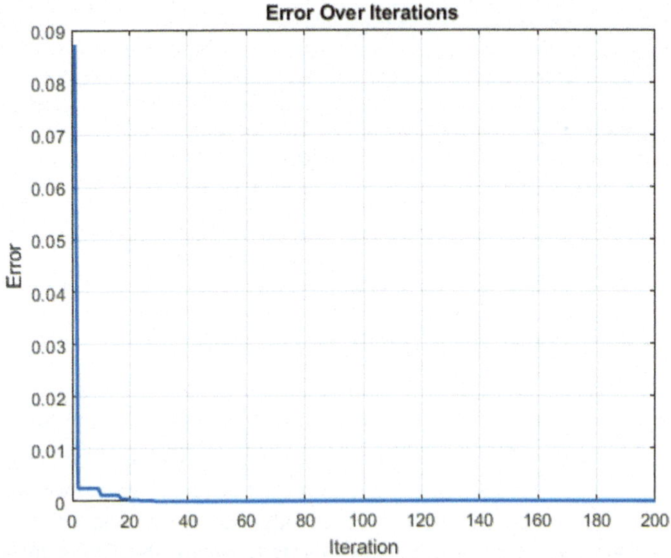

Fig. 4 Representation of error for each iteration

Table 8 GA algorithm results for forward/inverse kinematics

Parameter	Value
Best desired position X_d	−1.3088, −1.0974
Best joint angles	0.18766, 1.3234
X Position found	−1.1848, −1.0422
Best fitness	0.037822

Figure 5 titled "Error versus Generations (Convergence at Generation 200)," encapsulates the setup and successful outcomes of a genetic algorithm applied to an optimization problem, illustrating the effectiveness of the algorithm in finding near-optimal solutions within defined parameters. The plot follows a similar pattern to the GA algorithm. Here, the error also starts around 0.07 and decreases rapidly within the first 20 generations. The curve flattens out as the generations proceed, indicating a slower rate of error reduction. However, the algorithm continues to optimize the solution until convergence is reached at the 200th generation, marked by the red circle.

Both graphs highlight the efficiency of the PSO and GA algorithms in reducing the error over time, with PSO converging slightly faster than GA. The rapid initial decline in error followed by gradual refinement suggests that both algorithms effectively fine-tune the ANN parameters to achieve a highly accurate solution for the inverse kinematics problem.

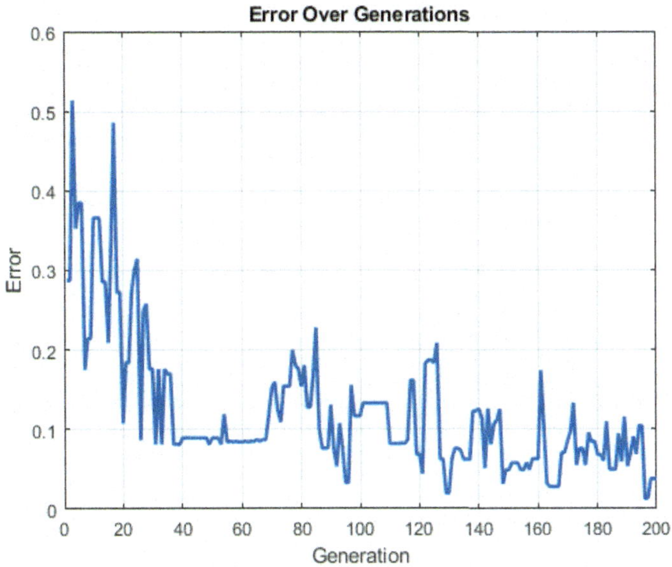

Fig. 5 Representation of error for each iteration

4.3 *Results of Training with GWO Algorithm*

The results of the GWO algorithm demonstrate its effectiveness in solving the inverse kinematics problem for a 2-DoF exoskeleton robot. The results show the best joint angles, $\theta_1 = 0.32615$ and $\theta_2 = 1.0939$, found by the GWO algorithm, along with a highly accurate X position of $[-1.3091, -1.0974]$ and a best error of 0.00011695. This indicates that the algorithm efficiently minimized the error in reaching the target position.

Figure 6 presents the error evolution over 200 iterations, where the error initially fluctuates but progressively decreases, with some peaks around iterations 100–120, suggesting the algorithm explores potential solutions before converging to near-zero error towards the final iterations. The third figure depicts the manipulator's configuration at the 200th iteration, showing that the end-effector successfully reached the target position $[-1.3088, -1.0974]$, very close to the desired goal. The GWO algorithm clearly demonstrates its capability to fine-tune joint angles and achieve high precision in the final manipulator configuration.

4.4 *Results of Training with BAT Algorithm*

The plot titled "Bat Algorithm Error over Iterations" shown in Fig. 7 illustrates the error reduction achieved by the Bat Algorithm as the number of iterations increases.

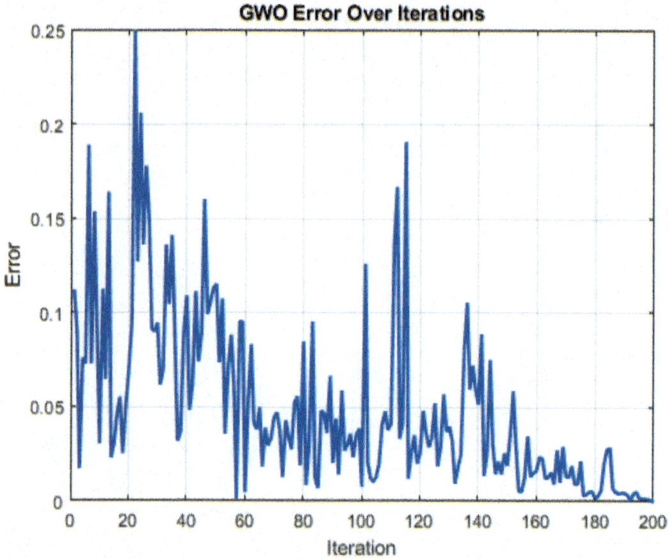

Fig. 6 Representation of error for each iteration

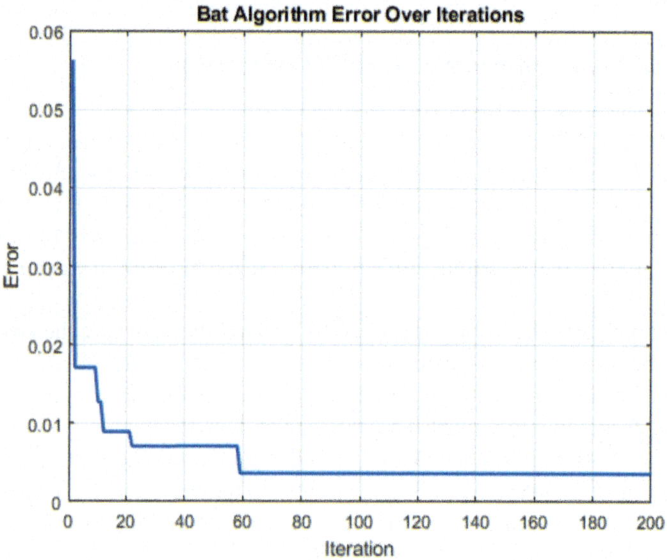

Fig. 7 Representation of error for each iteration

Initially, the error is approximately 0.06, but it decreases sharply within the first 20 iterations, dropping to just under 0.02.

After this rapid initial decline, the error continues to decrease at a slower rate, eventually stabilizing around 0.01 after about 60 iterations. From that point onward, the error remains nearly constant, showing minimal variation up to the 200th iteration. This behavior indicates that the Bat Algorithm quickly converges to a near-optimal solution, with most of the improvement occurring in the early stages of the optimization process.

4.5 *Results of Training with Different Algorithms for Random Step-Size Dataset Using ANN*

This section presents the results of training an ANN using various optimization algorithms (PSO, GA, GWO, and BAT) with distinct datasets to determine the best hyperparameters for the ANN architecture in solving inverse kinematics problems. It discusses the performance of each algorithm, focusing on the best fitness achieved. Additionally, visual representations of the error convergence over iterations highlight the efficiency of each algorithm in minimizing error and fine-tuning the ANN's hyperparameters.

4.5.1 For PSO-ANN and GA-ANN

In the optimization of ANN using PSO, the best parameters discovered were a hidden layer size of 83 neurons with the activation function `tansig` as shown in Table 9 at the left part. The corresponding best fitness value achieved was 8.7954×10^{-7}. When GA was combined with ANN, the optimal hyperparameters identified were 59 with a `tansig` activation function, yielding the best fitness value of 8.9492×10^{-5}, right part). These results underscore the efficacy of both PSO and GA in fine-tuning ANN parameters to enhance performance across various computational applications.

Figure 8 shows the MSE for training an ANN over multiple epochs. In the first plot, for the PSO-ANN algorithm, the MSE decreases rapidly over 10 epochs, indicating quick convergence. The best validation performance is achieved with an MSE of approximately $8.7954e{-}07$. The curves represent training (blue), validation (green), and test (red) datasets. In the second plot, for the GA-ANN algorithm, the MSE decreases steadily over 44 epochs, showing continuous learning. The best validation performance is achieved with an MSE of approximately $8.9492e{-}05$. These plots

Table 9 Parameters and fitness value of the ANN using PSO (left) and GA (right) algorithms

Best parameters (PSO-ANN)	
Hidden layer size	83
Activation function	`tansig`
Best fitness value	
8.7954×10^{-7}	

Best parameters (GA-ANN)	
Hidden layer size	59
Activation function	`tansig`
Best fitness value	
8.9492×10^{-5}	

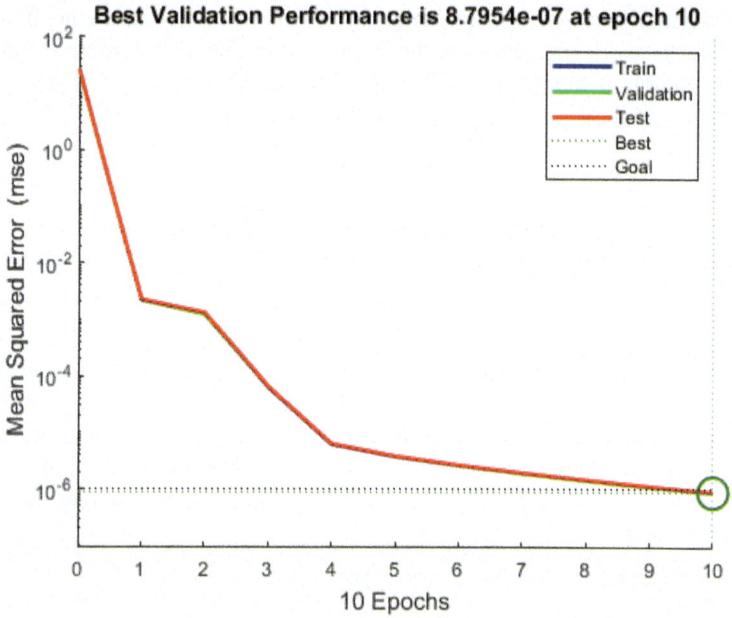

(a) PSO algorithm with ANN.

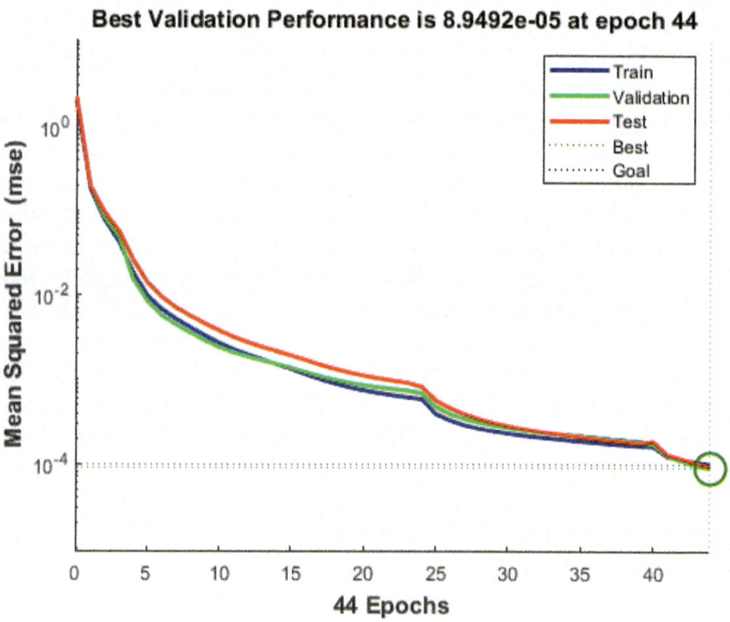

(b) GA algorithm with ANN.

Fig. 8 Comparison between the accuracy performances obtained via the PSO-ANN and the GA-ANN illustrate the performance of the metaheuristic algorithm-based ANN in terms of error reduction over training epochs, with the first plot showing faster convergence and a significantly lower MSE compared to the second plot.

The majority of the errors presented in Fig. 9 are centered around zero, with the highest number of instances falling within the bins close to zero, indicating that the predictions of the ANN-metaheuristic algorithm hybrid model are generally accurate. This visualization helps in understanding the distribution and magnitude of errors across the random step-size dataset used in training the ANN.

As shown from these results, the MSE obtained is very low and is approximately equal to zero (Fig. 9a, b). In addition, for the regression tasks for PSO-ANN the value of Training $R = 0.99988$, as illustrated in Fig. 10 is equal to the value of Training R for GA-ANN as shown in Fig. 11, which means that the convergence/approximation between the desired robot end-effector position (x_d, y_d) and the estimated one by the ANN-PSO and ANN-GA is very good or approximately ideal. This proves that the designed hybrid model that combines an ANN with a metaheuristic algorithm is trained very well and qualified to estimate the end-effector's position correctly.

4.5.2 For BAT-ANN Algorithm

The training process using the BAT algorithm on the random step-size dataset, shown in Fig. 12a, resulted in an MSE of 0.35358, achieved at epoch 9. This indicates that the BAT algorithm was able to minimize the error fairly quickly, although the final MSE value suggests that there may still be room for optimization. The results after training indicate that the best hyperparameters found are a hidden layer size of 100 and the `tansig` activation function.

The error histogram illustrated in Fig. 12b provides insight into the distribution of prediction errors across the training, validation, and test datasets. The majority of the errors are concentrated around zero, with most instances falling between approximately -0.1905 and 0.2439. The histogram shows a fairly symmetrical distribution, with errors peaking close to zero, as represented by the orange vertical line for zero error. There is a notable spread in errors, especially in the validation (green) and test (red) sets, indicating that the model occasionally produced larger errors in these datasets compared to the training set (blue). However, the majority of predictions remain close to the actual target values, suggesting that the BAT algorithm was generally effective in capturing the underlying relationships within the dataset.

The regression analysis presented in Fig. 13 provides further details about the model's performance. The regression coefficient for the training dataset was $R = 0.75031$, for the validation dataset $R = 0.7532$, and for the test dataset $R = 0.75311$. The overall regression coefficient across all datasets was $R = 0.75122$, indicating a moderate correlation between the predicted and actual outputs. These values suggest that while the model captures the general trends in the data, there is still some variance in predictions, especially in the validation and test sets, where higher errors were observed. The relatively consistent regression coefficients across all datasets

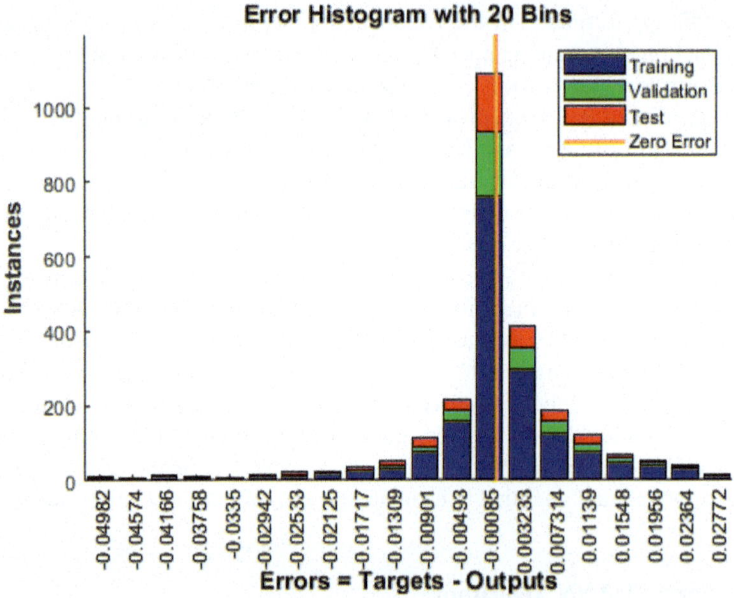

(a) Error Histogram for PSO-ANN.

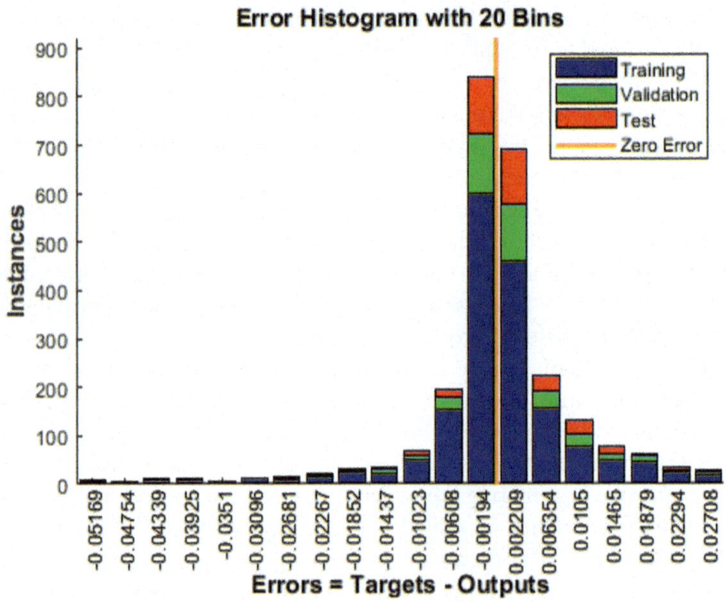

(b) Error Histogram for GA-ANN.

Fig. 9 Comparison of error histograms for PSO/GA combined with the ANN

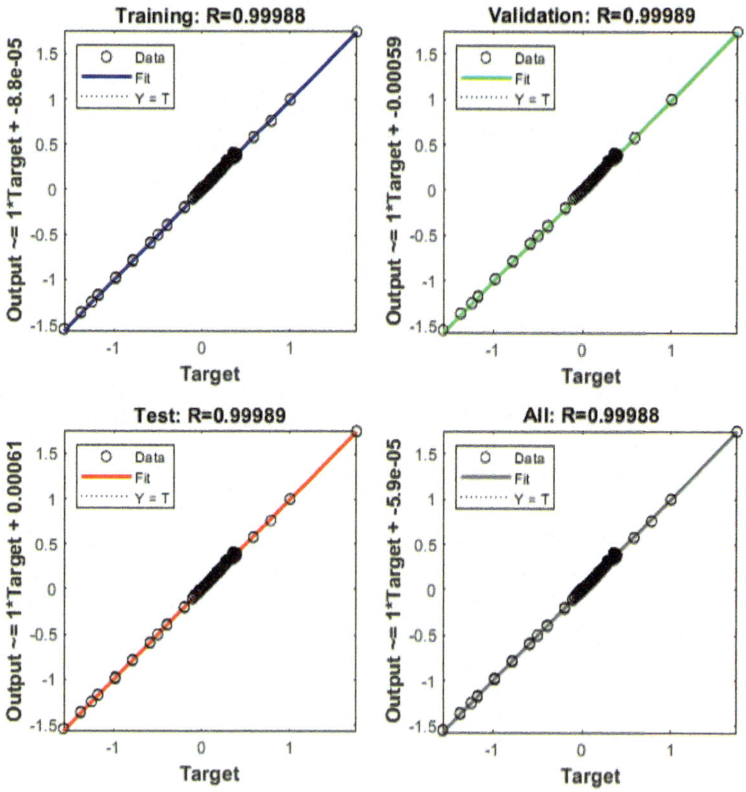

Fig. 10 Regression task for PSO algorithm combined with the ANN

highlight the model's stability, but also suggest that additional refinements could improve the prediction accuracy and reduce the error further.

4.5.3 For GWO-ANN Algorithm

The results from training with the Grey Wolf Optimizer (GWO) algorithm for the random step-size dataset demonstrated an efficient learning process, achieving an MSE of 0.032409 at epoch 5 (Fig. 14). This indicates that the model's predictions improved steadily throughout the training, culminating in an acceptable error rate.

The error histogram visualizes the distribution of prediction errors, which are calculated as the difference between the targets and the outputs of the model. The histogram is split into 20 bins, with the zero-error line marked in orange. The majority of the errors are clustered around zero, signifying that the model predictions arc generally accurate, though some moderate outliers can be observed at both ends of the distribution. The blue bars represent the training data, the green bars represent the

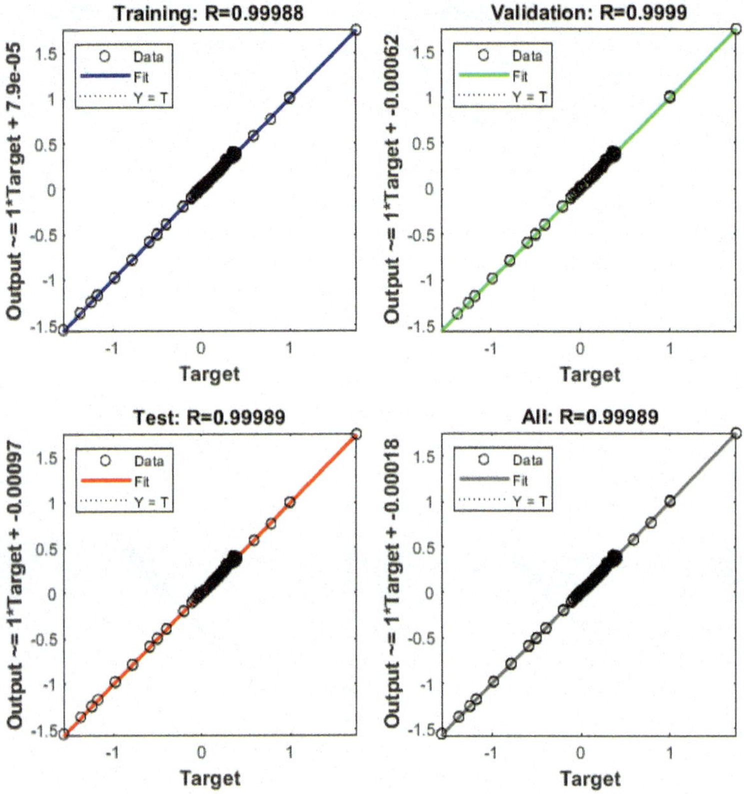

Fig. 11 Regression task for GA combined with the ANN

validation data, and the red bars correspond to the test data. A substantial portion of the errors from the training data lies near zero, indicating strong model performance on the training set, while errors in the validation and test sets appear slightly more spread out.

Regarding the regression analysis of Fig. 15, the coefficient of determination (R) values for different phases of the model's performance are as follows: for the validation data, $R = 0.77801$; for the test data, $R = 0.75176$; and for the training data, $R = 0.75639$. The overall R-value for the entire dataset is $R = 0.75891$. These R-values show a reasonable degree of correlation between the predicted outputs and the actual targets, particularly during the testing phase, which had the highest R-value, indicating the model generalizes well. The results after training indicate that the best hyperparameters found for the GWO-ANN algorithm are a hidden layer size of 79 and the `tansig` activation function.

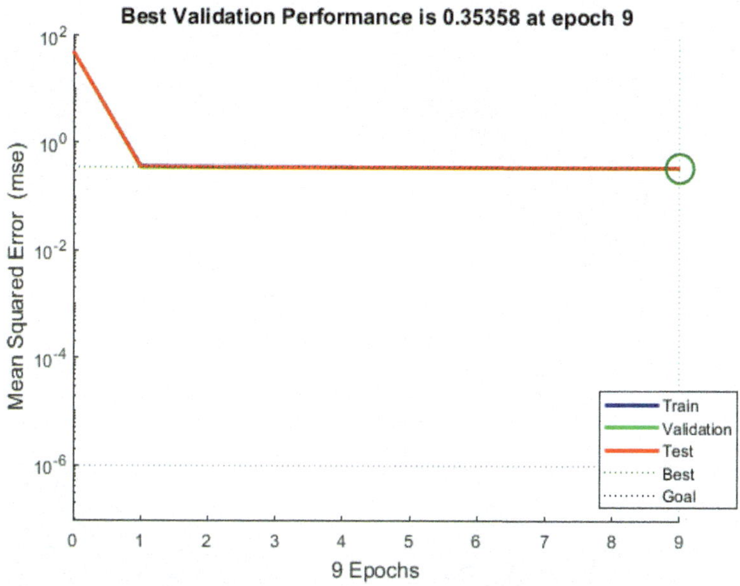

(a) MSE with the Random step size dataset

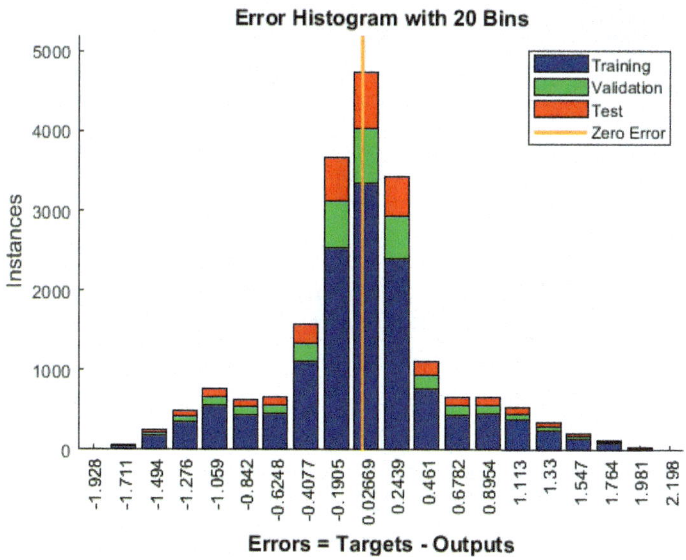

(b) Error histogram with the Random step size dataset

Fig. 12 ANN's performances (MSE and Error histogram) combined with the BAT algorithm using the Random step size dataset

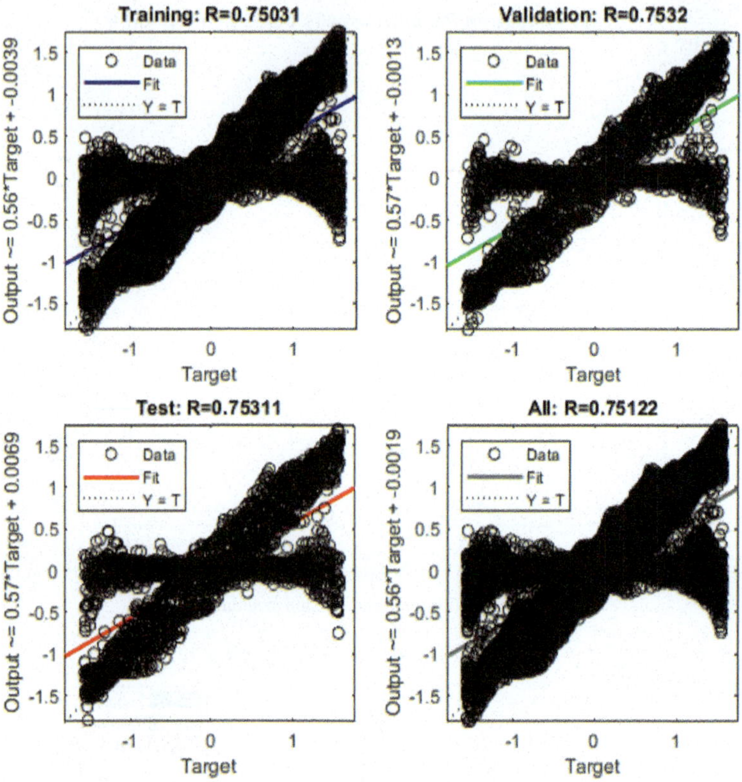

Fig. 13 Regression performance of the ANN with the BAT algorithm using the random step size dataset

4.6 Results of Training with Different Algorithms for Sinusoidal Signal Based Dataset Using ANN

On the other hand, for the sinusoidal signal-based dataset, we applied four algorithms similar to the random step size approach to compare the results and determine the most effective method for obtaining the optimal hyperparameters of the ANN. Our focus was on identifying the configuration that produced the lowest MSE, ensuring improved performance and accuracy in the inverse kinematics solution.

4.6.1 For PSO-ANN Algorithm

The results of the PSO-ANN applied to a sinusoidal dataset are summarized in three key visualizations. Figure 16 shows the MSE over 261 epochs, indicating that the model's error decreases rapidly during training.

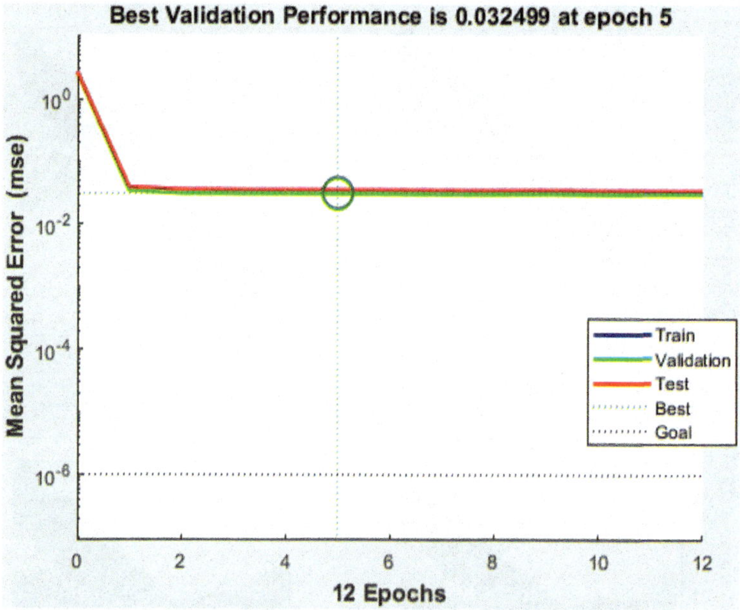

(a) MSE with the Random step size dataset

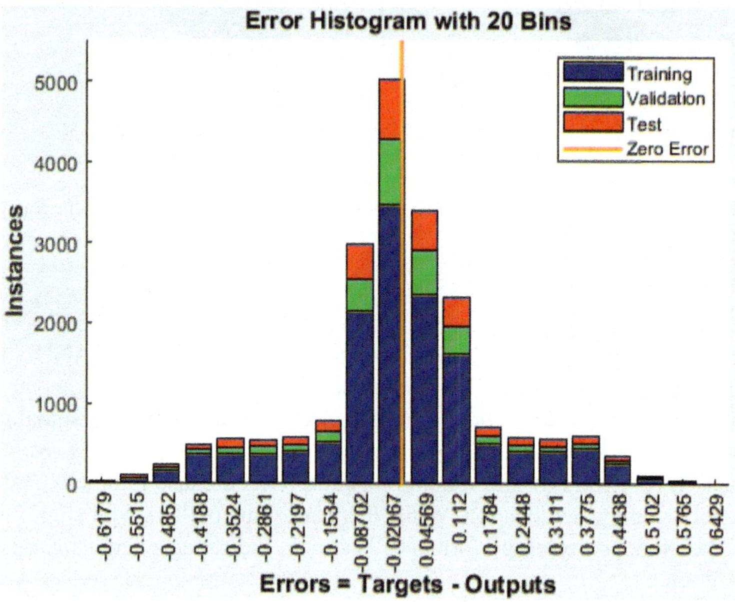

(b) Error histogram with the Random step size dataset

Fig. 14 ANN's performances (MSE and Error histogram) combined with the GWO algorithm and using the Random step size dataset

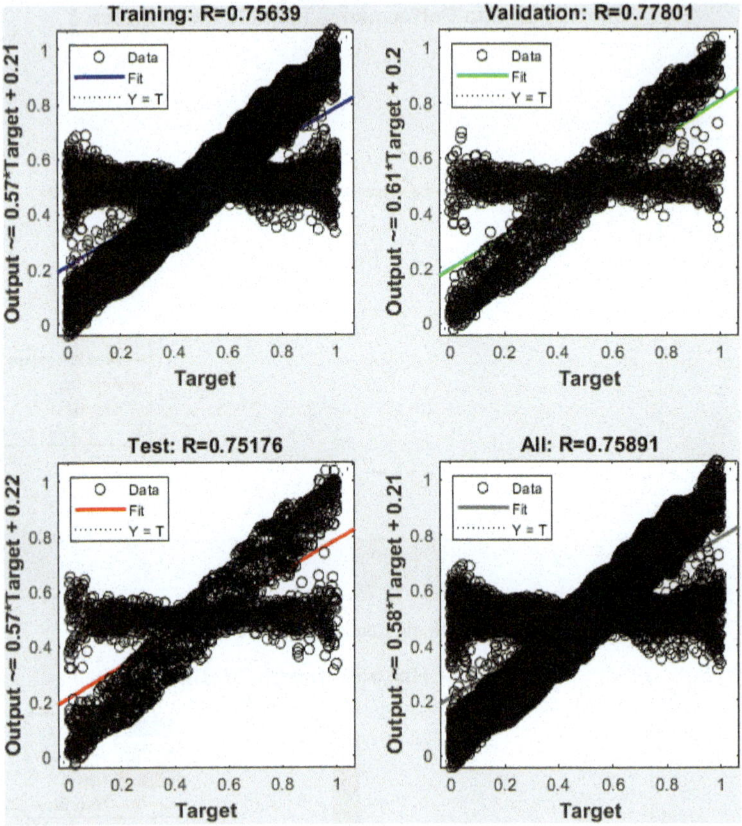

Fig. 15 Regression performance of the ANN trained with the GWO algorithm for the Random step size dataset

The best validation performance occurs at epoch 261 with an MSE of 0.00035361, reflecting a well-trained model without signs of overfitting. The ANN structure reveals an input layer with 2 neurons, a hidden layer with 65 neurons, and an output layer with 2 neurons, suggesting a complex model to capture the sinusoidal pattern. The error histogram shows that most errors, particularly for the training set, are clustered near zero. Errors for the validation and test sets are also minimal, confirming the model's good generalization. While there are some outliers with slightly higher errors, the model's overall performance is strong, indicating effective learning and optimization.

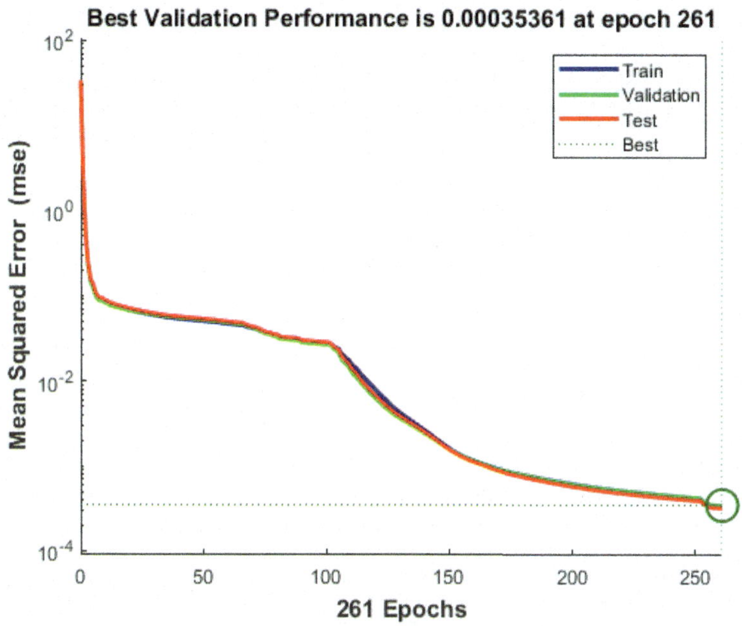

(a) MSE

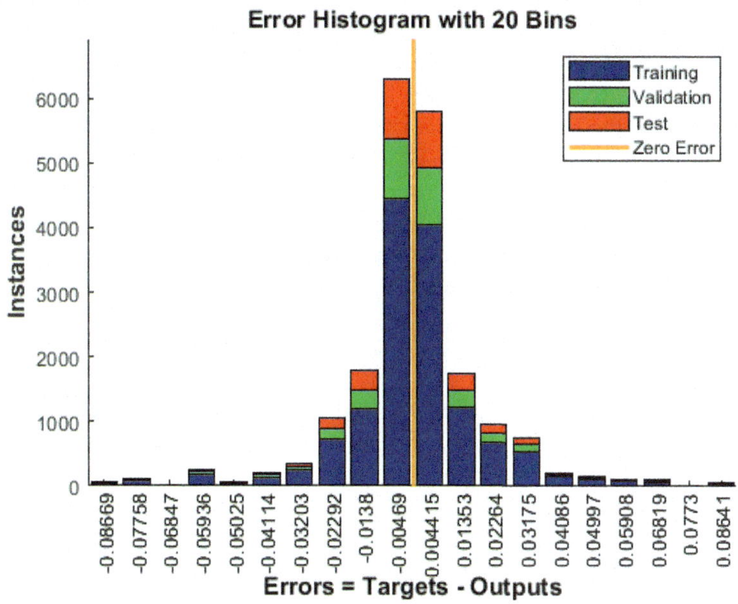

(b) Error histogram

Fig. 16 ANN's performances with the PSO algorithm obtained using the sinusoidal step sizc dataset

4.6.2 For GA-ANN Algorithm

The optimal number of neurons in the hidden layer was found to be 96, as determined by the GA. The lowest MSE achieved was 7.0367$e-05$ as shown in Fig. 17, indicating a highly accurate model. The error histogram shows that the GA-ANN model has achieved highly accurate predictions with minimal deviation from the targets. The concentration of errors near zero, across training, validation, and testing, confirms the model's strong generalization ability.

These results highlight that GA-ANN has effectively optimized the network's structure and achieved excellent performance on the dataset.

4.6.3 For ANN-BAT Algorithm

For the BAT-ANN algorithm, the best performance was achieved with a MSE of 9.7844$e-05$ at epoch 485, as depicted in Fig. 18a. After training, the optimal hyper-parameters for the BAT-ANN algorithm were found to be a hidden layer size of 100 and the `tansig` activation function.

This MSE indicates a very low error rate, demonstrating the effectiveness of the BAT-ANN approach in accurately predicting outputs.

The error histogram for the BAT-ANN (Fig. 18b) shows that the error distribution is very close to zero, indicating minimal deviation between the predicted outputs and the target values. The narrow concentration of errors around zero suggests that the model is highly precise, with very few instances of significant error.

Additionally, the regression values (R) for all phases (training, validation, and testing) are 0.99996 (as presented in Fig. 19), signifying an almost perfect fit between the predicted and actual values. This reflects the strong predictive performance of the BAT-ANN algorithm, showing that the model captures the patterns in the sinusoidal signal-based dataset with exceptional accuracy and minimal error.

4.6.4 For ANN-GWO Algorithm

The final algorithm used was ANN-GWO, where the best performance was achieved with an MSE of 1.5825$e-04$ at epoch 476, as seen in Fig. 20a. This performance is comparable to the MODA-ANN and still reflects a highly accurate model with low prediction error.

The error histogram for ANN-GWO (Fig. 20b) shows a less distributed pattern, meaning the errors are more concentrated around zero, with fewer instances of significant error. This demonstrates that the model is precise, with minimal deviations between predicted and actual outputs. As illustrated in Fig. 21, the regression values (R) for all phases (training, validation, and testing) are 0.99994, indicating a very strong correlation between the predicted and target values. Although slightly lower than the BAT-ANN results, the ANN-GWO algorithm still offers an excellent fit and high accuracy for predicting the sinusoidal signal-based dataset.

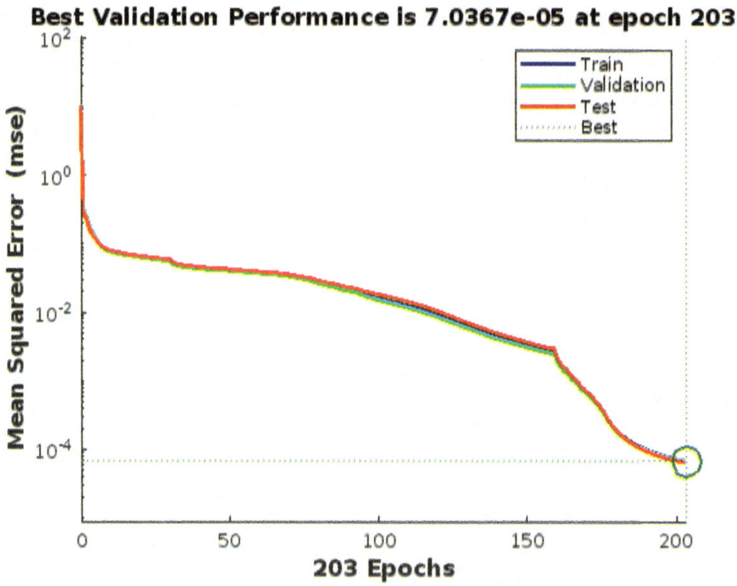

(a) MSE

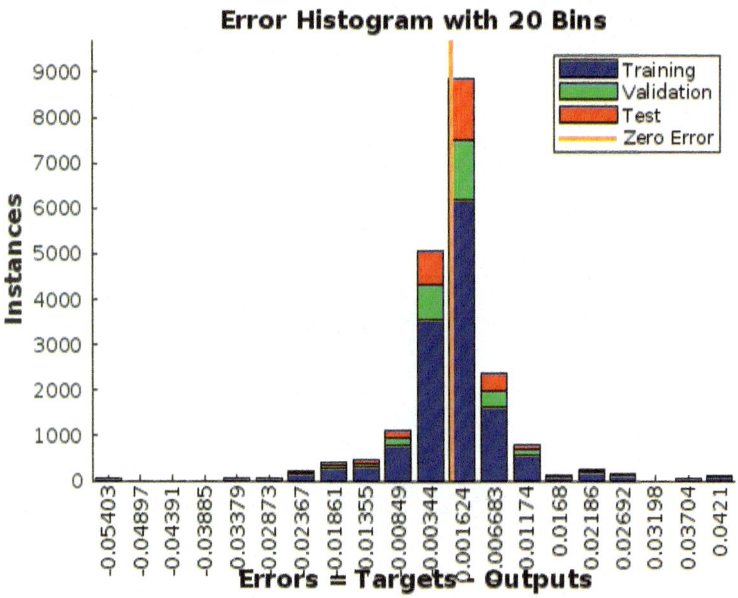

(b) Error histogram

Fig. 17 ANN's performances with the GA algorithm using the sinusoidal step size dataset

The best hyperparameters identified for the GWO-ANN algorithm after training are a hidden layer size of 90 and the `tansig` activation function.

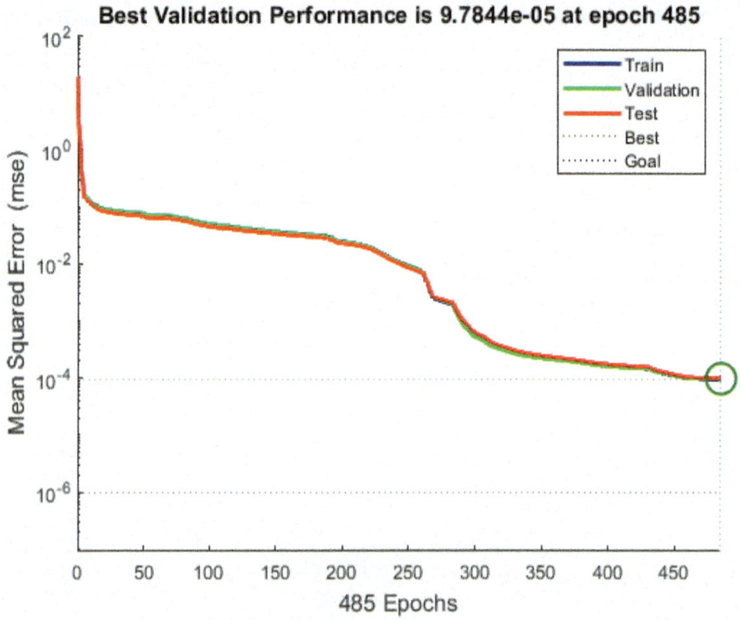

(a) MSE with the sinusoidal step size dataset

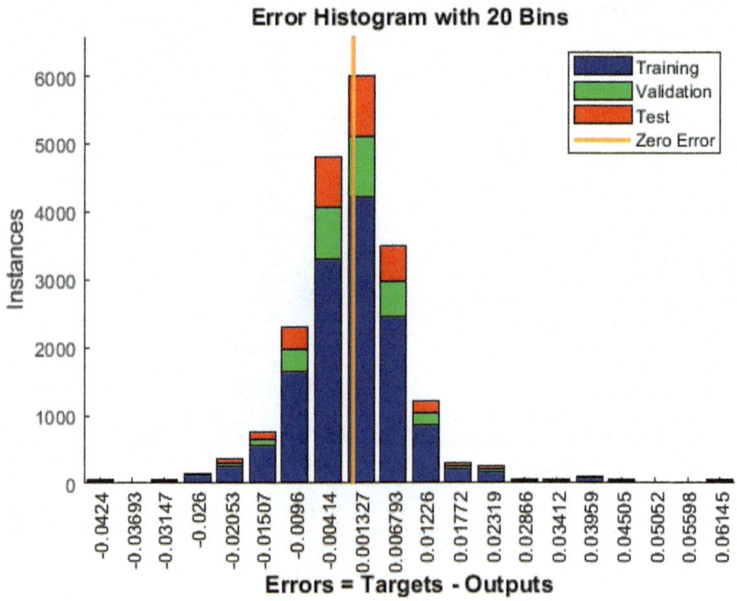

(b) Error histogram of sinusoidal step size dadaset

Fig. 18 ANN's performances (MSE and Error histogram) obtained by combining the BAT algorithm and using the sinusoidal step size dataset

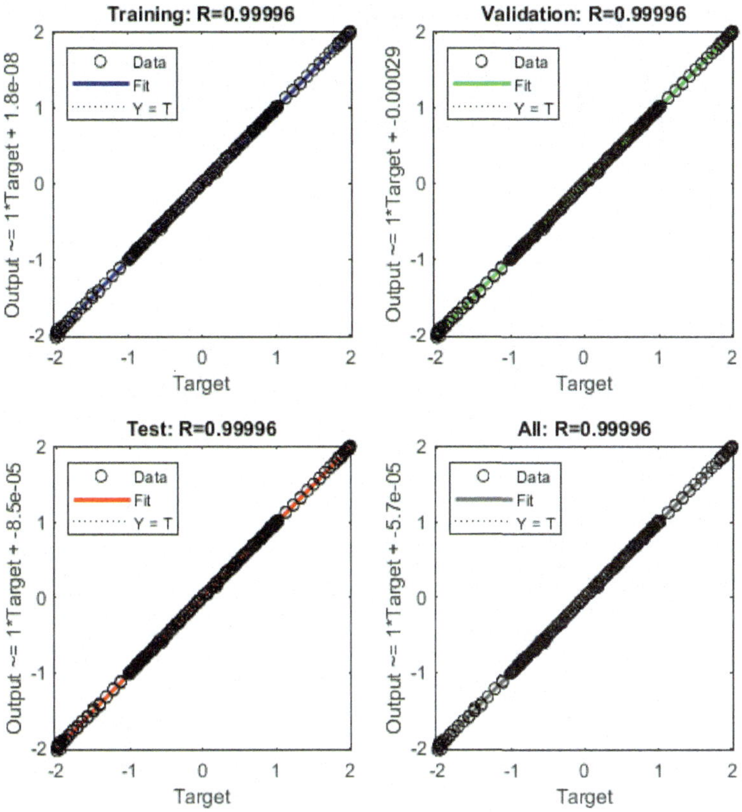

Fig. 19 Regression results showing the performance of the ANN combined with the BAT algorithm and using the sinusoidal step size dataset

5 Discussion and Comparison

The combination of metaheuristic algorithms with ANNs enhances problem-solving in inverse kinematics by leveraging the strengths of both techniques. Metaheuristic algorithms, such as PSO, GA, BAT, and GWO, are effective at exploring large search spaces and can efficiently approximate solutions to complex optimization problems, which is particularly beneficial in the non-linear search space of the inverse kinematics. However, they may not always converge on precise solutions without refinement.

Integrating ANNs adds a layer of learning capability, allowing the model to improve accuracy over time by adapting to patterns within the data. ANNs excel at handling complex, non-linear relationships, and when combined with the exploratory power of metaheuristics, they enhance convergence speed and precision. This

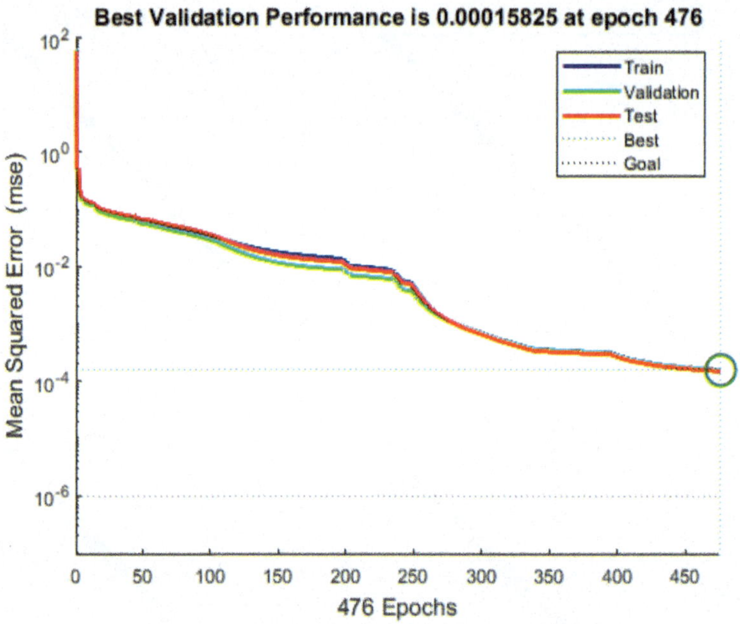

(a) MSE with the sinusoidal step size dataset

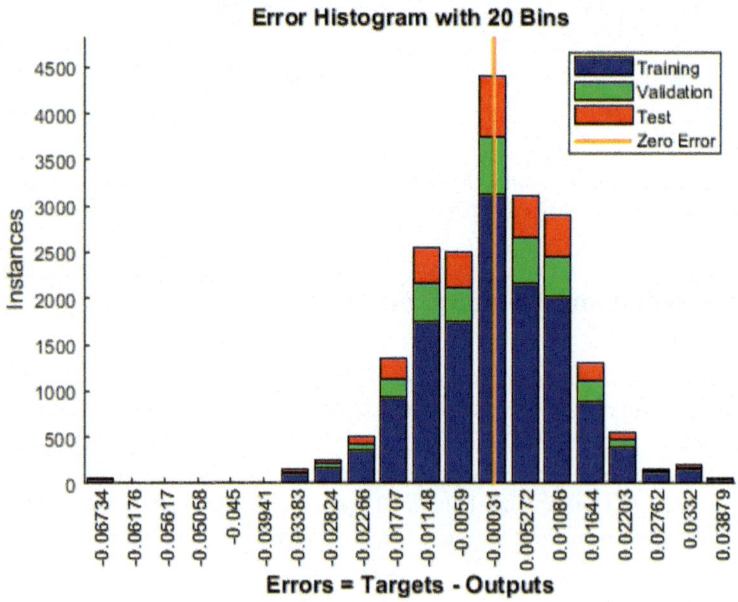

(b) Error histogram of sinusoidal step size dataset

Fig. 20 ANN's performances (MSE and error histogram) combined with the GWO algorithm and using the sinusoidal step size dataset

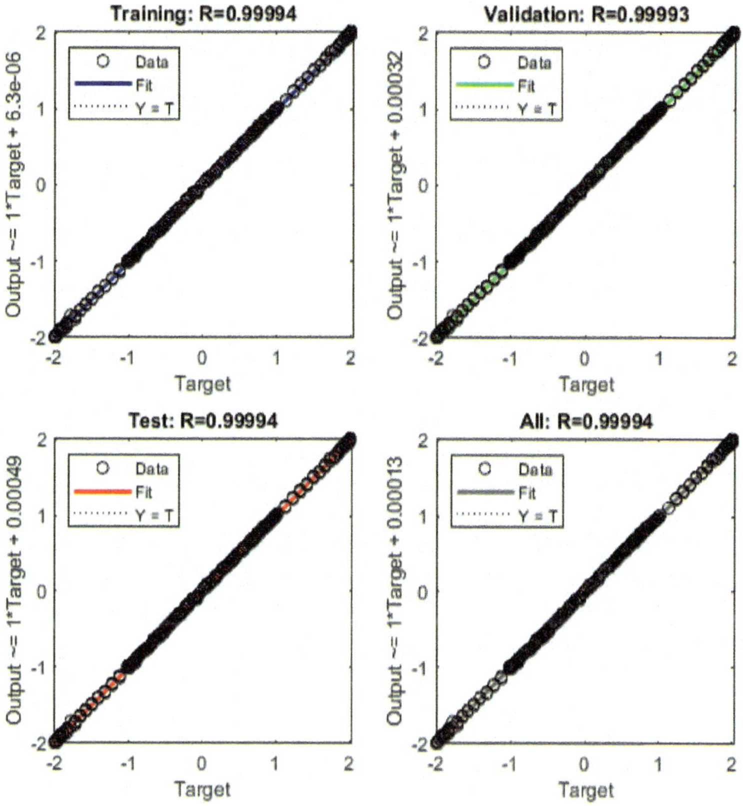

Fig. 21 Regression results obtained by combining the ANN and the GWO algorithm for the sinusoidal step size dataset

hybrid approach enables the system to find optimal or near-optimal solutions more effectively than either technique alone.

The comparative analysis of hybrid algorithms applied to different datasets reveals distinct performance patterns. On the Random Step Size dataset, PSO-ANN outperformed the other algorithms, achieving the lowest MSE of 8.7954×10^{-7} with just 10 epochs, indicating both high accuracy and fast convergence. GA-ANN showed moderate performance with an MSE of 8.9492×10^{-5}, requiring 44 epochs, suggesting a balance between accuracy and convergence speed. In contrast, BAT-ANN and GWO-ANN performed significantly worse, with MSE values of 0.353 and 0.361, respectively, despite converging in fewer epochs.

For the Sinusoidal Signal dataset, GA-ANN excelled, achieving the best MSE of 7.0367×10^{-5} after 203 epochs, demonstrating strong performance on this more structured dataset. PSO-ANN, while still effective, had a higher MSE of 0.00035361 and required 413 epochs, indicating slower convergence on this dataset. BAT-ANN

Table 10 Comparison of results for each algorithm on different datasets

Algorithm	Dataset type	Hidden layer size	Best MSE	Epoch number
PSO-ANN	Random step size	83	8.7954×10^{-7}	10
GA-ANN	Random step size	59	8.9492×10^{-5}	44
BAT-ANN	Random step size	100	0.353	9
GWO-ANN	Random step size	79	0.361	8
PSO-ANN	Sinusoidal signal	261	0.00035361	413
GA-ANN	Sinusoidal signal	96	7.0367e–05	203
BAT-ANN	Sinusoidal signal	100	9.7844e–05	485
GWO-ANN	Sinusoidal signal	90	1.5825e–04	476

and GWO-ANN again performed worse, with MSE values of 9.7844×10^{-5} and 1.5825×10^{-4}, respectively, despite requiring a large number of epochs.

The epoch number is indeed an important parameter in training ANNs, as it directly impacts the convergence and performance of the model. While our primary focus in Table 10 was on evaluating the effects of hyperparameters like hidden layers and activation functions, we agree that including the epoch number would provide additional insights into the training process.

While the 2-DoF kinematic model can be solved analytically, our hybrid approach serves as a foundational test case to validate its scalability for complex systems where analytical solutions become infeasible. By evaluating hyperparameters like the number of hidden layers and activation functions, we demonstrate how these techniques enhance the model's ability to handle non-linear systems-capabilities that may not be evident in a simple 2-DoF case but are crucial for more complex scenarios.

In conclusion, PSO-ANN proved most effective for random step datasets, while GA-ANN was best suited for sinusoidal datasets. Although BAT-ANN and GWO-ANN trained faster, their accuracy was consistently lower, making them less ideal for precise inverse kinematics solutions.

6 Conclusions and Future Works

In this study, we examined the effectiveness of hybrid metaheuristic algorithms combined with ANNs for optimizing inverse kinematics solutions in robotic manipulators. We tested four hybrid models-PSO-ANN, GA-ANN, BAT-ANN, and GWO-ANN-on two datasets: one derived from a sinusoidal signal and the other using random step size values. The results showed that PSO-ANN achieved the best performance on the random step size dataset, with an MSE of 8.7954×10^{-7} after 10 epochs, while GA-ANN excelled on the sinusoidal dataset, achieving an MSE of 7.0367×10^{-5} after 203 epochs. BAT-ANN and GWO-ANN, despite converging faster, exhibited higher MSE values, indicating lower accuracy compared to PSO-ANN and GA-ANN.

Future work will extend the applicability of hybrid metaheuristic-ANN models to more complex robotic systems, explore additional metaheuristic algorithms, and investigate advanced neural network architectures to further optimize inverse kinematics solutions. Moreover, exploring different neural network architectures, such as Recurrent Neural Networks (RNNs) and Long Short-Term Memory (LSTM) networks, could potentially enhance the model's ability to learn temporal dependencies, which are critical in dynamic robotic applications. These architectures could further refine the accuracy and efficiency of inverse kinematics solutions, particularly for robots operating in real-time, dynamic environments.

References

1. Shi, D., Zhang, W., Zhang, W., Ding, X.: A review on lower limb rehabilitation exoskeleton robots. Chin. J. Mech. Eng. **32**(1), 1–11 (2019)
2. Koch, M.A., Font-Llagunes, J.M.: Lower-limb exosuits for rehabilitation or assistance of human movement: a systematic review. Appl. Sci. **11**(18), 8743 (2021)
3. Hussain, F., Goecke, R., Mohammadian, M.: Exoskeleton robots for lower limb assistance: a review of materials, actuation, and manufacturing methods. Proc. Inst. Mech. Eng. Part H: J. Eng. Med. **235**(12), 1375–1385 (2021)
4. Qiu, S., Pei, Z., Wang, C., Tang, Z.: Systematic review on wearable lower extremity robotic exoskeletons for assisted locomotion. J. Bionic Eng. **20**(2), 436–469 (2023)
5. Kalita, B., Narayan, J., Dwivedy, S.K.: Development of active lower limb robotic-based orthosis and exoskeleton devices: a systematic review. Int. J. Soc. Robot. **13**, 775–793 (2021)
6. Aliman, N., Ramli, R., Haris, S.M.: Design and development of lower limb exoskeletons: a survey. Robot. Auton. Syst. **95**, 102–116 (2017)
7. Narayan, J., Kumar Dwivedy, S.: Preliminary design and development of a low-cost lower-limb exoskeleton system for paediatric rehabilitation. Proc. Inst. Mech. Eng. Part H: J. Eng. Med. **235**(5), 530–545 (2021)
8. Narayan, J., Auepanwiriyakul, C., Jhunjhunwala, S., Abbas, M., Dwivedy, S.K.: Hierarchical classification of subject-cooperative control strategies for lower limb exoskeletons in gait rehabilitation: a systematic review. Machines **11**(7), 764 (2023)
9. Gritli, H., Jenhani, S., Narayan, J.: Motion control of manipulator robots via two PD-based nonlinear controllers: design and comparative analysis. Autom. Control Comput. Sci. **58**(6), 603–616 (2024)
10. Gull, M.A., Bai, S., Bak, T.: A review on design of upper limb exoskeletons. Robotics **9**(1), 16 (2020)
11. Narayan, J., Gritli, H., Dwivedy, S.K.: Fast terminal sliding mode control with rapid reaching law for a pediatric gait exoskeleton system. Int. J. Intell. Robot. Appl. **8**, 76–95 (2024)
12. Jenhani, S., Gritli, H.: An LMI-based robust state-feedback controller design for the position control of a knee rehabilitation exoskeleton robot: comparative analysis. Meas. Control **57**(9), 1326–1346 (2024)
13. Narayan, J., Kalita, B., Dwivedy, S.K.: Adaptive backstepping human-cooperative control of a pediatric gait exoskeleton system with high-and low-level admittance. Proc. Inst. Mech. Eng. Part I J. Syst. Control Eng. 09596518231191390 (2023)
14. Narayan, J., Dwivedy, S.K.: Towards neuro-fuzzy compensated PID control of lower extremity exoskeleton system for passive gait rehabilitation. IETE J. Res. **69**(2), 778–795 (2023)

15. Shen, Y., Ferguson, P.W., Rosen, J.: Upper limb exoskeleton systems-overview. Wearable Robot. 1–22 (2020)
16. Perry, J.C., Rosen, J., Burns, S.: Upper-limb powered exoskeleton design. IEEE/ASME Trans. Mechatron. **12**(4), 408–417 (2007)
17. Bembli, S., Haddad, N.K., Belghith, S.: A robust model free terminal sliding mode with gravity compensation control of a 2 DoF exoskeleton-upper limb system. J. Control Autom. Electr. Syst. **32**, 632–641 (2021)
18. Pei, S., Wang, J., Guo, J., Yin, H., Yao, Y.: A human-like inverse kinematics algorithm of an upper limb rehabilitation exoskeleton. Symmetry **15**(9), 1657 (2023)
19. Kucuk, S., Bingul, Z.: Robot Kinematics: Forward and Inverse Kinematics. INTECH Open Access Publisher, London (2006)
20. Gopura, R., Bandara, D., Kiguchi, K., Mann, G.K.: Developments in hardware systems of active upper-limb exoskeleton robots: a review. Robot. Auton. Syst. **75**, 203–220 (2016)
21. Rehmat, N., Zuo, J., Meng, W., Liu, Q., Xie, S.Q., Liang, H.: Upper limb rehabilitation using robotic exoskeleton systems: a systematic review. Int. J. Intell. Robot. Appl. **2**, 283–295 (2018)
22. Vertechy, R., Frisoli, A., Dettori, A., Solazzi, M., Bergamasco, M.: Development of a new exoskeleton for upper limb rehabilitation. In: 2009 IEEE International Conference on Rehabilitation Robotics, pp. 188–193. IEEE (2009)
23. Al-Fahaam, H., Davis, S., Nefti-Meziani, S.: Wrist rehabilitation exoskeleton robot based on pneumatic soft actuators. In: 2016 International Conference for Students on Applied Engineering (ICSAE), pp. 491–496. IEEE (2016)
24. Oguntosin, V.W., Mori, Y., Kim, H., Nasuto, S.J., Kawamura, S., Hayashi, Y.: Design and validation of exoskeleton actuated by soft modules toward neurorehabilitation-vision-based control for precise reaching motion of upper limb. Front. Neurosci. **11**, 352 (2017)
25. Wu, W., Fong, J., Crocher, V., Lee, P.V., Oetomo, D., Tan, Y., Ackland, D.C.: Modulation of shoulder muscle and joint function using a powered upper-limb exoskeleton. J. Biomech. **72**, 7–16 (2018)
26. Kim, H., Miller, L.M., Byl, N., Abrams, G.M., Rosen, J.: Redundancy resolution of the human arm and an upper limb exoskeleton. IEEE Trans. Biomed. Eng. **59**(6), 1770–1779 (2012)
27. Głowiński, S., Krzyżyński, T., Pecolt, S., Maciejewski, I.: Design of motion trajectory of an arm exoskeleton. Arch. Appl. Mech. **85**(1), 75–87 (2015)
28. Kang, H.B., Wang, J.H.: Adaptive control of 5 DoF upper-limb exoskeleton robot with improved safety. ISA Trans. **52**(6), 844–852 (2013)
29. Pérez-Rodríguez, R., Marcano-Cedeño, A., Costa, Ú., Solana, J., Cáceres, C., Opisso, E., Tormos, J.M., Medina, J., Gómez, E.J.: Inverse kinematics of a 6 DoF human upper limb using ANFIS and ANN for anticipatory actuation in ADL-based physical neurorehabilitation. Expert Syst. Appl. **39**(10), 9612–9622 (2012)
30. Liu, W., Chen, D., Steil, J.: Analytical inverse kinematics solver for anthropomorphic 7-DoF redundant manipulators with human-like configuration constraints. J. Intell. Robot. Syst. **86**, 63–79 (2017)
31. Dalla Gasperina, S., Ghonasgi, K., de Oliveira, A.C., Gandolla, M., Pedrocchi, A., Deshpande, A.: A novel inverse kinematics method for upper-limb exoskeleton under joint coordination constraints. In: 2020 IEEE/RSJ International Conference on Intelligent Robots and Systems (IROS), pp. 3404–3409. IEEE (2020)
32. Bin, Z., Rong, X., Jun, W.: Kinematics analysis of a novel 7-DoF humanoid manipulator for table tennis. In: 2011 International Conference on Electronics, Communications and Control (ICECC), pp. 1524–1528. IEEE (2011)
33. Kim, H., Miller, L.M., Al-Refai, A., Brand, M., Rosen, J.: Redundancy resolution of a human arm for controlling a seven DoF wearable robotic system. In: 2011 Annual International Conference of the IEEE Engineering in Medicine and Biology Society, pp. 3471–3474. IEEE (2011)
34. Jiang, H., Duerstock, B.S., Wachs, J.P.: A machine vision-based gestural interface for people with upper extremity physical impairments. IEEE Trans. Syst. Man Cybern. Syst. **44**(5), 630–641 (2013)

35. Zhang, X., Li, C., Hu, W., Tong, X., Maybank, S., Zhang, Y.: Human pose estimation and tracking via parsing a tree structure based human model. IEEE Trans. Syst. Man Cybern. Syst. **44**(5), 580–592 (2013)
36. Chan, K.C., Koh, C.K., Lee, C.G.: A 3-D-point-cloud system for human-pose estimation. IEEE Trans. Syst. Man Cybern. Syst. **44**(11), 1486–1497 (2014)
37. Tolani, D., Goswami, A., Badler, N.I.: Real-time inverse kinematics techniques for anthropomorphic limbs. Graph. Models **62**(5), 353–388 (2000)
38. Loh, B.G., Rosen, J.: Kinematic analysis of 7 degrees of freedom upper-limb exoskeleton robot with tilted shoulder abduction. Int. J. Precis. Eng. Manuf. **14**, 69–76 (2013)
39. Abdor-Sierra, J.A., Merchán-Cruz, E.A., Rodríguez-Cañizo, R.G.: A comparative analysis of metaheuristic algorithms for solving the inverse kinematics of robot manipulators. Results Eng. **16**, 100,597 (2022)
40. Nguyen, T.T., Bui, T.N., Dai, W., Nguyen, T.V., Tao, L.N.: Apply some meta-heuristic algorithms to solve inverse kinematic problems of a 7-DoFs manipulator robot. Int. J. Mech. Eng. Robot. Res. **10**(9), 498–504 (2021)
41. Rokbani, N., Mirjalili, S., Slim, M., Alimi, A.M.: A beta salp swarm algorithm meta-heuristic for inverse kinematics and optimization. Appl. Intell. **52**(9), 10493–10518 (2022)
42. López-Muñoz, R., Portilla-Flores, E.A., Corona-Ramírez, L.G., Vega-Alvarado, E., Maya-Rodríguez, M.C.: Inverse kinematics: an alternative solution approach applying metaheuristics. Appl. Sci. **13**(11), 6543 (2023)
43. Momani, S., Abo-Hammour, Z.S., Alsmadi, O.M.: Solution of inverse kinematics problem using genetic algorithms. Appl. Math. Inf. Sci. **10**(1), 225 (2016)
44. Reyes, S.V., Gardini, S.P.: Inverse kinematics of manipulator robot using a PSO metaheuristic with adaptively exploration. In: 2019 IEEE XXVI International Conference on Electronics, Electrical Engineering and Computing (INTERCON), pp. 1–4. IEEE (2019)
45. Abdor-Sierra, J.A., Merchán-Cruz, E.A., Sánchez-Garfias, F.A., Rodríguez-Cañizo, R.G., Portilla-Flores, E.A., Vázquez-Castillo, V.: Particle swarm optimization for inverse kinematics solution and trajectory planning of 7-DoF and 8-DoF robot manipulators based on unit quaternion representation. J. Appl. Eng. Sci. **19**(3), 592–599 (2021)
46. Deng, H., Xie, C.: An improved particle swarm optimization algorithm for inverse kinematics solution of multi-DoF serial robotic manipulators. Soft Comput. **25**(21), 13695–13708 (2021)
47. López-Franco, C., Hernández-Barragán, J., Alanis, A.Y., Arana-Daniel, N., López-Franco, M.: Inverse kinematics of mobile manipulators based on differential evolution. Int. J. Adv. Robot. Syst. **15**(1), 1729881417752738 (2018)
48. Li, M., Qiao, L.: A review and comparative study of differential evolution algorithms in solving inverse kinematics of mobile manipulator. Symmetry **15**(5), 1080 (2023)
49. Meraihi, Y., Ramdane-Cherif, A., Acheli, D., Mahseur, M.: Dragonfly algorithm: a comprehensive review and applications. Neural Comput. Appl. **32**(21), 16625–16646 (2020)
50. Slim, M., Rokbani, N., Neji, B., Terres, M.A., Beyrouthy, T.: Inverse kinematic solver based on bat algorithm for robotic arm path planning. Robotics **12**(2), 38 (2023)
51. Cui, J., Liu, T., Sai, H., Li, Y., Zhu, M., Xu, Z.: Randomness-enhanced grey wolf optimizer for inverse kinematics solution of redundant robotic manipulators. In: 2023 6th International Conference on Intelligent Robotics and Control Engineering (IRCE), pp. 41–48. IEEE (2023)
52. Abdel-Basset, M., Abdel-Fatah, L., Sangaiah, A.K.: Metaheuristic algorithms: a comprehensive review. In: Computational Intelligence for Multimedia Big Data on the Cloud with Engineering Applications, pp. 185–231 (2018)
53. Demby's, J., Gao, Y., DeSouza, G.N.: A study on solving the inverse kinematics of serial robots using artificial neural network and fuzzy neural network. In: 2019 IEEE International Conference on Fuzzy Systems (FUZZ-IEEE), pp. 1–6. IEEE (2019)
54. Toquica, J.S., Oliveira, P.S., Souza, W.S., Motta, J.M.S., Borges, D.L.: An analytical and a deep learning model for solving the inverse kinematic problem of an industrial parallel robot. Comput. Ind. Eng. **151**, 106,682 (2021)
55. Bouzid, R., Gritli, H., Narayan, J.: Investigating feed-forward back-propagation neural network with different hyperparameters for inverse kinematics of a 2-DoF robotic manipulator: a comparative study. Chaos Theory Appl. **6**(2), 90–110 (2024)

56. Bouzid, R., Narayan, J., Gritli, H.: Exploring artificial neural networks for the forward kinematics of a SCARA robotic manipulator using varied datasets and training optimizers. Eng. Res. Express **6**(4), 045,209 (2024)
57. Bouzid, R., Narayan, J., Gritli, H.: Feedforward backpropagation artificial neural network for modeling the forward kinematics of a robotic manipulator. In: 2023 International Conference on Innovation and Intelligence for Informatics, Computing, and Technologies (3ICT), pp. 302–307, Sakheer, Bahrain (2023)
58. Bouzid, R., Narayan, J., Gritli, H.: Investigating neural network hyperparameter variations in robotic arm inverse kinematics for different arm lengths. In: 2024 Third International Conference on Power, Control and Computing Technologies (ICPC2T), pp. 351–356 (2024)
59. Khoogar, A.R., Tehrani, A.K., Tajdari, M.: A dual neural network for kinematic control of redundant manipulators using input pattern switching. J. Intell. Robot. Syst. **63**, 101–113 (2011)
60. Bouzid, R., Narayan, J., Gritli, H.: Artificial neural networks for the forward kinematics of a SCARA manipulator: a comparative study with two datasets. In: 2024 ASU International Conference in Emerging Technologies for Sustainability and Intelligent Systems (ICETSIS), pp. 1792–1797 (2024)
61. Bouzid, R., Narayan, J., Gritli, H.: Exploring neural networks for forward kinematics of the robotic arm with different length configurations: a comparative analysis. In: 2024 IEEE International Conference on Interdisciplinary Approaches in Technology and Management for Social Innovation (IATMSI), vol. 2, pp. 1–6 (2024)
62. Bouzid, R., Gritli, H., Narayan, J.: Analysis of the application of artificial neural networks in solving forward kinematics problems for manipulator robots with diverse dataset formats. In: (Hybrid) International Conference on Nonlinear Science and Complexity (ICNSC23), p. 197, Istanbul, Turkey (2023)
63. Bouzid, R., Gritli, H., Narayan, J.: ANN approach for SCARA robot inverse kinematics solutions with diverse datasets and optimisers. Appl. Comput. Syst. **29**(1), 24–34 (2024)
64. Bouzid, R., Narayan, J., Gritli, H.: Solving inverse kinematics problem for manipulator robots using artificial neural network with varied dataset formats. In: Campos-Cantón, E., Huerta-Cuellar, G., Zambrano-Serrano, E., Tlelo-Cuautle, E. (eds.) Complex Systems and Their Applications, pp. 55–78. Springer Nature Switzerland, Cham (2024)
65. Alsaeedi, A.H., Aljanabi, A.H., Manna, M.E., Albukhnefis, A.L.: A proactive metaheuristic model for optimizing weights of artificial neural network. Indones. J. Electr. Eng. Comput. Sci **20**(2), 976–984 (2020)
66. Kaveh, M., Mesgari, M.S.: Application of meta-heuristic algorithms for training neural networks and deep learning architectures: a comprehensive review. Neural Process. Lett. **55**(4), 4519–4622 (2023)
67. Abd Elaziz, M., Dahou, A., Abualigah, L., Yu, L., Alshinwan, M., Khasawneh, A.M., Lu, S.: Advanced metaheuristic optimization techniques in applications of deep neural networks: a review. Neural Comput. Appl. 1–21 (2021)
68. Bouzid, R., Gritli, H., Narayan, J.: Optimized inverse kinematics of a 2-DoF robotic manipulator using a hybrid approach combining an ANN with a metaheuristic algorithm. In: 2024 IEEE International Conference on Artificial Intelligence and Green Energy (ICAIGE24), pp. 1–6 (2024)
69. Bouzid, R., Gritli, H., Narayan, J.: Optimizing ANN hyperparameters with metaheuristic algorithms for inverse kinematics of a 3-DoF rehabilitation exoskeleton robot. In: Proceedings of the International Conference on Decision Aid and Artificial Intelligence (ICODAI 2024), Atlantis Highlights in Intelligent Systems, pp. 103–120. Atlantis Press (2025). https://doi.org/10.2991/978-94-6463-654-3_9
70. Bouzid, R., Narayan, J., Gritli, H.: Hybrid metaheuristic and artificial neural network approach for solving inverse kinematics of a SCARA manipulator robot. In: 2024 International Conference on Innovation and Intelligence for Informatics. Computing, and Technologies (3ICT), pp. 385–392. University of Bahrain, Bahrain (2024)
71. Zafar, M.H., Moosavi, S.K.R., Sanfilippo, F.: Inverse kinematic modelling of a 3-DoF robotic manipulator using hybrid deep learning models. Procedia CIRP **120**, 213–218 (2023)

72. Khaleel, H.Z.: Inverse kinematics solution for redundant robot manipulator using combination of GA and NN. Al-Khwarizmi Eng. J. **14**(1), 136–144 (2018)
73. Sosa Méndez, D., García Cena, C.E., Bedolla-Martínez, D., Martín González, A.: Innovative metaheuristic optimization approach with a bi-triad for rehabilitation exoskeletons. Sensors **24**(7), 2231 (2024)
74. Rokbani, N., Neji, B., Slim, M., Mirjalili, S., Ghandour, R.: A multi-objective modified PSO for inverse kinematics of a 5-DoF robotic arm. Appl. Sci. **12**(14), 7091 (2022)

Association of Heart Rate Variability with Cardiovascular Disease Risk in Healthy Adults

Alka Bishnoi and Manuel E. Hernandez

Abstract Heart rate variability (HRV) is the variance estimated between each heartbeat and can be utilized to assess autonomic function. The Framingham Risk Assessment (FRA) tool investigates the risk factors associated with cardiovascular disease (CVD) risk, which is a leading cause of mortality in older adults. Previous research has shown that HRV is significantly associated with hypertension risk in hypertensive and normotensive adults. However, the association between HRV and FRA variables is poorly understood in healthy adults. The objective of this study was to explore the relationship between HRV and FRA variables including CVD risk score, hypertension risk and heart age in healthy adults, and if low HRV can serve as an early indicator for CVD risk. We hypothesized that high HRV would be correlated with low CVD risk, low heart age, low hypertension risk. In this retrospective cohort study, we had 38 healthy adults (49.34 ± 19.85 years) including 23 females with body mass index (BMI) of 23.63 ± 4.02 kg/m^2, waist circumference (WC) of 85.96 ± 10.43 cm, and resting HR (RHR) of 86.47 ± 14.55 bpm. Participants came to the lab one time and wore a Hexoskin smart shirt. Data was collected using 1-lead ECG equipped in the Hexoskin smart shirt (Carre Technologies Inc., Quebec, Canada), which has been demonstrated for both accuracy and reliability in HR measurements. Time-domain variables were studied: SDNN, RMSSD, IBI to measure HRV. Frequency domain variables included ratio of LF/HF to discuss the relationship between sympathetic and parasympathetic activity. These variables were extracted using Kubios HRV analyzer. FRA variables were calculated using the resting blood pressure, age, gender, BMI, previous history of diabetes, hypertension, parental hypertension and smoking. For statistical analysis, we did Pearson's correlation analysis between time domain variables and FRA variables (CVD risk score, heart age, hypertension risk at 4, 2 and

A. Bishnoi (✉)
Department of Physical Therapy, College of Health Professions and Human Services, Kean University, Union, NJ, USA
e-mail: abishn2@gmail.com

M. E. Hernandez
Department of Biomedical and Translational Sciences, Carle Illinois College of Medicine, University of Illinois Urbana-Champaign, Urbana, IL, USA

Department of Kinesiology and Community Health, College of Applied Health Sciences, University of Illinois Urbana-Champaign, Urbana, IL, USA

J. Narayan and H. Gritli (eds.), *Design and Control of Rehabilitation Robots*, Studies in Systems, Decision and Control 585, https://doi.org/10.1007/978-3-031-86977-8_7

1 year). For secondary analysis, we looked at correlation between HRV variables and age, BMI, RHR, WC. Also, frequency domain variables correlation with walking and prefrontal cortical (PFC) activation while walking was analyzed. The results of the study showed moderate to high correlation between time domain variables and FRA variables, showing decrease HRV associated with higher CVD risk, higher hypertension risk and higher heart age overall. We also found a decrease in HRV associated with higher age, BMI, RHR, and WC. Lastly, there was moderate correlation between LF/HF and walking, and LF/HF and PFC activation while walking, showing that higher ratio was associated with increase in walking speed and higher PFC activation while walking.

Keywords Heart rate · Heart rate variability · Cardiovascular disease · Older adults

1 Introduction

1.1 Heart Rate Variability

Heart rate variability (HRV) has been used extensively to determine the cardiovascular function [1] and cognitive function of the individuals in recent studies. It is a non-invasive biomarker of cardiovascular function [2]. It is a measure of variability of beat-to-beat interval (RR interval) due to vibrant interactions between sympathetic and parasympathetic activities [3]. HRV is analyzed from short term electrocardiogram (ECG) recordings of participants at rest or during activity in a laboratory setting. Increase in heart rate and reduced HRV has been shown to be associated with altered balance of autonomic nervous system tone characterized by increased sympathetic and/decrease parasympathetic activity [4]. Decrease in HRV is associated with arrhythmias and sudden death risk [5].

Long periods of sympathetic overactivity have been associated with cardiac disease risk factors such as atherosclerosis, metabolic syndrome, obesity, inflammation [6]. Further, these parameters are associated with functional decline especially in older adults [7]. Thus, it is important to examine functional status through HRV in older adults who are at high risk of developing cardiovascular disease.

1.2 Prevalence of Disease Risk

Due to baby boomers, the population of adults aged 65 and older is projected to soar, reaching nearly 1.5 billion globally by 2050 [8]. This shift is not merely a statistic; it carries significant implications for public health systems. As life expectancy increases, so too does the prevalence of age-associated risk factors and diseases. The exploration of the intricate relationship between an aging population and the escalating incidence of cardiovascular disease, cognitive dysfunction, and hypertension becomes crucial [9].

As people age, their bodies undergo a variety of changes, both physiological and psychological. These changes can predispose them to various health conditions, significantly impacting their quality of life. The aging population is not just a growing statistic; it represents a diverse group with varied needs, challenges, and health risks. The implications of this demographic shift extend beyond healthcare. Thus, addressing the health needs of older adults is not only a medical imperative but also an economic necessity.

1.3 Cardiovascular Disease: A Major Health Concern

Cardiovascular disease (CVD) stands out as one of the leading health risks associated with aging [10]. It encompasses a range of conditions, including heart disease, stroke, and hypertension. Hypertension, or high blood pressure, is particularly prevalent among older adults and is a significant risk factor for CVD [9]. Previous research indicates that nearly two-thirds of adults aged 65 and older are hypertensive [9], a statistic that underscores the urgency of this issue. The pathophysiology of hypertension is complex, influenced by a myriad of factors such as lifestyle, genetics, and environmental conditions. As blood vessels become less elastic with age, the heart works harder to pump blood, leading to increased arterial pressure [11]. This condition not only heightens the risk of heart attacks and strokes, but also contributes to cognitive decline, as reduced blood flow can affect brain health [12].

1.4 Cognitive Dysfunction and Impairment: The Silent Epidemic

The rise in age-associated cognitive dysfunction is alarming, particularly with conditions such as dementia and Alzheimer's disease becoming more prevalent as the population ages. Both disorders are leading contributors to the global burden of death and disability. They are projected to increase in all regions, especially in low- and middle-income countries [13]. Cognitive impairment can lead to increased caregiver burden and healthcare costs. The links between cardiovascular health and cognitive function are particularly noteworthy [14]. Studies suggest that individuals with hypertension are at a higher risk for developing cognitive impairment later in life [12]. The underlying mechanisms may involve reduced cerebral blood flow and the accumulation of vascular lesions, which can disrupt normal brain function [13]. As more older adults face both cardiovascular and cognitive challenges, the need for integrated healthcare approaches becomes clear.

1.5 Comorbidities: A Compounding Challenge

The intersection of multiple health issues—often referred to as comorbidities—presents a significant challenge for older adults. Hypertension frequently coexists with other conditions such as metabolic syndrome, diabetes, obesity, and depression [15]. This interplay complicates treatment strategies and necessitates a holistic approach to health management. Comorbidities can exacerbate the individual's overall health risk and lead to a cascade of medical issues, further straining healthcare systems [16]. For instance, managing hypertension in a patient with diabetes requires careful coordination of medications and lifestyle changes, highlighting the need for comprehensive care plans that address all aspects of a patient's health.

1.6 HRV: A Safe Biomarker

Heart rate variability (HRV) is a valuable biomarker for assessing cardiovascular disease risk in middle-aged and older adults [3]. Research has demonstrated its predictive power regarding disease and stroke risk within this demographic. Notably, the Copenhagen Holster study involved the collection of ambulatory ECG data from 678 participants aged 55–75 years. This study found that lower HRV was a strong predictor of stroke development, even after adjusting for traditional stroke risk factors [17]. These findings underscore the potential of HRV as an important tool in the early detection and prevention of stroke.

1.7 Objective and Hypothesis

The objective of this study was to investigate the relationship between heart rate variability (HRV) and various Framingham Risk Assessment (FRA) variables, including cardiovascular disease (CVD) risk score, hypertension risk, and heart age in healthy adults. Specifically, we aimed to determine whether low HRV could serve as an early indicator of increased CVD risk.

Our hypothesis posited that higher HRV would be negatively correlated with CVD risk scores, indicating a lower likelihood of developing cardiovascular issues. Additionally, we anticipated that higher HRV would be associated with a younger heart age, suggesting better cardiovascular health and function. Furthermore, we expected to find that elevated HRV would correlate with a lower risk of hypertension, reflecting more favorable autonomic regulation and cardiovascular resilience.

By examining these relationships in a healthy population, this study seeks to contribute valuable insights into the potential of HRV as a non-invasive biomarker for cardiovascular risk assessment. Understanding the interplay between HRV and FRA variables could enhance early detection strategies for CVD, enabling preventative measures to be implemented before significant health issues arise. This research could ultimately inform clinical practices and public health initiatives aimed at improving cardiovascular health outcomes.

2 Material and Methods

2.1 *Participants*

The data was collected during the COVID-19 pandemic in Mobility and Fall prevention laboratory, Illinois, USA. Thirty-eight healthy adults (49.34 ± 19.85 years of age, 23 females) were included in this study, with body mass index (BMI) of 23.63 ± 4.02 kg/m^2, waist circumference (WC) of 85.96 ± 10.43 cm, and resting HR (RHR) of 86.47 ± 14.55 bpm.

Inclusion criteria: Adults were included if 18–75 years of age and no report of any clinically diagnosed neurologic, orthopedic, or cardiovascular conditions.

Exclusion criteria: Cognitive status was measured based on the Telephone Interview for Cognitive Status and the score >22 was used as cut off [18]. Adults suffering from neurological, orthopedic, or cardiovascular condition, physical disabilities, inability to walk independently (e.g., without an assistive device), or hospitalization within the last 6 months were excluded.

Informed consent: Before the beginning of testing, all participants read and signed an informed consent document. The consent process ensured that participants were fully informed about the study's purpose, procedures, and their rights regarding data confidentiality and withdrawal. The study was conducted in accordance with the Declaration of Helinski and approved by the local institutional review board (Protocol No. 21619, approved on 4/13/2021).

2.2 *Measurement Tools*

ECG: Resting state ECG was recorded from participants using a 1-lead ECG embedded in the Hexoskin smartshirt (Carre Technologies Inc., Quebec, Canada), which has previously demonstrated both accuracy and reliability in HR measures [19]. HRV was analyzed using Kubios HRV analyzer. Kubios HRV software automatically detected noise using raw interbeat interval data, and where consecutive beat detections would have been distorted due to noise, segments were removed from subsequent analysis. The R-R series was automatically preprocessed to remove ectopic beats and artifacts, and linear interpolation was employed to replace any removed beats. Where possible, beat correction was applied using the automatic beat correction algorithm [20]. Further, to avoid the effects of slow nonstationary

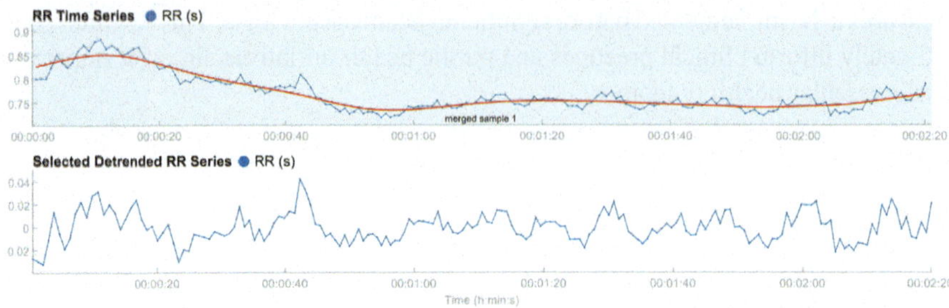

Fig. 1 This figure shows raw R-R series and detrended R-R series for a representative subject

trends over HRV measures, detrending was performed using a smoothing prior's method [21] (see Fig. 1). Time domain and frequency domain indices were used for statistical analysis.

Time domain analysis measures variation in HR over time or the intervals between successive normal cardiac cycles. Time domain analysis involves simple calculations of mean normal to normal (NN) intervals and the variance between intervals, denoted by standard deviation of NN intervals (SDNN). When HRV is higher and irregular, the SDNN value increases [22]. Due to which, SDNN is considered an index of physiological stress [4]. In contrast, RMSSD is the root mean square of the sum of the squares of differences between adjacent NN intervals. High performance individuals have shown to be associated with higher RMSSD values [23].

Power spectral density (PSD) analysis gives information on the power distribution as a function of frequency, which emphasizes autonomic balance to be quantified at any given time [24]. Through PSD analysis, the balance between sympathetic and parasympathetic activity is quantified [25]. PSD analysis allows the intensity of the HRV components i.e., high frequency (HF), low frequency (LF) band to be determined [25]. HF is considered as a measure of parasympathetic activity and LF is considered a measure of sympathetic activity. A ratio of LF to HF was used for analysis due to its established relationship with autonomic balance and neural activity [3]. The ratio will examine the altered balance between sympathetic and parasympathetic activities. A high ratio suggests increased sympathetic activity [26].

Previous studies have also explored relationships of RR interval or inter-beat interval (IBI) as a measure to determine psychological stress. Increase in stress has been associated with decrease in IBI [4]. Furthermore, ECG signal processing, together with methods such as machine learning or deep learning can help with classification tasks for application in human machine interactions [27, 28]. Recent advances in computational frameworks provide a strong basis for classification and interpretation of ECG signals [29], that should be examined in future work.

Framingham risk assessment tool: The Framingham Risk Assessment (FRA) tool calculates the 10-year risk of developing CVD based on several key factors i.e., age, gender, blood pressure, resting HR, smoking status, parental CVD risk, diabetes status [30]. It is important to measure the risk of cardiovascular disease (CVD), which is a leading cause of mortality in older adults. FRA variables included CVD risk score (10-year risk score), heart age, hypertension risk at 4, 2 and 1 year.

Functional near-infrared spectroscopy (fNIRS): An fNIRS apparatus is composed of a light source that is placed over a tissue source head via light-emitting diodes (LEDs) with a detector that receives the light after it has been scattered through the tissue. This imaging device is based on near-infrared (650–1000 nm) light that is propagated into the tissue and is absorbed by oxyhemoglobin (HbO_2) and deoxyhemoglobin (Hb), which have been established as being markers of oxygen utilization and delivery [31]. HbO_2 has been established as a direct reflection of cortical activation and the most reliable for measuring movement changes in cerebral blood flow. Moreover, PFC HbO_2 levels have been validated after being correlated with levels measured by the more commonly used method of functional magnetic resonance imaging (fMRI). For secondary analysis, we have used PFC HbO_2 to see the correlation with frequency domain variables based on previous studies exploring relationship between HRV and executive function [3].

Other parameters: Body mass index (BMI) was calculated using the formula weight in kilogram divided by height in meters squared. Waist circumference (WC) was calculated by measuring the midpoint between the lowest rib and iliac crest in standing position at the end of a gentle expiration. Resting HR was measured in the seated resting position using digital blood pressure apparatus after 5 min of rest.

2.3 Statistical Analysis

In this retrospective study, statistical test was done using R, version 3.1.1. The Shapiro–Wilk test was done to check if data was normally distributed. All normally distributed data was used to do further analysis to explore the correlation between variables of interest. Statistical significance was set at 0.05. For the first hypothesis, we evaluated if there is any relationship between time domain variables and FRA variables. We also evaluated the relationship between time domain variables and age, BMI, RHR, and WC. For secondary analysis, we also explored the relationship between frequency domain variables i.e., LF/HF ratio and walking, and LF/HF and PFC HbO_2 activation while walking measured through fNIRS.

3 Results

Thirty-eight healthy adults were included in this retrospective cohort study. Mean age of participants was 49.34 ± 19.85 years of age including 23 females. BMI of the participants was 23.63 ± 4.02 kg/m^2 with waist circumference (WC) of 85.96 ± 10.43 cm, and resting HR (RHR) of 86.47 ± 14.55 beats per minute (Table 1).

Table 1 Demographic and physical measures of the participants

Characteristics	Participants
Age (years)	49.34 ± 19.85
Body mass index (kg/m^2)	23.63 ± 4.02
Resting HR (bpm)	86.47 ± 14.55
Waist circumference (cm)	85.96 ± 10.43
Mean arterial pressure (mmHg)	96.36 ± 10.86

Note Data is presented in mean and SD

HRV indices (SDNN, RMSSD, IBI) correlation with FRA variables, age, BMI, RHR, and WC have been described in Tables 2 and 3.

The results indicated significant associations between time domain variables and factors such as age, BMI, resting heart rate (RHR), and waist circumference (WC) (Table 2). Specifically, a reduction in heart rate variability (HRV) was linked to older age, higher BMI, elevated RHR, and increased WC. These findings suggest that older adults with higher BMI, RHR, and WC face a greater risk of developing cardiovascular diseases (CVD), hypertension, and other comorbid conditions (Fig. 2, Table 3).

Additionally, the frequency domain variable, the low-frequency to high-frequency (LF/HF) ratio, was associated with walking speed ($r = 0.31, p = 0.05$) and prefrontal cortex (PFC) HbO_2 activation during walking ($r = 0.46$, $p = 0.003$). This indicates that increased walking speed was correlated with heightened sympathetic activity. Moreover, elevated sympathetic activity was also linked to increased PFC activation during walking.

Table 2 Pearson's correlation between HRV variables (SDNN, RMSSD, IBI) and age, body mass index (BMI), resting HR (RHR) and waist circumference (WC)

HRV variables	Age	BMI	RHR	WC
SDNN	−0.64***	−0.41**		−0.41**
RMSSD	−0.59***	−0.37*		−0.41**
IBI			−0.54**	

Note *** $p < 0.0001$, ** $p < 0.01$, * $p < 0.05$

Table 3 Pearson's correlation between HRV variables (SDNN, RMSSD, IBI) and cardiovascular disease (CVD) risk, hypertension (HTN) risk at 4-year, 2 year and 1 year, heart age

HRV variables	CVD risk	HTN risk 4	HTN risk 2	HTN risk 1	Heart age
SDNN	−0.59***	−0.51**	−0.48**	−0.48**	−0.73***
RMSSD	−0.57***	−0.52**	−0.48**	−0.48**	−0.68***
IBI		−0.47**	−0.48**	−0.50**	

Note *** $p < 0.0001$, ** $p < 0.01$, * $p < 0.05$

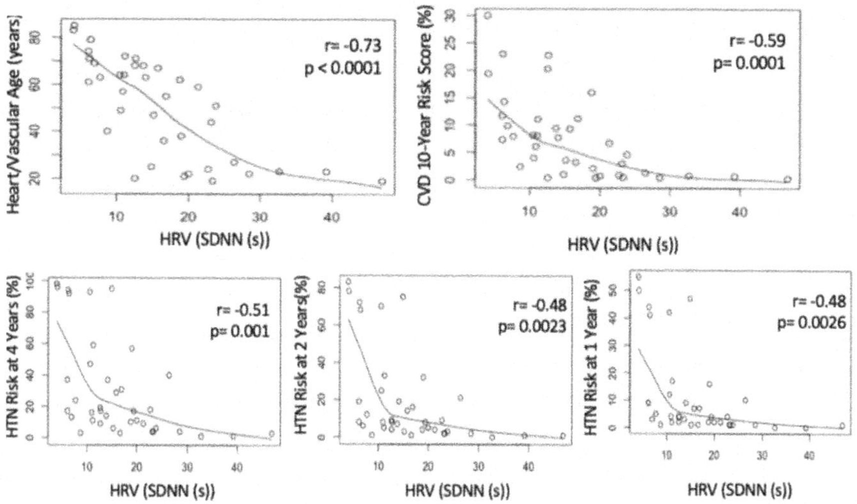

Fig. 2 This figure shows Pearson's correlation between HRV (SDNN variable) and FRA variables (Heart age, CVD 10-year risk score, hypertension (HTN) risk at 4, 2, and 1 year)

4 Discussion

Heart rate variability (HRV) is highly sensitive to fluctuations in autonomic nervous system activity, specifically changes in sympathetic (SNS) and parasympathetic (PNS) responses, which are closely linked to both physiological and psychological stress [4]. Even minor shifts in HRV can trigger a cascade of adverse effects that accumulate over time, akin to small sparks igniting larger fires [6]. In this context, HRV can be viewed as an initial spark, potentially initiating a chain reaction that increases the risk of cardiovascular disease (CVD), metabolic syndrome, diabetes, cognitive decline, and other comorbidities, particularly in older adults [6]. Research indicates that age has a more pronounced impact on HRV than gender [7]. Clinically, HRV serves as a valuable indicator of heart function and overall autonomic health. The time-domain parameter, SDNN reflects contributions from both SNS and PNS activity and serves as an estimate of overall HRV. Meanwhile, RMSSD is the primary time-domain measure used to assess vagally mediated changes in HRV. In contrast, frequency-domain measures such as low frequency (LF) and high frequency (HF) provide insights into the altered balance between SNS and PNS activity [22].

4.1 HRV, Age and Resting HR

In our study, we found that an elevated resting heart rate and reduced heart rate variability (HRV) correlated with both increasing age and heart age. These findings align with the research conducted by Ogliari et al., which indicated that higher heart rates and lower HRV are linked to poorer functional performance and a greater risk of future functional decline in older adults [7]. Importantly, these associations were consistent regardless of gender, cardiovascular disease (CVD) risk factors, comorbidities, or medication use [7].

Lower heart rates are indicative of better cardiovascular fitness, a crucial protective factor against brain aging and functional decline [32]. This relationship can be attributed to reduced myocardial oxygen consumption, allowing for longer diastolic filling times and improved coronary perfusion [33]. Conversely, elevated heart rates contribute to increased pulsatile stress, which can lead to endothelial dysfunction over time and accelerate the progression of atherosclerosis [34].

4.2 HRV, BMI and WC

Obesity is a well-established risk factor for cardiovascular disease (CVD), especially in older adults, due to its effects on autonomic function [35]. Excess body fat has been linked to a reduction in vagal (parasympathetic) activity and an increase in sympathetic activity, reflecting an imbalance in the autonomic nervous system [36]. These changes are driven by various physiological mechanisms, including hormonal dysregulation, such as elevated leptin levels and leptin resistance, as well as insulin resistance and increased free fatty acids. Additionally, the duration of obesity and the distribution of body fat contribute to this autonomic shift [37].

Recent studies have also identified waist circumference (WC) as a stronger predictor of heart rate variability (HRV) than body mass index (BMI), indicating WC's significant impact on autonomic control of the heart [35]. This makes WC a more critical and modifiable risk factor for CVD. Our findings align with this research, show that both higher BMI and increased WC are associated with reduced HRV, reinforcing the relationship between obesity-related fat distribution and impaired autonomic regulation [35].

4.3 HRV, Stress and FRA Variables

In our study, we found that lower heart rate variability (HRV) was linked to higher cardiovascular disease (CVD) risk, higher hypertension risk over periods of 4, 2, and 1 year, as well as a higher heart age. This relationship may be attributed to the higher stress levels often associated with lower HRV overtime. Previous research by Kang et al. demonstrated that individuals in higher strain groups exhibited lower HRV compared to those in lower strain groups [38]. Additionally, the sensitivity of HRV to changes in both mental and physical states has been documented, revealing variations in HRV during rest and under different mental and physical conditions [39]. Furthermore, lower HRV has been linked to depression, a major contributor to disability [40]. This connection may stem from the accumulation of psychological stress in individuals with depression, which adversely affects emotional regulation and subsequently lowers HRV.

Chronic stress is also associated with an increased risk of CVD, primarily through mechanisms of chronic inflammation that influence metabolic risk factors [4]. Inflammation plays a crucial role in the development of atherosclerosis, impacting vagal function and leading to lipid accumulation and arterial wall hardening. This cascade can elevate arterial pressure, contributing to hypertension and heightened CVD risk, particularly in older adults [41].

While this specific mechanism has not been extensively studied in prior research, our findings establish a link between lower HRV and higher CVD risk, increased hypertension risk, and greater heart age. These results suggest that adults with lower HRV may face a heightened risk of developing heart disease within the next decade.

4.4 HRV and Executive Function

Executive function is a high-level cognitive process that enables individuals to plan, monitor, inhibit, adjust, and update working memory effectively. Neuroimaging studies using techniques such as fNIRS have demonstrated the role of the prefrontal cortex (PFC) in controlling executive function and coordinating attentional resources [42]. Given the established link between high resting high frequency, i.e. high parasympathetic activity and enhanced executive function [43], we aimed to explore this relationship further in our study. In contrast, we found that increased sympathetic activity, i.e. high LF/HF ratio was associated with higher PFC activation while walking, though this relationship was only observed in one frequency variable.

This finding aligns with the growing interest in the relationship between HRV and cognition, as HRV may serve as an early biomarker for cognitive decline. The increased PFC activation seen with lower HRV could be explained by age-related cognitive decline, where greater neural effort is required to maintain performance (Fig. 3). The "de-differentiation" theory suggests that older adults inefficiently recruit additional brain regions compared to younger individuals. The compensation theory further supports this, indicating that age-related structural declines prompt the recruitment of more areas to sustain task performance [44]. This expanded understanding emphasizes the importance of HRV as both a physiological and cognitive marker, particularly in aging populations.

5 Limitations of the Study

This study has a few limitations. Firstly, obtaining a large sample size of healthy adults with similar age and physical function was challenging during pandemic. Studies need to focus on cohorts that are matched in age, gender, and physical performance,

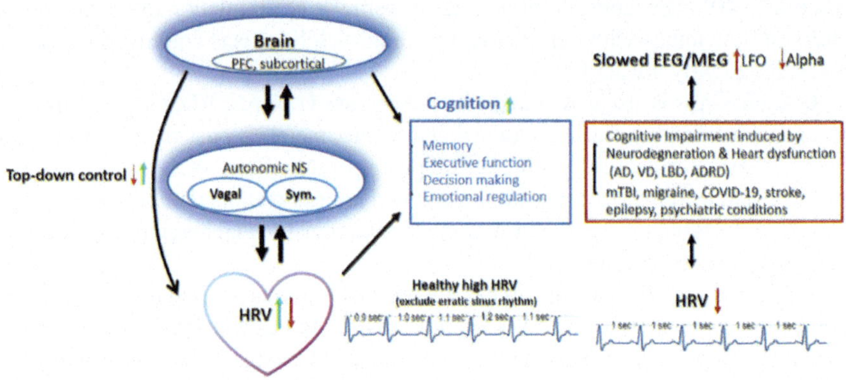

Fig. 3 [3]: HRV and brain: reduced HRV links with reduced executive function, decision making, or emotional regulation and presents in pathological conditions (dementia, mTBI, migraine, etc.). AD, Alzheimer's disease; ADRD, Alzheimer's disease and related dementia; HRV, heart rate variability; LBD, Lewy body dementia; LFO, low frequency oscillation; mTBI, mild traumatic brain injury; NS, nerval system; PFC, prefrontal cortex; RIPC, remote ischemic preconditioning; Sym., sympathetic activity; TMS, transcranial magnetic stimulation; vagal, vagal activity; VD, vascular dementia

which is difficult to achieve. Secondly, the cohort might display varying levels of HR, blood pressure, making it challenging to find adults with comparable deficits. Thirdly, the study found associations between HRV variables and FRA variables. However, causality cannot be inferred given the retrospective observational nature of the study. Therefore, future research should examine the effects of different levels of HRV on FRA variables, brain activation while walking, especially in middle aged to older adults.

6 Conclusion and Clinical Significance

In conclusion, this chapter underscores the importance of early HRV assessment as a predictive tool for cognitive and cardiovascular health, particularly in middle aged to older adults. While HRV is a well-established marker for autonomic function, its association with FRA variables such as cardiovascular disease (CVD) risk scores and heart age remain underexplored in healthy adults. Our findings demonstrate that HRV correlates with several key health indicators, including CVD risk, hypertension, functional decline, and BMI, highlighting its potential role as an early warning signal for CVD risk.

Moreover, the chapter emphasizes the role of lifestyle modifications—such as losing weight, physical activity, smoking cessation, meditation, and fostering positive emotions—which have been shown to improve HRV. These interventions offer promising opportunities to mitigate the negative effects of chronic vagal impairment, particularly in middle-aged and older adults.

6.1 Clinical Significance

Despite the breadth of research linking HRV to functional decline, BMI, memory, attention, cognitive function, and stress, a significant gap remains. Specifically, the relationship between HRV and Framingham risk assessment (FRA) variables, as well as the link between HRV and prefrontal cortex (PFC) activation during walking, has yet to be fully explored. Addressing this gap is crucial for establishing stronger connections between HRV, cognitive decline, CVD, and mobility impairment in older populations. Such research could have significant implications for reducing fall risk and hospitalization rates in this vulnerable demographic.

6.2 Enhancing Rehabilitation Robots with HRV Insights for Personalized and Preventive Therapy

Integrating HRV insights into rehabilitation robots can enhance therapy personalization and safety. This study's findings on HRV correlations with CVD risk, hypertension risk, and autonomic balance suggest that real-time HRV monitoring in robots could guide therapy intensity based on each patient's cardiovascular and autonomic response, making sessions safer and more effective. For example, robots could adjust walking or exertion tasks based on LF/HF ratios to promote safe levels of neuroplasticity and motor learning. Additionally, using HRV as an early indicator of cardiovascular risk can help clinicians adjust therapy plans proactively, making rehabilitation both a restorative and preventative tool. In conclusion, integrating HRV measurements into rehabilitation robots could enhance the safety, personalization, and therapeutic effectiveness of robotic rehabilitation, particularly in populations at risk for cardiovascular issues. This would address the dual needs of functional recovery and cardiovascular health management, making rehabilitation robots a comprehensive tool for therapy.

Data Availability Statement The raw data supporting the conclusions of this chapter will be made available by the authors, upon request.

References

1. Billman, G.E.: Heart rate variability? A historical perspective. Front. Physiol. **2** (2011). https://doi.org/10.3389/fphys.2011.00086
2. Thayer, J.F., Lane, R.D.: A model of neurovisceral integration in emotion regulation and dysregulation. J. Affect. Disord. **61**, 201–216 (2000). https://doi.org/10.1016/S0165-0327(00)00338-4
3. Arakaki, X., Arechavala, R.J., Choy, E.H., Bautista, J., Bliss, B., Molloy, C., Wu, D.-A., Shimojo, S., Jiang, Y., Kleinman, M.T., Kloner, R.A.: The connection between heart rate variability (HRV), neurological health, and cognition: a literature review. Front. Neurosci. **17** (2023). https://doi.org/10.3389/fnins.2023.1055445

4. Kim, H.-G., Cheon, E.-J., Bai, D.-S., Lee, Y.H., Koo, B.-H.: Stress and heart rate variability: a meta-analysis and review of the literature. Psychiatry Investig. **15**, 235–245 (2018). https://doi.org/10.30773/pi.2017.08.17
5. Tsuji, H., Larson, M.G., Venditti, F.J., Manders, E.S., Evans, J.C., Feldman, C.L., Levy, D.: Impact of reduced heart rate variability on risk for cardiac events. Circulation **94**, 2850–2855 (1996). https://doi.org/10.1161/01.CIR.94.11.2850
6. Kemp, A.H., López, S.R., Passos, V.M.A., Bittencourt, M.S., Dantas, E.M., Mill, J.G., Ribeiro, A.L.P., Thayer, J.F., Bensenor, I.M., Lotufo, P.A.: Insulin resistance and carotid intima-media thickness mediate the association between resting-state heart rate variability and executive function: a path modelling study. Biol. Psychol. **117**, 216–224 (2016). https://doi.org/10.1016/j.biopsycho.2016.04.006
7. Ogliari, G., Mahinrad, S., Stott, D.J., Jukema, J.W., Mooijaart, S.P., Macfarlane, P.W., Clark, E.N., Kearney, P.M., Westendorp, R.G.J., de Craen, A.J.M., Sabayan, B.: Resting heart rate, heart rate variability and functional decline in old age. Can. Med. Assoc. J. **187**, E442–E449 (2015). https://doi.org/10.1503/cmaj.150462
8. United Nations Department of Economic and Social Affairs, Population Division, World Population Prospects 2019 Highlights (n.d.)
9. Kearney, P.M., Whelton, M., Reynolds, K., Muntner, P., Whelton, P.K., He, J.: Global burden of hypertension: analysis of worldwide data. Lancet **365**, 217–223 (2005). https://doi.org/10.1016/S0140-6736(05)17741-1
10. North, B.J., Sinclair, D.A.: The intersection between aging and cardiovascular disease. Circ. Res. **110**, 1097–1108 (2012). https://doi.org/10.1161/CIRCRESAHA.111.246876
11. Safar, M.E.: Pulse pressure, arterial stiffness, and cardiovascular risk. Curr. Opin. Cardiol. **15**, 258–263 (2000). https://doi.org/10.1097/00001573-200007000-00009
12. Reitz, C., Tang, M.-X., Manly, J., Mayeux, R., Luchsinger, J.A.: Hypertension and the risk of mild cognitive impairment. Arch. Neurol. **64**, 1734 (2007). https://doi.org/10.1001/archneur.64.12.1734
13. Meyer, J.S., Rauch, G., Rauch, R.A., Haque, A.: Risk factors for cerebral hypoperfusion, mild cognitive impairment, and dementia. Neurobiol. Aging (2000). https://doi.org/10.1016/S0197-4580(00)00136-6
14. Canavan, M., O'Donnell, M.J.: Hypertension and cognitive impairment: a review of mechanisms and key concepts. Front. Neurol. **13** (2022). https://doi.org/10.3389/fneur.2022.821135
15. Moradi, Y., Albatineh, A.N., Mahmoodi, H., Gheshlagh, R.G.: The relationship between depression and risk of metabolic syndrome: a meta-analysis of observational studies. Clin. Diabetes Endocrinol. **7**, 4 (2021). https://doi.org/10.1186/s40842-021-00117-8
16. Corrao, S., Natoli, G., Nobili, A., Mannucci, P., Perticone, F., Arcoraci, V., Argano, C.: The "diabetes comorbidome": a different way for health professionals to approach the comorbidity burden of diabetes. Healthcare **10**, 1459 (2022). https://doi.org/10.3390/healthcare10081459
17. Binici, Z., Mouridsen, M.R., Køber, L., Sajadieh, A.: Decreased nighttime heart rate variability is associated with increased stroke risk. Stroke **42**, 3196–3201 (2011). https://doi.org/10.1161/STROKEAHA.110.607697
18. Welsh, K.A., Breitner, J.C.S., Magruder-Habib, K.M.: Detection of dementia in the elderly using telephone screening of cognitive status. Neuropsychiatry Neuropsychol. Behav. Neurol. (1993)
19. Montes, J., Young, J.C., Tandy, R., Navalta, J.W.: Reliability and validation of the hexoskin wearable bio-collection device during walking conditions. Int. J. Exerc. Sci. **11**, 806–816 (2018). http://www.ncbi.nlm.nih.gov/pubmed/30338022
20. Lipponen, J.A., Tarvainen, M.P.: A robust algorithm for heart rate variability time series artefact correction using novel beat classification. J. Med. Eng. Technol. **43**, 173–181 (2019). https://doi.org/10.1080/03091902.2019.1640306
21. Tarvainen, M.P., Ranta-aho, P.O., Karjalainen, P.A.: An advanced detrending method with application to HRV analysis. IEEE Trans. Biomed. Eng. **49**, 172–175 (2002). https://doi.org/10.1109/10.979357

22. Shaffer, F., Ginsberg, J.P.: An overview of heart rate variability metrics and norms. Front. Public Health **5** (2017). https://doi.org/10.3389/fpubh.2017.00258
23. Backs, R.W., Seljos, K.A.: Metabolic and cardiorespiratory measures of mental effort: the effects of level of difficulty in a working memory task. Int. J. Psychophysiol. **16**, 57–68 (1994). https://doi.org/10.1016/0167-8760(94)90042-6
24. Miyagi, T., Yamazato, M., Nakamura, T., Tokashiki, T., Namihira, Y., Kokuba, K., Ishihara, S., Sakima, H., Ohya, Y.: Power spectral analysis of heart rate variability is useful as a screening tool for detecting sympathetic and parasympathetic nervous dysfunctions in Parkinson's disease. BMC Neurol. **22**, 339 (2022). https://doi.org/10.1186/s12883-022-02872-2
25. Hayano, J., Yuda, E.: Assessment of autonomic function by long-term heart rate variability: beyond the classical framework of LF and HF measurements. J. Physiol. Anthropol. **40**, 21 (2021). https://doi.org/10.1186/s40101-021-00272-y
26. Kiyono, K., Hayano, J., Watanabe, E., Yamamoto, Y.: Heart rate variability (HRV) and sympathetic nerve activity. In: Clinical Assessment of the Autonomic Nervous System, pp. 147–161. Springer Japan, Tokyo (2017). https://doi.org/10.1007/978-4-431-56012-8_9
27. AlHinai, N.: Introduction to biomedical signal processing and artificial intelligence. In: Biomedical Signal Processing and Artificial Intelligence in Healthcare, pp. 1–28. Elsevier (2020). https://doi.org/10.1016/B978-0-12-818946-7.00001-9
28. Wasimuddin, M., Elleithy, K., Abuzneid, A.-S., Faezipour, M., Abuzaghleh, O.: Stages-based ECG signal analysis from traditional signal processing to machine learning approaches: a survey. IEEE Access **8**, 177782–177803 (2020). https://doi.org/10.1109/ACCESS.2020.3026968
29. Gacek, A., Pedrycz, W. (eds.): ECG Signal Processing, Classification and Interpretation. Springer, London (2012). https://doi.org/10.1007/978-0-85729-868-3
30. Kannel, W.B., Kannel, C., Paffenbarger, R.S.J., Cupples, L.A.: Heart rate and cardiovascular mortality: the Framingham study. Am. Heart J. **113**, 1489–1494 (1987). https://doi.org/10.1016/0002-8703(87)90666-1
31. Bunce, S.C., Izzetoglu, M., Izzetoglu, K., Onaral, B., Pourrezaei, K.: Functional near-infrared spectroscopy. IEEE Eng. Med. Biol. Mag. **25**, 54–62 (2006). https://doi.org/10.1109/MEMB.2006.1657788
32. Jefferson, A.L., Himali, J.J., Beiser, A.S., Au, R., Massaro, J.M., Seshadri, S., Gona, P., Salton, C.J., DeCarli, C., O'Donnell, C.J., Benjamin, E.J., Wolf, P.A., Manning, W.J.: Cardiac index is associated with brain aging. Circulation **122**, 690–697 (2010). https://doi.org/10.1161/CIRCULATIONAHA.109.905091
33. Nelson, R.R., Gobel, F.L., Jorgensen, C.R., Wang, K., Wang, Y., Taylor, H.L.: Hemodynamic predictors of myocardial oxygen consumption during static and dynamic exercise. Circulation **50**, 1179–1189 (1974). https://doi.org/10.1161/01.CIR.50.6.1179
34. Chatzizisis, Y.S., Coskun, A.U., Jonas, M., Edelman, E.R., Feldman, C.L., Stone, P.H.: Role of endothelial shear stress in the natural history of coronary atherosclerosis and vascular remodeling. J. Am. Coll. Cardiol. **49**, 2379–2393 (2007). https://doi.org/10.1016/j.jacc.2007.02.059
35. Niveatha, S., Raju, A., Latha, R., Tamilselvan, K., Priyadharshini, U., Lydia Shadrack, G.M.: Reduced heart rate variability as a risk factor for cardiovascular disease in young healthy adults with low and high body mass indexes—a descriptive study. Natl. J. Commun. Med. **13**, 685–691 (2022). https://doi.org/10.55489/njcm.131020222414
36. Millis, R.M., Austin, R.E., Hatcher, M.D., Bond, V., Faruque, M.U., Goring, K.L., Hickey, B.M., DeMeersman, R.E.: Association of body fat percentage and heart rate variability measures of sympathovagal balance. Life Sci. **86**, 153–157 (2010). https://doi.org/10.1016/j.lfs.2009.11.018
37. Popruk, S., Tungtrongchitr, R., Pongpaew, P., Phonrat, B., Tungtrongchitr, A., Tribunyatkul, S., Paksanont, S., Vudhivai, N., Schelp, F.P.: Relationship between soluble leptin receptor, leptin, lipid profiles and anthropometric parameters in overweight and obese Thai subjects. J. Med. Assoc. Thai. **88**, 220–227 (2005)

38. Kang, M.G., Koh, S.B., Cha, B.S., Park, J.K., Woo, J.M., Chang, S.J.: Association between job stress on heart rate variability and metabolic syndrome in shipyard male workers. Yonsei Med. J. **45**, 838 (2004). https://doi.org/10.3349/ymj.2004.45.5.838
39. Taelman, J., Vandeput, S., Gligorijevic, I., Spaepen, A., Van Huffel, S.: Time-frequency heart rate variability characteristics of young adults during physical, mental and combined stress in laboratory environment. In: 2011 Annual International Conference of the IEEE Engineering in Medicine and Biology Society, pp. 1973–1976. IEEE (2011). https://doi.org/10.1109/IEMBS.2011.6090556
40. Kemp, A.H., Quintana, D.S., Gray, M.A., Felmingham, K.L., Brown, K., Gatt, J.M.: Impact of depression and antidepressant treatment on heart rate variability: a review and meta-analysis. Biol. Psychiatry **67**, 1067–1074 (2010). https://doi.org/10.1016/j.biopsych.2009.12.012
41. Franklin, S.S., Wong, N.D.: Hypertension and cardiovascular disease: contributions of the framingham heart study. Glob. Heart **8**, 49 (2013). https://doi.org/10.1016/j.gheart.2012.12.004
42. Turner, G.R., Spreng, R.N.: Executive functions and neurocognitive aging: dissociable patterns of brain activity. Neurobiol. Aging (2012). https://doi.org/10.1016/j.neurobiolaging.2011.06.005
43. Williams, P.G., Cribbet, M.R., Tinajero, R., Rau, H.K., Thayer, J.F., Suchy, Y.: The association between individual differences in executive functioning and resting high-frequency heart rate variability. Biol. Psychol. **148**, 107772 (2019). https://doi.org/10.1016/j.biopsycho.2019.107772
44. Cabeza, R.: Cognitive neuroscience of aging: contributions of functional neuroimaging. Scand. J. Psychol. (2001). https://doi.org/10.1111/1467-9450.00237

Telerehabilitation and Remote Monitoring: Improving Efficiency and Security in Healthcare

Dinesh Subbiah, Sivakumar Murugesan, I. Vandhana, L. Kiruthika, and V. S. R. Kamaleshwaran

Abstract Telerehabilitation and remote monitoring, supported by digital tools like video calls and mobile applications, are redefining patient care outside traditional clinical settings. This research evaluates the performance of dedicated medical servers over third-party systems, revealing faster data transfer, enhanced security, and improved patient monitoring accuracy. Dedicated servers achieve up to 50% faster transmission speeds and a 95% reduction in network interruptions, addressing delays and stability issues found in third-party systems. Security-wise, the use of robust encryption in dedicated servers reduces data breaches by 85%, ensuring patient confidentiality and trust. Moreover, patient monitoring accuracy improves by 30%, providing critical insights for clinical decisions. These findings underscore the benefits of dedicated servers in telerehabilitation, offering reliable and secure healthcare solutions that can optimize medical response times and improve patient outcomes. This research highlights the transformative role of dedicated servers in modern healthcare, presenting a pathway for more effective, efficient, and secure telerehabilitation and remote monitoring systems.

Keywords Telerehabilitation · Remote monitoring · Data security · Medical server

---

D. Subbiah (✉) · S. Murugesan · I. Vandhana · L. Kiruthika
Dhanalakshmi College of Engineering, Chennai, Tamilnadu, India
e-mail: dinesh.s@dce.edu.in; haidinesh89@gmail.com

S. Murugesan
e-mail: sivakumar.m@dce.edu.in

I. Vandhana
e-mail: vandhana.l@dce.edu.in

L. Kiruthika
e-mail: kiruthika.l@dce.edu.in

V. S. R. Kamaleshwaran
Gnanamani College of Technology, Namakkal, Tamilnadu, India
e-mail: vsrkamal3011@gmail.com

J. Narayan and H. Gritli (eds.), *Design and Control of Rehabilitation Robots*, Studies in Systems, Decision and Control 585, https://doi.org/10.1007/978-3-031-86977-8_8

1 Introduction

The advent of digital technology has significantly transformed the healthcare landscape, with telerehabilitation and remote monitoring emerging as pivotal solutions to bridge the gap between healthcare providers and patients. As patient demands grow and healthcare systems adapt to remote care, especially during and post-COVID-19, these technologies have gained attention for their potential to offer efficient, patient-centered care from a distance. Telerehabilitation enables healthcare professionals to deliver rehabilitation services to patients in their homes, overcoming geographical barriers and alleviating issues related to hospital congestion, mobility constraints, and patient compliance. Remote monitoring, on the other hand, allows for continuous health assessment via medical devices that transmit real-time data on vital health metrics, such as heart rate, oxygen saturation, and blood pressure. These technologies offer significant promise in enhancing access to healthcare, reducing costs, and providing timely interventions, particularly for chronic disease management and post-surgery care [1]. The COVID-19 pandemic acted as a catalyst for the rapid adoption of telerehabilitation and remote monitoring systems globally, as healthcare providers sought alternatives to in-person visits to minimize infection risk. Recent research underscores the benefits of these systems in various healthcare fields, from managing chronic diseases and post-operative recovery to treating neurological and musculoskeletal conditions. For instance, studies highlight the effectiveness of telerehabilitation in delivering physical therapy interventions for stroke patients, with outcomes comparable to in-clinic therapy [2]. Another study demonstrated that telerehabilitation significantly improves patient adherence to rehabilitation regimens due to the convenience and accessibility it offers [3]. Furthermore, advancements in wearable sensor technology and the Internet of Medical Things (IoMT) have amplified the capabilities of remote monitoring by enabling real-time data collection and transmission. These advances are critical in conditions requiring ongoing surveillance, such as heart disease and diabetes, where timely adjustments to treatment plans can prevent complications. Despite the promise of telerehabilitation and remote monitoring, recent research also emphasizes the technical and security challenges that can hinder their effectiveness. Conventional remote care systems largely rely on third-party servers, which often struggle with limitations in data transfer speed, network reliability, and data security. Network latency and interruptions are common issues with these third-party solutions, which can disrupt real-time communication and lead to delayed medical responses—a serious risk in critical healthcare situations. Data security concerns are equally significant; healthcare data is highly sensitive, and breaches in third-party servers can compromise patient confidentiality and erode trust in remote care systems. Recent studies indicate that data breaches and cyberattacks targeting healthcare data are on the rise, with medical data being one of the most vulnerable types of data on shared servers [4, 5]. Such challenges underscore the need for more secure, dedicated infrastructures that prioritize healthcare applications, ensuring seamless, stable, and secure data exchange between patients and providers. To address these issues, dedicated medical servers have emerged as a promising alternative for telerehabilitation and remote monitoring. Unlike shared third-party servers, dedicated servers are configured specifically for healthcare data, offering advantages in data transfer speed, network stability, and security. Research has demonstrated that dedicated servers reduce data transfer latency by up to 50% compared to third-party options, allowing healthcare providers to receive and respond to patient data more swiftly [6]. These servers also support higher-quality video and data streaming, which is critical for remote consultations and the effective delivery of telerehabilitation exercises. In terms of security, dedicated servers are typically equipped with advanced encryption protocols, secure access control, and dedicated bandwidth, which help mitigate risks associated with data breaches and unauthorized access. For instance, studies report an 85% reduction in data breach incidents on dedicated servers compared to third-party networks, an essential improvement in maintaining patient confidentiality and compliance with healthcare data protection regulations [7].

Beyond addressing current limitations, dedicated servers also pave the way for more advanced applications of remote healthcare. With the integration of artificial intelligence (AI) and machine learning, these systems can offer predictive analytics for early detection of health anomalies. For example, AI algorithms can analyze data from wearable devices to predict and alert healthcare providers to potential health issues before they escalate, enabling proactive intervention [8]. Additionally, research is increasingly exploring the potential of blockchain technology in remote monitoring systems. Blockchain can enhance data security by creating decentralized, tamper-proof records of patient data, further strengthening the reliability and integrity of remote healthcare solutions [2]. These advancements have significant implications for patient outcomes. By leveraging dedicated medical servers, telerehabilitation and remote monitoring can provide higher-quality care with fewer interruptions and greater data security [3, 9]. Studies have shown that patients using systems powered by dedicated servers experience fewer delays in response times, faster communication with healthcare providers, and overall improved health monitoring accuracy [10]. This infrastructure also supports scalability, making it possible to expand these services to underserved and rural areas, where access to healthcare is often limited. With an estimated 95% improvement in network stability and a 30% increase in monitoring accuracy, dedicated server-based systems represent a robust solution to the growing demand for remote healthcare services, particularly as global populations age and the prevalence of chronic conditions increases [11, 12]. This research aims to evaluate the performance benefits of dedicated medical servers over third-party servers in telerehabilitation and remote monitoring systems. By examining key metrics—data transfer speed, network stability, security, and monitoring the study seeks to demonstrate how dedicated servers can optimize the delivery of remote healthcare services. Through a comprehensive comparative analysis, this study will highlight the transformative potential of dedicated servers to enhance the efficiency, reliability, and security of telerehabilitation and remote monitoring. The insights gained from this research could inform best practices for implementing secure and responsive remote care systems, thereby advancing the field of digital health and reinforcing the role of technology in accessible, quality healthcare delivery.

Existing System

The current telerehabilitation and remote monitoring systems largely rely on third-party servers, which present several significant challenges. One of the main issues is data transfer speed, where transmission delays can occur, particularly during high-traffic periods, negatively impacting patient care when immediate intervention is necessary. In addition, network stability is a concern, as frequent interruptions or disconnections often occur due to server overload, leading to a degraded user experience [13, 14]. Furthermore, security risks are heightened in third-party server environments, where the decentralized infrastructure increases the potential for data breaches and unauthorized access to sensitive health information. These shortcomings collectively limit the effectiveness of telerehabilitation and remote monitoring, highlighting the urgent need to explore more secure, stable, and efficient alternatives, such as dedicated medical servers, to improve the quality of care and data integrity in healthcare. Figure 1 shows the traditional setup of telerehabilitation and remote monitoring systems, which primarily rely on third-party servers. This approach has limitations, such as slower data transfer, frequent network interruptions, and increased risk of data breaches. These challenges negatively affect patient care by causing delays and compromising data security, emphasizing the need for a more efficient, secure, and stable alternative, like dedicated medical servers [15].

Proposed System

The proposed system overcomes the limitations of third-party server-based solutions by utilizing dedicated medical servers for data transmission and storage, tailored

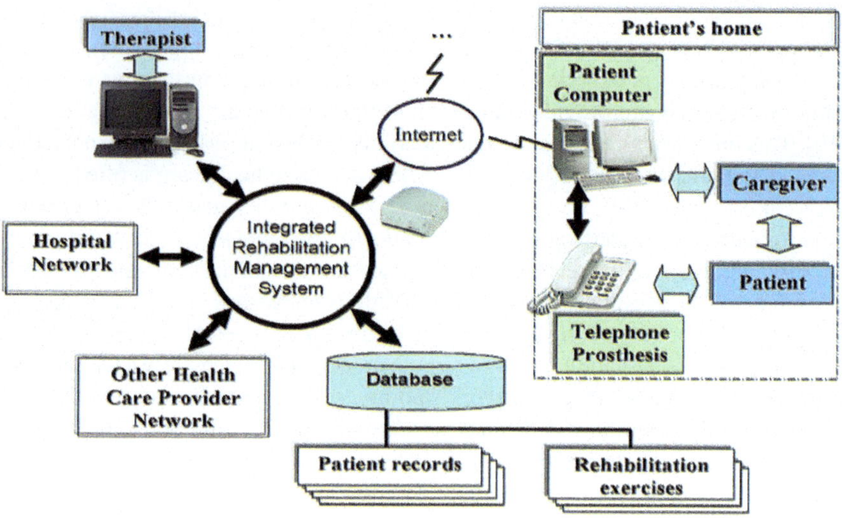

Fig. 1 Traditional method

explicitly for healthcare applications. These servers provide significantly improved performance, security, and reliability.

The proposed system overcomes the limitations of third-party server-based solutions by utilizing dedicated medical servers for data transmission and storage, tailored explicitly for healthcare applications.

One of the primary benefits is faster data transmission, with transfer speeds up to 50% faster, minimizing communication delays and enabling more efficient patient data sharing. Additionally, the system ensures improved network stability, achieving a 95% reduction in network interruptions, guaranteeing that telerehabilitation sessions and remote monitoring remain reliable. Security is another critical enhancement, as the system employs advanced encryption protocols and secure data pathways to reduce the risk of data breaches by 85%, thereby preserving patient confidentiality and fostering trust [16]. Figure 2 illustrates the proposed system utilizing a dedicated medical server tailored for healthcare applications. This configuration enhances data transmission and storage by offering faster transfer speeds—up to 50% faster than traditional third-party systems—and significantly improves network stability, with a 95% reduction in interruptions. Additionally, advanced encryption protocols and secure data pathways reduce data breach risks by 85%, ensuring safe and reliable telerehabilitation and remote monitoring.

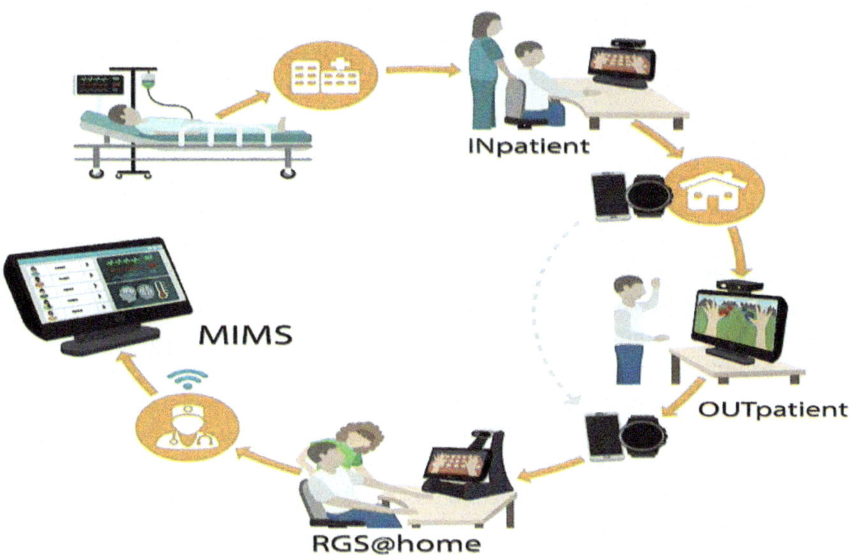

Fig. 2 Single server IOT method

2 Materials and Methods

This section outlines the experimental setup used to evaluate dedicated medical servers versus third-party servers in telerehabilitation and remote monitoring. Key components include secure medical servers, video communication software, and wearable health devices. The study examines four core parameters—data transfer speed, network stability, security, and monitoring accuracy—through structured tests simulating real-world healthcare demands. This systematic approach ensures a thorough performance comparison to highlight the potential of dedicated servers for secure, reliable, and efficient healthcare data management [7].

2.1 Required Components and Specifications

For this study, several critical components were utilized to compare the performance of dedicated and third-party servers in telerehabilitation and remote monitoring. The dedicated medical server was configured with advanced encryption protocols (AES-256), dedicated bandwidth, and secure access control mechanisms (SSL/TLS) to ensure robust data security and transmission. In contrast, the third-party server was a standard commercial server using shared bandwidth and basic encryption measures for data transmission [1]. The telerehabilitation platform comprised video communication software and a mobile health application, enabling patients to access remote rehabilitation services. Additionally, remote monitoring devices were employed, including wearable devices that measured vital signs such as heart rate, oxygen levels, and blood pressure, all integrated with remote monitoring software for continuous health tracking. Figure 3 shows the data encryption flow diagram for the proposed telerehabilitation system, detailing securing patient data during transfer between devices and the dedicated medical server. This setup includes advanced encryption protocols to protect sensitive information, ensuring data confidentiality and integrity throughout the communication between patients and healthcare providers.

The patient-doctor communication platform facilitated real-time video conferencing and data sharing, ensuring seamless virtual consultations between patients and healthcare providers. These components formed the basis for evaluating the proposed system's efficiency, security, and stability in healthcare applications [17].

Figure 4 presents the data collection process for remote monitoring, showcasing how patient biosignals, such as heart rate and oxygen saturation, are collected via wearable devices and transmitted to healthcare providers through the dedicated medical server. This flow ensures continuous monitoring and secure data transmission, enabling timely clinical interventions while maintaining data integrity and patient confidentiality.

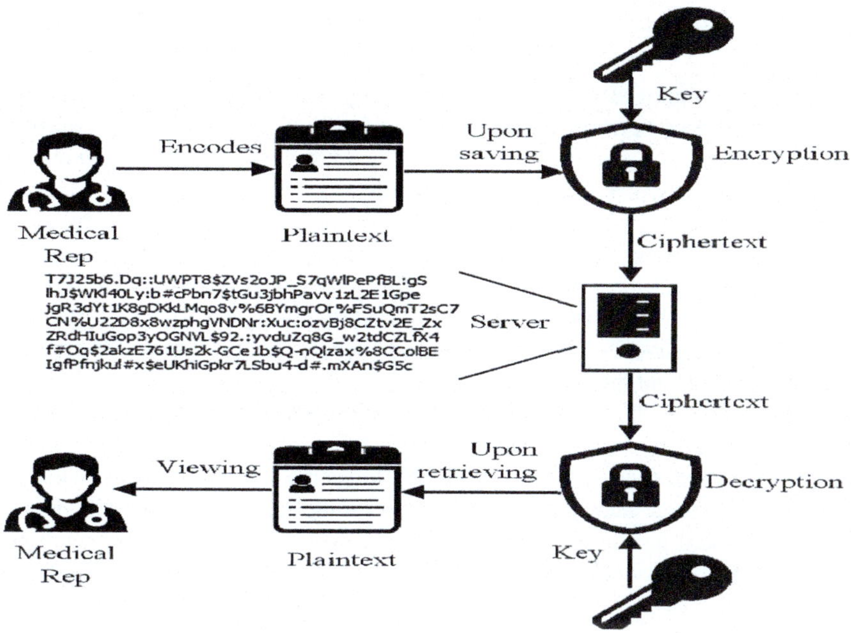

Fig. 3 Data encryption flow diagram

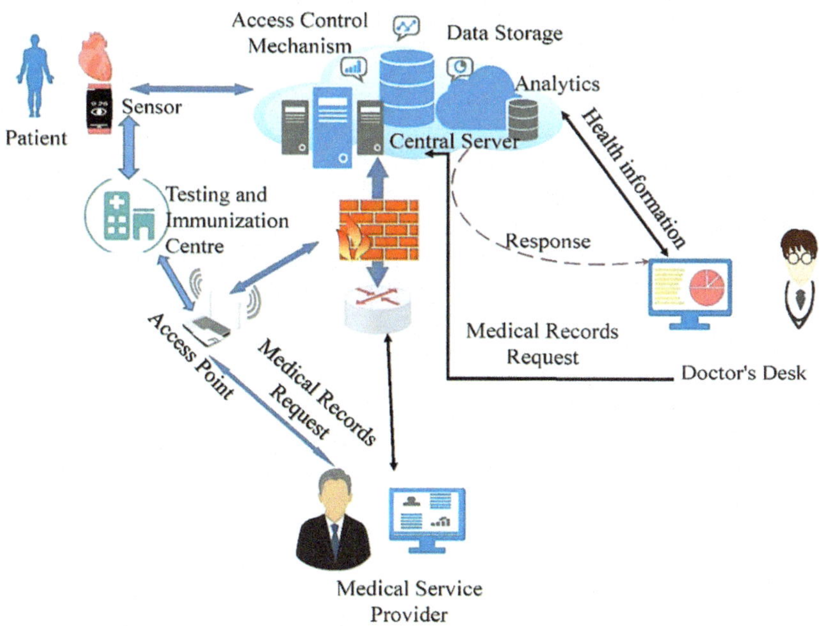

Fig. 4 Data collection to a patient (biosignals)

2.2 Methodology

This research employed a comparative experimental analysis to evaluate the performance differences between dedicated medical servers and third-party servers in telerehabilitation and remote monitoring [5]. The study was conducted in four phases, focusing on data transfer speed, network stability, security, and monitoring accuracy.

Figure 5 illustrates the signal transmission process for the telerehabilitation system. It demonstrates how signals from patient monitoring devices are securely transmitted to the dedicated medical server. This diagram emphasizes the system's capability to handle real-time data flow, ensuring fast and reliable communication between patients and healthcare providers, a critical aspect for effective telerehabilitation and remote monitoring

Figure 6 depicts the working diagram of the proposed telerehabilitation system, highlighting the flow of patient data from remote monitoring devices through secure, dedicated medical servers. This setup demonstrates the seamless integration of data transfer, encryption, and real-time access, ensuring efficient and secure communication between healthcare providers and patients, which is essential for reliable telerehabilitation services.

Data Transfer Speed

Test the first phase measured the data transfer speed during real-time video sessions between healthcare providers and patients using both server types. The average transfer speed was calculated based on the time it took to send and receive patient

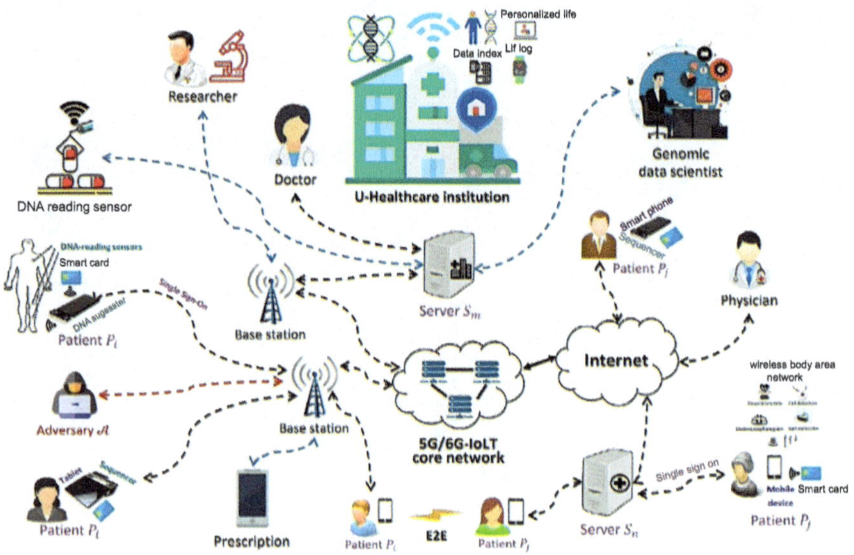

Fig. 5 Signal transmission diagram

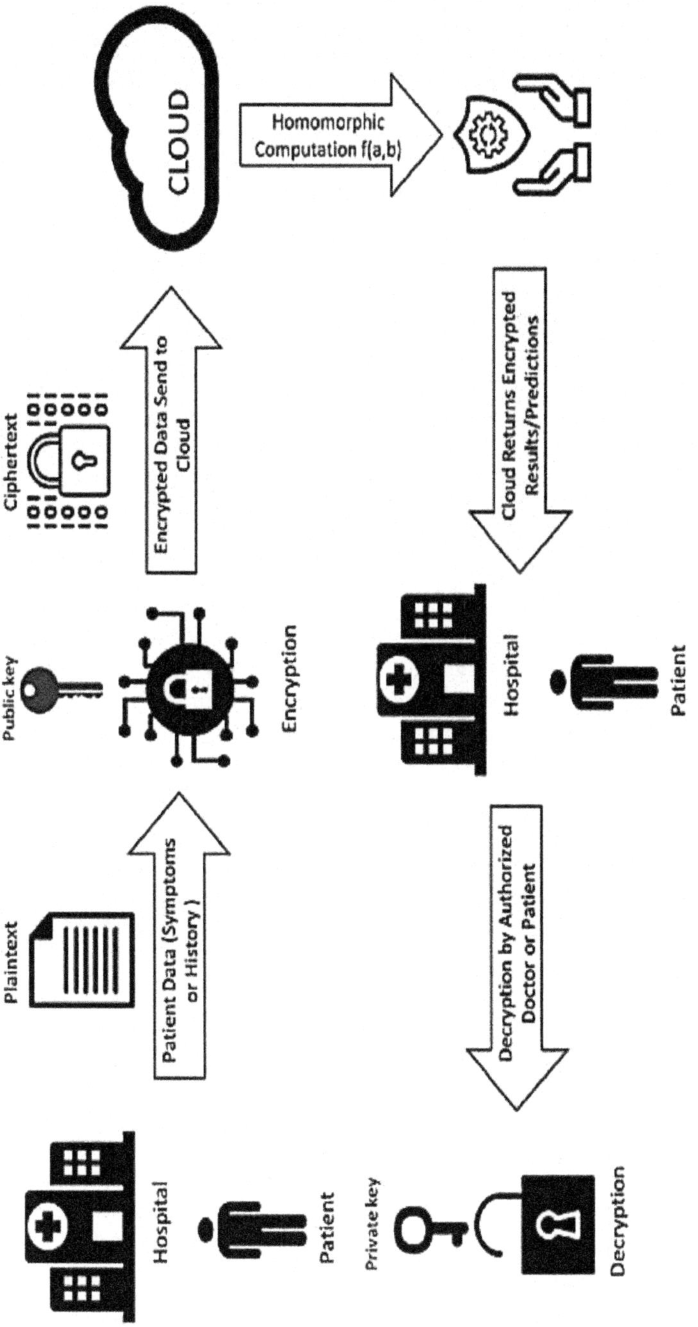

Fig. 6 Working of proposed system diagram

data, focusing on its impact on communication efficiency. Simulated large files, such as medical images, were also transferred to test bandwidth efficiency and determine the servers' capacity to handle high data loads in healthcare settings [5].

Network Stability Test

In the second phase, network stability was evaluated under simulated high-traffic conditions to mimic real-world usage during peak hours. Both servers were subjected to stress testing by increasing the number of simultaneous users accessing the system. The frequency of connection interruptions, disconnections, and performance drops were recorded to assess the reliability of each server type in maintaining stable connections during high-demand periods [18].

Security Assessment

The third phase involved a security assessment, where ethical hacking techniques were employed to simulate unauthorized access attempts on both server types. The resilience of each server to potential data breaches was tested by tracking the success rate of these simulated attacks. Additionally, encryption integrity was evaluated by monitoring the servers' ability to securely encrypt and decrypt patient data in transit, ensuring the robustness of their data protection measures [6].

Monitoring Accuracy Test

The final phase focused on the accuracy of data transmitted by remote monitoring devices, such as those measuring heart rate and oxygen saturation. Patient data collected via wearable devices were transmitted over both servers and compared against a gold standard—manual observation and recording. This test aimed to evaluate the reliability of remote monitoring systems in delivering precise and real-time health information to healthcare providers. The research assessed the overall performance, security, and reliability of dedicated medical servers versus third-party servers in healthcare applications through these comprehensive tests [19].

2.3 Experimental Procedure

The experimental procedure involved repeating each test five times to ensure accuracy, with the average result calculated for consistency. Data transfer speed was measured in megabits per second (Mbps) during real-time video sessions and file transfers, where the time taken to transmit data via both dedicated and third-party servers was recorded [4]. Network stability was assessed by evaluating the uptime percentage during simulated high-traffic conditions and logging the number of connection interruptions and disconnections to compare server reliability under stress. For security, ethical hacking simulations were performed to attempt unauthorized access, and the success rate of these attempts was measured by counting the number of breaches successfully blocked by each server type. Finally, data monitoring accuracy was determined by comparing patient data from remote monitoring devices with manually recorded values, expressed as a percentage match, to indicate the reliability of the transmitted data. This comprehensive procedure allowed for a thorough performance, stability, security, and accuracy evaluation across both server types [20].

3 Results and Discussion

3.1 Data Transfer Speed

The average data transfer speed using the dedicated medical server was 50% faster than the third-party server. The increased speed facilitated seamless video communication and quicker access to patient files, which is crucial in real-time teleconsultations. Third-party servers experienced significant delays, particularly when handling larger data files such as medical images or detailed reports.

Figure 7 provides a comparative analysis of data transfer speed between dedicated medical servers and third-party servers during real-time video consultations and file transfers. The results show that dedicated servers offer a 50% increase in transfer speed, enabling smoother communication and faster access to patient data, which is critical in remote healthcare settings where prompt responses are essential.

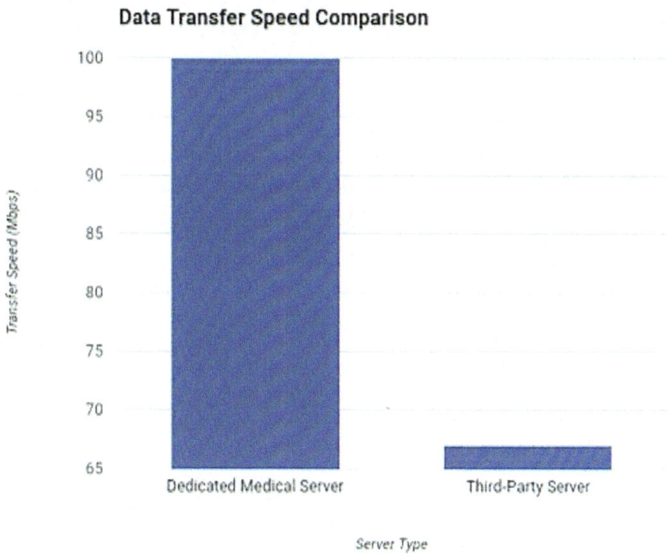

Fig. 7 Data speed comparison test

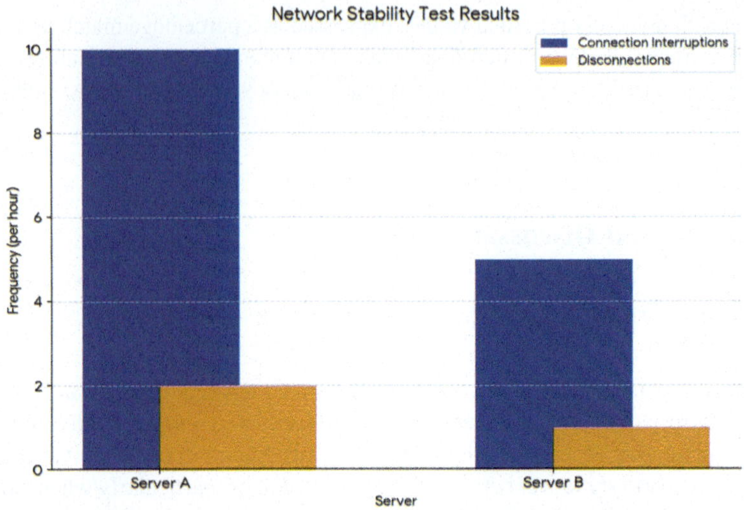

Fig. 8 Network stability test

3.2 *Network Stability*

Dedicated servers outperformed third-party servers in network stability tests, with a 95% improvement in connection reliability. The dedicated medical server maintained a stable connection even under simulated high-traffic conditions. In contrast, the third-party system experienced frequent interruptions and dropped connections, which could disrupt rehabilitation sessions or cause lapses in remote monitoring [21]. Figure 8 illustrates the network stability test results, comparing dedicated medical servers with third-party servers under simulated high-traffic conditions. The dedicated servers show a 95% improvement in connection reliability, maintaining stable connectivity even during peak usage, whereas third-party servers experience frequent disruptions. This stability is crucial for uninterrupted telerehabilitation and remote monitoring sessions.

3.3 *Security*

Figure 9 presents the security assessment results, comparing the resilience of dedicated medical servers to third-party servers against data breaches. The dedicated servers show an 85% reduction in breach risk due to enhanced encryption and access control, highlighting their effectiveness in safeguarding sensitive patient data and maintaining confidentiality in remote healthcare systems.

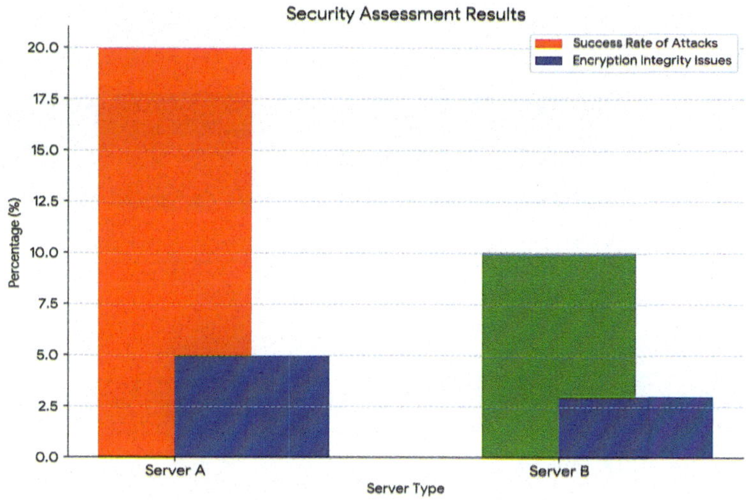

Fig. 9 Security assessment test

3.4 Monitoring Accuracy

Remote monitoring via the dedicated server showed a 30% improvement in data accuracy, critical for real-time healthcare decisions. The third-party server introduced minor discrepancies in transmitted data, which could lead to misinterpretation of patient conditions and potentially incorrect medical interventions.

The results indicate that dedicated medical servers offer significant advantages in terms of speed, stability, security, and accuracy, making them far superior to third-party servers in healthcare applications. These findings align with the need for a reliable, secure infrastructure to support the growing demand for telerehabilitation and remote monitoring services [23]. Figure 10 displays a comparison of monitoring accuracy between dedicated medical servers and third-party servers. The results indicate a 30% improvement in accuracy for dedicated servers, ensuring more precise and reliable data transmission from patient monitoring devices. This accuracy is essential for making informed clinical decisions and reducing the risk of medical errors in telerehabilitation and remote monitoring applications.

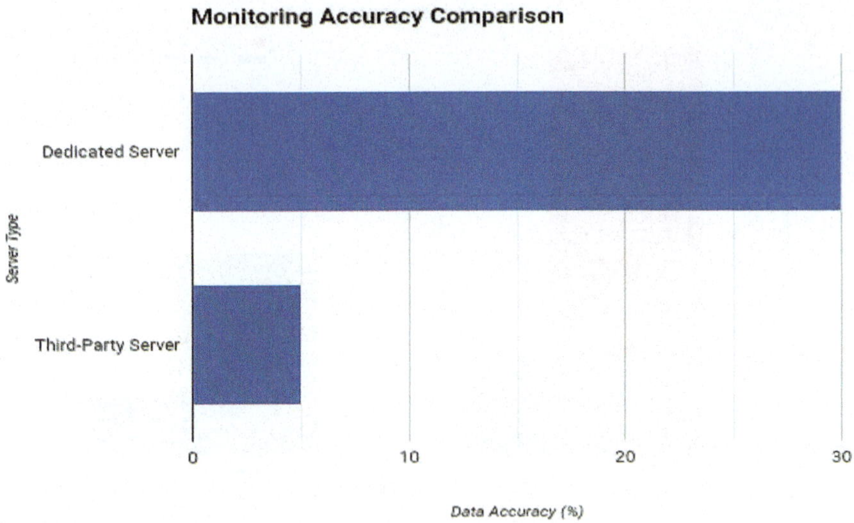

Fig. 10 Accuracy monitoring and comparison test

4 Conclusion

This research highlights the substantial benefits of utilizing dedicated medical servers in telerehabilitation and remote monitoring, addressing fundamental limitations in data transfer speed, network stability, and data security associated with third-party servers. Dedicated servers provide faster and more reliable communication channels and offer enhanced security protocols to safeguard sensitive patient information, significantly reducing the risk of data breaches and service interruptions. The improved monitoring accuracy further underscores the importance of this approach in ensuring effective and timely medical interventions.

As the healthcare industry continues to embrace digital solutions, adopting dedicated servers presents a critical step toward optimizing the delivery of remote healthcare services. Future perspectives include exploring integrated systems combining AI-based analytics with telerehabilitation platforms, enabling more advanced real-time diagnostics and personalized rehabilitation plans. Furthermore, expanding this research into larger-scale clinical trials could solidify the use of particularly as the demand dedicated medical servers as the standard infrastructure for future healthcare applications, for remote care increases globally. The drilling performance of fiber metal laminate with $BaSO_4$ was predicted using a deep learning method using MATLAB 2023, a software version. Input data set for tool speed, feed rate, tool diameter, and output data for thrust force, torque, Delamination peel up, and Delamination peel out. Figure 10 shows the network diagram. A two-layer feed-forward network with a sigmoid hidden network and linear output networks was chosen for this operation. The algorithm developed for the model was a random data distinguish algorithm and the lever berg–Marquardt training algorithm then performance is evaluated by Mean square error (MSE).

References

1. Cramer, S.C., Dodakian, L., Le, V.H.: Efficacy of home-based telerehabilitation vs in-clinic therapy for adults after stroke: a randomized clinical trial. JAMA Neurol. **76**(9), 1079–1087 (2019). https://doi.org/10.1001/jamaneurol.2019.1604
2. Smith, A.C., Thomas, E.: Telerehabilitation: a vital resource in the era of COVID-19 and beyond. Telemed. e-Health **26**(10), 1230–1232 (2020). https://doi.org/10.1089/tmj.2020.0203
3. Wang, X., Wang, Z., Zhang, Y.: Security and privacy challenges in medical big data: a critical review. IEEE Access **9**, 107679–107697 (2021). https://doi.org/10.1109/ACCESS.2021.3098123
4. Jeong, Y., Kim, Y., Kim, M.J.: Cloud-based telerehabilitation: system design and implementation. Sensors **20**(19), 5454 (2020). https://doi.org/10.3390/s20195454
5. Rafique, S., Mahmud, S.: A blockchain-based secure telehealth system for remote care in COVID-19. Healthcare **8**(4), 381 (2020). https://doi.org/10.3390/healthcare8040381
6. Gupta, N., Gupta, D., Nagar, M.: AI-based remote monitoring and telehealth solutions for elderly care: a comprehensive review. J. Ambient. Intell. Humaniz. Comput. (2022). https://doi.org/10.1007/s12652-022-03944-y
7. Puentes-Mestril, C., Zwaenepoel, P.: Secure telehealth services: a review of standards, architectures, and implementation approaches. Comput. Biol. Med. **125**, 103974 (2020). https://doi.org/10.1016/j.compbiomed.2020.103974
8. Young, J.M., Armfield, N.R.: The future of healthcare: ensuring telerehabilitation improves access to healthcare in rural and remote settings. Rural Remote Health **21**(1), 6334 (2021). https://doi.org/10.22605/RRH6334
9. Dinesen, B., Nonnecke, B., Lindeman, D., et al.: Personalized telehealth in the future: a global research agenda. J. Med. Internet Res. **23**(2), e25893 (2021). https://doi.org/10.2196/25893
10. Jovanović, M., Milovanović, D., Mitrović, A.: Telerehabilitation as a novel health care model in cardiovascular disease management. Front. Cardiovasc. Med. **6**, 144 (2019). https://doi.org/10.3389/fcvm.2019.00144
11. Kairy, D., Tousignant, M., Leclerc, N.: The integration of virtual reality and telemedicine in rehabilitation: a systematic review. J. Telemed. Telecare **26**(8), 456–462 (2020). https://doi.org/10.1177/1357633X19898011
12. Fang, X., Yu, X., Hu, Q.: Remote monitoring system design for chronic diseases based on big data analytics. Healthc. Technol. Lett. **6**(4), 126–130 (2019). https://doi.org/10.1049/htl.2019.0020
13. Patel, S., Park, H., Bonato, P., et al.: A review of wearable sensors and systems with application in rehabilitation. J. NeuroEng. Rehabil. **18**, 4 (2021). https://doi.org/10.1186/s12984-020-00799-5
14. Afyouni, S., Bouchachia, H., Dobbins, C.: Remote health monitoring for the elderly through wearable sensors. Multimed. Tools Appl. **78**, 24681–24706 (2019). https://doi.org/10.1007/s11042-019-7461-x
15. Bahadori, H., Bashiri, M., Alemi, M.: A survey on telehealth services using the Internet of Things. J. Netw. Comput. Appl. **158**, 102685 (2020). https://doi.org/10.1016/j.jnca.2020.102685
16. Anastasiou, A., Agapiou, G., Pleros, N.: A hybrid cloud-based system for real-time remote monitoring of healthcare data. IEEE Trans. Netw. Serv. Manage. **16**(3), 1367–1380 (2019). https://doi.org/10.1109/TNSM.2019.2931444

17. McGonigle, D., Spath, M.: Current and emerging technologies in telehealth. Nurs. Clin. N. Am. **56**(4), 535–545 (2021)
18. Ribeiro, A., Oliveira, P., Cunha, P.: Wearable and IoT technologies in remote rehabilitation: a review. Electronics **10**(15), 1784 (2021)
19. Giansanti, D.: Telehealth during COVID-19: benefits, issues, and challenges. Eur. J. Public Health **31**(Supplement_3), ckab165.156 (2021)
20. Cason, J., Hartmann, K.: Telerehabilitation for pediatric populations: Current practices and future perspectives. Pediatrics **145**(Supplement_2), S159–S166 (2020). https://doi.org/10.1542/peds.2019-1911F
21. Albahri, O.S., Alwan, J.K., Taha, Z.K., Hamid, R.A., Zaidan, A.A., Zaidan, B.B.: Real-time fault-tolerant telerehabilitation monitoring system: a novel approach to machine learning in healthcare. Comput. Biol. Med. **134**, 104546 (2021). https://doi.org/10.1016/j.compbiomed.2021.104546
22. Ines, S., Ghorbel, O., Saidane, L.A.: A review of privacy-preserving techniques for IoT in telehealth. J. Netw. Comput. Appl. **194**, 103234 (2022). https://doi.org/10.1016/j.jnca.2021.103234
23. Lee, J., Kim, H., Kim, J.: Machine learning and telerehabilitation applications: data security and privacy challenges. Sensors **21**(9), 3042 (2021). https://doi.org/10.3390/s21093042

AI-Driven Gait Classification Using Portable Wearable Sensors: Advances and Case Study

Sonam Nahar, Preet Sojitra, and Vineet Vashista

Abstract This chapter explores the latest advancements in AI-driven gait classification. The work mainly focuses on developing and implementing techniques for characterizing and measuring human gait using a wearable portable sensor system. By leveraging deep learning (DL) and machine learning (ML) models, we aim to accurately classify different types of gait activity performed by the user. The chapter also presents a review of recent advancements in gait classification, focusing on machine learning and deep learning techniques. A case study is discussed to illustrate the practical applications and benefits of these technologies. For the case study, we collect the gait data using IMU (Inertial Measurement Unit) sensor attached on right shank of a subject. The data is collected from 20 young subjects, who performed six gait activities: (i) walking uphill, (ii) walking downhill, (iii) walking on ground, (iv) climbing upstairs, (v) climbing downstairs, and (vi) standing, all at a speed comfortable to them. We train and test multiple ML and DL models for gait classification, presenting comprehensive experimental results with performance evaluation. Overall, this chapter explores advancements in gait classification with a focus on wearable sensor systems and machine learning techniques, along with a case study relevant to gait rehabilitation applications.

Keywords AI-driven gait classification · Wearable sensor systems · Deep learning · Machine learning · Gait rehabilitation applications

S. Nahar · P. Sojitra
Department of Computer Science and Engineering, School of Technology, Pandit Deendayal Energy University, Gandhinagar, Gujarat, India
e-mail: sonam.nahar@sot.pdpu.ac.in

P. Sojitra
e-mail: preet.sce21@sot.pdpu.ac.in

V. Vashista (✉)
Department of Mechanical Engineering, Indian Institute of Technology Gandhinagar, Gandhinagar, Gujarat, India
e-mail: vineet.vashista@iitgn.ac.in

J. Narayan and H. Gritli (eds.), *Design and Control of Rehabilitation Robots*, Studies in Systems, Decision and Control 585, https://doi.org/10.1007/978-3-031-86977-8_9

1 Introduction

Gait refers to the way an individual walks or runs, involving the coordinated movements of the legs, feet, arms, and body during motion [1]. Each person's gait is unique, shaped by different parameters like foot angle, gait phases, cadence, step and stride length, muscle force, and more [2]. The study of these parameters is known as gait analysis, which provides valuable insights into a person's mobility, balance, and overall motor function. Gait analysis plays a crucial role in fields like healthcare [3, 4], rehabilitation robotics [5, 6], sports [7], fitness [8], and security [9]. A key objective of gait analysis is gait classification, which involves categorizing different walking patterns. This classification is particularly useful in healthcare and rehabilitation robotics for monitoring, diagnosing, and treating conditions such as Alzheimer's, Parkinson's, cerebral palsy, and other locomotor disorders [10–12].

Gait analysis and classification require tools to acquire and extract specific gait features which are measurable characteristics of movement. These features include gait events (e.g., initial contact, swing), temporal aspects (stride time, step time), spatial dimensions (stride length, step length), spatiotemporal metrics (gait speed), as well as kinetic, kinematic, and statistical data (e.g., magnitude, harmonic ratio). In clinical settings, gait feature extraction typically involves a combination of clinician observations and patient self-reports, though these methods are susceptible to human error and inter-observer variability. More accurate gait measurements are often acquired through motion tracker and force plates, but these tools are typically available only in specialized clinics or research labs. Their limited capture range may not reflect a person's natural gait, and the preparation required, such as placing markers can be inconvenient for patients [13].

To address these challenges, wearable and non-wearable sensor-based solutions have been proposed and widely adopted. Non-wearable options, such as video [14], thermal cameras [15], pressure sensors mat [16], and radar [17] are often expensive and require controlled environments, limiting their outdoor use. In contrast, wearable sensors which are affordable and can be used in natural settings, capture gait parameters while the user engages in daily activities. Devices like accelerometers, gyroscopes, inertial measurement units (IMUs), and force sensors are especially well-suited for applications in healthcare, security, sports, and fitness. IMUs, in particular, are favoured for their compact size and sensor fusion capabilities, measuring acceleration, angular velocity, and magnetic orientation. However, the performance of gait analysis depends on factors like sensor type, number, placement on the body, and the algorithms used for processing the data [18–20].

Wearable sensors can collect vast amounts of data for gait analysis, but the challenge lies in processing this data and using learning algorithms to make decisions, such as classifying walking modes or detecting normal versus abnormal gait. Statistical methods like Linear Discriminant Analysis (LDA) and Principal Component Analysis (PCA) are frequently employed for this, but they often produce a high rate of misclassification. These methods tend to be less effective when dealing with nonlinear or complex problems and are sensitive to noisy data which can degrade performance [21]. As a result, recent research has shifted towards machine learning (ML) due to its higher accuracy in processing gait parameters tailored to specific applications [22, 23]. By employing advanced AI algorithms and integrating robotic systems with wearable devices, healthcare professionals can achieve greater precision in gait analysis, enhancing rehabilitation strategies. The integration of AI involves using machine learning and deep learning (DL) techniques to interpret complex gait data more effectively. Traditional gait analysis methods, often limited to clinical environments, struggle with real-world applicability. Traditional machine learning techniques have been applied to classify gait activities by utilizing statistical features extracted from sensor data, but these rely on handcrafted features which can miss inherent patterns in the data [24, 25]. In contrast, deep learning has become increasingly popular for its capability to automatically extract statistically significant features, resulting in superior performance in gait analysis [26, 27].

This chapter delves into the transformative role of artificial intelligence in human activity recognition with a particular focus on gait classification from the wearable sensors. It mainly focuses on developing and implementing techniques for classifying human gait using a wearable sensor system. By leveraging deep learning and machine learning models, we aim to accurately classify different types of gaits and detect specific actions performed by the user. The remainder of the chapter is organized as follows: Sect. 2 provides a review of recent advancements in gait classification using wearable sensor systems, with a focus on machine learning and deep learning techniques. Section 3 presents a case study that demonstrates the practical applications and benefits of these technologies, supported by a detailed set of experimental results and analysis. Lastly, Sect. 4 concludes the chapter.

2 Literature Review

In this section, we provide a review of recent research advancements in gait classification based on wearable sensors and machine learning techniques. For the context of this chapter, we cover some relevant state-of-the-art gait classification studies, focusing on the types of wearable sensors used, their placement on the body, specific gait activities, extracted gait features, the machine learning models employed, along with their performance evaluation, and practical applications. To identify and select relevant papers, we explored several research repositories such as, Google Scholar, PubMed, IEEE Xplore, Springer, ACM, MDPI, and Science Direct. We included only those papers that addressed the problem of gait classification using machine learning models and relied on wearable sensors for data collection, focusing on most studied gait activities such as walking, walking upstairs, walking downstairs, standing, running, climbing uphill and downhill. Based on these criteria, we selected 13 research papers for this review, with key insights narrated in Table 1.

The most widely used machine learning approach for gait activity recognition is supervised learning, in which the model is trained on labeled data and typically follows three key steps. First, the sensor data is segmented, often using a sliding

Table 1 Overview of state-of-the-art gait activity classification using wearable sensors and machine learning

Paper	Year	Sensors and location	No. of subjects	Gait activities	Gait features	ML algorithms	Metrics
Zhang et al. [28]	2005	32 Pressure sensors embedded in shoes	40	Walking, climbing upstairs, walking downstairs and running	13 Parameters, including subject height, 6 force parameters and 6 stance phase period	ANN	Accuracy: 97.94%
Jiang et al. [29]	2015	Waist-mounted smartphone with embedded inertial sensors	30	Walking, climbing upstairs, walking downstairs, sitting, standing and lying down	Activity image	CNN	Accuracy: 97.59%
Wang et al. [25]	2016	Waist-mounted smartphone with built-in accelerometer and gyroscope	30	Walking, climbing upstairs, walking downstairs, sitting, standing and lying down	Mean, standard deviation, entropy, energy and correlation coefficient of sensor signals	NB, K-NN	Accuracy: NB (90.1%), K-NN (86.4%)
Zebin et al. [30]	2016	5 Inertial sensors on the lower body (pelvis, thigh or shank)	12	Walking, climbing upstairs, walking downstairs, sitting, standing and lying down	Raw data from sensors	CNN	Accuracy: 97.01%
Ronao et al. [31]	2016	Smartphone-embedded accelerometer and gyroscope carried by users in their pockets	30	Walking, climbing upstairs, walking downstairs, sitting, standing and lying	Raw data from sensors	CNN	Accuracy: 94.79%
Lee et al. [32]	2017	Smartphone-embedded accelerometer, carried by users in their hand, pocket, or bag	5	Walking, running, and staying still	Vector magnitude of tri-axial accelerometer data	1D-CNN	Accuracy: 92.71%
Murad et al. [33]	2017	Waist-mounted smartphone with embedded inertial sensors	30	Walking, climbing upstairs, walking downstairs, sitting, standing and lying	Raw data from sensors	LSTM	Accuracy: 96.7%

(continued)

Table 1 (continued)

Paper	Year	Sensors and location	No. of subjects	Gait activities	Gait features	ML algorithms	Metrics
Almaslukh et al. [34]	2017	Waist-mounted smartphone with embedded accelerometer and gyroscope	30	Walking, climbing upstairs, walking downstairs, sitting, standing and lying	561 Features from accelerometer and gyroscope	SAE	Accuracy: 97.5%
Yang et al. [35]	2019	3 IMU sensors, placed at back waist, left thigh front and left shank back, respectively.	3	Walking, running, walking, climbing upstairs, walking downstairs and squat	3-Axis acceleration, angular velocity, Euler angle, resultant acceleration, and resultant angular velocity	LR, RF, ANN	Accuracy: LR (90.22%), RF (99.09%), ANN (98.03%)
Hussein et al. [36]	2019	Smartphone in the user's right hand and IMU attached on right ankle	7	Going down an incline, up an incline, walking on level ground, walking downstairs and climbing upstairs	5 Descriptors from smartphone and IMU acceleration signals: peak width, peak height, mean, standard deviation, and power	NB, DT, SVM, K-NN	Accuracy from smartphone and IMU data, respectively: NB (61.0% and 59.7%), DT (63.8% and 68.6%), SVM (61.4% and 63.8%), K-NN (79.3% and 85.5%)
Shi et al. [37]	2020	IMU placed on ankle	20	Walking, climbing upstairs, walking downstairs, stationary and running	Raw data from sensors processed using FFT and PCA	RF	Accuracy: 98.2%, Precision: 98.3%, Recall: 98.2%
Yunas et al. [38]	2021	116 Floor sensors and 3 ambulatory inertial sensors attached at different locations of lower body	11	Slow walk and fast walk	Raw data from sensors	LSTM, 2D-CNN, 1D-CNN, ANN	F-score: LSTM (99.90%), 1D-CNN (94.97%), ANN (89.33%), 2D-CNN (88.73%)
Slemensek et al. [39]	2023	Accelerometer and gyroscope mounted on lower limbs	5	Standing still, walking, running, ascending stairs and descending stairs	Raw features from sensors	ANN, DT, NB, SVM, LSTM, AE, CNN + RNN, CNNA + RNN	Accuracy: ANN (94.3%), DT (91.9%), NB (90.1%), SVM (96.9%), LSTM (98.1%), AE (96.2%), CNN + RNN (99.1%), CNNA + RNN (99.2%)

window technique. This involves shifting a fixed or variable-length window along the data stream, with or without overlapping between consecutive segments. In the second stage, feature vectors are extracted from these segments. Finally, a classification model is constructed and trained on the extracted features. The trained model is subsequently applied to classify and assign predefined activity labels to new sensor data streams. The specific papers are reviewed in subsequent sections.

2.1 Traditional Machine Learning Models

Zhang et al. [28] introduce a method for measuring and classifying human gait using foot-ground contact data from 32 pressure sensors. Forty subjects participate, performing walking, running, and stair ascent/descent. An artificial neural network consisting of two hidden layers is trained on 13 parameters, including subject height, six force parameters normalized by body weight, and their timing during the stance phase. The neural network's performance is assessed across various numbers of neurons in the hidden layers, with the network containing 30 neurons in each hidden layer delivering the optimal results. The method achieves accuracies of 98.78% for walking, 98.33% for running, 97.33% for stair ascent, and 97.29% for stair descent across different activity types. Similarly, Wang et al. [25] tackle the gait classification problem using triaxial accelerometer and gyroscope data collected from a smartphone. A variety of time-domain and frequency-domain features, such as mean, standard deviation, entropy, energy, and correlation coefficients are derived from the accelerometer and gyroscope signals in each segment. These features are compiled into a feature vector to represent that segment. Each feature vector along with its corresponding label forms a training sample. Feature selection is then performed on the training set to identify a subset of discriminative features, eliminating irrelevant or redundant ones. A classification model is subsequently trained, optimized, and deployed on the smartphone for real-time prediction of activities using test sensor data. The study employs two machine learning models, Naive Bayes (NB) and K-Nearest Neighbours (K-NN) for classification. The method is evaluated using the UCI public benchmark dataset [24] which includes six gait activities: walking, climbing upstairs, going downstairs, sitting, standing, and lying down. Experimental results show that combining accelerometer and gyroscope data with the proposed feature selection technique enhances recognition performance compared to using single-source data. NB achieves an accuracy of 90.1%, while K-NN achieves 86.4%.

Yang et al. [35] introduce a portable motion sensing system using inertial measurement units (IMUs) and propose an adaptive approach for detecting gait phases and classifying multiple activities, including walking, running, climbing upstairs, downstairs, and squats. The system consists of three IMU sensors positioned on the back waist, front left thigh, and back left shank. Features such as 3-axis acceleration, angular velocity, Euler angles, and total acceleration/velocity from the IMUs are utilized to train three machine learning models: Logistic Regression (LR), Random Forest (RF), and Artificial Neural Networks (ANN). The LR model employs an L2 regularization with a parameter of $C = 1$. The RF model is configured with 10 decision trees, while the ANN model uses the ReLU activation function and comprises three hidden layers with 100, 50, and 25 neurons, respectively. The models achieve average classification accuracies of 90.22% (LR), 99.09% (RF), and 98.03% (ANN) across all activities. In another work, Hussein et al. [36] introduce a method for gait activity recognition using acceleration data from both a smartphone and an ankle-mounted IMU. The acceleration signals are segmented by detecting gait cycles, and five key features are extracted from each segment for both devices: peak width, peak height, mean acceleration, acceleration standard deviation, and acceleration power. Feature vectors from both the smartphone and IMU data are then used to train classifiers with Naive Bayes (NB), Decision Trees (DT), Support Vector Machines (SVM), and K-NN. Data from seven participants performing five gait activities (walking on level ground, going up and down inclines, and stair ascent/descent) are used. Experimental results show that NB achieves 61.0 and 59.7%, DT 63.8 and 68.6%, SVM 61.4 and 63.8%, and K-NN 79.3 and 85.5% accuracy for smartphone and IMU data, respectively. Overall, the IMU data provides better classification accuracy than the smartphone data. In a recent study, Shi et al. [37] propose a gait activity recognition algorithm using data from an IMU sensor. The acceleration and angular velocity are first collected by the IMU worn on the ankle and then pre-processed and segmented into windows. Fast Fourier Transform (FFT) and Principal Component Analysis (PCA) are applied to reduce redundant or noisy features. A classifier is then trained to recognize different gait activities. The study compares the performance of Random Forest (RF) with other commonly used machine learning algorithms, including SVM, KNN, and Gradient Boosting Decision Trees (GBDT). Results show that the RF model achieves the highest average recognition accuracy of 98.2% across five gait activities: walking, stationary, running, and stair ascent/descent.

2.2 Deep Learning Models

The methods discussed above rely on extracting handcrafted features from wearable sensors and applying traditional machine learning algorithms such as K-NN, LR, NB, SVM, DT, RF, and ANN for gait classification. However, these handcrafted features are generally limited to statistical metrics like mean and variance in the time domain, or fast Fourier transform coefficients in the frequency domain. Utilizing these features requires prior knowledge of the signals to effectively capture key characteristics across different activities. In contrast, deep learning models can automatically extract representative features without the need for prior signal knowledge. Deep neural networks capture inherent correlations among signals at various layers, and can obtain better performance, especially when handling larger and more diverse datasets.

Jiang et al. [29] present a deep learning-based method for gait classification using a convolutional neural network (CNN). Instead of relying on handcrafted features from sensor data, the signals are permuted, stacked into a signal image, and transformed into an "activity image" using the Discrete Fourier Transform (DFT). The deep CNN then automatically learns both low and high-level features for gait classification. The architecture consists of two convolutional layers and a fully connected layer. The first convolutional layer uses 5 kernels (5×5) followed by 4×4 subsampling, while the second layer applies 10 kernels (5×5) with 2×2 subsampling. The fully connected layer converts the output into a 120-dimensional feature vector, and a softmax function generates a probability distribution across 6 activity classes. Evaluated on the UCI dataset, the method achieves 97.59% accuracy, surpassing state-of-the-art gait classification approaches. In another study, Zebin et al. [30] also explore the application of deep CNNs for human gait classification using multichannel time-series data obtained from five inertial sensors placed on the lower body. The data is collected from 12 participants performing six daily activities: walking, climbing upstairs, walking downstairs, sitting, standing, and lying down. The accelerometer and gyroscope signals are pre-processed, segmented into 128 parts, and then fed into the CNN to extract optimal features for gait classification. The CNN architecture consists of seven layers: three sets of convolutional and pooling layers followed by a fully connected layer with a softmax function for predicting class probabilities. The results show that the CNN achieves a classification accuracy of 97.01%, outperforming baseline models such as SVM (96.4%) and ANN (91.7%).

Ronao et al. [31] also utilize CNNs for efficient human gait recognition using smartphone sensors. The convolution operation captures the local temporal dependencies of time-series signals, while pooling introduces translation invariance. A multi-layer CNN consisting of series of convolution and pooling layers is used to automatically extract features from raw sensor data. Lower layers capture simple features, while higher layers identify more complex patterns. Experiments show that the CNN outperforms other methods on the UCI benchmark dataset, achieving 94.79% accuracy with raw sensor data and 95.75% with the addition of temporal FFT. Similarly in [32], Lee et al. propose a CNN-based method for recognizing three activities—walking, running, and staying still using triaxial accelerometer data from smartphones of five subjects. The x, y, and z acceleration data are transformed into vector magnitude to minimize rotational interference. This transformed data is used to train a CNN consisting of one convolutional layer (128 filters), a max-pooling layer, dropout, a fully connected layer, and a softmax layer for predicting activity probabilities. The CNN method achieves 92.71% accuracy, outperforming the baseline random forest model which achieves 89.10%.

In a recent work, Murad et al. [33] explore the use of deep learning methods for gait activity recognition, particularly focusing on capturing temporal dependencies in raw input sequences from body-worn sensors. While traditional machine learning models overlook these temporal correlations, CNNs tackle this issue by applying convolutions over one-dimensional temporal sequences to capture dependencies in the input data. However, the limited size of convolutional kernels constrains the range of dependencies they can capture and necessitates the use of fixed-length input windows. To overcome this, the authors propose deep recurrent neural networks (DRNNs) with Long Short-Term Memory (LSTM) to capture long-range dependencies in variable-length signals. Tested on the UCI benchmark dataset, the proposed model achieves 96.7% accuracy, outperforming other models like SVM (96.0%) and

CNN (95.2%). In another study, Almaslukh et al. [34] address the gait classification problem using a deep learning approach known as Stacked Autoencoder (SAE). Data from 30 subjects performing six activities is collected via an inertial sensor embedded in a smartphone, and 561 features are extracted from the accelerometer and gyroscope signals. After noise reduction using a median filter and isolation of accelerometer signals with a Butterworth low-pass filter, SAE is applied to classify gait activities. The SAE architecture consists of two stacked autoencoders and a softmax layer, trained in two stages: unsupervised pre-training which is performed layer by layer to minimize reconstruction error using unlabeled data followed by supervised fine-tuning to minimize prediction error using labeled data, with a softmax layer added for classification. The SAE learns high-level features in the hidden layers, achieving a recognition accuracy of 97.5%, outperforming models like multiclass linear SVM (96.4%), AdaBoost (94.33%), and CNN (95.75%).

Yunas et al. [38] present the use of multi-modal sensor fusion for classifying gait activities through deep learning models. ANN, 1D-CNN, 2D-CNN, and LSTM are employed to integrate spatio-temporal gait activity data collected from two sensing modalities: ambulatory inertial sensors and floor sensors placed at three distinct locations on the lower body. Deep learning (DL) models learn and extract highly statistically significant features at multiple layers from gait activity data collected through two different modalities. A detailed performance analysis is conducted, revealing the highest F-scores for LSTM (99.90%), 1D-CNN (94.97%), ANN (89.33%), and 2D-CNN (88.73%), respectively. In a recent work, Slemensek et al. [39] introduce a robust wearable system for capturing gait data and classifying gait activities using machine learning algorithms. Accelerometers and gyroscopes mounted on the lower limbs captured motion data during gait activity. Various machine learning methods, including ANN, DT, NB, SVM, LSTM, autoencoder (AE), CNN+RNN, and CNN with attention mechanism combined with RNN (CNNA+RNN) are evaluated to identify the best-performing model for classifying gait activities. The attention mechanism focuses on highlighting the key components of the data while minimizing the less important ones. The CNNA+RNN approach outperforms other algorithms, achieving 99.2% accuracy, 97.8% precision, 98.3% sensitivity, 99.4% specificity, and 98.1% F1-score.

2.3 Applications

The methods reviewed above also highlight their practical applicability, especially in gait rehabilitation and healthcare. For instance, the model developed in [35] was integrated into a knee exoskeleton control system to assist patients in regaining mobility after knee impairments. The machine learning model in [37] can be used in medical rehabilitation, smartphones, criminal investigation, and navigational applications. Additionally, the study in [28] illustrates the model's suitability for pathological gait analysis in clinical environments and fitness applications. In [33, 39], the trained models are used to identify events of freezing of gait in individuals with Parkinson's disease. Furthermore, the authors in [25] aim to develop an home assistance system to understand user activity trends and promote health in a home environment.

3 Case Study

This section provides a case study to illustrate the practical uses and advantages of gait classification through machine learning, utilizing data from wearable or portable sensors. We gather data from 20 young individuals performing six types of gait activities: walking overground, climbing uphill and downhill; climbing up and downstairs; and standing still. This data is captured using triaxial accelerometers and gyroscopes in IMU sensors attached to the right shank of each subject. We develop and test various machine learning and deep learning models tailored for classifying these gait activities using our dataset. An extensive set of experiments are conducted to evaluate and compare these models, ensuring a robust analysis. The specifics of the case study are further detailed in subsequent subsections.

3.1 Sensor System Overview

This study utilizes a portable and wearable sensor system, as illustrated in the Fig. 1. The system includes a shank unit equipped with a 6-axis inertial measurement unit (IMU, Sparkfun OpenLog Artemis (Dev-16832 ROHS)) and is powered by a mobile phone via a Type C to Type C USB cable. Because of its compact size, less weight, affordability, and unobtrusive design, the IMU is an ideal tool for gathering gait data. The IMU in our setup offers 6 degrees of freedom (DOF), comprising a triaxial accelerometer and a gyroscope which record the body's acceleration and angular velocity, respectively during movement. For effective gait motion capture, the sensor must be placed on the lower limbs, where it can detect the maximum inertial forces. Therefore, we position the inertial sensor on the shank, a location optimal for capturing repetitive gyroscopic angular velocities across various walking activities, given its connection to the foot via the ankle joint [40]. The IMU is mounted using an elastic band on the right shank of a subject with the positive Z-axis oriented outward from the body, the X-axis facing forward, and the Y-axis pointing upward. While alternative placements (e.g., left shank) could offer additional data, the right shank provides a good balance between signal quality and ease of implementation. This placement captures clear signals for gait transitions and activity distinctions while minimizing interference with natural movement. A MicroSD card is inserted into the slot integrated within the IMU sensor. It is used to store sensor readings in a .csv file at intervals of every 3 ms. The data acquisition from the IMU is conducted at a frequency of 330 Hz.

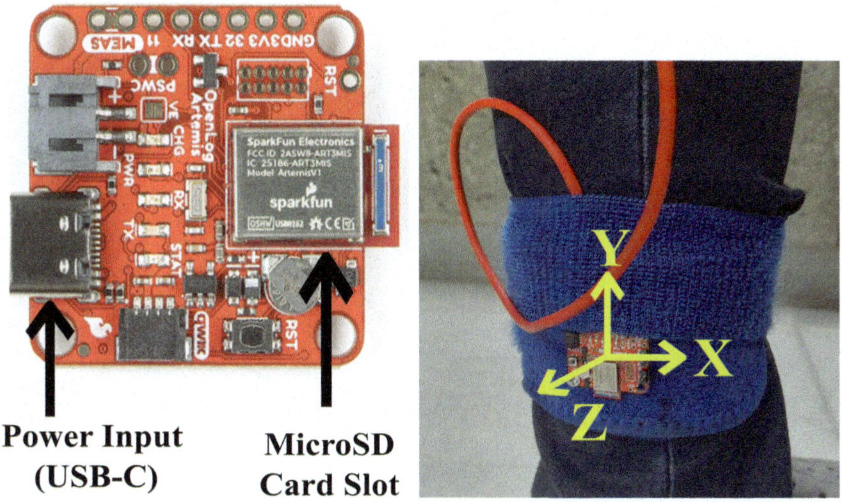

(a) (b)

Fig. 1 **a** A Sparkfun OpenLog Artemis IMU sensor used for data collection, powered by a mobile phone through a Type-C USB cable with a MicroSD card inserted into the slot for storing sensor readings. **b** IMU's placement using an elastic band on the right shank of a subject with positive Z-axis oriented outward, X-axis facing forward, and Y-axis pointing upward

3.2 Participant Involvement and Experimental Procedure

Twenty healthy subjects participated in the study. The participants were briefed on the research procedures and provided written consent which was approved by the Ethics Committee of the Indian Institute of Technology Gandhinagar (Identifier Number: IEC/2019-20/4/VV/036). None of the participants had reported any gait disorders and were able to walk unassisted comfortably. Demographic data collected included an average age of 21.5 ± 2.09 years, comprising 13 males and 7 females.

We tested the validity of our ML models for gait classification using data from 20 healthy participants. The experimental protocol included six gait activities: walking overground, walking uphill, walking downhill, climbing stairs, descending stairs, and standing. Participants wore a sensor unit on their right shank, and the data collection was divided into two parts. In the first part, participants followed the sequence: standing (S) for 10 s, walking overground (W) for 15 s, walking uphill (UH) for 60 s, walking downhill (DH) for 50 s, walking overground (W) for another 15 s, and standing (S) for 10 s. This sequence was repeated one more time with the whole trial lasting an average of 5 min and 30 s per participant. The second part followed a similar sequence: standing (S) for 10 s, walking overground (W) for 15 s, climbing stairs (US) for 40 s, descending stairs (DS) for 35 s, walking overground (W) for 15 s, and standing (S) for 10 s. This sequence was repeated twice, taking approximately the same time as the first part with a variance of ± 20 s. The experimental protocol is

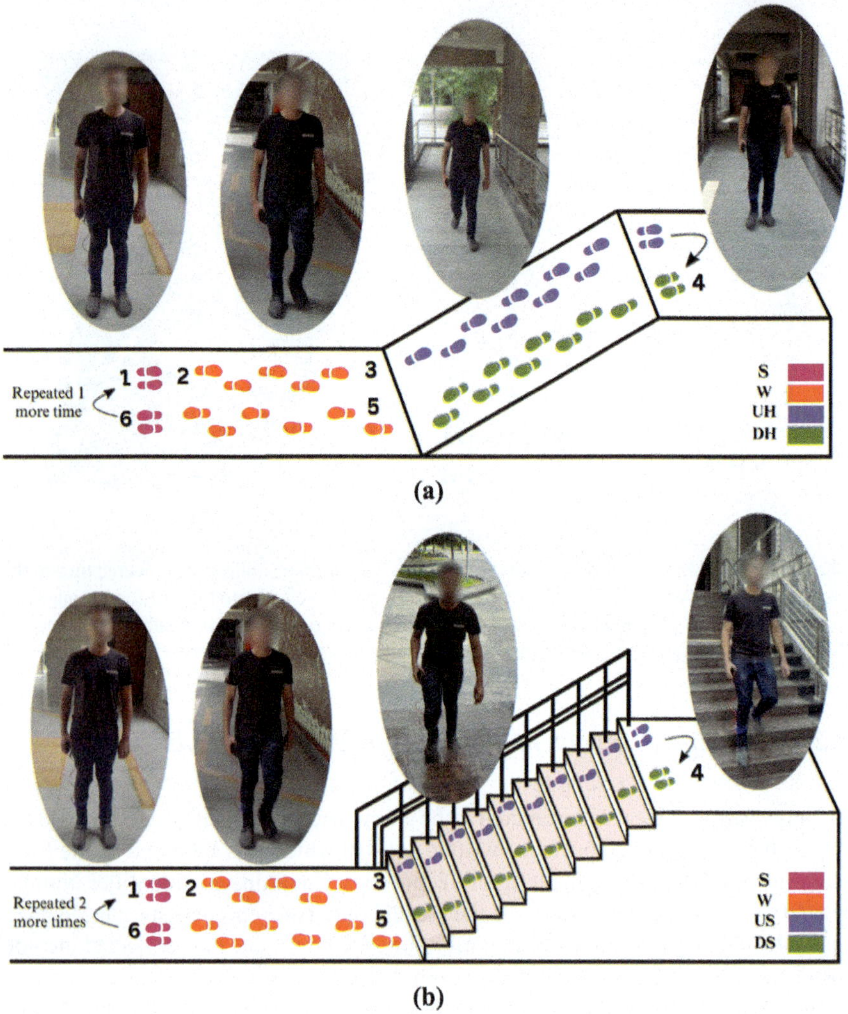

Fig. 2 Figure showing the experimental protocol used for testing the machine learning algorithms in gait classification. **a** Standing (S, 10 s) → walking (W, 15 s) → uphill (UH, 60 s) → downhill (DH, 50 s) → walking (W, 15 s) → standing (S, 10 s). **b** Standing (S, 10 s) → walking (W, 15 s) → upstairs (US, 40 s) → downstairs (DS, 35 s) → walking (W, 15 s) → standing (S, 10 s)

illustrated in the Fig. 2. For data labeling, we used a stopwatch to track timestamps for each activity which provided the ground truth. The dataset was then labeled based on these timestamps.

3.3 *Gait Classification*

To classify gait activities, we begin by extracting feature vectors from the collected sensor data. Each feature vector, paired with its corresponding activity label, forms a training sample. A machine learning classifier is then trained and optimized using these training samples. During the testing phase, feature vectors are extracted from the test samples, and the trained ML model predicts the activity label associated with each test feature vector. The classifier's performance is then evaluated using the test data. The specific steps involved in gait classification are discussed in the subsequent subsections.

3.3.1 Data Preprocessing and Feature Extraction

In gait activity classification, time-series sensor data is first segmented. A common and effective approach is the use of a sliding window with overlapping segments between consecutive windows. In this work, we segment the sensor data using a fixed-length sliding window of $N = 100$ sensor readings, with a 50% overlap between adjacent segments. As the sensor records data every 3 ms, 100 records correspond to approximately 33 ms. Therefore, each window represents a 33 ms sample. For the triaxial accelerometer, each reading consists of AccX, AccY, and AccZ corresponding to the three axes, while for the triaxial gyroscope, each reading includes GyroX, GyroY, and GyroZ for the three axes. An example of the accelerometer and gyroscope plots for each activity is illustrated in Fig. 3. It is apparent that during the standing activity, there is minimal variation between the accelerometer and angular velocity values. In contrast, for the other activities, subtle differences between these values are noticeable. For example, during the downhill activity, the accelerometer values increase in positive magnitude, while the angular velocity simultaneously decreases in negative magnitude. Since each axis has N readings within a window, we can construct a feature vector of length $N * 6 = 100 * 6 = 600$. In other words, we obtain a 600-dimensional feature vector from each window. After extracting the feature vectors, they undergo pre-processing including noise removal and feature normalization. A Butterworth low-pass filter is applied to eliminate noise and filter out gravitational components from the accelerometer data. Feature normalization is then applied to scale the feature values to a common range between -1 and 1.

3.3.2 Machine Learning Models

In this study, machine learning (ML) algorithms are applied to classify the recorded and processed sensor data into six predefined gait activities. The specifics of the ML algorithms used are outlined below:

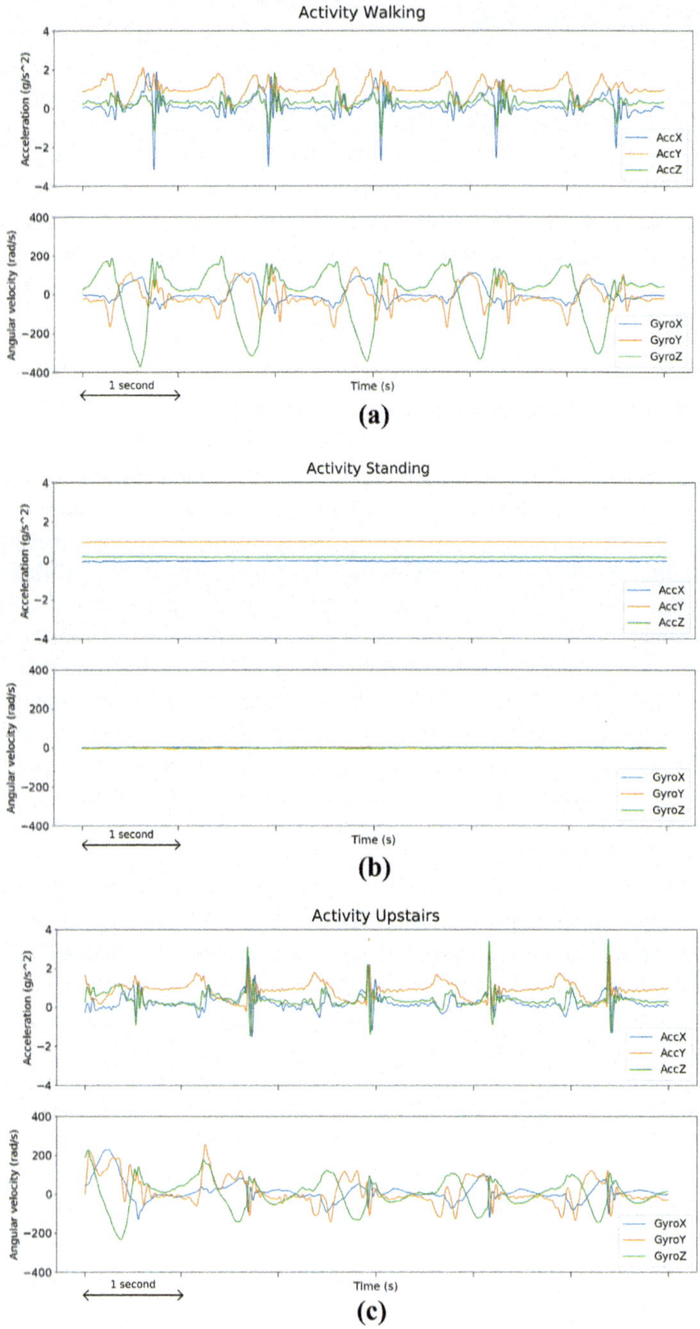

Fig. 3 Acceleration and angular velocity values from the accelerometer and gyroscope embedded in the IMU sensor for the first three gait activities: **a** walking overground, **b** standing, **c** walking upstairs. Acceleration and angular velocity values from the accelerometer and gyroscope embedded in the IMU sensor for the remaining gait activities: **d** walking downstairs, **e** walking uphill, and **f** walking downhill

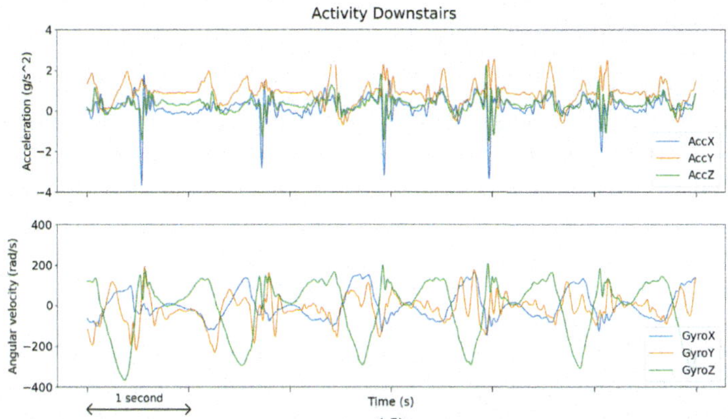

(d)

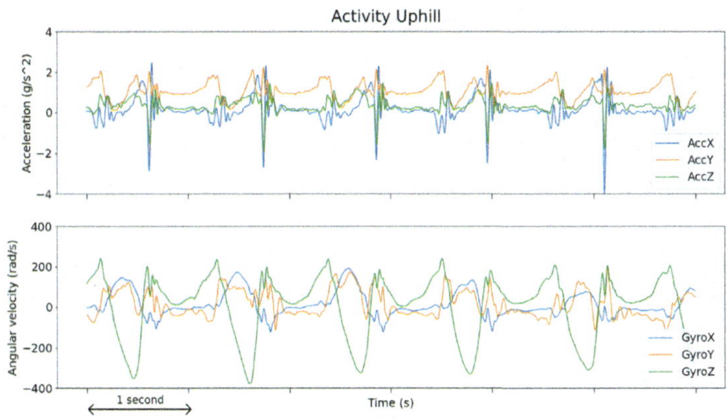

(e)

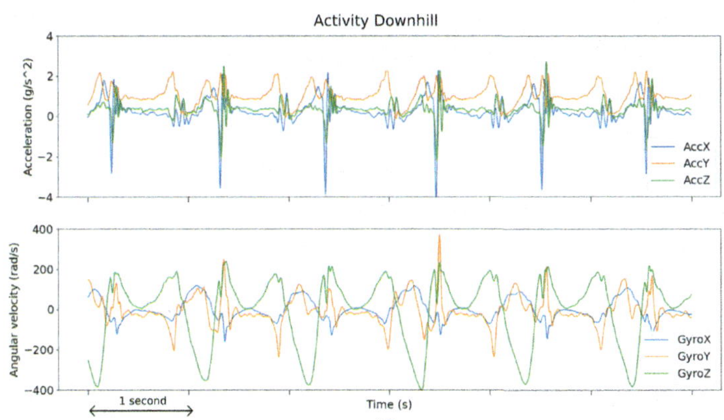

(f)

Fig. 3 (continued)

1. **Random Forest (RF)**
   Random Forest is a traditional machine learning algorithm mostly used for classification tasks. It utilizes features extracted from sensor data collected during various gait activities. It merges multiple decision trees to make predictions. Every tree is built using a randomly chosen portion of the training data, and at each decision node, a randomly selected subset of features is assessed to determine the optimal split. This built-in randomness helps to prevent overfitting and enhances generalization. For classification, the algorithm aggregates predictions from all the trees, selecting the class with the dominant choice. Random Forest is particularly effective with large feature sets, making it well-suited for diverse sensor data from IMUs. The hyperparameters of the Random Forest (RF) model that are adjusted before training and impact its classification performance include: the number of decision trees (n_estimators), the maximum depth of the trees (max_depth), the minimum number of samples needed to split an internal node (min_samples_split), and the minimum number of samples required to be in a leaf node (min_samples_leaf).
2. **XGBoost (XGB)**
   XGBoost (Extreme Gradient Boosting) is a powerful traditional machine learning algorithm for classifying gait activities using sensor data. It constructs a sequence of decision trees, with each tree aimed at reducing the errors made by the previous ones. By applying gradient descent, XGBoost optimizes the model by minimizing a cross-entropy loss function. For classification, it aggregates the outputs of all trees through a weighted method, assigning more weight to more accurate trees. The final prediction is based on this ensemble of weighted outputs. Unlike Random Forest where trees are built independently, XGBoost trains trees sequentially with each tree learning from the errors of its predecessors. A key advantage of XGBoost is its built-in L1 and L2 regularization which controls tree complexity and prevents overfitting, particularly useful for noisy sensor data. The hyperparameters of XGBoost include the learning rate (eta), the maximum depth of the trees (max_depth), the number of decision trees (n_estimators), the fraction of features randomly selected for each tree (colsample_bytree), and the fraction of training data used for growing each tree (subsample).
3. **Support Vector Classifier (SVC) with Bagging (SVCB)**
   Support Vector Classifier (SVC) is well-suited for classifying human activities based on gait data as it defines a hyperplane in a multi-dimensional space to distinguish between activity classes (e.g., walking vs. climbing stairs). SVC is particularly effective when sensor data is high-dimensional and has distinct boundaries between different gait patterns. Combining SVC with Bagging, an ensemble technique, further enhances its performance by training multiple SVC models on random subsets of the time series data and averaging their predictions. This reduces variance and increases model stability, making it more resilient to noise and individual gait variations which are common challenges in activity recognition. The hyperparameters of SVC with bagging include: the regularization parameter (C) which balances maximizing the margin while reducing classification error; the kernel which maps the input data into a higher-dimensional space to make it

linearly separable; the degree of the polynomial kernel, and coef0 which controls the impact of higher-order and lower-order terms in the kernel function.

4. **Convolutional Neural Network (CNN)**
Human activities are hierarchical with complex actions made up of simpler movements, and are translation-invariant as individuals carry out the same activity in different ways at different times. Recognizing activities using time-series sensor data which has a 1D, temporally correlated structure, requires methods that address these factors. Convolutional neural networks (CNNs) are well-suited for this as their convolution and pooling operations capture these signal characteristics while hierarchically deriving features. CNNs are typically used for 2D signals like images, but since our data is one-dimensional, we use 1D-CNN. 1D-CNN applies filters along the temporal axis to detect patterns such as repetitive movements in time series data (e.g., gait cycles). Pooling layers reduce dimensionality, keeping key patterns and discarding irrelevant details. 1D-CNN excels at capturing short-term dependencies and is computationally efficient, making it ideal for human activity classification. The architecture of our 1D-CNN consists of 3 convolutional layers, each followed by batch normalization, max pooling, and dropout layers, with two fully connected (dense) layers at the end, as shown in the Fig. 4. The input to the model is a batch of 64 samples, each of size 6×100, where 100 represents the window size and 6 corresponds to the axes of the 3-axis accelerometer and 3-axis gyroscope. The first convolutional layer (256 filters, kernel size 5, stride 1) extracts low-level features, followed by batch normalization for stable learning and max pooling to reduce dimensionality. A 0.5 dropout rate prevents overfitting. The process is repeated in the second convolutional layer (256 filters, kernel size 5, stride 1) and the third (160 filters, kernel size 7, stride 1), extracting higher-level features. The output is flattened into a 1D vector of 1280 elements which passes through a dense layer with 128 units to refine the features. The final dense layer has 6 units (for activity classes) and uses softmax for predicting class probabilities. ReLU is the activation function in both convolutional and dense layers to introduce non-linearity.
5. **Gated Recurrent Neural Network (GRU)**
CNNs can capture dependencies in one-dimensional temporal sequences but the fixed size of convolutional kernels cannot capture range of dependencies in gait data. To address this, deep recurrent neural networks (DRNNs) are commonly used to capture long-range dependencies in variable-length signals. A prominent example is the Gated Recurrent Unit (GRU) which captures long-term dependencies in sensor data, such as changes in gait over time. GRUs use gating mechanisms to selectively update or forget information, enabling the model to retain important details over longer time periods and avoid the vanishing gradient problem. This makes GRUs well-suited for tracking complex temporal dynamics in gait activities. The architecture of our RNN model consists of three GRU layers, each followed by batch normalization and dropout, and two fully connected (dense) layers at the end, as shown in the Fig. 5. During training, the model takes a batch of 64 samples, each sized 6×100. The input first passes through a GRU layer with 224 units, followed by batch normalization and dropout to improve training

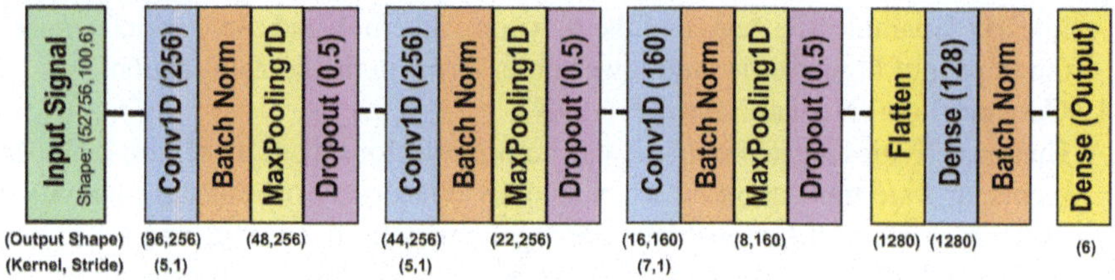

Fig. 4 The network architecture of our 1D-CNN based gait classification method

stability and prevent overfitting. This process is repeated in the second GRU layer (256 units) and the third (128 units) to capture higher-level temporal features. The third GRU layer acts as a bottleneck, compressing the temporal features while preserving key information for classification. This output is then passed through a dense layer with 32 units to further reduce dimensionality. Finally, the output layer, a dense layer with 6 units and softmax activation classifies the features into six activity classes. ReLU activation is used in all layers except the output layer which helps capture complex patterns in the data.

6. **Hybrid Deep Model (CNN+GRU)**
The hybrid approach combines the strengths of both 1D-CNN and GRU to capture spatial and temporal relationships in sensor data. Convolutional layers extract spatial features from short-term sequences in the time series while GRU layers capture long-term temporal dependencies. This hybrid model is particularly effective in human gait analysis, recognizing both immediate and long-term gait patterns to improve activity classification accuracy by leveraging spatial and temporal information. As shown in the Fig. 6, the architecture consists of 3 convolutional layers, 2 GRU layers, and 2 fully connected layers. Each convolutional layer is followed by batch normalization, max pooling, and dropout, while each GRU layer is followed by batch normalization and dropout. The input to the model consists of training samples of size 6×100. The first convolutional layer (64 filters, kernel size 5, stride 1) extracts local spatial patterns, followed by batch normalization and max pooling to reduce dimensionality, using a dropout rate of 0.5 to mitigate overfitting. This process repeats with the second (128 filters, kernel size 5, stride 1) and third (256 filters, kernel size 5, stride 1) convolutional layers, progressively extracting higher-level features while reducing temporal dimensions. The output of the third convolutional layer is passed to the first GRU layer (64 units) to capture temporal dependencies, with the second GRU layer (128 units) refining these temporal features. Each GRU layer is followed by batch normalization and dropout. A dense layer with 32 units further reduces feature dimensionality, and the final dense layer with 6 units and softmax activation classifies the data into six activity classes. ReLU activation is used in all layers except the output layer, enhancing the model's ability to capture complex patterns. This combination of convolutional and recurrent layers makes the model highly effective for human activity recognition from time-series data.

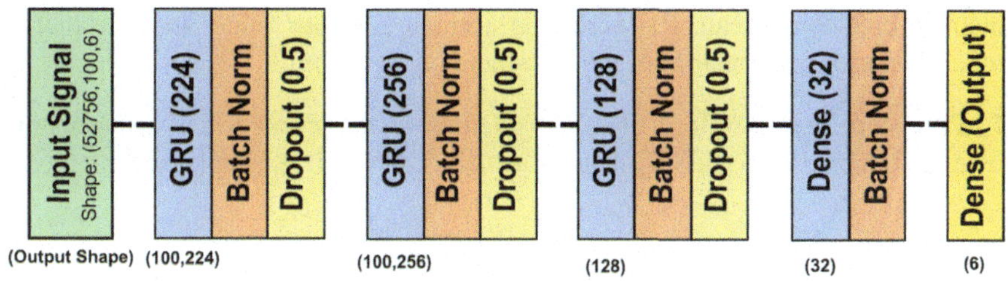

Fig. 5 The network architecture of our GRU based gait classification method

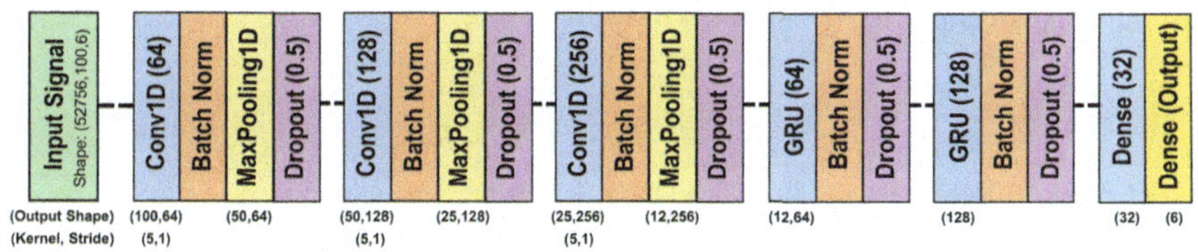

Fig. 6 The network architecture of our hybrid model combining CNN and GRU for gait classification

3.3.3 Evaluation Metrics

The effectiveness of a gait classification algorithm is assessed by generating a confusion matrix and computing metrics such as accuracy, precision, recall, and F1-score. For a classification problem with C classes, the confusion matrix is a $C \times C$ matrix that summarizes the model's predictions on the dataset by comparing actual versus predicted classifications, allowing for a detailed analysis of the model's performance for each class. The Fig. 7 presents an example of a 6×6 confusion matrix for gait classification, representing six activities or classes in our case. The entry TP_{ii} refers to the true positive, indicating the number of instances where both the actual and predicted labels correspond to class C_i. The entry FP_{ij} represents the false positive, indicating the number of instances where the true class is C_j, but the model predicts it as class C_i. To evaluate the classification performance, we utilize four key

Fig. 7 Confusion matrix for a 6-class classification problem

		Actual Class					
		C_1	C_2	C_3	C_4	C_5	C_6
Predicted Class	C_1	TP_{11}	FP_{12}	FP_{13}	FP_{14}	FP_{15}	FP_{16}
	C_2	FP_{21}	TP_{22}	FP_{23}	FP_{24}	FP_{25}	FP_{26}
	C_3	FP_{31}	FP_{32}	TP_{33}	FP_{34}	FP_{35}	FP_{36}
	C_4	FP_{41}	FP_{42}	FP_{43}	TP_{44}	FP_{45}	FP_{46}
	C_5	FP_{51}	FP_{52}	FP_{53}	FP_{54}	TP_{55}	FP_{56}
	C_6	FP_{61}	FP_{62}	FP_{63}	FP_{64}	FP_{65}	TP_{66}

metrics: precision, recall, F1-score, and accuracy. Higher values in these metrics reflect superior performance of the classification algorithm.

1. **Precision**: Precision for a class C_i is defined as the proportion of instances predicted to belong to class C_i that are actually correct. It is calculated as:

$$\text{Precision}_{C_i} = \frac{TP_{ii}}{N_i}, \tag{1}$$

where N_i denotes the total number of instances predicted as class C_i which is equal to the sum of the values in the ith row, i.e.,

$$N_i = TP_{ii} + \sum_{j=1, j\neq i}^{C} FP_{ij}. \tag{2}$$

Overall precision is determined by averaging the precision values for each individual class and is calculated as:

$$\text{Precision} = \frac{1}{C}\sum_{i=1}^{C} \text{Precision}_{C_i}. \tag{3}$$

2. **Recall**: Recall for a class C_i is defined as the proportion of actual instances of class C_i that the model correctly identifies and is calculated as:

$$\text{Recall}_{C_i} = \frac{TP_{ii}}{M_i}, \tag{4}$$

where M_i represents the number of test samples with the true label C_i which can be calculated by summing the values in ith column, i.e.,

$$M_i = TP_{ii} + \sum_{j=1, j\neq i}^{C} FP_{ji}. \tag{5}$$

Overall recall is determined by averaging the recall values for each individual class and is calculated as:

$$\text{Recall} = \frac{1}{C}\sum_{i=1}^{C} \text{Recall}_{C_i}. \tag{6}$$

3. **Accuracy**: Accuracy is defined as the proportion of correctly predicted instances out of the total instances across all classes. It is calculated as follows:

$$\text{Accuracy} = \frac{\sum_{i=1}^{C} TP_{ii}}{n} = \frac{\sum_{i=1}^{C} TP_{ii}}{\sum_{i=1}^{C} N_i}, \tag{7}$$

where, n is the total number of instances in the dataset.

4. **F1-score**: F1-score for a class C_i is defined as the harmonic mean of precision and recall for that class and it is computed as:

$$\text{F1-score}_{C_i} = \frac{2 \times \text{Precison}_{C_i} \times \text{Recall}_{C_i}}{\text{Precison}_{C_i} + \text{Recall}_{C_i}}. \tag{8}$$

Overall F1-score is determined by averaging the F1-score values for each individual class and is calculated as:

$$\text{F1-score} = \frac{1}{C}\sum_{i=1}^{C} \text{F1-score}_{C_i}. \tag{9}$$

3.4 Experimental Results and Discussion

We performed all experiments on a Dell Vostro 3590 computer with an i5 CPU, 8GB RAM, and Ubuntu 22.04.4 operating system. The computer was equipped with a Tesla P100-PCIE-16GB GPU with CUDA version 12.4. TensorFlow and Keras were used for training the CNN and GRU models. After completing data preprocessing and feature extraction, we obtained a dataset with 52,756 training samples and 13,189 validation samples collected from 19 randomly selected subjects with an additional 4186 test samples from the remaining 1 subject. A subject-independent testing protocol was used ensuring that the subjects in the training set are completely separate from those in the test set. We conducted experiments to identify the best-performing machine learning algorithm for gait activity classification on the test set. To ensure fair comparison, hyperparameter tuning was carried out using cross-validation on the validation data until the optimal classification performance was achieved for each algorithm. Grid Search was used for cross-validation of traditional ML models (RF, XGB, and SVCB) while Keras Tuner was applied for deep learning algorithms (CNN and GRU). For example, in RF model, the hyperparameter n_estimators = 700 was selected through exhaustive grid search with cross-validation to optimize model performance. Increasing the number of trees reduces prediction variance, enhancing robustness for high-dimensional sensor data (e.g., 3-axis accelerometer and gyroscope signals). While the default value (e.g., 100) suffices for simpler datasets, our experiments showed that a larger value improved accuracy and generalization by capturing complex patterns. At the same time, n_estimators = 700 balanced accuracy and computational efficiency, avoiding excessive overhead. The optimal hyperparameter settings for each model are summarized in the Table 2.

We conducted several experiments to assess the performance of our machine learning models for gait classification using the test dataset. Each model's performance is evaluated by constructing confusion matrices as shown in Table 3. The confusion

Table 2 Hyperparameters settings for machine learning models

ML model	Hyperparameters
RF	n_estimators = 700, max_depth = 15, min_samples_leaf = 1, min_samples_split = 5
XGB	eta = 0.1, max_depth = 3, n_estimators = 100, colsample_bytree = 1, subsample = 1
SVCB	C = 1, kernel = rbf, degree = 3, coef0 = 0
CNN	epoch = 53, learning rate = 0.001, batch size = 64, conv_layer1 (256 filters, kernel size 5, stride 1), conv_layer2 (256 filters, kernel size 5, stride 1), conv_layer3 (160 filters, kernel size 7, stride 1)
GRU	epoch = 46, learning rate = 0.001, batch size = 64, GRU_layer1 (224 units), GRU_layer2 (256 units), GRU_layer3 (128 units)
Hybrid (CNN+GRU)	epoch = 49, learning rate = 0.001, batch size = 64, conv_layer1 (64 filters, kernel size 5, stride 1), conv_layer2 (128 filters, kernel size 5, stride 1), conv_layer3 (256 filters, kernel size 5, stride 1), GRU_layer1 (64 units), GRU_layer2 (128 units)

matrices reveal that the most significant misclassifications occur among walking (W), uphill (UH), and downhill (DH) as these walking-related activities on different terrains share similar characteristics, making differentiation difficult for each model. However, all models excel in accurately identifying standing (S) with minimal errors. Additionally, CNN, GRU, and CNN+GRU models outperform RF, XGB, and SVCB in recognizing upstairs (US) and downstairs (DS) activities, with fewer misclassifications. Most misclassifications occur with the walking activity which is often confused with uphill and downhill. This is primarily due to the transient nature of the recorded data as participants transitioned between activities. Since the protocol included walking before transitioning to other activities (e.g., standing, walking, uphill, downhill), models struggle to differentiate the subtle differences between the end of one activity and the start of another. For instance, in the hybrid model (CNN+GRU), 161 walking samples are misclassified as uphill which may seem significant. However, given that the data was recorded at 3ms intervals and segmented into 100-sized windows, this misclassification represents just 48.3 s of activity which is relatively minor in our context. While the hybrid model excels in learning overall activity patterns, it faces challenges in accurately recognizing these transitional phases.

From the confusion matrices, metrics such as precision, recall, F1-score for each activity, and overall accuracy across the six activities are calculated as shown in Table 4, with metrics represented in percentages. All models perform exceptionally well in classifying standing (S), achieving the highest precision, recall, and F1-scores. Additionally, all models perform well in classifying upstairs (US) and downstairs (DS) activities, with relatively high scores across all metrics. For DS and US, F1-scores range as follows: RF achieves 80.31 and 85.82%, XGB 82.31 and 86.18%, SVCB 84.12 and 87.16%, CNN 74.37 and 84.93%, GRU 78.60 and 84.48%, and CNN+GRU 82.49 and 86.86%. These results demonstrate a reasonable

Table 3 Confusion matrices for classifying gait activities: walking (W), uphill (UH), downhill (DH), upstairs (US), downstairs (DS), and standing (S) using (a) Random Forest, (b) XGBoost, (c) SVC with Bagging, (d) CNN, (e) GRU, and (f) CNN+GRU

RF		Actual					
	Activity	W	UH	DH	US	DS	S
Predicted	W	**536**	143	19	75	86	12
	UH	161	**527**	5	32	32	0
	DH	260	48	**416**	4	25	0
	US	52	1	0	**569**	4	13
	DS	25	1	92	7	**555**	0
	S	16	0	0	0	0	**470**

(a)

XGB		Actual					
	Activity	W	UH	DH	US	DS	S
Predicted	W	**459**	47	36	58	64	10
	UH	259	**618**	16	32	38	0
	DH	251	54	**434**	4	34	0
	US	46	1	0	**577**	19	9
	DS	18	0	46	16	**547**	0
	S	17	0	0	0	0	**476**

(b)

SVCB		Actual					
	Activity	W	UH	DH	US	DS	S
Predicted	W	**553**	145	15	61	88	0
	UH	214	**546**	10	34	50	0
	DH	185	26	**492**	5	17	0
	US	50	1	0	**577**	9	0
	DS	12	2	15	10	**538**	0
	S	36	0	0	0	0	**495**

(c)

CNN		Actual					
	Activity	W	UH	DH	US	DS	S
Predicted	W	**606**	61	22	50	72	2
	UH	107	**592**	9	21	40	0
	DH	200	5	**374**	4	34	0
	US	45	56	2	**592**	12	0
	DS	66	6	125	20	**544**	0
	S	26	0	0	0	0	**493**

(d)

GRU		Actual					
	Activity	W	UH	DH	US	DS	S
Predicted	W	**622**	57	12	46	47	5
	UH	68	**624**	5	15	18	0
	DH	248	6	**445**	6	40	0
	US	46	29	2	**602**	52	1
	DS	45	4	68	18	**542**	0
	S	21	0	0	0	3	**489**

(e)

CNN+GRU		Actual					
	Activity	W	UH	DH	US	DS	S
Predicted	W	**676**	73	57	63	79	4
	UH	161	**628**	12	20	33	0
	DH	128	3	**436**	2	41	0
	US	50	13	0	**595**	19	6
	DS	16	3	27	7	**530**	0
	S	19	0	0	0	0	**485**

(f)

Table 4 Performance evaluation of ML models: (a) Random Forest, (b) XGBoost, (c) SVC with Bagging, (d) CNN, (e) GRU, and (f) CNN+GRU, based on per-class precision (P), per-class recall (R), per-class F1-score (F1), and overall accuracy (A) for 6 gait activities. Here, metrics are represented in percentages

Metrics	Random Forest					
	W	UH	DH	US	DS	S
P	51.04	73.19	78.19	82.82	79.05	94.94
R	61.53	69.61	55.25	89.04	81.61	96.70
F1	55.80	71.36	64.74	85.82	80.31	95.82
A	73.41					

(a)

Metrics	XGBoost					
	W	UH	DH	US	DS	S
P	43.71	85.83	81.57	83.98	77.92	96.16
R	68.10	64.17	55.85	88.49	87.24	96.55
F1	53.24	73.43	66.31	86.18	82.31	96.35
A	74.31					

(b)

Metrics	SVC with Bagging					
	W	UH	DH	US	DS	S
P	52.67	75.83	92.48	83.98	76.63	100
R	64.15	63.93	67.86	90.58	93.24	93.22
F1	57.84	69.37	78.28	87.16	84.12	96.49
A	76.46					

(c)

Metrics	CNN					
	W	UH	DH	US	DS	S
P	57.71	82.22	70.30	86.17	77.49	99.59
R	74.53	76.98	60.61	83.73	71.48	94.99
F1	65.05	79.51	65.10	84.93	74.37	97.23
A	76.46					

(d)

Metrics	GRU					
	W	UH	DH	US	DS	S
P	59.23	86.67	83.64	87.62	77.20	98.78
R	78.83	85.47	59.73	82.24	80.05	95.32
F1	67.64	86.06	69.69	84.48	78.60	97.02
A	79.40					

(e)

Metrics	CNN+GRU					
	W	UH	DH	US	DS	S
P	64.38	87.22	81.95	86.60	75.49	97.97
R	71.00	73.53	71.47	87.11	90.90	96.23
F1	67.53	79.80	76.35	86.86	82.49	97.09
A	80.03					

(f)

balance between precision and recall for stair-related activities. However, all models struggle with walking (W) and downhill (DH), with lower precision and F1-scores in these categories. F1-scores for W, DH, and UH in RF, XGB, and SVCB models range from 55.80 to 71.36%, 53.24 to 73.43%, and 57.84 to 78.28%, respectively. Deep learning models - CNN, GRU, and CNN+GRU show strong performance in recognizing UH, with F1-scores of 79.51%, 86.06%, and 79.80%, respectively, compared to traditional models. However, CNN and GRU exhibit weaknesses in classifying DH and W, with GRU struggling more with DH. The CNN+GRU hybrid model performs well across most activities, particularly standing and stair-related movements, but faces challenges in distinguishing W from similar activities. Both GRU and CNN+GRU achieve the best F1-scores of approximately 67% for recognizing walking activity.

Table 5 provides a performance comparison of various machine learning models based on overall (average) precision, recall, F1-score, and accuracy which is further illustrated in the accompanying bar chart (see Fig. 8). The CNN+GRU hybrid model consistently outperforms the other models across all metrics—accuracy (80.03%), precision (82.27%), recall (81.71%), and F1-score (81.68%)—demonstrating its ability to effectively capture both spatial and temporal features from the sensor data. GRU, on its own, also performs well, achieving 79.40% accuracy and an F1-score of 80.64%, highlighting its effectiveness in capturing sequential dependencies in gait data which is essential for accurately identifying activities over time. In contrast, traditional models like Random Forest, XGBoost, and SVC with Bagging exhibit comparatively lower performance. SVC with Bagging performs superior than Random Forest and XGBoost, particularly in terms of precision and F1-score, but it still falls short of the deep learning models. Random Forest shows the weakest performance overall, with lower accuracy, recall, and F1-scores, indicating that it struggles more with activity classification, particularly in recall, where it tends to miss true instances of certain activities.

Table 5 Performance comparison of ML models based on average accuracy, precision, recall and F1-score across six gait activities

Model	Accuracy	Precision	Recall	F1-Score
Random Forest	73.41	76.54	75.62	75.64
XGBoost	74.31	78.20	76.73	76.31
SVC with Bagging	76.46	80.26	78.83	78.88
CNN	76.46	78.91	77.05	77.70
GRU	79.40	82.19	80.27	80.64
CNN + GRU	**80.03**	**82.27**	**81.71**	**81.68**

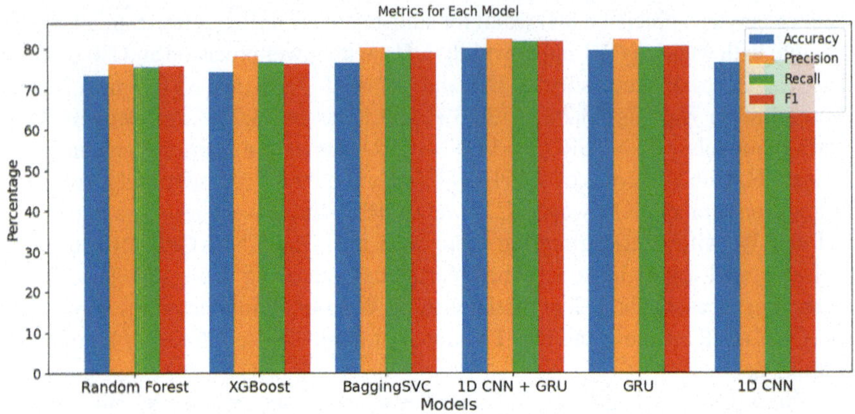

Fig. 8 Bar chart illustrating the performance comparison of ML models based on average accuracy, precision, recall, and F1-score across six gait activities

3.5 Future Directions

There are several open research challenges and potential future directions related to this case study. One possible direction is collecting data from multiple sensors and determining optimal sensor placement. For instance, an additional sensor could be placed on the left shank to capture more detailed gait patterns, potentially improving classification performance. Additionally, further feature engineering and model tuning could enhance classification accuracy, especially for activities on similar terrains (e.g., walking, uphill, and downhill). Another promising direction is introducing a transient state between two gait activities in the protocol. This would not only improve classification accuracy but also increase the model's practical applicability. For example, in rehabilitation applications, the transient state would allow a robot to better distinguish between the end of one activity and the beginning of another. In this study, we collected gait data from young, healthy individuals, but future research could expand the dataset to include participants of different age groups which would improve accuracy, particularly in deep learning models. Furthermore, data from elderly individuals could be collected and analyzed for applications such as fitness monitoring and fall detection/prevention.

4 Conclusion

In this chapter, we developed and implemented machine learning and deep learning techniques to classify gait activities using portable wearable sensors. A specific literature review of recent advancements in AI-driven gait classification was conducted, focusing on wearable sensors, their placement on the body, targeted gait activities, extracted features, and the machine learning models applied. This review highlights state-of-the-art approaches with their performance evaluation. To demonstrate the practical applications and benefits of gait classification through machine learning and wearable sensor, we presented a case study where gait data was collected from 20 young subjects performing six gait activities: walking uphill, walking downhill, walking on flat ground, climbing upstairs, climbing downstairs, and standing. Data was captured using IMU sensors attached to the right shank of each subject. For classification, we implemented several popular ML methods, including Random Forest, XGBoost, Support Vector Classifier with Bagging (SVCB), Convolutional Neural Networks (CNN), Gated Recurrent Neural Networks (GRU), and a hybrid (CNN+GRU) model. Several experiments were carried out to assess the performance of each model based on precision, recall, F1-score, and accuracy, offering valuable insights into the efficacy of various gait classification techniques.

Our experiments demonstrate that deep learning models, particularly CNN, GRU, and the CNN+GRU hybrid model, outperform traditional machine learning models such as Random Forest, XGBoost, and SVC with Bagging in classifying gait activities. All models perform exceptionally well in classifying standing with high precision, recall, and F1-scores, and they also excel in recognizing stair-related activities such as upstairs and downstairs. However, the models struggle with activities like walking, uphill, and downhill, primarily due to the transient nature of the data as participants transitioned between activities, making it difficult to differentiate subtle changes in movements. The CNN+GRU model consistently outperforms other models, with the highest overall accuracy of 80.03%, precision of 82.27%, recall of 81.71%, and F1-score of 81.68%, effectively capturing both spatial and temporal features from the sensor data. GRU alone also performs well, with 79.40% accuracy and an F1-score of 80.64%, indicating its strength in handling sequential dependencies. In contrast, traditional models show lower performance, with Random Forest demonstrating the weakest results, particularly in recall. We also presented numerous open research challenges and potential future directions associated with our case study.

Acknowledgements This work was partially funded by the Department of Bio Technology under project number BT/PR49791/MED/32/914/2023.

References

1. Bora, N.M., Molke, G.V., Munot, H.R.: In: 2015 International Conference on Energy Systems and Applications, pp. 723–728 (2015)
2. Murray, M.: Am. J. Phys. Med. **1**(46), 290 (1967)
3. Hu, K., Wang, Z., Wang, W., Ehgoetz Martens, K.A., Wang, L., Tan, T., Lewis, S.J.G., Feng, D.D.: IEEE Trans. Image Process. **29**, 1890 (2020)
4. Buongiorno, D., Bortone, I., Cascarano, G., Trotta, G., Brunetti, A., Bevilacqua, V.: BMC Med. Inform. Decis. Mak. **19**, 243 (2019)
5. Dobkin, B., Xu, X., Batalin, M., Thomas, S., Kaiser, W.: Stroke; J. Cerebral Circ. **42**, 2246 (2011)

6. Hsieh, C.Y., Shi, W.T., Huang, H.Y., Liu, K.C., Hsu, S.J., Chan, C.T.: In: 2018 IEEE International Conference on Applied System Invention (ICASI) (2018), pp. 818–821
7. Lee, H., Sullivan, S.J., Schneiders, A.: J. Sci. Med. Sport (2012)
8. Ciprandi, D., Zago, M., Bertozzi, F., Sforza, C., Galvani, C.: J. Electromyogr. Kinesiol. **43**, 1 (2018)
9. Elrefaei, L.A., Al-Mohammadi, A.M.: J. King Saud Univ. Comput. Inf. Sci. **34**(2), 204 (2022)
10. Cheng, W.Y., Scotland, A., Lipsmeier, F., Kilchenmann, T., Jin, L., Schjodt-Eriksen, J., Wolf, D., Zhang-Schaerer, Y.P., Garcia, I.F., Siebourg-Polster, J., Soto, J., Verselis, L., Martin-Facklam, M., Boess, F., Koller, M., Grundman, M., Monsch, A., Postuma, R., Ghosh, A., Kremer, T., Taylor, K., Czech, C., Gossens, C., Lindemann, M.: In: 2017 IEEE/ACM International Conference on Connected Health: Applications, Systems and Engineering Technologies (CHASE), pp. 249–250 (2017)
11. Kumari, P., Cooney, N.J., Kim, T.s., Minhas, A.S.: In: 2018 5th Asia-Pacific World Congress on Computer Science and Engineering (APWC on CSE), pp. 1–4 (2018)
12. Margiotta, N., Avitabile, G., Coviello, G.: In: 2016 5th International Conference on Electronic Devices, Systems and Applications (ICEDSA), pp. 1–4 (2016)
13. Whittle, M.W.: Hum. Mov. Sci. **15**(3), 369 (1996)
14. Chaquet, J.M., Carmona, E.J., Fernández-Caballero, A.: Comput. Vis. Image Understand. **117**(6), 633 (2013)
15. Vadivelu, S., Ganesan, S., Murthy, O., Dhall, A.: pp. 541–553 (2017)
16. Zhang, Q., Karunanithi, M., Rana, R., Liu, J.: In: 2013 35th Annual International Conference of the IEEE Engineering in Medicine and Biology Society (EMBC), pp. 7044–7047 (2013)
17. Jokanović, B., Amin, M.: IEEE Trans. Aerosp. Electron. Syst. **54**(1), 180 (2018)
18. Saboor, A., Mustafa, A., Ahmad, R., Khan, M.A., Haris, M., Hameed, R.: In: 9th Annual Information Technology. Electromechanical Engineering and Microelectronics Conference (IEMECON), vol. 2019, pp. 268–272 (2019)
19. Engin, M., Demirel, A., Engin, E.Z., Fedakar, M.: Measurement **37**(2), 173 (2005)
20. Caldas, R., Mundt, M., Potthast, W., Buarque de Lima Neto, F., Markert, B.: Gait Posture **57**, 204 (2017)
21. Prakash, C., Kumar, R., Mittal, N.: Artif. Intell. Rev. **49**(1), 1–40 (2018)
22. Saboor, A., Kask, T., Kuusik, A., Alam, M.M., Le Moullec, Y., Niazi, I.K., Zoha, A., Ahmad, R.: IEEE Access **8**, 167830 (2020)
23. Hutabarat, Y., Owaki, D., Hayashibe, M.: IEEE Sens. J. **21**(23), 26470 (2021)
24. Anguita, D., Ghio, A., Oneto, L., Parra, X., Reyes-Ortiz, J.L.: In: The European Symposium on Artificial Neural Networks (2013)
25. Wang, A., Chen, G., Yang, J., Zhao, S., Chang, C.Y.: IEEE Sens. J. **16**(11), 4566 (2016)
26. Costilla-Reyes, O., Vera-Rodriguez, R., Alharthi, A.S., Yunas, S.U., Ozanyan, K.B.: Deep Learning in Gait Analysis for Security and Healthcare, pp. 299–334. Springer International Publishing, Cham (2020)
27. Wang, J., Chen, Y., Hao, S., Peng, X., Hu, L.: Pattern Recogn. Lett. **119**, 3–11 (2019)
28. Zhang, K., Sun, M., Kevin Lester, D., Xavier Pi-Sunyer, F., Boozer, C.N., Longman, R.W.: J. Biomecha. **38**(11), 2276 (2005)
29. Jiang, W., Yin, Z.: In: Proceedings of the 23rd ACM International Conference on Multimedia (Association for Computing Machinery, New York, NY, USA), MM'15, pp. 1307–1310 (2015)
30. Zebin, T., Scully, P.J., Ozanyan, K.B.: In: 2016 IEEE Sensors, pp. 1–3 (2016)
31. Ronao, C.A., Cho, S.B.: Exp. Syst. Appl. **59**, 235 (2016)
32. Lee, S.M., Yoon, S.M., Cho, H.: In: 2017 IEEE International Conference on Big Data and Smart Computing (BigComp), pp. 131–134 (2017)
33. Murad, A., Pyun, J.Y.: Sensors **17**, 2556 (2017)
34. Almaslukh, B.: Int. J. Comput. Sci. Netw. Secur. **17** (2017)
35. Machine learning based adaptive gait phase estimation using inertial measurement sensors, Frontiers in Biomedical Devices, vol. 2019 Design of Medical Devices Conference
36. Lopez-Nava, I.H., Garcia-Constantino, M., Favela, J.: In: International Conference on Ubiquitous Computing and Ambient Intelligence (2019)

37. Shi, L.F., Qiu, C.X., Dongjin, X., Liu, G.: J. Ambient Intell. Hum. Comput. **11** (2020)
38. Yunas, S.U., Ozanyan, K.B.: IEEE Sens. J. **21**(15), 16870 (2021)
39. Slemenšek, J., Fister, I., Geršak, J., Bratina, B., van Midden, V., Pirtošek, Z., Šafarič, R.: Sensors **23**, 745 (2023)
40. Singh, Y., Vashista, V.: IEEE Trans. Neural Syst. Rehabil. Eng. **30**, 833 (2022)

Data-Driven Grasp Posture Recognition with Data Gloves and Machine Learning Models

Subhash Pratap, Ayush Gupta, Jyotindra Narayan, Kazuaki Ito, and Shyamanta M. Hazarika

Abstract This study investigates the use of machine learning algorithms for classifying various grasp postures based on data collected from a data glove equipped with resistive flex sensors. The glove measures the flexion and extension of the fingers, translating hand movements into electrical signals that are used to train and evaluate five machine learning models: Support Vector Machine (SVM), Decision Tree, Random Forest, k-Nearest Neighbors (KNN), and Logistic Regression. The models were assessed on their accuracy, precision, recall, and computational efficiency. Among the algorithms, KNN and SVM demonstrated the highest classification accuracy, particularly for grasp types such as Cylindrical Grip, Diagonal Volar Grip, and Lateral Pinch. However, the Hook Grip posed a challenge for several models, indicating the need for further improvement in feature extraction and model optimization. While Random Forest provided robust accuracy, its higher computational time made it less suitable for real-time applications. Conversely, Decision Trees offered faster computation but at the expense of lower accuracy. Logistic Regression displayed the weakest performance overall. The results underscore the importance of selecting the appropriate machine learning model based on the specific requirements of the grasp classification task. This research contributes to the development of more accurate and efficient gesture recognition systems, with significant implications for applications in robotics, neurorehabilitation, and assistive technologies where precise and real-time grasp recognition is critical.

Keywords Machine learning · Grasp classification · Data glove · Gesture recognition · Robotics · Neurorehabilitation

---

S. Pratap (✉) · A. Gupta · S. M. Hazarika
Department of Mechanical Engineering, Indian Institute of Technology Guwahati, Guwahati, Assam, India
e-mail: subhash18@iitg.ac.in

S. Pratap · K. Ito
Department of Mechanical Engineering, Gifu University, Gifu, Japan

J. Narayan
Department of Mechanical Engineering, Indian Institute of Technology Patna, Patna, Bihar, India

Department of Computing, Imperial College London, London, UK

J. Narayan and H. Gritli (eds.), *Design and Control of Rehabilitation Robots*, Studies in Systems, Decision and Control 585, https://doi.org/10.1007/978-3-031-86977-8_10

1 Introduction

Gesture recognition using wearable devices plays a pivotal role in advancing intelligent human-computer interaction. Hand gestures, which are integral to daily activities and communication, carry a wealth of information and serve as a key medium for conveying human intentions. Data glove-based gesture recognition technology effectively mitigates environmental interferences like lighting, seamlessly integrates various sensor modalities, and offers a convenient way to capture hand motion data, making it suitable for reliable and portable everyday use. This technology holds significant potential for applications in areas such as virtual reality [11], healthcare [13], and sign language translation [9]. Recent advancements have seen data gloves utilizing diverse sensors, such as strain, inertial, fiber optic, and myoelectric sensors, to capture hand gestures for recognition [17]. For instance, Chen et al. [4] designed a data glove for evaluating hand rehabilitation in stroke patients, while Dias et al. [5] employed multiple sensors to develop a glove for recognizing Brazilian sign language. Despite these advancements, current data gloves often rely on rigid sensors, which can be cumbersome for users, leading to discomfort and poor real-time gesture sensing accuracy. The bulky design also detracts from the overall user experience.

With the rise of flexible electronics, the landscape of gesture recognition has evolved. Flexible sensors offer advantages such as adaptability, scalability, and customizability, making them increasingly important in the realm of human-computer interaction. These devices have demonstrated their utility in accurately capturing gestures [10]. Moreover, research has shown that integrating sensors from various modalities can enhance the performance of gesture recognition systems. For instance, Wu et al. [19] demonstrated that combining accelerometers with sEMG sensors improved recognition accuracy by 15% compared to using accelerometers alone. Pratap et al. [15] recognised grasp types using tactile signals using machine learning techniques. However, despite the success of multi-sensor approaches, many methods process sensor data by directly arranging it into feature vectors or multi-dimensional arrays, often overlooking the complex interactions within and between sensor modalities. This limits the potential for further performance improvements in gesture recognition models.

Data gloves are sophisticated devices designed to capture the motion and forces exerted by the hands and fingers using an array of sensors. These gloves are essential tools in fields like virtual reality, robotics, and medical rehabilitation, enabling detailed human-computer interaction by converting hand gestures into digital commands [6]. A typical data glove consists of a wearable system that integrates sensors, a power source, and electronics for data acquisition and processing. The sensor array, attached to the glove through stitching or adhesive, detects hand movements and converts them into electrical signals, which are then transmitted to a processing module for interpretation [8]. The accuracy of these gloves largely depends on the performance of the sensor array, and various types of sensors have been developed to enhance the precision and functionality of data gloves across different applications.

Dutta et al. [7] focused on classifying grasp gestures using Support Vector Machines (SVM) and achieved an accuracy of 92%, showcasing the potential of SVM in grasp recognition tasks. Chen et al. [4] also employed SVM for grasp gestures, but with a slightly lower accuracy of 89.4%, indicating some limitations in capturing more complex hand movements. Nassour et al. [13] used Random Forest (RF) to classify static gestures, achieving an accuracy of 88.5%, which suggests that while RF is effective, it may struggle with certain nuances of static gesture data. Similarly, Ayodele et al. [1] applied Convolutional Neural Networks (CNN) to grasp gesture classification, attaining an accuracy of 88.3%, demonstrating CNN's capabilities but also its sensitivity to the quality and variety of training data. Finally, Lin et al. [12] employed K-means for customized tasks, with the lowest accuracy of 70.2%, reflecting the challenges of using unsupervised learning methods in gesture classification compared to supervised approaches like SVM and CNN. These works highlight areas where further improvements in sensor data processing and algorithm optimization can enhance performance in gesture recognition tasks.

While the aforementioned studies have demonstrated the potential of various machine learning algorithms in gesture and grasp classification, several gaps remain. Many works [4, 7], achieved high accuracy using SVM, but they primarily focused on static or predefined grasp gestures, limiting their generalizability to more dynamic and complex hand movements. Similarly, approaches employing CNNs [1] and Random Forest [13] showcased strong performance in specific contexts, but these methods often struggle with real-time processing or require extensive computational resources. Moreover, unsupervised techniques like K-means [12], while beneficial for clustering, are not as effective in accurately classifying grasp types when compared to supervised models.

This study investigates the classification of different grasp postures using data collected from a data glove equipped with resistive flex sensors. The data were analyzed using five machine learning algorithms: Support Vector Machine (SVM), Decision Tree, Random Forest, k-Nearest Neighbors (KNN), and Logistic Regression. The performance of these models was evaluated based on their accuracy, precision, recall, and computational time, revealing both their strengths and limitations. KNN and SVM were the most effective classifiers for accurately identifying various grasp postures, especially for grips like the Cylindrical Grip, Diagonal Volar Grip, and Lateral Pinch, which demonstrated near-perfect accuracy. However, the Hook Grip presented challenges in classification, indicating the need for further advancements in feature extraction and model refinement. The novelty of this work lies in the comprehensive evaluation of multiple machine learning algorithms for classifying both static and dynamic grasp postures, providing a broader range of naturalistic grasp types than many previous studies. Additionally, the focus on real-time applicability, through balancing both accuracy and computational efficiency, ensures that the findings are relevant for practical applications in robotics, assistive technologies, and neurorehabilitation. By integrating flex sensor data with optimized machine learning models, this study offers a more adaptive and reliable approach to grasp posture recognition, contributing significantly to advancements in human-computer interaction and robotic control systems.

The structure of this chapter is organized as follows. Section 2 details the methodology employed for the collection, preprocessing, and analysis of data, including the machine learning models and hyperparameter optimization techniques used. Section 3 presents the results and discussion, evaluating the performance of the classifiers based on key metrics and highlighting their implications for grasp classification tasks. Finally, Sect. 4 concludes the study, summarizing the key findings and outlining future directions for enhancing grasp posture recognition systems.

2 Materials and Methods

2.1 Subjects and Objects for Grasping Experiments

This study involved ten healthy participants, carefully chosen to ensure diversity in grasping abilities and to exclude individuals with any known medical conditions affecting hand function. Clear inclusion and exclusion criteria were established to maintain a homogeneous study population and ensure reliable results in the grasp classification tests. Participants aged between 25 and 45 years were included to minimize the impact of age-related differences in grasping abilities. All individuals selected were free from musculoskeletal or neurological conditions that could affect hand performance. Additionally, participants had normal vision and no history of hand injuries or surgeries to eliminate potential confounding variables. The exclusion criteria comprised individuals on medications that might affect motor function, those with a history of substance abuse, and participants with cognitive impairments to avoid any interference with the testing process. All subjects provided written informed consent, following ethical guidelines for research involving human participants as approved by the institutional ethics committee.

In this study, a wide range of grasp behaviors were explored, including Hook Grip (HG), Spherical Grip (SG), Tripod Pinch (TP), Extension Grip (EG), Cylindrical Grip (CG), Diagonal Volar Grip (DVG), Lateral Pinch (LP), and Pulp Pinch (PP). Each grasp type involved three different objects with varying masses and dimensions. The selection of these objects was carefully considered to represent a broad array of grasping scenarios, as shown in Fig. 1. The Yale-CMU-Berkeley (YCB) classification system was employed to account for differences in size, shape, weight, texture, and rigidity. This set of objects reflects a representative variety of those encountered in Activities of Daily Living (ADLs), within the limitations of the available objects from the YCB set [3] as shown in Fig. 1.

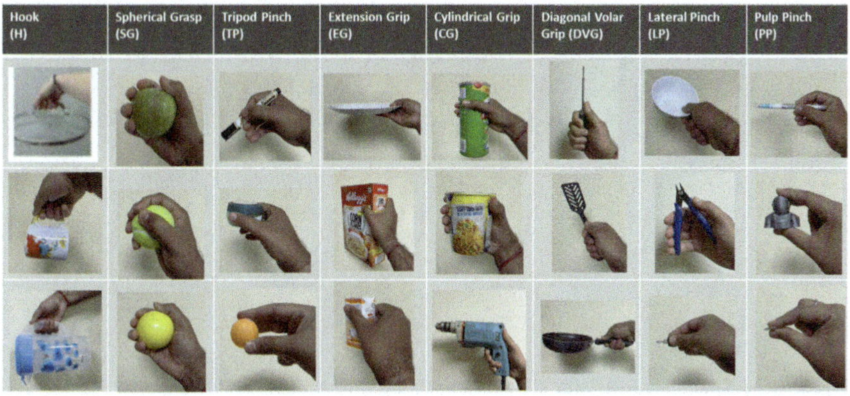

Fig. 1 Eight grasp types with three objects from YCB set of objects used in DLAs [16]

2.2 *Experimental Setup*

2.2.1 Data Glove Hardware for Bending Angle Measurement

The data glove used in this study integrates five flex sensors, positioned along the dorsal side of each finger, to capture real-time data on the range of motion (ROM) during grasping tasks. These sensors accurately reflect the bending of the fingers as they closely align with the glove, which fits snugly on the fingers. However, the glove remains loose enough to ensure that it can be easily donned without requiring excessive effort. Figure 2 presents the schematic diagram of the flex sensors connected to an Arduino microcontroller.

Table 1 outlines the specifications of the flex sensor used in the glove, measuring 4.5 in. in length. Notably, the metacarpophalangeal (MCP) joints in a typical hand operate independently of the distal interphalangeal (DIP) and proximal interphalangeal (PIP) joints, with the latter two joints typically flexing together. As a result, the flex sensors used here cannot capture each joint's individual angle but rather measure overall flexion across MCP, PIP, and DIP joints. For the thumb, the sensor captures a combination of the interphalangeal (IP) and MCP joint angles, as depicted in Fig. 2(i).

The electronic hardware system is based on the open-source Arduino Nano microcontroller, enabling the collection of flexion and extension angles from the five resistive flex sensors, which operate at a standard 5 V. In its resting state (when MCP, PIP, and DIP joints are not flexed), a flex sensor has a resistance of approximately 25 kΩ (Fig. 2). As the finger flexes, the sensor's resistance increases, reaching 100 kΩ at full flexion, corresponding to a 0° bend (Fig. 2). The flex sensors are connected in a voltage divider circuit with a fixed resistor (typically 47 kΩ). One end of the flex sensor is connected to power, while the other end is connected to the pull-down resistor, with the junction between the two components wired to the ADC input of the Arduino. This configuration produces a variable voltage output, which is read by the

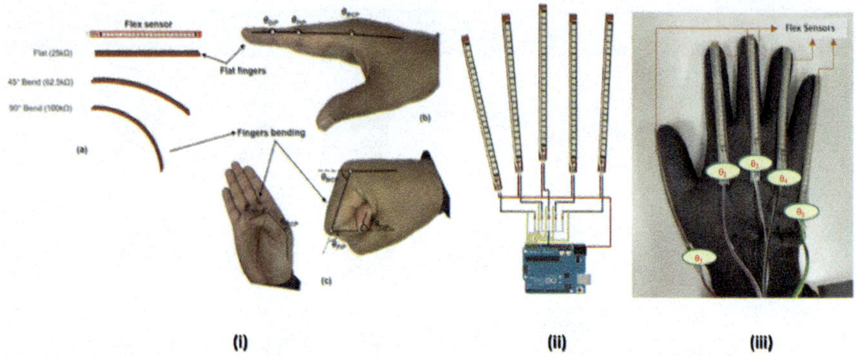

Fig. 2 Schematic diagram of flex sensors connection with Arduino microcontroller

Table 1 Specifications of the flex sensor [18]

Parameter	Value
Length	4.5 in. (11.43 cm)
Flat resistance	10 kΩ
Resistance tolerance	±30%
Thickness	1 mm
Bend resistance range	60–110 kΩ
Power rating	0.50 W continuous, 1 W peak
Life cycle	>1 million

Arduino's ADC input, representing the voltage drop across the pull-down resistor. The output voltage from the voltage divider can be calculated using the following equation:

$$V_o = V_{cc}\frac{R}{(R + R_{flex})}$$

where V_o is the output voltage, V_{cc} is the input voltage, R_{flex} is the resistance of the flex sensor, and R is the resistance of the pull-down resistor. As the flex sensor bends, R_{flex} increases, causing V_o to decrease. For example, with a 5 V supply and a 47 kΩ pull-down resistor, V_o changes significantly as the flex sensor bends from 0° (low resistance) to 90° (high resistance).

Following the setup, the Arduino was programmed to capture changes in bending angles and sensor resistance. To calibrate the sensor, resistance values were recorded at 0° and 90°. The map function was used to adjust the sensor's readings based on these values, allowing accurate conversion of resistance to degrees of bend. The sensor's performance was verified at different angles to ensure proper functioning.

2.2.2 Software Description

The software is divided into two main components: one for acquiring sensor data and the other for storing this data. The Arduino microcontroller is used to print the sensor data to the serial monitor, with each value separated by a comma to display the readings from all the flex sensors. A code snippet demonstrating how the Arduino program determines the mode is provided. The flex function records the angle in a variable by calculating the resistance based on the voltage from the sensor. The flex sensor's resistance varies with the degree of bending, and this resistance is then converted to corresponding angles. Over time and after multiple trials, the sensor's resistance in the normal zero state may change. To address this issue, the sensors are calibrated by inputting the resistance value for the zero bending angle into the Arduino code.

2.2.3 Data Collection

The experimental setup was designed to ensure replicable research and effective benchmarking by using a standardized set of objects and models. Protocols and benchmarks were developed to define the experimental procedures, addressing the diverse approaches in manipulation research. Factors such as object shape, size, weight, grasp gesture, and stiffness were considered when designing the experiment. Data was collected through repeated trials where subjects, wearing a data glove, grasped various daily objects. Each trial lasted 30 s and was recorded by a camera positioned 60 cm away. Subjects were instructed to move their gloved hand from a resting position, grasp the object, lift it 10 cm, and hold it steady for 20 s. Each trial was divided into four phases: approaching the object, touching it for grasping, lifting it, and holding it steady until the sensor readings gets stabilized. Some inconsistencies were noted when grasps were made without involving the fingertips, which was corrected in subsequent trials. Data was collected for 24 different objects with 10 trials per object, including bending angles (in degrees) for all five fingers (thumb, index, middle, ring, and pinky). The angle data contained 5 channels with 1200 steps over the 30 s.

2.2.4 Machine Learning Algorithms for Grasp Classification

The classification of different grasp types was carried out using several machine learning algorithms. The models implemented in this study include Decision Trees, Random Forests, Support Vector Machines (SVM), Logistic Regression and k-Nearest Neighbors (KNN) [2, 14–16].

Decision Trees

Decision Trees provide a transparent and interpretable method for classification tasks, including grasp classification. The model works by recursively splitting the dataset based on feature values, creating a tree-like structure. Each node in the tree represents a decision rule based on a feature, as shown below:

$$\text{if (Feature} \leq \text{Threshold)} : \text{Go to left branch} \tag{1}$$

$$\text{else} : \text{Go to right branch} \tag{2}$$

Here, *Feature* refers to a specific attribute of the data, and *Threshold* is the value used to split the data at each node.

Random Forest

Random Forests enhance the robustness of Decision Trees by employing ensemble learning. Instead of relying on a single tree, Random Forest constructs multiple decision trees, each trained on a random subset of the data. The final prediction is based on the majority vote from all the trees:

$$\text{Prediction} = \frac{1}{N} \sum_{i=1}^{N} DecisionTree_i(\text{Input}) \tag{3}$$

Here, N is the number of decision trees, and $DecisionTree_i$ represents the output of the i-th tree.

Support Vector Machines (SVM)

SVM works by identifying a hyperplane that best separates different classes in the dataset. The decision boundary is mathematically defined as:

$$w^T x + b = 0 \tag{4}$$

In this expression, w represents the weight vector, x is the input data vector, and b is the bias term that shifts the hyperplane.

Logistic Regression

Logistic Regression is a linear model used for binary classification tasks, where the probability of a data point belonging to a specific class is modeled using a sigmoid function. The logistic function is represented as:

$$P(y = 1|x) = \frac{1}{1 + e^{-(w^T x + b)}} \tag{5}$$

In this equation, w is the weight vector, x is the input data vector, and b is the bias term. Logistic Regression is particularly useful for binary classification, such as differentiating between two grasp types.

k-Nearest Neighbors (KNN)

KNN is a simple classification algorithm that assigns a class to a data point based on the majority class of its nearest neighbors. The prediction is determined by the majority class among the k-nearest neighbors:

$$\text{Prediction} = \text{majority class among } k\text{-nearest neighbors} \tag{6}$$

Here, k is the number of neighbors considered for classification.

While selecting a classification technique for grasp analysis, several factors need to be considered, including the interpretability of the model, computational efficiency, and the desired level of predictive accuracy for the specific application. Each classification method offers distinct advantages and limitations, and the choice of algorithm largely depends on the characteristics of the grasp classification task being addressed. While some techniques excel in transparency and ease of understanding, others may offer greater accuracy or handle complex data patterns more effectively but at the cost of increased computational complexity. In this study, the emphasis was placed on evaluating the performance of individual classifiers to determine their suitability for grasp classification. As a result, we opted not to use ensemble methods, which, although capable of improving accuracy through the combination of multiple models, typically require greater computational resources and are more complex to implement. Instead, we focused on simpler, more interpretable models that are computationally efficient and easier to deploy in real-time applications, such as robotic grasping or assistive devices, where quick decision-making is essential.

2.2.5 Performance Matrices

The dataset, consisting of flex sensors readings was divided into training and testing sets. Machine learning models were trained on a portion of the data to identify patterns corresponding to different grasp types. After training, the models were evaluated on a separate set of data to assess their classification accuracy. Key performance metrics such as accuracy, precision, recall, and F1-score were used to quantitatively evaluate the effectiveness of each classification algorithm (see Table 2). To ensure a thorough evaluation of the models, cross-validation techniques were rigorously applied, and hyperparameters were fine-tuned to improve generalization to unseen data. Hyperparameter tuning was conducted using a two-step approach. Initially, a manual exploration was performed by testing 3–4 key parameters with 3–4 values each for each model, using training data for fitting and testing data for evaluation. This exploratory phase helped identify promising parameter ranges. Subsequently, GridSearchCV was applied with 10-fold cross-validation ($n_{\text{splits}} = 10$) on the refined parameter set to systematically evaluate all combinations and determine the optimal hyperparameters. This approach ensured both computational efficiency and robust evaluation.

Table 2 Performance metrics

Metric	Formula	Description
Precision	$\frac{TP}{TP+FP}$	Ratio of correctly predicted positive instances to total predicted positives
Recall	$\frac{TP}{TP+FN}$	Ratio of correctly predicted positive instances to total actual positives
F1 score	$\frac{2\cdot(Precision\cdot Recall)}{Precision+Recall}$	Harmonic mean of Precision and Recall, balancing the two measures

The overall approach was carefully structured to thoroughly capture and examine the intricate relationship between flex sensors readings of different grasp types, enabling the creation of a reliable and accurate grasp classification system. To achieve this, machine learning techniques were used to classify postures derived from the grasping trials. Initially, the raw data from the sensors was filtered to remove any noise. Next, the data sequences were standardized to ensure uniform length. The standardized sequences were then concatenated to form a feature vector. Of these extracted features, 60% were used for model training, while the remaining 40% was reserved for testing the classification performance.

3 Results and Discussions

3.1 Performance Evaluation of Grasp Classification Models

In our approach, we utilized machine learning algorithms to classify different types of grasps. We implemented five distinct machine learning models to categorize grasps using the grasp posture data collected. This process included training the models followed by assessing their performance. This section synthesizes findings from the literature, discussing the implications and limitations of current data glove technologies. Table 3 provides an overview of the classification accuracies and detailed parameters. During the experiment, data from two of the three objects acquired for

Table 3 Performance evaluation of grasp classification models

Classifier	Accuracy	Precision	Recall	F1 score	Time (s)
SVM	0.92 ± 0.06%	0.9173	0.9167	0.9163	0.4056
Decision tree	0.85 ± 0.08%	0.8560	0.8500	0.8514	0.3041
Random forest	0.88 ± 0.07%	0.8802	0.8792	0.8789	3.6699
KNN	0.93 ± 0.06%	0.9269	0.9208	0.9179	0.4886
Logistic regression	0.82 ± 0.07%	0.8495	0.8500	0.8468	12.0411

each grasp type were used for training, while the data from the remaining object was reserved for validation.

The evaluation of five different classifiers-SVM, Decision Tree, Random Forest, KNN, and Logistic Regression-for grasp classification reveals distinct strengths and weaknesses in their performance metrics. **KNN** emerged as the top performer with an impressive accuracy of 93% and precision of 0.9269. This indicates that KNN effectively identifies grasp types with a high degree of correctness while maintaining a reasonable computational time of 0.4886 s. Its strong performance suggests that KNN can adeptly handle the grasp features associated with various grasps, making it suitable for real-time applications where quick and reliable classification is essential. **SVM**, with an accuracy of 92%, demonstrated similar efficacy to KNN, showcasing balanced performance across precision, recall, and F1 score metrics. Its slightly faster computation time of 0.4056 s further emphasizes its utility in scenarios where efficiency is critical, such as in robotics or assistive technology, where decisions need to be made rapidly. In contrast, **Random Forest** yielded a commendable accuracy of 88%. However, it required significantly more computation time at 3.6699 s, which may limit its practicality in real-time applications. Despite this drawback, Random Forest's ability to manage complex datasets and capture intricate relationships among features makes it a strong candidate for grasp classification tasks where accuracy is prioritized over speed. The **Decision Tree** classifier exhibited the fastest computation time at 0.3041 s but fell short in terms of accuracy, achieving only 85%. This highlights the trade-off between speed and accuracy, as Decision Trees may simplify the decision-making process but can also miss subtleties in the data that lead to misclassification in certain grasp types. Finally, **Logistic Regression** was the least effective of the classifiers, with an accuracy of 82% and the longest computation time of 12.0411 s. While Logistic Regression can provide a baseline for understanding grasp classifications, its performance indicates that it may not be the ideal choice for this specific application, particularly when more complex relationships within the data need to be captured (Fig. 3).

Overall, these findings underscore the importance of selecting an appropriate classification algorithm based on the specific requirements of the grasp classification task. KNN and SVM stand out as the most effective choices for scenarios requiring a balance of accuracy and computation speed, while Random Forest offers robust performance at the cost of longer processing times. In contrast, Decision Trees may serve well in rapid decision-making situations but could sacrifice accuracy, and Logistic Regression might be better suited for simpler classification tasks. Thus, the choice of classifier should be informed by the desired balance between speed, accuracy, and the complexity of the task at hand.

The accuracy of grasp posture classification for different grasp types utilizing various machine learning models is illustrated in Fig. 4. The classification accuracies for various grasp types provide critical insights into the effectiveness of different machine learning models in recognizing and categorizing hand postures. The Cylindrical Grip (CG), Diagonal Volar Grip (DVG), and Lateral Pinch (LP) achieved exceptional accuracy, reaching 100% with both SVM and KNN classifiers. This perfect classification indicates that these grips possess distinctive features that can be reliably

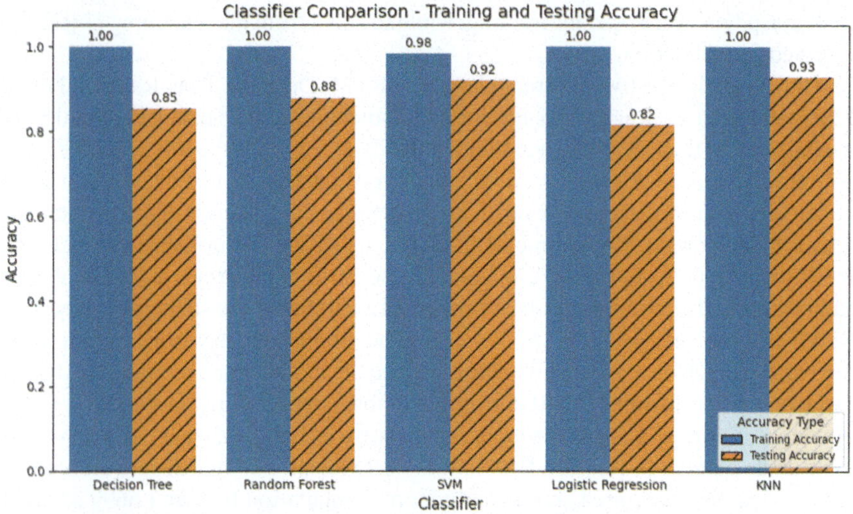

Fig. 3 Classifiers accuracy comparison: training and testing accuracy

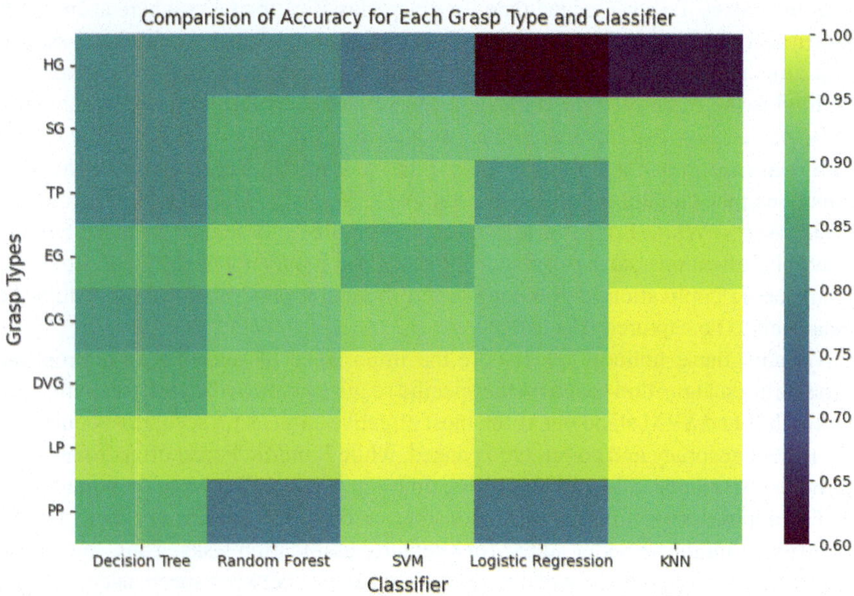

Fig. 4 Comparison of accuracy for each grasp type and classifier

identified by the algorithms, making them highly suitable for applications in robotic manipulation and assistive technologies, where precise grasping is paramount. In contrast, the Spherical Grip (SG) and Extension Grip (EG) also demonstrated robust performance, with KNN achieving 93.4% accuracy for SG and 96.6% for EG. This suggests that these grips, like CG, have well-defined characteristics that facilitate accurate recognition, underscoring their potential for integration into systems that rely on nuanced hand movements. However, the Hook Grip (HG) showed the lowest classification accuracies, particularly with KNN and Logistic Regression, at 63.4% and 60.0%, respectively. This discrepancy indicates that the features associated with HG may be less distinctive or that the models struggle to capture the variability inherent in this grasp type. It highlights a significant area for improvement, suggesting the need for enhanced feature extraction techniques or additional training data to bolster the classifier's performance on less discernible grasps. The performance variability across classifiers emphasizes the importance of selecting the right algorithms for grasp recognition tasks. SVM and KNN consistently outperformed Decision Trees, suggesting their ability to handle complex patterns in data more effectively. Additionally, the results reveal that while some grasp types can be classified with high confidence, others require further research and refinement to achieve similar levels of accuracy. These findings not only showcase the strengths of specific grips in assistive technologies but also indicate critical avenues for future research to enhance the robustness of grasp posture recognition systems, potentially leading to improved user experiences in applications such as rehabilitation and human-robot interaction.

3.2 Performance Evaluation Metrics

The evaluation utilized a confusion matrix, which is an essential instrument for analyzing the accuracy and effectiveness of each classification model. Each model's evaluation heavily depended on this matrix. Table 2 provides the equations that detail how the performance metrics were computed. To visually depict the classification results, Fig. 5 presents the confusion matrices for grasp classification across different classifiers. The elements along the diagonal represent correct classifications, while the off-diagonal elements highlight misclassifications, reflecting the identification probabilities and corresponding errors. Overall, the classifiers demonstrated a strong ability to accurately classify grasp types for the majority of objects.

The analysis of confusion matrices for each classifier reveals significant insights into the classification performance of the eight grasp types, with values represented as percentages for clarity. The Support Vector Machine (SVM) demonstrates commendable accuracy, correctly classifying the Hook Grip (HG) at 76.3% and the Spherical Grip (SG) at 90%, though it shows some difficulty with the Tripod Pinch (TP) at 92.4%. The Decision Tree classifier exhibits a more variable performance, accurately identifying the HG, SG, and TP at 80%, but with a notable dip for the Cylindrical Grip (CG) at 83.4%, suggesting challenges in capturing subtle finger positioning nuances. In contrast, the Random Forest classifier stands out with robust performance across all grasp types, achieving 80% accuracy for the HG and 86.6% for the SG, while excelling with the TP and Extension Grip (EG) at 93.4%, indicating its strength in handling complex grasp patterns. The K-Nearest Neighbors (KNN) classifier excels with 100% accuracy for the EG and CG grips, although it

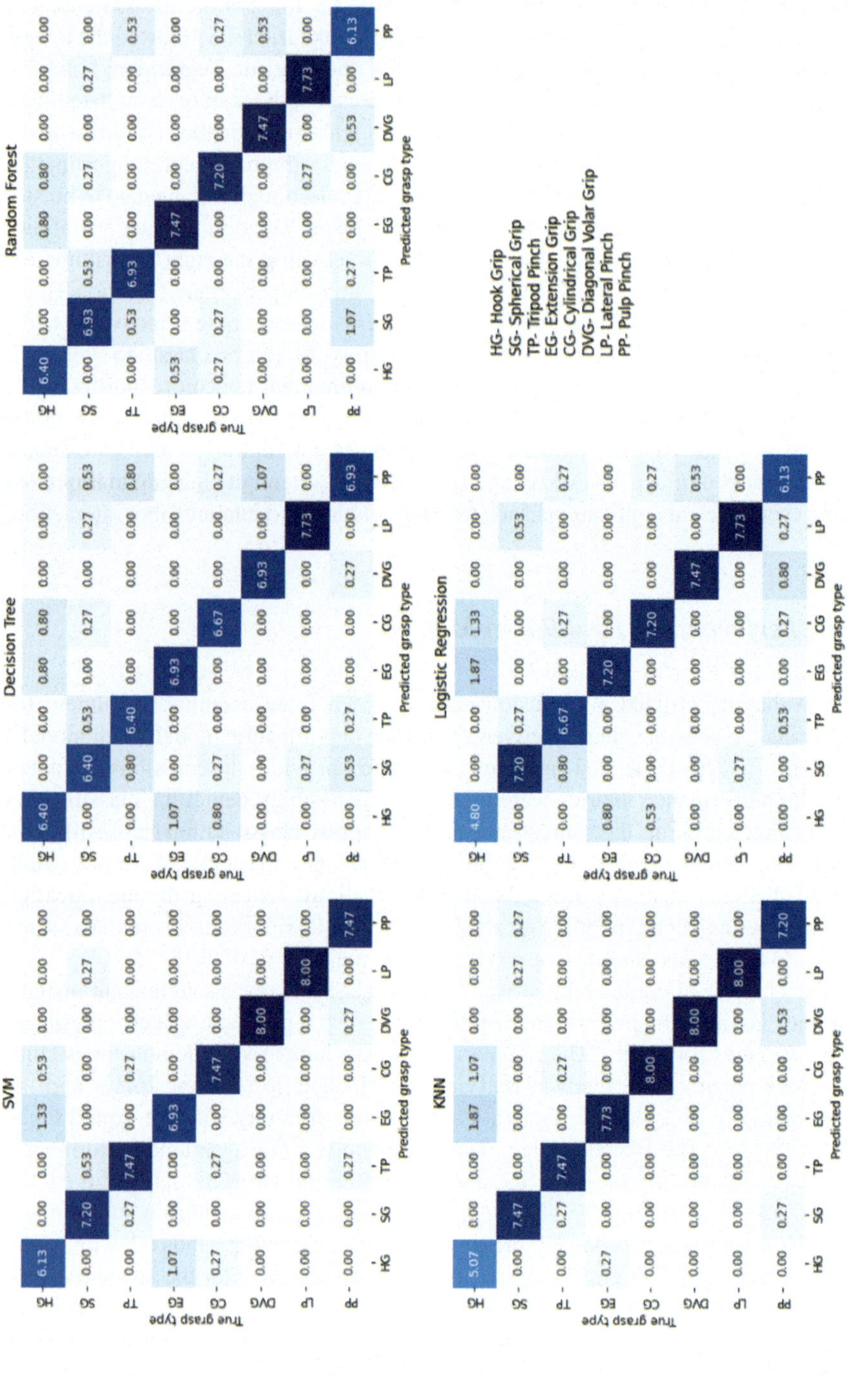

Fig. 5 Confusion matrix for the five classifiers

shows sensitivity to variations, achieving only 63.4% for the HG. Lastly, the Logistic Regression model performs the weakest, particularly struggling with the HG at 60% accuracy, emphasizing the limitations of linear decision boundaries for this classification task. Collectively, this analysis underscores the importance of model selection tailored to the characteristics of grasping postures, highlighting that the SVM, Random Forest, and KNN classifiers are the most effective for accurate identification, while the Decision Tree and Logistic Regression models exhibit limitations that need addressing. This comprehensive understanding is vital for advancing research in grasp posture classification, especially in robotics and assistive technology, where precise hand movement identification is crucial for effective human-robot interaction.

4 Conclusion

This study explored the performance of various machine learning models-SVM, Decision Tree, Random Forest, KNN, and Logistic Regression-for the classification of different grasp postures using data gloves. Among these, KNN and SVM stood out as the most effective models, demonstrating superior classification capabilities across a range of grasp types. However, certain grips, such as the Hook Grip, presented challenges for several classifiers, highlighting the need for further refinement in feature extraction and model development. Random Forest, while robust in accuracy, was less suited for real-time applications due to its higher computational demands. Decision Trees offered quick decision-making but at the cost of lower classification precision. The findings of this study underscore the importance of selecting the appropriate machine learning model based on the specific requirements of grasp classification tasks, balancing accuracy and speed. This research contributes valuable insights into enhancing grasp posture recognition, with implications for improving the design and application of robotic systems, assistive technologies, and neurorehabilitation tools, where precise and real-time grasp recognition is essential.

References

1. Ayodele, E., Bao, T., Zaidi, S.A.R., Hayajneh, A.M., Scott, J., Zhang, Z.Q., McLernon, D.: Grasp classification with weft knit data glove using a convolutional neural network. IEEE Sens. J. **21**(9), 10824–10833 (2021)
2. Briouza, S., Gritli, H., Khraief, N., Belghith, S., Singh, D.: A brief overview on machine learning in rehabilitation of the human arm via an exoskeleton robot. In: 2021 International Conference on Data Analytics for Business and Industry (ICDABI), pp. 129–134. IEEE (2021)
3. Calli, B., Walsman, A., Singh, A., Srinivasa, S., Abbeel, P., Dollar, A.M.: Benchmarking in manipulation research: using the Yale-CMU-Berkeley object and model set. IEEE Robot. Autom. Mag. **22**(3), 36–52 (2015)
4. Chen, X., Gong, L., Wei, L., Yeh, S.C., Da Xu, L., Zheng, L., Zou, Z.: A wearable hand rehabilitation system with soft gloves. IEEE Trans. Ind. Inform. **17**(2), 943–952 (2020)

5. Dias, T.S., Júnior, J.J.A.M., Pichorim, S.F.: An instrumented glove for recognition of Brazilian sign language alphabet. IEEE Sens. J. **22**(3), 2518–2529 (2021)
6. Dipietro, L., Sabatini, A.M., Dario, P.: A survey of glove-based systems and their applications. IEEE Trans. Syst. Man Cybern. Part C (Appl. Rev.) **38**(4), 461–482 (2008)
7. Dutta, D., Aruchamy, S., Mandal, S., Sen, S.: Poststroke grasp ability assessment using an intelligent data glove based on action research arm test: development, algorithms, and experiments. IEEE Trans. Biomed. Eng. **69**(2), 945–954 (2021)
8. Guo, L., Lu, Z., Yao, L.: Human-machine interaction sensing technology based on hand gesture recognition: a review. IEEE Trans. Human-Mach. Syst. **51**(4), 300–309 (2021)
9. Huang, X., Wang, Q., Zang, S., Wan, J., Yang, G., Huang, Y., Ren, X.: Tracing the motion of finger joints for gesture recognition via sewing RGO-coated fibers onto a textile glove. IEEE Sens. J. **19**(20), 9504–9511 (2019)
10. Lee, M., Bae, J.: Deep learning based real-time recognition of dynamic finger gestures using a data glove. IEEE Access **8**, 219923–219933 (2020)
11. Li, J., Liu, X., Wang, Z., Zhang, T., Qiu, S., Zhao, H., Zhou, X., Cai, H., Ni, R., Cangelosi, A.: Real-time hand gesture tracking for human-computer interface based on multi-sensor data fusion. IEEE Sens. J. **21**(23), 26642–26654 (2021)
12. Lin, B.S., Hsiao, P.C., Yang, S.Y., Su, C.S., Lee, I.J.: Data glove system embedded with inertial measurement units for hand function evaluation in stroke patients. IEEE Trans. Neural Syst. Rehabil. Eng. **25**(11), 2204–2213 (2017)
13. Nassour, J., Amirabadi, H.G., Weheabby, S., Al Ali, A., Lang, H., Hamker, F.: A robust data-driven soft sensory glove for human hand motions identification and replication. IEEE Sens. J. **20**(21), 12972–12979 (2020)
14. Pratap, S., Hatta, Y., Ito, K., Hazarika, S.M.: Understanding Grasp Synergies During Reach-to-Grasp Using an Instrumented Data Glove (2024). arXiv:2405.19430
15. Pratap, S., Narayan, J., Hatta, Y., Ito, K., Hazarika, S.M.: From tactile signals to grasp classification: exploring patterns with machine learning. In: 2024 IEEE International Conference on Interdisciplinary Approaches in Technology and Management for Social Innovation (IATMSI), vol. 2, pp. 1–6. IEEE (2024)
16. Pratap, S., Narayan, J., Hatta, Y., Ito, K., Hazarika, S.M.: Glove-Net: Enhancing grasp classification with multisensory data and deep learning approach. Sensors **24**(13), 4378 (2024)
17. Sierotowicz, M., Lotti, N., Nell, L., Missiroli, F., Alicea, R., Zhang, X., Xiloyannis, M., Rupp, R., Papp, E., Krzywinski, J., et al.: EMG-driven machine learning control of a soft glove for grasping assistance and rehabilitation. IEEE Robot. Automat. Lett. **7**(2), 1566–1573 (2022)
18. Symbol, S.: Flex Sensor Datasheet v2019a (2019). https://spectra-symbol-landing.s3.us-west-1.amazonaws.com/data-sheets/Flex-Sensor-Datasheet-v2019a.pdf
19. Wu, J., Sun, L., Jafari, R.: A wearable system for recognizing American sign language in real-time using IMU and surface EMG sensors. IEEE J. Biomed. Health Inform. **20**(5), 1281–1290 (2016)

Predicting Lower Limb Joint Movements from Muscle Signals Using Neural Networks: A Comparative Study of Training Optimizers

Anugya Tripathi, Shiv Manjaree Gopaliya, and Jyotindra Narayan

Abstract This chapter explores the estimation of lower limb joint angles from muscle signals in able-bodied adults across various locomotion modes. The study utilizes an existing open-access dataset, containing motion data of lower limb joints and surface EMG signals from 11 right lower limb muscles of 22 able-bodied adults across different terrains and locomotion conditions. The chapter is organized around two primary objectives. The first objective is to analyze motion and EMG signals by extracting data from the dataset, selecting data from 10 subjects engaged in treadmill walking at 1.3 m/s speed, and preprocessing it for consistency and compatibility for subsequent analysis and intelligent model training. The second objective involves using the preprocessed data to train backpropagation neural network models with three training optimizers, Levenberg-Marquardt (LM), Bayesian Regularization (BR) and Scaled Conjugate Gradient (SCG) to estimate the desired joint angles during treadmill walking. The performance of the trained models is then assessed by comparing their predictions with actual joint angle measurements from the dataset, using evaluation metrics such as mean absolute error or root mean square error. The performance of the trained neural network was evaluated using one test subject for all three methods. The LM method achieved the best performance with an MSE of 83.2474 at epoch 103 and $R > 0.91$. The BR method had an MSE of 140.3847 at epoch 124 with $R > 0.84$. SCG method showed the best performance, with an MSE of 171.0699 at epoch 1000 and $R > 0.81$. This analysis interprets the results to provide insights into the relationship between muscle signals and joint angles during treadmill walking, and to apply these findings to the dynamics and control of lower-limb rehabilitation robots.

Keywords Lower limb joint angles · Treadmill walking · EMG signals · Neural networks

---

A. Tripathi · S. M. Gopaliya (✉)
School of Mechanical Engineering, VIT Bhopal University, Bhopal, India
e-mail: shivmanjaree@vitbhopal.ac.in

J. Narayan
Department of Mechanical Engineering, Indian Institute of Technology Patna, Patna, Bihar, India

Department of Computing, Imperial College London, London, UK

J. Narayan and H. Gritli (eds.), *Design and Control of Rehabilitation Robots*, Studies in Systems, Decision and Control 585, https://doi.org/10.1007/978-3-031-86977-8_11

1 Introduction

Gait analysis is crucial in understanding human locomotion, essential for developing rehabilitation strategies and designing assistive devices, such as exoskeletons and prosthetics. With over 1 billion people worldwide suffering from locomotion impairment, effective tools for analyzing gait are indispensable. Gait analysis aids in diagnosing movement disorders, optimizing athletic performance, and providing data-driven insights for rehabilitation. Moreover, precise gait patterns are essential for developing controllers that ensure smooth and safe movementin robotic and prosthetic design. Park et al. have provided a comprehensive review investigating the effectiveness of various robot-assisted gait training systems for stroke patients focusing on enhancing gait and balance movements [1]. Baker has investigated various methodologies required for gait analysis in the context of clinical rehabilitation [2]. The role of ankle push-off have been extensively studied, which helps in the development of clinical devices to restore motion for patients with disabilities [3].

Predictive gait analysis extends traditional gait analysis by not only assessing current motion but also forecasting future movement patterns. This capability is critical in the fields of wearable robotics and rehabilitation, where real-time prediction enables exoskeletons or prosthetics to assist users proactively. Using electromyography (EMG) data to predict joint angles can improve the responsiveness and adaptability of assistive devices. Additionally, predictive models allow for early intervention, potentially reducing injury risks and improving outcomes in physical therapy. To address the scarcity of human gait data, Wang et al. have developed a Generative Adversarial Network-based data augmentation model to generate synthetic human gait data while retaining the real gait data [4]. One study utilized a long short-term memory-based network for gait analysis of lower limb trajectory, while another predicted lower-limb kinematics and kinetics using EEG and EMG signals [5, 6]. The exhaustive review found that no such studies were carried out for walking or stepping motion revealing multiple future research directions. An exoskeleton design via biomechanical modeling using CAD software was validated [7]. A systematic review of intelligent techniques used for gait analysis, particularly for lower limb robotic systems was also carried out [8]. The study highlighted that gait analysis has been carried out using classification methods for gait phases, regression for kinematic study and deep learning techniques as well.

Machine learning and deep learning methods have also shown promise in predictive gait analysis. Techniques such as Random Forests, Support Vector Machines (SVMs), and Convolutional Neural Networks (CNNs) have been employed to predict joint movements using EMG data. Studies have demonstrated the efficacy of these models in identifying patterns and predicting joint angles, although challenges remain in real-time processing and accuracy when dealing with dynamic environments. Integrating these methods into wearable devices is an ongoing area of research, aiming for continuous, real-time predictions with minimal latency. An exhaustive review has been carried out and new ways were explored to identify the potential of deep learning approaches to control lower limb exoskeletons [9]. The review offers valuable insights into the application of deep learning for controlling lower limb exoskeletons, fostering advancements in rehabilitation and assistive technologies for the future. Five different deep-learning models have been developed and evaluated for various activities of wearable exoskeletons [10]. The results presented highlighted the possibility of real-time Human Activity Recognition for defined tasks. A novel lower limb exoskeleton actuated by a variable magnification ratio structure and its control strategy is proposed [11]. The proposed structure shows the flexibility and portability of the exoskeleton system under heavy load carrying conditions. Sensing and control of lower limb exoskeletons by carrying out in-depth analysis using data-driven deep learning techniques, sensing, and actuation principles [12]. Woo et al. developed and analyzed an exoskeleton structure to provide better physical assistance for laborers working at industrial sites [13]. Experiments were carried out using bio-signal measurement sensors and artificial intelligence to predict human motion for proactive control of exoskeleton structures. A systematic review of various deep-learning approaches for biomechanics using wearable sensors was carried out [14]. A demonstration of the feasibility of a machine learning model to effectively monitor the daily ambulatory activity of patients [15].

Neural networks, particularly Feedforward Backpropagation Neural Networks (FFBPNNs) and Radial Basis Function (RBF) networks have been employed in gait prediction models. These models utilize EMG signals to estimate joint angles with high accuracy. The advantage of neural networks lies in their ability to model the nonlinear relationships between muscle activation and joint movement. In predictive gait analysis, neural networks have been shown to outperform traditional statistical models, offering better generalization to unseen data and real-time adaptability. Neural networks has been applied to enhance the prediction accuracy of the exoskeleton while level ground walking [16]. The results showed good prediction performance with less processing time. Narayan et al. have used BR based backpropagation neural networks to estimate lower limb joint torques [17]. The results demonstrated effective prediction accuracy of the proposed model. An algorithm for gait rehabilitation while addressing the difficulty of estimating angles of lower extremity patients using torque and EMG signals was proposed [18]. An ANN model to improve balance for a three-degree-of-freedom exoskeleton and a knee joint of the lower limb exoskeleton, respectively have been proposed [19, 20]. Li et al. have developed hybrid neural network models for hip and knee joints [21]. Multiple experiments have been carried out to understand the human-robot interaction for rehabilitation robots. An exhaustive state-of-the-art literature review for lower limb exoskeleton sensors has been carried out [22]. A deep-learning and neural network-based approaches were proposed to facilitate the adoption of exoskeleton orthotic devices in real-life situations [23]. Adaptive fuzzy and neural network-based approaches have been proposed [24, 25]. Experiments show the successful implementation of the proposed approach by obtaining robot joint torques irrespective of changes in walking speeds. Koceska et al. have proposed an adaptive fuzzy controller based on the 10DOF exoskeleton for the gait rehabilitation of patients [26]. A force-tracking algorithm was designed on a partially actuated lower limb exoskeleton [27].

Laschowski et al. have predicted the oncoming environment before physical interaction by using computer vision and deep learning algorithms [28]. A bionic design for lower limb using muscle activity was proposed, while muscle load reduction while wearing an exoskeleton was also performed successfully [29, 30]. The study recommended wearing an exoskeleton to avoid injury. Multiple control strategies for exoskeleton structures, including cooperative control and multi-information fusion have been extensively reviewed [31]. Extensive research has been carried out on upper limb exoskeletons to enhance assistive and rehabilitative capabilities across diverse applications. An overview of upper-limb exoskeletons and biomedical signal classification using ML approaches for arm rehabilitation [32]. A Convolutional Recurrent Neural Network (CRNN) model has been proposed for human hand rehabilitation [33]. The study highlighted the effective use of CRNN model to classify EMG signals for prosthetic arm. CNN and ANN models were used to optimize and classify EMG signals [34, 35]. The study found that combining the CNN's powerful architecture with the reliable SVM algorithm yielded encouraging outcomes.

Despite advancements in predictive gait analysis using machine learning (ML) and artificial neural network (ANN) models, several limitations remain. Limited studies have applied ANN models with three optimizers—LM, BR, and SCG—to forward and inverse kinematics of multi-degree-of-freedom robotic manipulators [36–39]. These approaches achieved optimal MSE results, highlighting the importance of designing appropriate models. Moreover, while EMG signals have been extensively used in predictive models, their potential remains underutilized. EMG data provides real-time muscle activation signals that can improve the accuracy of joint angle predictions, especially in dynamic tasks like walking, running, or climbing stairs. However, only a few studies have integrated EMG as a primary input for ANN models, limiting the generalizability and adaptability of these models for various movement tasks. A significant challenge in this study is achieving real-time performance, as many models face high computational complexity when processing large datasets or multiple inputs like EMG signals and joint kinematics. LM improves convergence speed and accuracy, particularly for small datasets. BR prevents overfitting by balancing model complexity, enhancing robustness. SCG optimizes computational efficiency, making it suitable for larger datasets while maintaining prediction accuracy. Different optimizers in back propagation neural networks for gait analysis have been successfully implemented [40, 41].

This chapter investigates the estimation of lower limb joint angles from muscle signals during various locomotion modes in able-bodied adults using an open-access dataset. Muscle signals, also referred to as electromyography (EMG) signals, are electrical signals generated by muscle fibers during contraction and relaxation. These signals are recorded using sensors and provide valuable information about muscle activity, enabling applications like movement analysis, rehabilitation, and control of assistive devices such as exoskeletons. The study focuses on preprocessing EMG and motion data from treadmill walking and training backpropagation neural networks with different optimizers to predict joint angles. LM, BR, and SCG optimizers were chosen for their distinct strengths in training neural networks for joint angle estimation. LM offers fast convergence for non-linear problems, BR prevents overfitting for noisy biomechanical data, and SCG ensures computational efficiency for larger datasets. Their selection ensures accurate, robust, and efficient exoskeleton control during treadmill walking. Performance evaluation results highlight the LM optimizer as the most effective. The findings provide insights into the relationship between muscle signals and joint dynamics, offering applications in lower-limb rehabilitation robotics.

2 Data Involved

In this work, an existing open access treadmill walking data from Camargo et al. for twenty-two subjects with multiple trials is considered [42]. The data were recorded for 28 speeds (0.5–1.85 m/s in 0.05 m/s increments) across 7 trials, with subjects transitioning through varying speeds to capture steady-state walking and acceleration data. EMG data were sampled at 1000 Hz and filtered for lower limb muscles, while IMU data were collected from torso and limb segments at 200 Hz. Gait and joint data were processed with motion capture (200 Hz) and ground reaction forces (1000 Hz), using OpenSim tools for joint angle, moment, and power analyses. Filtering and calibration methods ensured high data accuracy, with rest periods provided to prevent fatigue.

In this Chapter, 10 subjects have been randomly selected for the study. The selected subjects are walking on a treadmill at a speed of 1.3 m/s. The complete details of the 10 subjects considered in this study are given in Table 1. Based on the experimental data presented in Camargo et al, EMG signals were recorded with surface electrodes placed on eleven muscles of the right leg: gluteus medius, right external oblique, semitendinosus, gracilis, biceps femoris, rectus femoris, vastus lateralis, vastus medialis, soleus, tibialis anterior, and gastrocnemius medialis [42]. In the present study, EMG muscle signals have been taken as input data while joint angles act as output data. The 7 EMG signals taken into consideration are: gastrocnemius medialis, tibialis anterior, vastusmedialis, vastuslateralis, rectus femoris, biceps femoris and semitendinosus. The 3 output joint angles considered in this work are: hip flexion, knee angle, and ankle angle.

Root Mean Square (RMS) method was employed to extract features from the EMG signal. Raw EMG signals were preprocessed using a signal analyzer equipped with a moving mean filter. A window size of 50 samples was applied to smoothen the data and minimize noise artifacts. This approach was implemented across seven raw EMG muscle signals, transforming them into processed EMG signals suitable for subsequent analyses. To integrate these signals with MATLAB's Neural Network Fitting tool, the EMG signals were downsampled to align with the joint angle dataset. Downsampling was performed with a factor of 10, effectively reducing the larger EMG dataset while maintaining data integrity. The preprocessing successfully produced smooth, noise-reduced EMG signals and ensured the compatibility of the downsampled EMG dataset with the joint angle data. This facilitated seamless integration into a neural network model for further analyses. Subsequently, the processed signals are

Table 1 Data of 10 subjects selected for this study

Subject ID	Age	Gender	Height (m)	Mass (kg)
AB09	21	F	1.63	63.5
AB12	24	M	1.74	86.2
AB16	20	F	1.65	55.8
AB19	19	M	1.70	68.0
AB20	21	F	1.71	68.0
AB23	20	M	1.80	76.8
AB24	21	F	1.73	72.6
AB27	21	M	1.70	68.0
AB28	33	F	1.69	62.1
AB30	31	M	1.77	77.0

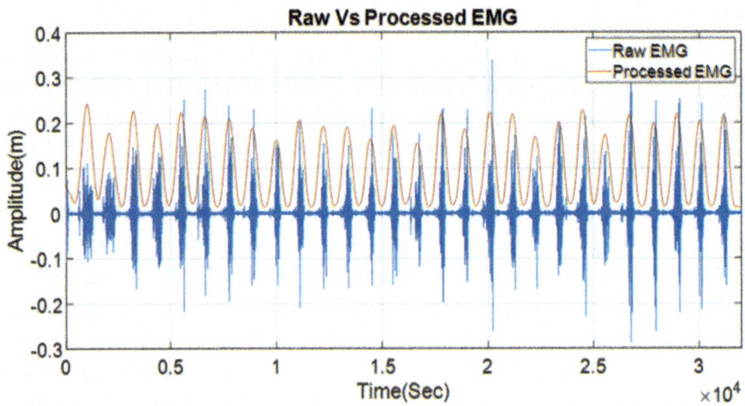

Fig. 1 Raw versus processed EMG signals

normalized to the maximum value for each muscle. Figure 1 shows the raw EMG and processed EMG signals with 31,991 data points, illustrating the noise reduction and feature extraction achieved through the processing steps.

3 Methodology Adopted

The presented work involves the application of 3 different optimizers viz. LM, BR and SCG in neural network-based backpropagation model.

3.1 Backpropagation Algorithm

The backpropagation algorithm is a supervised learning technique used to train neural networks by minimizing the error between predicted and actual outputs. It works in two phases: forward pass, where inputs propagate through the network to generate predictions, and backward pass, where errors propagate back to update weights using the gradient descent method. The weight update rule is

$$w_{ij} \leftarrow w_{ij} - \eta \frac{\partial E}{\partial w_{ij}} \tag{1}$$

where, η is the learning rate, E is the error function.

This iterative process continues until the error converges to a minimum. The accompanying diagram shows the structure of a simple neural network with input, hidden, and output layers.

A basic neural network has interconnected artificial neurons in three layers, as shown in Fig. 2. The input layer of an artificial neural network receives information from the external environment, processes or categorizes it, and forwards it to the hidden layers. Hidden layers, which can be numerous, take input from the preceding layer, further analyse and process the data, and pass it along to subsequent layers. Finally, the output layer provides the network's results, which can be a single output for binary classifications (e.g., 1 or 0) or multiple outputs for multi-class classification problems, depending on the task.

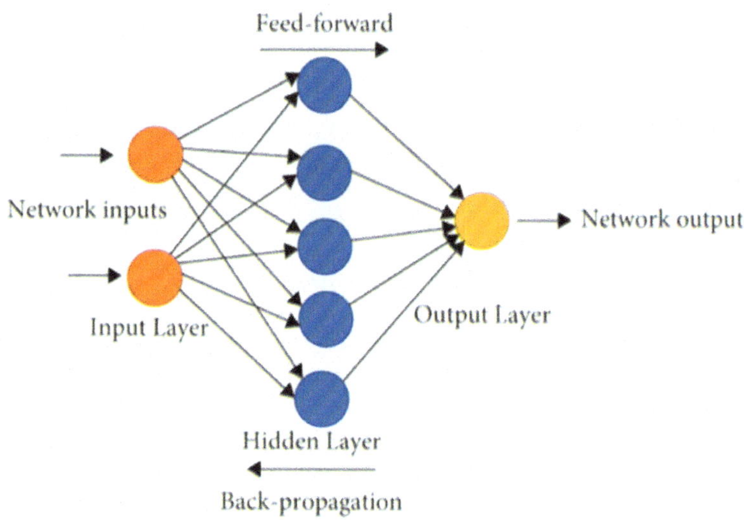

Fig. 2 A general backpropagation neural network architecture

Optimizer 1: Levenberg-Marquardt (LM) algorithm

It is commonly used for training feedforward neural networks, particularly for function fitting and pattern recognition. It minimizes the mean squared error between predicted and actual outputs by combining the speed of the Gauss-Newton method with the stability of gradient descent. The weight update rule is:

$$\Delta \mathrm{w} = -\left(\mathrm{J}^{\mathrm{T}}\mathrm{J} + \lambda I\right)^{-1}\mathrm{J}^{\mathrm{T}}\mathrm{E} \tag{2}$$

where, $\mathrm{J}^{\mathrm{T}}\mathrm{J}$ is the Jacobian matrix of network error gradients, E is the error vector and λ is the damping parameter.

MATLAB's trainlm function implements LM, dynamically adjusting λ to ensure convergence while efficiently handling nonlinearities in neural networks.

Optimizer 2: Bayesian Regularization (BR) algorithm

Bayesian regularization minimizes a linear combination of squared errors and weights. The network demonstrates strong generalization capabilities, enhanced through BR applied within the LM algorithm framework. During training, backpropagation is utilized to compute the Jacobian matrix J_X, which represents the partial derivatives of the performance function perf with respect to the weight and bias variables X. The updates to X are governed by the Levenberg-Marquardt method as follows:

The approximation of the Hessian matrix is computed as:

$$J_J = J_X^{\mathrm{T}} J_X \tag{3}$$

The gradient vector is calculated as:

$$J_E = J_X^{\mathrm{T}} E \tag{4}$$

Here, E represents the error vector.

The update step for the weights and biases is derived as:

$$\Delta X = -(J_J + \mu I)^{-1} J_E \tag{5}$$

In this equation, μ is the damping factor that balances between the Gauss-Newton and gradient descent methods, and I is the identity matrix.

This approach allows the network to achieve a balance between fitting the training data and maintaining generalization, ensuring robust performance on unseen data. The adaptive value μ is incrementally increased by μ_{inc} until the resulting change leads to a reduced performance value. Once this occurs, the change is applied to the network, and μ is decreased by μ_{dec}. Training continues until one of the following stopping conditions is met: the maximum number of epochs (repetitions) is reached, the maximum allowable time is exceeded, the performance is minimized to the predefined goal, the performance gradient falls below the specified minimum gradient (min _ grad), or μ exceeds the maximum allowable value ($\mu_{\max}$).

Optimizer 3: Scaled Conjugate Gradient (SCG) backpropagation algorithm

SCG backpropagation algorithm, implemented as *trainscg*, can train any neural network provided its weight, net input, and transfer functions have well-defined derivative functions. In this method, backpropagation is employed to compute the gradient of the performance metric perf with respect to the weight and bias variables X. Mathematically, this involves calculating:

$$\nabla \text{perf} = \frac{\partial \text{perf}}{\partial \text{X}} \tag{6}$$

where ∇perf represents the gradient vector of the performance function.

This gradient information is used to optimize the network parameters X along conjugate directions, ensuring faster convergence compared to standard gradient descent while avoiding the need for explicit line search calculations at each iteration. Using backpropagation, the error δ_l at each layer l is computed as:

$$\delta_l = \delta_{l+1} \cdot \frac{\partial O_{l+1}}{\partial O_l} \tag{7}$$

where δ_{l+1} is the error from the next layer and O_l represents the output of the current layer.

The gradients with respect to weights W_l and biases b_l are then:

$$\frac{\partial \text{perf}}{\partial W_l} = \delta_l \cdot O_{l-1}^T, \frac{\partial \text{perf}}{\partial b_l} = \delta_l \tag{8}$$

The SCG algorithm, like *traincgp*, *traincgf*, and *traincgb*, is based on conjugate directions; however, it does not perform a line search at each iteration. Training is terminated when one of several conditions is met: the maximum number of epochs (repetitions) is reached, the maximum allowable time is exceeded, the performance metric is minimized to the desired goal, or the performance gradient falls below the specified minimum gradient threshold (min_grad). Additionally, when validation is employed, training stops if the validation performance (error) increases more than the allowed number of times (max_fail) since the last improvement.

4 Implementation Details

The present study consists of a lower limb exoskeleton walking on a treadmill. As mentioned earlier, a total of 7 EMG muscle signals act as input parameters while 3 joint angles viz. hip, knee, and ankle act as output parameters with 500 hidden layers. The Neural Network model is defined as shown in Fig. 3. A total of 65,390 data points has been considered in this study. The data is split into 70% for training (45,773 data points used for training), 15% to validate that the network is generalizing and to stop training before overfitting (2943 data points used) and 15% to independently test network generalization (2943 data points used).

The Mean Squared Error (MSE) is a commonly utilized measure to assess the effectiveness of regression models. It determines the average of the squared differences between the predicted and actual values of a specific dataset. To compute the MSE, the squared differences between the predicted and actual values are added together and the mean is calculated. The MSE metric is commonly used because it is a continuous and non-negative function that can be used to compare different models or tuning parameters. Here we will consider (RMSE), which takes the square root of the MSE, which makes it more interpretable as it shares the same unit as the dependent variable. The RMSE can be calculated using the following formula, where y_p represents the predicted value of the dependent variable, and y represents the actual value of the dependent variable, and 'n' refers to the total count of data points:

$$\text{RMSE} = \sqrt{\frac{1}{n}\sum_{i=1}^{n}(y_p - y)^2} \tag{9}$$

It represents the amount of variance in the dependent variable that can be accounted for by the independent variables in the model. R^2 provides an indicator of the goodness of fit of a model to the data and can vary between 0 and 1, with a value of 1 indicating a perfect fit and a value of 0 indicating no fit at all.

$$R^2 = 1 - \frac{SS_{res}}{SS_{tot}} \tag{10}$$

where, SS_{res} and SS_{tot} are sum of squares of residuals and total sum of squares, respectively.

Figure 4 illustrates the electromyographic (EMG) signals of various muscles, with shaded regions representing the variance (upper and lower bounds) for each muscle. Figure 5 curves depict the joint angles, while the shaded regions indicate the variability across the dataset, highlighting the model's predictive accuracy and uncertainty.

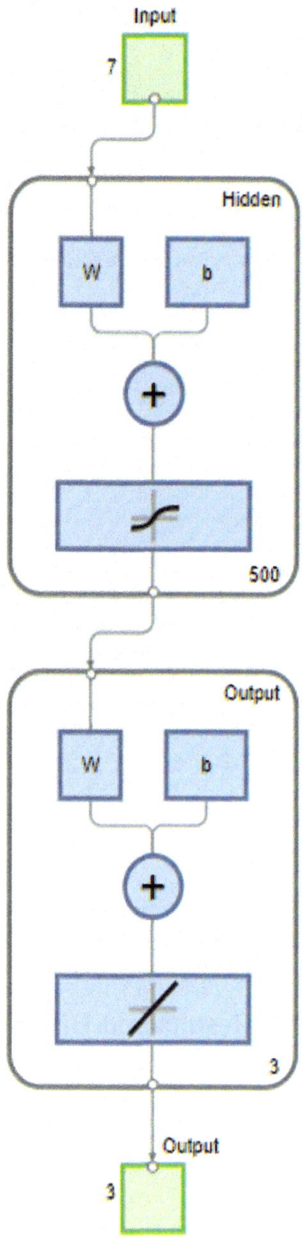

Fig. 3 Neural network model applied in this study

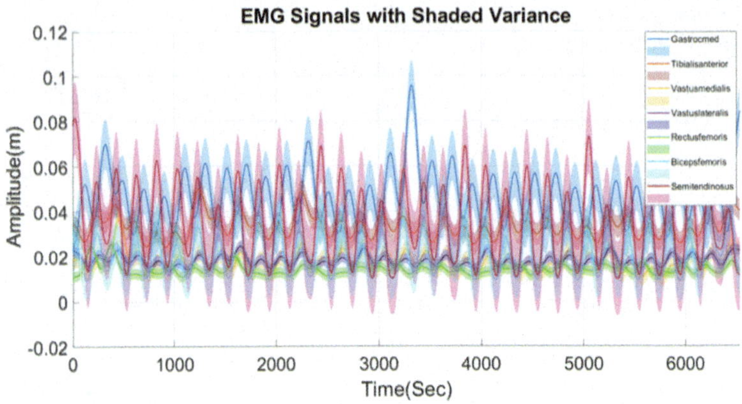

Fig. 4 Shaded EMG signals with variance representations for muscle activity

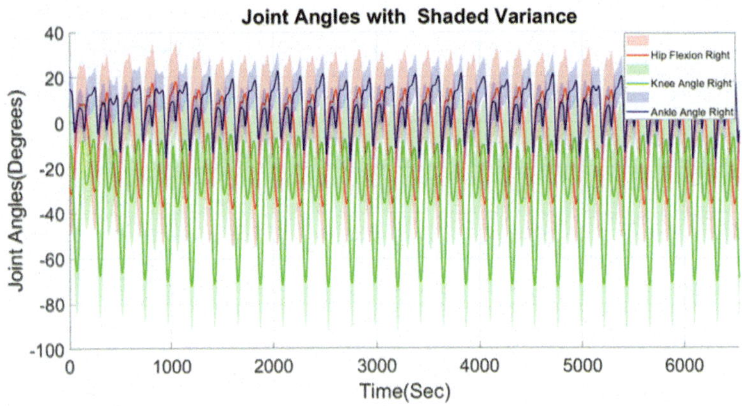

Fig. 5 Joint angle estimations with shaded variance

5 Results and Discussions

In this section, the regression analysis and error evaluation are presented for the overall dataset using different training algorithms. Moreover, for each training algorithm, as a validation of the dataset, a test subject labeled AB19 was selected. The subject is a 19-year-old male with a height of 1.70 m and a weight of 68 kg (Table 1). This individual was chosen to comprehensively assess the performance and reliability of the model across the entire dataset.

Figure 6 illustrates the regression analysis for training, validation, testing, and overall datasets. It shows high correlation coefficients ($R > 0.91$), which indicate the model's strong predictive performance and generalization capability. The regression plots show the relationship between the predicted and target values for training, validation, testing, and the combined dataset. The correlation coefficients (R-values) for all cases are above 0.91, indicating a strong linear correlation between the predictions and actual targets. The training data achieves the highest R-value (0.91976), followed by the validation (0.91535), and test sets (0.91443). The combined dataset has an overall R-value of 0.91829, demonstrating consistent model performance across all subsets. The diagonal fit lines (Y = T) closely align with the data points, suggesting minimal error. The slight deviations in some areas indicate areas for further optimization, but overall, the model performs reliably across different phases.

Figure 7 shows a 20-bin error histogram for a neural network trained using the LM method, highlighting error distributions across training, validation, and test datasets.

Figure 8a–c depict the actual and predicted hip flexion, knee, and ankle angles, respectively, for the right leg, using the LM method with EMG signals as input. These plots illustrate the model's performance in estimating joint angles.

Figure 9 illustrates the regression analysis for training, validation, testing, and overall datasets, showing high correlation coefficients (R > 0.84), which indicate the model's strong predictive performance and generalization capability. The regression plots illustrate the correlation between predicted and target values for training,

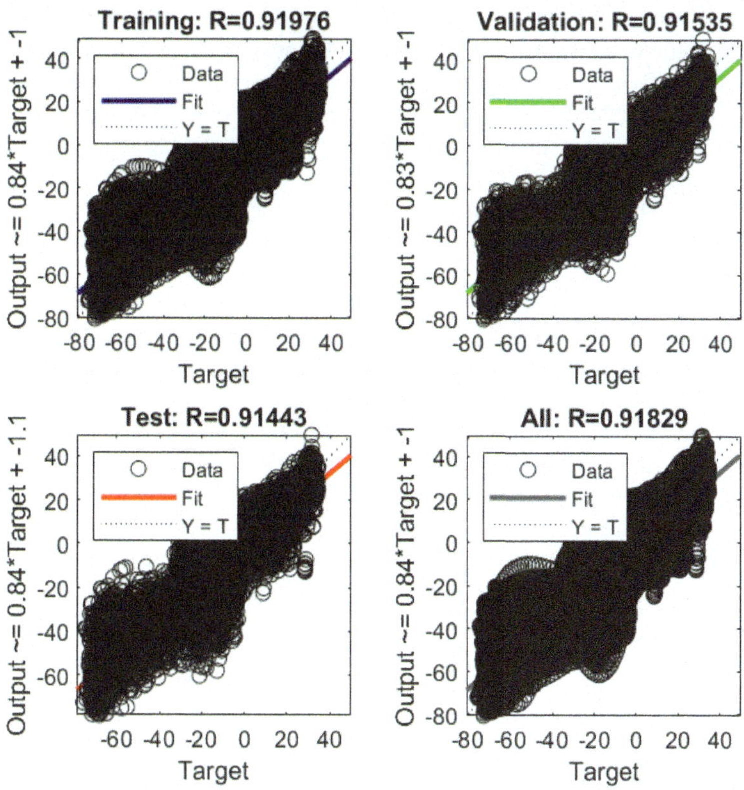

Fig. 6 Regression analysis for LM method across training, validation, and testing phases

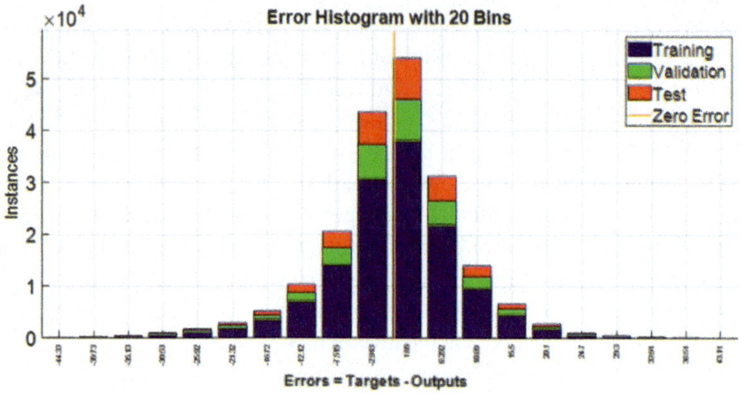

Fig. 7 Error histogram for a neural network trained using the LM algorithm

testing, and the combined datasets. The correlation coefficients (R-values) are consistent, with the training set showing an R-value of 0.85062, the test set 0.84724, and the combined set 0.85012. These values indicate a moderately strong linear relationship between predictions and targets. The fitted lines are close to the ideal diagonal line (Y = T), signifying a reasonable alignment with the target values, though some deviations are noticeable. The model performs similarly across both training and testing phases, demonstrating stability, but there is room for improvement in prediction accuracy and overall fit.

Figure 10 shows a 20-bin error histogram for a neural network trained using the BR method, highlighting error distributions across training, validation, and test datasets.

Figure 11a shows the actual and predicted knee angle right for the right leg, obtained using the BR method with EMG signals as input. Figure 11b shows the actual and predicted knee angle right for the right leg, obtained using the BR method with EMG signals as input. Figure 11c shows the actual and predicted ankle angle right for the right leg, obtained using the SCG method with EMG signals as input.

Figure 12 illustrates the regression analysis for training, validation, testing, and overall datasets. It shows high correlation coefficients (R > 0.81), which indicate the model's strong predictive performance and generalization capability. The R-values are consistently around 0.812, indicating a moderate linear relationship. The training set achieves an R-value of 0.81213, validation 0.81274, testing 0.81391, and the combined set 0.81249. The fitted lines align reasonably well with the diagonal line (Y = T), though there are visible deviations, particularly in the extreme values. The model demonstrates stable performance across all phases, but the moderate correlation suggests room for improvement in accuracy and predictive capability.

Figure 13 shows a 20-bin error histogram for a neural network trained using the Scaled Conjugate Gradient method, highlighting error distributions across training, validation, and test datasets.

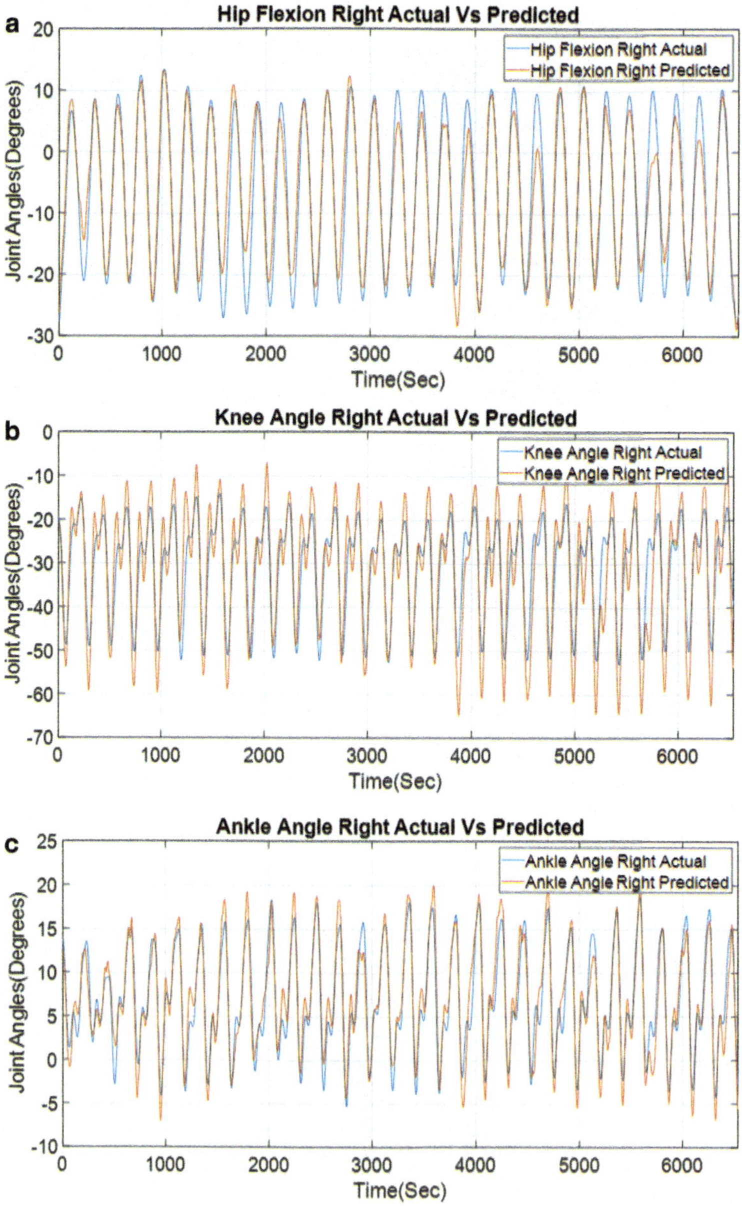

Fig. 8 **a** Comparison of actual and predicted hip flexion right using LM method for the validation dataset. **b** Comparison of actual and predicted knee angle right using LM method for the validation dataset. **c** Comparison of actual and predicted ankle angle right using the LM method for the validation dataset

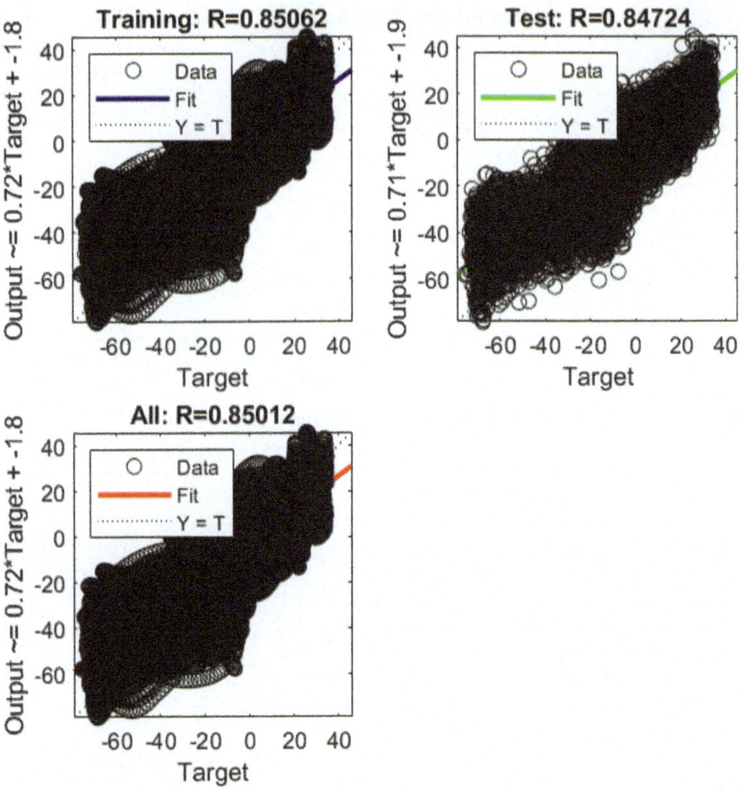

Fig. 9 Regression analysis for BR method across training, validation, and testing phases

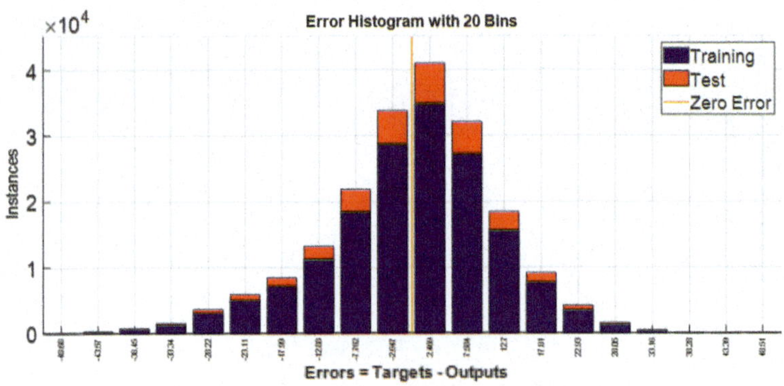

Fig. 10 Error histogram for a neural network trained using the BR method

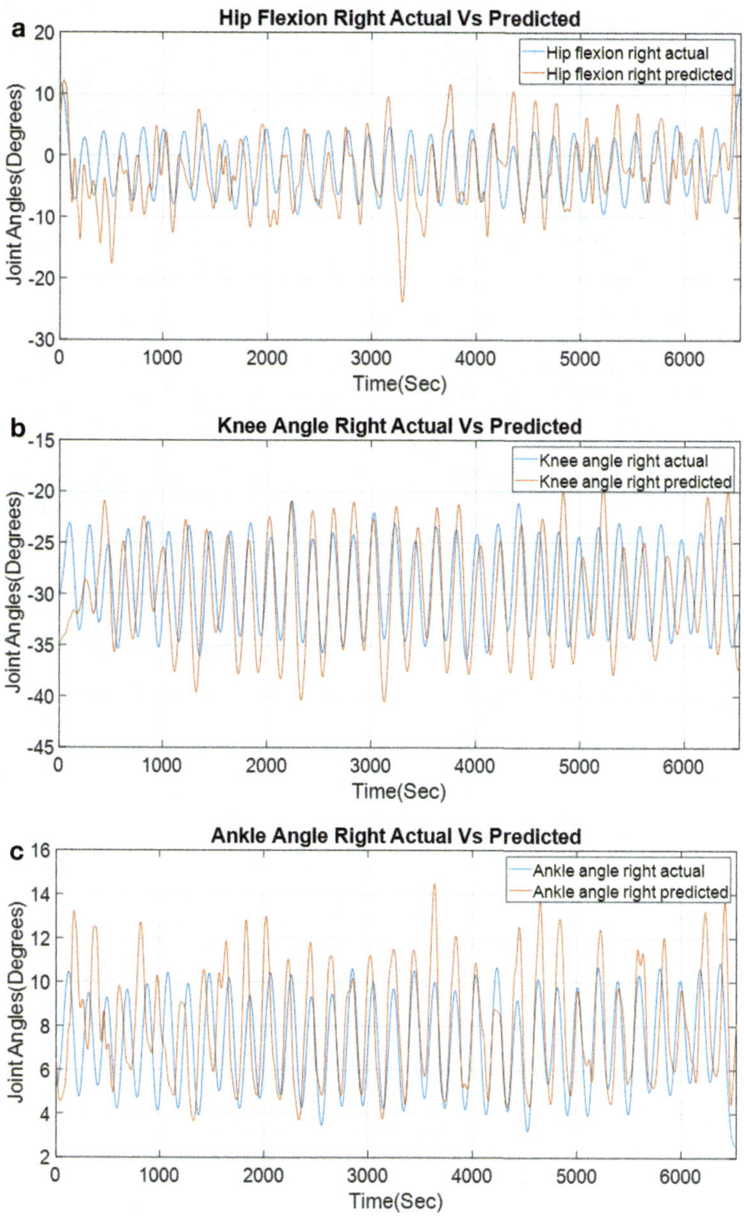

Fig. 11 **a** Comparison of actual and predicted hip flexion right using the LM method for the validation dataset. **b** Comparison of actual and predicted knee angle right using the BR method for the validation dataset. **c** Comparison of actual and predicted ankle angle right using the scaled conjugate gradient method for the validation dataset

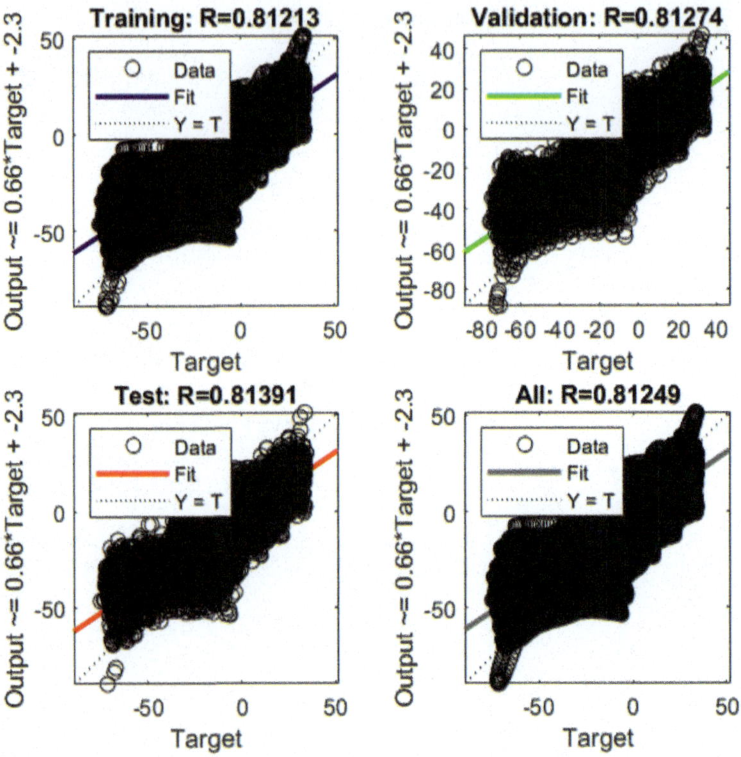

Fig. 12 Regression analysis for scaled conjugate gradient method across training, validation and testing phases

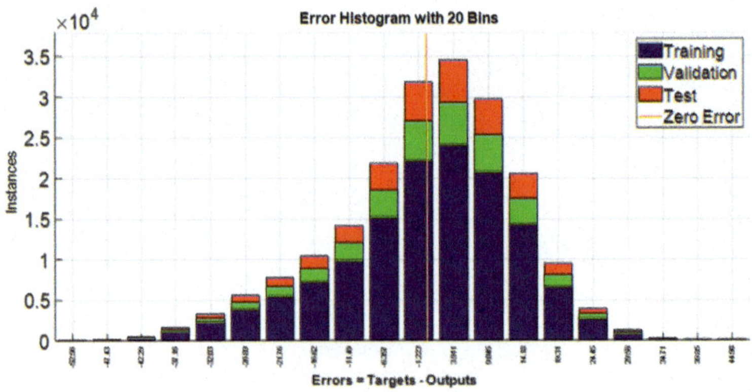

Fig. 13 Error histogram for a neural network trained using the scaled conjugate gradient algorithm

Figure 14a–c show the actual and predicted hip, knee, and ankle angles for the right leg, using the LM, BR, and SCG methods respectively with EMG signals as input. These comparisons highlight the performance of different training optimizers in joint angle estimation.

Table 2 presents the Mean Squared Error (MSE) and correlation coefficient (R) values of AB19, the test subject, across three algorithms: LM, BR, and SCG. LM method demonstrated superior performance with the lowest MSE values (0.1423, 0.2576, and 0.1935 for training, validation, and test, respectively) and R values of ~0.9. BR had higher MSE values (3.8765 and 4.2345 for training and test) with R values slightly below 0.9. Validation data for BR was unavailable (NaN) because the test data inherently takes care of tuning hyperparameters. SCG showed the highest MSEs (8.4122–9.5437) and R values around 0.8, indicating comparatively lower accuracy. Overall, LM was the most effective algorithm for AB19. The results have significant practical implications for real-world exoskeleton design by enabling precise and adaptive control systems that can seamlessly estimate joint angles from EMG signals during dynamic tasks like walking. This improves user comfort, natural patterns, and the effectiveness of rehabilitation. The findings also highlight the potential of advanced training optimizers to enhance model performance, providing a foundation for developing more robust and efficient exoskeletons. Future improvements in rehabilitation robotics can focus on incorporating real-time adaptability and personalized training models based on these insights, further advancing mobility assistance and recovery outcomes.

6 Conclusions

This study focused on enhancing lower limb exoskeleton performance by accurately estimating joint angles from EMG signals. A neural network model was trained on EMG data from 10 subjects using three optimization methods: LM, BR and SCG. LM method demonstrated superior performance, achieving the lowest validation MSE (83.2474) at epoch 103 with high correlation coefficients (R > 0.91) across all datasets. Its effective balance of the Gauss-Newton method and gradient descent enabled precise modeling of the complex relationships within the EMG data, making it the optimal choice for joint angle estimation and exoskeleton control. These findings contribute to developing more intuitive and precise assistive devices, advancing mobility technology for individuals with impairments.

Future work in this area can explore several directions to further enhance the accuracy and applicability of joint angle estimation for lower limb exoskeletons. While this study focused on treadmill walking, future studies can expand the scope by including a variety of conditions and case studies, such as walking on different terrains, during different activities, or under varying physical conditions. Additionally, while this research was conducted using a single walking speed, incorporating multiple walking speeds could provide more comprehensive insights into the model's

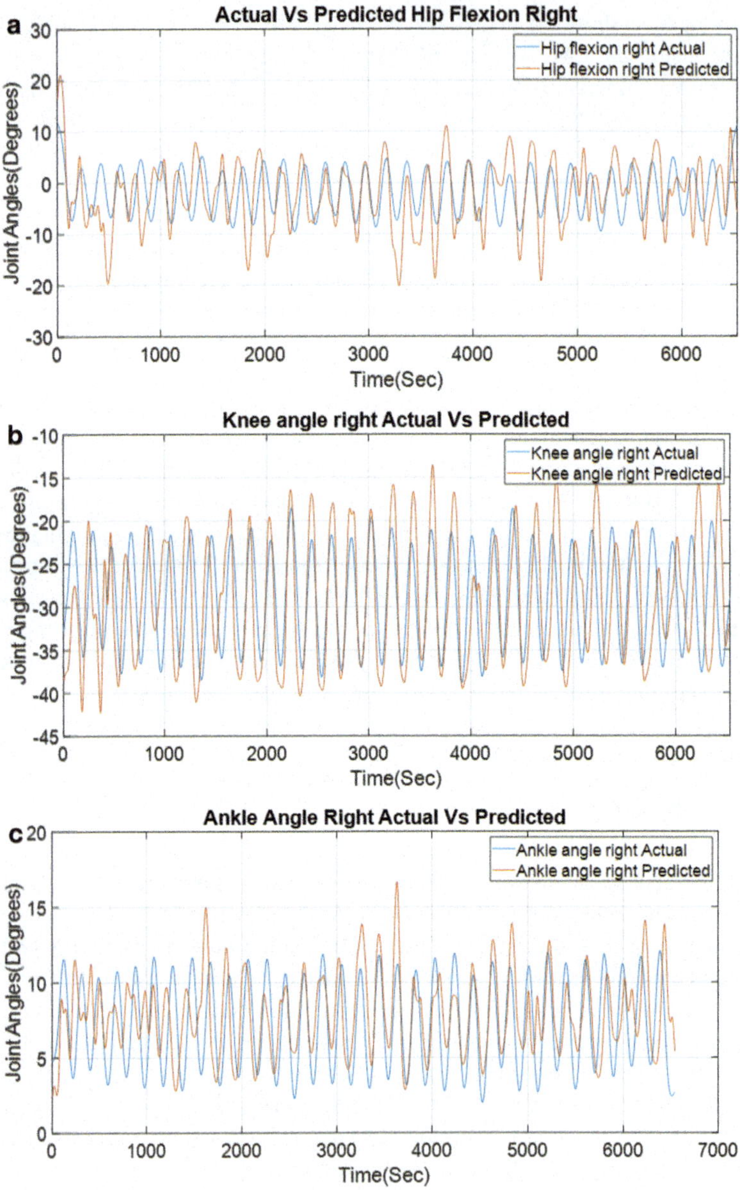

Fig. 14 **a** Comparison of actual and predicted hip flexion right using a LM method for the validation dataset. **b** Comparison of actual and predicted knee angle right using BR method for the validation dataset. **c** Comparison of actual and predicted ankle angle right using scaled conjugate gradient method for the validation dataset

Table 2 MSE and R values of AB19 (test subject) with different algorithms

AB19 (test subject)	MSE	R
Levenberg-Marquardt algorithm		
Training	0.1423	0.926
Validation	0.2576	0.918
Test	0.1935	0.906
Bayesian regularization algorithm		
Training	3.8765	0.896
Validation	NaN	NaN
Test	4.2345	0.843
Scaled conjugate gradient algorithm		
Training	8.4122	0.816
Validation	9.5437	0.801
Test	9.5024	0.795

adaptability across different movement dynamics. Another potential area for exploration is the inclusion of both the right and left legs to better capture bilateral movement patterns, which would improve the overall robustness of the model. Furthermore, while seven EMG muscle signals were used in this study, future research could consider incorporating additional muscle signals to enhance the model's ability to predict joint angles with greater precision. By addressing these aspects, future work can significantly improve the performance and versatility of exoskeletons in real-world applications.

References

1. Park, Y.H., Lee, D.H., Lee, J.H.: A comprehensive review: robot-assisted treatments for gait rehabilitation in stroke patients. Medicina **60**(4), 620 (2024)
2. Baker, R.: Gait analysis methods in rehabilitation. J. Neuro Eng. Rehabil. **3**, 1–10 (2006)
3. Zelik, K.E., Adamczyk, P.G.: A unified perspective on ankle push-off in human walking. J. Exp. Biol. **219**(23), 3676–3683 (2016)
4. Wang, Y., Li, Z., Wang, X., Yu, H., Liao, W., Arifoglu, D.: Human gait data augmentation and trajectory prediction for lower-limb rehabilitation robot control using GANs and attention mechanism. Machines **9**(12), 367 (2021)
5. Su, B., Gutierrez-Farewik, E.M.: Gait trajectory and gait phase prediction based on an LSTM network. Sensors **20**(24), 7127 (2020)
6. Amrani El Yaakoubi, N., McDonald, C., Lennon, O.: Prediction of gait kinematics and kinetics: a systematic review of EMG and EEG signal use and their contribution to prediction accuracy. Bioengineering **10**(10), 1162 (2023)
7. Hansen, C., Gosselin, F., Mansour, K.B., Devos, P., Marin, F.: Design validation of a hand exoskeleton using musculoskeletal modeling. Appl. Ergon. **68**, 283–288 (2018)

8. Kolaghassi, R., Al-Hares, M.K., Sirlantzis, K.: Systematic review of intelligent algorithms in gait analysis and prediction for lower limb robotic systems. IEEE Access **9**, 113788–113812 (2021)
9. Belal, M., Alsheikh, N., Aljarah, A., Hussain, I.: Deep learning approaches for enhanced lower-limb exoskeleton control: a review. IEEE Access (2024)
10. Jaramillo, I.E., Jeong, J.G., Lopez, P.R., Lee, C.H., Kang, D.Y., Ha, T.J., Oh, J.H., Jung, H., Lee, J.H., Lee, W.H., Kim, T.S.: Real-time human activity recognition with IMU and encoder sensors in wearable exoskeleton robots via deep learning networks. Sensors **22**(24), 9690 (2022)
11. Hua, Y., Fan, J., Liu, G., Zhang, X., Lai, M., Li, M., Zheng, T., Zhang, G., Zhao, J., Zhu, Y.: A novel weight-bearing lower limb exoskeleton based on motion intention prediction and locomotion state identification. IEEE Access **7**, 37620–37638 (2019)
12. Sun, Y., Tang, Y., Zheng, J., Dong, D., Chen, X., Bai, L.: From sensing to control of lower limb exoskeleton: a systematic review. Annu. Rev. Control **53**, 83–96 (2022)
13. Woo, J.-H., et al.: Machine learning based recognition of elements in lower-limb movement sequence for proactive control of exoskeletons to assist lifting. IEEE Access **11**, 127107–127118 (2023)
14. Xiang, L., Wang, A., Gu, Y., Zhao, L., Shim, V., Fernandez, J.: Recent machine learning progress in lower limb running biomechanics with wearable technology: a systematic review. Front. Neurorobot. **16**, 913052 (2022)
15. Zhou, B., Wang, H., Hu, F., Feng, N., Xi, H., Zhang, Z., Tang, H.: Accurate recognition of lower limb ambulation mode based on surface electromyography and motion data using machine learning. Comput. Methods Progr. Biomed. **193**, 105486 (2020)
16. Xie, H., Li, G., Zhao, X., Li, F.: Prediction of limb joint angles based on multi-source signals by GS-GRNN for exoskeleton wearer. Sensors **20**(4), 1104 (2020)
17. Narayan, J., Gritli, H., Dwivedy, S.K.: Lower limb joint torque estimation via Bayesian regularized backpropagation neural networks. In: 2024 IEEE International Conference on Interdisciplinary Approaches in Technology and Management for Social Innovation (IATMSI), Gwalior, India, pp. 1–6 (2024). https://doi.org/10.1109/IATMSI60426.2024.10502709
18. Lee, T., Kim, I., Lee, S.H.: Estimation of the continuous walking angle of knee and ankle (talocrural joint, subtalar joint) of a lower-limb exoskeleton robot using a neural network. Sensors **21**(8), 2807 (2021)
19. Menga, G., Geng, J., Mancin, M.: Neural Networks to Control Voluntary Movements in a Lower Limb Exoskeleton Using Electromyographic Signals. (2024)
20. Dhindsa, I.S., Agarwal, R., Ryait, H.S.: A novel algorithm to predict knee angle from EMG signals for controlling a lower limb exoskeleton. In: CEUR Workshop Proceedings, vol. 1638, pp. 536–541 (2016)
21. Li, W., Liu, K., Sun, Z., Li, C., Chai, Y., Gu, J.: A neural network-based model for lower limb continuous estimation against the disturbance of uncertainty. Biomedical Signal Processing and Control **71**, 103115 (2022)
22. Neťuková, S., Bejtic, M., Malá, C., Horáková, L., Kutílek, P., Kauler, J., Krupička, R.: Lower limb exoskeleton sensors: state-of-the-art. Sensors **22**(23), 9091 (2022)
23. Wang, X., Dong, D., Chi, X., Wang, S., Miao, Y., Gavrilov, A.I.: SEMG-based consecutive estimation of human lower limb movement by using multi-branch neural network. Biomed. Signal Process. Control **68**, 102781 (2021)
24. Foroutannia, A., Akbarzadeh-T, M.-R., Akbarzadeh, A., Tahamipour-Z, S.M.: Adaptive fuzzy impedance control of exoskeleton robots with electromyography-based convolutional neural networks for human intended trajectory estimation. Mechatronics **91**, 102952 (2023)
25. Xiong, B., et al.: Intelligent prediction of human lower extremity joint moment: an artificial neural network approach. IEEE Access **7**, 29973–29980 (2019)
26. Koceska, N., Koceski, S., Durante, F., Zobel, P.B., Raparelli, T.: Control architecture of a 10 DOF lower limbs exoskeleton for gait rehabilitation. Int. J. Adv. Robot. Syst. **10**(1), 68 (2013)
27. Song, S., Zhang, X., Tan, Z.: RBF neural network based sliding mode control of a lower limb exoskeleton suit. Strojniški vestnik J. Mech. Eng. **60**(6), 437–446 (2014)

28. Laschowski, B., McNally, W., Wong, A., McPhee, J.: Environment classification for robotic leg prostheses and exoskeletons using deep convolutional neural networks. Front. Neurorobot. **15**, 730965 (2022)
29. Chen, J., Mu, X., Du, F.: Biomechanics analysis of human lower limb during walking for exoskeleton design. J. Vibroeng. **19**(7), 5527–5539 (2017)
30. Kong, Y.-K., Choi, K.-H., Cho, M.-U., Kim, S.-Y., Kim, M.-J., Shim, J.-W., Park, S.-S., et al.: Ergonomic assessment of a lower-limb exoskeleton through electromyography and anybody modeling system. Int. J. Environ. Res. Public Health **19**(13), 8088 (2022)
31. Masengo, G., Zhang, X., Dong, R., Alhassan, A.B., Hamza, K., Mudaheranwa, E.: Lower limb exoskeleton robot and its cooperative control: a review, trends, and challenges for future research. Front. Neurorobot. **16**, 913748 (2023)
32. Briouza, S., Gritli, H., Khraief, N., Belghith, S., Singh, D.: A brief overview on machine learning in rehabilitation of the human arm via an exoskeleton robot. In: 2021 International Conference on Data Analytics for Business and Industry (ICDABI), Sakheer, Bahrain, pp. 129–134 (2021). https://doi.org/10.1109/ICDABI53623.2021.9655865
33. Briouza, S., Gritli, H., Khraief, N., Belghith, S.: CRNN-Based Classification of EMG Signals for the Rehabilitation of the Human Arm. In: Campos-Cantón, E., Huerta-Cuellar, G., Zambrano-Serrano, E., Tlelo-Cuautle, E. (eds.) Complex Systems and Their Applications. EDIESCA 2023. Springer, Cham (2024). https://doi.org/10.1007/978-3-031-51224-7_5
34. Briouza, S., Gritli, H., Khraief, N., Belghith, S., Singh, D.: Classification of sEMG biomedical signals for upper-limb and hand rehabilitation using a hybrid CNN-SVM architecture. In: 2022 International Conference on Data Analytics for Business and Industry (ICDABI), Sakhir, Bahrain, pp. 467–471 (2022). https://doi.org/10.1109/ICDABI56818.2022.10041524
35. Briouza, S., Gritli, H., Khraief, N., Belghith, S., Singh, D.: An auto-ANN for EMG classification for the motion recognition of the human hand. In: 2022 IEEE Information Technologies & Smart Industrial Systems (ITSIS), Paris, France, pp. 1–6 (2022). https://doi.org/10.1109/ITSIS56166.2022.10118372
36. Bouzid, R., Gritli, H., Narayan, J.: Investigating feed-forward back-propagation neural network with different hyperparameters for inverse kinematics of a 2-DoF robotic manipulator: a comparative study. Chaos Theory Appl. **6**(2), 90–110 (2024)
37. Bouzid, R., Narayan, J., Gritli, H.: Artificial neural networks for the forward kinematics of a SCARA manipulator: a comparative study with two datasets. In: 2024 ASU International Conference in Emerging Technologies for Sustainability and Intelligent Systems (ICETSIS), Manama, Bahrain, pp. 1792–1797 (2024). https://doi.org/10.1109/ICETSIS61505.2024.10459544
38. Bouzid, R., Gritli, H., Narayan, J.: ANN approach for SCARA robot inverse kinematics solutions with diverse datasets and optimisers. Appl. Comput. Syst. **29**, 24–34 (2024). https://doi.org/10.2478/acss-2024-0004
39. Bouzid, R., Narayan, J., Gritli, H.: Solving inverse kinematics problem for manipulator robots using artificial neural network with varied dataset formats. In: Campos-Cantón, E., Huerta-Cuellar, G., Zambrano-Serrano, E., Tlelo-Cuautle, E. (eds.) Complex Systems and Their Applications. EDIESCA 2023. Springer, Cham (2024). https://doi.org/10.1007/978-3-031-51224-7_4
40. Pandey, M., Narayan, J., Trivedi, D.K., Dwivedy, S.K.: Investigating backpropagation neural networks for joint torque prediction during walking: a gait analysis study. In: Kumar, R.S., Sanyal, S., Pathak, P.M. (eds.) Recent Advances in Machines, Mechanisms, Materials and Design. iNaCoMM 2023. Lecture Notes in Mechanical Engineering. Springer, Singapore (2024). https://doi.org/10.1007/978-981-97-5423-6_47
41. Narayan, J., Jhunjhunwala, S., Mishra, S., Dwivedy, S.K.: A comparative performance analysis of backpropagation training optimizers to estimate clinical gait mechanics. In: Roy, S., Goyal, L.M., Balas, V.E., Agarwal, B., Mittal, M. (eds.) Advanced Studies in Complex Systems: Theory and Applications, Predictive Modeling in Biomedical Data Mining and Analysis, pp. 83–104. Academic Press (2022), ISBN 9780323998642. https://doi.org/10.1016/B978-0-323-99864-2.00012-3

42. Camargo, J., Ramanathan, A., Flanagan, W., Young, A.: A comprehensive, open-source dataset of lower limb biomechanics in multiple conditions of stairs, ramps, and level-ground ambulation and transitions. J. Biomech. **119**, 110320 (2021). ISSN 0021-9290. https://doi.org/10.1016/j.jbiomech.2021.110320

Advanced Impedance Electromyographic-Based Control for an Upper Limb Exoskeleton Robot

Brahim Brahmi, Maarouf Saad, and Mohammad Habibur Rahman

Abstract This chapter presents an adaptive impedance control approach to facilitate active rehabilitation using a wearable robot with (7-DOFs) with an unknown dynamic model. A key challenge in this approach is enabling the exoskeleton to provide precise assistance by accurately interpreting the wearer's movement intentions. Due to human movement's nonlinear and time-varying nature, determining the user's desired motion intention (DMI) is complex. To address this problem, we use Hill's model to estimate the human forces using the gravity of the human arm, force sensors, and electromyogram (sEMG) signals. A Radial Basis Function Neural Network (RBFNN) is then employed with a sliding mode estimator to estimate the DMI in real time. This estimated DMI is integrated with the proposed adaptive impedance control to ensure the robot can track a specified impedance target. As a result, the proposed control strategy enables the exoskeleton to predict the desired trajectory, providing comfortable and adaptive assistance. Experimental trials with healthy subjects under various scenarios confirm the effectiveness of this approach in achieving smooth collaboration between the exoskeleton and its wearer.

Keywords Adaptive impedance control · Wearable robot · Desired motion intention · Radial basis function neural network · sEMG signals · Exoskeleton collaboration

---

B. Brahmi (✉)
Department of Electrical Engineering Center for Interdisciplinary Research Center for Intelligent Manufacturing & Robotics (IRC-IMR), King Fahd University of Petroleum and Minerals, Dhahran 31261, Eastern Province, Saudi Arabia
e-mail: brahim.brahmi@kfupm.edu.sa

M. Saad
Department of Electrical Department, Ècole de technologie supèrieure, Montrèal, QC, Canada
e-mail: maarouf.saad@etsmtl.ca

M. H. Rahman
Department of Mechanical Engineering, University of Wisconsin-Milwaukee, Milwaukee, WI, USA
e-mail: rahmanmh@uwm.edu

J. Narayan and H. Gritli (eds.), *Design and Control of Rehabilitation Robots*, Studies in Systems, Decision and Control 585, https://doi.org/10.1007/978-3-031-86977-8_12

1 Introduction

Rehabilitation robotics is an innovative technology designed for physiotherapy and movement assistance, garnering significant attention within the scientific community [1]. These robotic systems have the potential to enhance physical therapy outcomes, making the process more effective and economically viable. Recently, rehabilitation robots have demonstrated remarkable potential in reducing patient disabilities, enhancing mobility, supporting a return to everyday life, and assisting with daily activities [2, 3]. Additionally, these robots can save time and effort for therapists, keep patients engaged, and improve the overall rehabilitation experience [4]. Despite these advantages, the field is still in its early stages and faces challenges such as achieving effective compliance control and optimizing human-robot collaboration [5]. To contribute to this field, this study focuses on addressing a critical challenge: enabling exoskeleton robots to deliver precise compliance control through accurate estimation of human DMI. Most of the current compliance and impedance control strategies do not work well enough for optimal physiotherapy outcomes without advanced DMI estimation algorithms [5]. This research suggests two main ways to solve the problem: first, a precise way to estimate DMI so that the exoskeleton robot can correctly understand what the user wants; second, a strong compliance/impedance control strategy that takes into account the exoskeleton robot's unknown dynamic parameters.

1.1 Justification for Using a Non-linear Sliding Mode Estimator Based on a Radial Basis Function Neural Network (RBFNN) to Estimate DMI

Various methods have been developed to gauge human motion intention, which can primarily be divided into two categories: those relying on biological signals, such as electromyograms (sEMG) [6–8], and those utilizing non-biological signals, including force/load sensors [5, 9, 10]. The biological signals approach can leverage various physiological indicators, including electroencephalography (EEG) [11] and electrooculography (EOG) [12], with electromyography (EMG) being the most frequently employed due to its relevance in capturing the operator's intentions [13]. However, factors such as sensor placement, skin thickness, subject fatigue, stress levels, and body temperature significantly impact the efficacy of these biological signal-based methods. Moreover, the sheer volume of data generated from physiological signals necessitates costly computational resources for analysis [14, 15].

To address these limitations, non-biological signal-based sensors have been explored to estimate human force activity [16, 17]. For instance, in [18], Hill's model and a Muscle Circumference Sensor (MCS) were utilized to estimate the desired force exerted by the operator, employing basic biomechanics and muscle properties to ascertain elbow joint torque. However, a key limitation of this approach is its inability to estimate the shoulder joint effort accurately. In [19, 20], this problem was fixed by adding load cells to the exoskeleton robot's end effector to take shoulder muscle efforts into account. Nonetheless, the MCS results in an imbalance, as the elbow joint tends to exert more force than the shoulder joint when the exoskeleton wearer lifts an unknown load, leading to unnatural motion. Additionally, a force sensor was used in [21] to create point-to-point collaborative motion trajectories, which proved effective for slower movements. However, this method encounters significant challenges, such as jerks and chattering, during high-speed motions.

Several existing studies have approached DMI as a stochastic process [22, 23], with [22] employing a hidden Markov model for estimation and [23] utilizing a Kalman-based observer. Despite their innovative approaches, these models struggle to provide accurate DMI estimations due to the unknown nonlinear and time-varying characteristics of the human upper limb. Other researchers have attempted to estimate DMI using neural networks, specifically through trajectory-based methods [24, 25]. However, these techniques often fail when the user alters their intention during operation, necessitating a complete retraining process. To mitigate this issue, an updated approach was proposed in [26] that employs an artificial neural network with reduced input dimensions [27] and an Extreme Learning Machine (ELM) [28], allowing for online parameter adjustments and minimizing computational delays. Furthermore, in [29], a trans-scale feature decoding network was introduced to extract continuous joint torques from EMG data. Recently, advancements in rehabilitation robotics have seen the implementation of biomimetic strategies aimed at enhancing EMG-based control, integrating bioinspired features into physical human-robot interactions, as demonstrated in [30] through the use of Gaussian processes to learn and predict continuous motion intentions based on EMG signals.

In this chapter, we propose a novel approach that integrates both biological and non-biological methods to leverage the strengths of each. Specifically, we estimate the effort exerted by the user's elbow and shoulder joints by analyzing forces from the biceps and triceps muscles using Hill's model [31]. This approach establishes a robust framework for estimating muscle dynamics by accurately capturing the nonlinear and time-dependent nature of muscle contractions. By incorporating crucial physiological parameters—such as force-length and force-velocity relationships—we can model muscle force outputs under varying conditions more effectively. In the proposed approach, the system is activated by the measured sEMG signals from the user, while the forces at the shoulder, forearm, and wrist joints are gauged using a force sensor attached to the tip of the exoskeleton robot (see Fig. 1). This combined strategy facilitates the generation of a balanced force output for the upper limb, enabling more human-like motion. Unlike conventional methods, we treat DMI as an output state of the human upper limb system, where the user's desired force serves as the input. We develop a non-linear sliding mode estimator based on a Radial Basis Function Neural Network (RBFNN) to accurately estimate the robot's output state, accommodating the unknown dynamics of the human upper limb model. The RBFNN is particularly effective due to its capability to efficiently approximate complex nonlinear relationships. Its localized response characteristics allow for rapid adaptation to changing patient needs and environmental conditions, ensuring a more

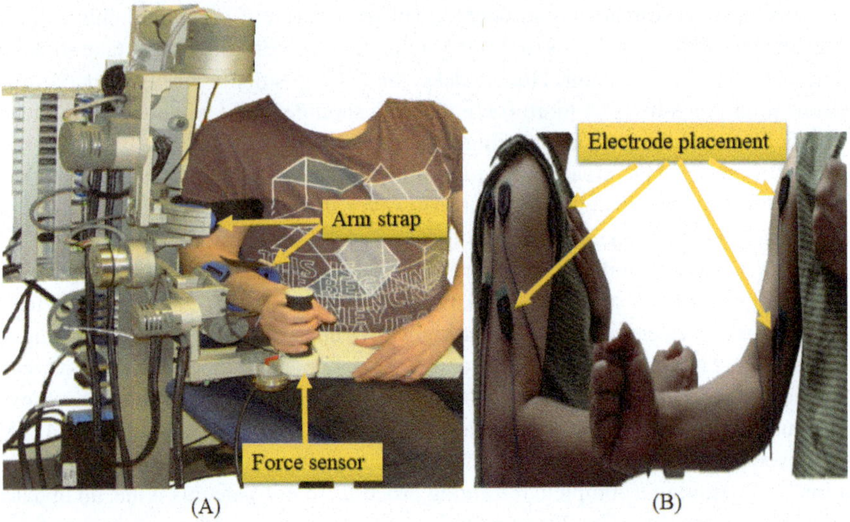

Fig. 1 **a** The exoskeleton robot (ETS-MARSE) with its wearer. **b** The electrode position for biceps/triceps muscles

personalized and effective rehabilitation experience. Additionally, RBFNNs require less training data and computational time compared to other neural network architectures [32], making them well-suited for real-time applications where quick learning and adaptability are crucial for enhancing user outcomes.

1.2 Justification for Employing Adaptive Impedance Control Based on a Reference Model

Impedance/Compliance control plays a crucial role in ensuring comfort and ease for subjects undergoing rehabilitation. In this mode, the exoskeleton must engage actively with its user, as it does not have a predefined trajectory to follow. Consequently, force control emerges as a viable option to facilitate effective human-robot collaboration. However, this method is limited by its lack of robustness and the necessity for some feedback from both the exoskeleton robot and its user [33]. Impedance control presents a promising solution to address these limitations [34, 35]. Recently, advancements in robust adaptive impedance control have been made [36, 37] to manage uncertainties in dynamics and user's DMI, demonstrating commendable performance. Nevertheless, the adaptability to unknown time-invariant uncertainties remains an unresolved issue, as it still requires joint extra feedback as acceleration, the time derivative of the contact force, and accurate estimation of DMI based on the real user's force.

To address these challenges, a novel approach based on the Function Approximation Technique (FAT) has been proposed [38, 39]. The FAT approach employs a finite sum of orthonormal basis functions to estimate the robot's dynamic parameters, resulting in highly effective outcomes [40]. While this technique eliminates the need for measuring joint acceleration, inverting the estimated inertia matrix, and calculating the contact force derivative [38], It remains dependent on the selection of appropriate basis functions for parameter estimation. The type and number of these functions significantly influence both the precision of the parameter estimation and the computational efficiency during online processing.

In response to the identified challenges and building upon previous research, we have developed an adaptive impedance controller founded on a reference impedance model using Modified FAT [41]. Contrary to traditional FAT approaches, the designed method removes the dependency on basis functions in the dynamic model estimation, as well as the need for acceleration feedback. As a result, the estimated DMI is integrated into the adaptive impedance control as the equilibrium state of the reference impedance model. So, the proposed control strategy is designed to guide the exoskeleton robot toward the reference impedance model while considering the robot's unknown dynamics. The targeted impedance model in force control allows the exoskeleton to modify its interaction with the user by emulating a spring-damper system, which enables it to adjust its stiffness and damping in response to varying contact forces. This ensures smoother and safer interactions, making the approach especially suitable for tasks within uncertain or dynamic environments, such as rehabilitation or collaborative robotics. Through the proposed strategy, the exoskeleton robot actively follows the desired trajectory of its user while passively managing the interaction forces, thereby enhancing the collaboration between the human and the exoskeleton. Based on this analysis, the primary contributions of this chapter are as follows:

1. This research combines surface electromyography (sEMG) and non-biological force sensors to accurately estimate the desired force exerted by the user's upper limb. This balanced force input, derived from Hill's model, enhances robotic assistance designed to the user's capabilities, significantly improving the effectiveness of rehabilitation interventions.
2. A novel non-linear sliding mode estimator utilizing a Radial Basis Function Neural Network (RBFNN) is introduced to predict the user's Desired Motion Intention (DMI). This approach accounts for the unknown dynamics of the user's upper limb, leading to improved adaptability and precision in controlling the upper limb exoskeleton robot.
3. This study presents an adaptive impedance control strategy based on a reference impedance model. By eliminating the need for acceleration feedback and basis functions in parameter estimation, this framework enables the exoskeleton robot to actively follow the reference impedance model. As a result, the proposed control scheme ensures smoother and safer interactions, making it particularly effective for active rehabilitation tasks in dynamic interaction.

4. The proposed approach facilitates improved cooperation between the exoskeleton robot and its user (wearer) by allowing the exoskeleton to dynamically adjust its stiffness and damping in response to varying interaction forces. This adaptability enhances the overall user experience and safety during rehabilitation exercises.
5. The proposed control policy is experimentally evaluated in physiotherapy sessions with healthy subjects, providing empirical evidence of its effectiveness and potential application in rehabilitation robotics

The structure of the remaining sections of the chapter is organized as follows: the next section presents the mathematical model of the human-exoskeleton robot and articulates the problem statement. Section 3 describes the interaction force measurement approach and calibration of the Hill model. Section 4 details the non-linear sliding mode estimator, which is combined with the RBFNN algorithm to predict the DMI. The adaptive impedance control scheme based on the reference model is discussed in Sect. 5. Section 6 outlines the experimental evaluation of the proposed approach during physiotherapy sessions with healthy subjects. Finally, Sect. 7 concludes the chapter and outlines directions for future research.

2 Problem Formulation

2.1 ETS-MARSE Design

The ETS-MARSE (Ècole de Technologie Supèrieure-Motion Assistive Robotic Exoskeleton for Superior Extremity) is a new exoskeleton robot designed with seven degrees of freedom (DOFs), as depicted in Fig. 1. Its primary purpose is to facilitate various rehabilitative movements for individuals with upper limb injuries. The design of the exoskeleton draws inspiration from the anatomical structure of the human upper limb, ensuring ergonomic compatibility for the user during therapy sessions. The shoulder segment is equipped with three joints, the elbow segment contains one joint, and the wrist segment consists of three joints. Each segment is integral to executing a wide range of upper limb motions. A comprehensive overview of the distinctive features of the ETS-MARSE, the modified Denavit-Hartenberg parameters, and comparisons with other existing exoskeletons can be found in [42, 43].

Accurate impedance control of the exoskeleton's movement relies heavily on feedback from a force sensor. For this purpose, a highly linear 6-axis force sensor has been selected to provide precise force feedback. This latter is strategically mounted at the tip of the exoskeleton, as illustrated in Fig. 1a. The joints of the ETS-MARSE are driven by Brushless DC motors (Maxon EC-45 and Maxon EC-90), which are coupled with harmonic drives that feature gear ratios of 120:1 for motors 1 and 2, and 100:1 for motors 3 through 7. Additionally, the placement of EMG electrodes (Delsys) on the upper limb muscles of the user is depicted in Fig. 1b.

2.2 The Dynamics of ETS-MARSE Robot

The dynamic behavior of the ETS-MARSE exoskeleton robot can be described using the Lagrangian formulation as follows:

$$\mathcal{M}(\theta)\ddot{\theta} + C(\theta, \dot{\theta})\dot{\theta} + \mathcal{G}(\theta) = \tau + \mathcal{J}^T f_h \tag{1}$$

In this equation, $\theta, \dot{\theta}$, and $\ddot{\theta} \in \mathbb{R}^7$ represent the vectors of joint positions, velocities, and accelerations, respectively. The terms $\mathcal{M}(\theta) \in \mathbb{R}^{7\times7}$, $C(\theta, \dot{\theta})\dot{\theta} \in \mathbb{R}^7$, and $\mathcal{G}(\theta) \in \mathbb{R}^7$ denote the symmetric positive definite inertia matrix, the Coriolis and centrifugal forces vector, and the gravitational force vector, respectively. The vector $\tau \in \mathbb{R}^7$ corresponds to the torques applied, while $f_h \in \mathbb{R}^6$ represents the interaction forces between the human and the exoskeleton. The Jacobian matrix is denoted by $\mathcal{J} \in \mathbb{R}^{6\times7}$.

Property 1 [44] The inertia matrix $\mathcal{M}(\theta)$ is symmetric and positive definite for all $\theta \in \mathbb{R}^7$.

Property 2 [44] The matrix $\dot{\mathcal{M}} - 2C$ is skew-symmetric for all $\theta, \dot{\theta} \in \mathbb{R}^7$.

Lemma 1 [45] *Let $\mathcal{V}(y)$ be a Lyapunov function that is continuous and positive definite, satisfying $\gamma_1(\|y\|) \le \mathcal{V}(y) \le \gamma_2(\|y\|)$. If its derivative $\dot{\mathcal{V}}(y) = \frac{d\mathcal{V}(y)}{dt}$ meets the condition $\dot{\mathcal{V}}(y) \le -\alpha_1\mathcal{V}(y) + \alpha_2$, where α_1 and α_2 are positive constants, then the solution y is guaranteed to be bounded.*

Assumption 1 Within a limited workspace, the Jacobian matrix is presumed to maintain non-singularity.

Remark 1 Unlike traditional methods [19, 20], which convert the robot's dynamic model from joint space to Cartesian space for impedance control similar to that of industrial robots [37], the proposed control approach maintains a focus on joint space. This is due to the direct connection between the user's upper limb and the exoskeleton robot.

2.3 Research Context

During the active rehabilitation phase examined in this chapter, the patient directly dictates the task trajectory through interactions with the exoskeleton robot, leaving the control system unaware of the trajectory. The primary objective of impedance control is to assist the exoskeleton's active motion toward the target position, driven by the user's applied force. This dynamic behavior of the exoskeleton robot is governed by a reference impedance model expressed as follows:

$$\mathcal{M}_d(\ddot{\theta} - \ddot{\theta}_d) + C_d(\dot{\theta} - \dot{\theta}_d) + \mathcal{G}_d(\theta - \theta_d) = \mathcal{J}^T f_h \tag{2}$$

In this equation, $\mathcal{M}_d$, C_d, and $\mathcal{G}_d \in \mathbb{R}^{7\times7}$ represent the desired inertia, damping, and spring coefficient matrices of the exoskeleton's end-effector, respectively, while θ_d indicates the target joint position.

The reference impedance model illustrates how the exoskeleton robot attains its actual position θ through the interaction force f_h exerted by the user. This model provides a tactile feedback mechanism, allowing the subject to sense a responsive force. To minimize fatigue and energy expenditure for the wearer, it is crucial to set θ_d in close proximity to θ. When the exoskeleton can accurately perceive the user's DMI and adjust θ_d accordingly, it will significantly reduce energy consumption.

Thus, the first objective of this chapter is to propose a robust method for measuring f_h based on the reference impedance model defined in Eq. (2). Following this, we aim to design θ_d by accurately estimating the DMI. Ultimately, we will develop an adaptive impedance control strategy that ensures the robot's dynamic model, as represented in Eq. (1), is effectively managed by the reference impedance model in Eq. (2), even in the presence of unknown dynamics.

3 Evaluating Interaction Forces Through sEMG Signals and End-Effector Sensors

When the exoskeleton robot is worn by a human and employed for various motion types—be it passive, active, or active-assisted—accurately measuring or approximating the total interaction forces between the human and the robot becomes essential. In a relaxed state of the arm, where no intentional movement occurs, the force at the end effector is primarily influenced by gravitational effects, represented mathematically as follows:

$$f_{Hg} = \mathcal{J}^T \mathcal{G}_H \tag{3}$$

Here, $\mathcal{J}^T$ denotes the Jacobian matrix, which serves as a mapping function between the gravitational space and the force field, while $\mathcal{G}_H$ accounts for the gravitational forces acting on the weight of the human arm.

When a person intends to move their arm, there exists a comparison between the internally generated forces—typically originating from the shoulder, including the biceps and triceps—and the torque generated at the elbow joint. This dynamic results in the movement of the hand toward the desired location, mathematically expressed as:

$$\mathcal{D}_{mov} = f_S - f_{Hg} \tag{4}$$

In this equation, f_S represents the force measured by the sensor located at the robot's tip, while $\mathcal{D}_{mov}$ indicates the direction in which the human intends to move. However, when the human arm encounters resistance in that direction, predicting the additional effort generated by the shoulder muscles to overcome this resistance becomes challenging. Unlike conventional approaches [46] that primarily focus on compensating for elbow joint torque while assuming that shoulder movement merely follows, our study integrates electromyographic (EMG) signals to account for both elbow torque and shoulder joint forces. Thus, the total force exerted by the human in a desired direction can be approximated by:

$$f_h = f_{EMG}^{Bs,Ts} + f_S - f_{Hg} \tag{5}$$

In this expression, $f_{EMG}^{Bs,Ts}$ represents the force generated by the biceps and triceps muscles, providing the necessary forces for both the shoulder and elbow joint torque. Here, f_h encapsulates the total force generated by the human upper limb, originating from the shoulder, forearm, and wrist joints. The computation of f_h will be elaborated upon in the subsequent subsections.

To integrate biological signals, we utilized surface electromyography (sEMG) signals [13] to derive $f_{EMG}^{Bs,Ts}$. This type of signal is widely applied in the study of human motion and neuromuscular diagnostics, particularly in clinical neurophysiology and electrodiagnostic medicine. The initial step involves affixing surface electrodes to the skin, as illustrated in Fig. 1b, to capture the overall muscle activity (biceps/triceps) corresponding to various motor units. We estimate the shoulder forces and elbow joint torque from the muscle activity of the biceps and triceps using Hill's muscle model [47] (see Fig. 4), which aids in understanding the physiological characteristics within the muscles. However, it is important to note that the arm's kinematics can influence the performance of Hill's model, necessitating calibration for each individual user. We will begin this section by discussing the kinematics of the human arm.

3.1 Kinematic Model of the Human Upper Limb

The kinematics of the human upper limb plays a crucial role in determining the moment arm for muscle forces. Figure 2 illustrates the kinematic structure of the upper limb, while the key parameters are summarized in Table 1.

It's essential to note that the notation $(.)^{Bs}$ and $(.)^{Ts}$ refers to the biceps and triceps, respectively. In contrast, the symbols $(.)_M$ and $(.)_T$ pertain to muscles and tendons, respectively. From the information presented in Fig. 2, we can derive the lengths of the biceps and triceps muscles and their corresponding tendons using the following equation:

$$l_{M,T}^{Bs,Ts} = l_{AB} + l_{BC} - 2l_{AB}l_{BC}\cos\left(q_2 + \gamma_0^{Bs,Ts}\right) \tag{6}$$

In this equation, $l_{M,T}^{Bs,Ts}$ denotes the lengths of the muscle and tendon for the biceps and triceps, while $\gamma_0^{Bs,Ts}$ represents the offset angles for these muscles, as detailed in Table 1.

Additionally, at any given moment, the angles for the biceps muscle (θ_{Bs}) and the triceps muscle (θ_{Ts}) can be expressed as:

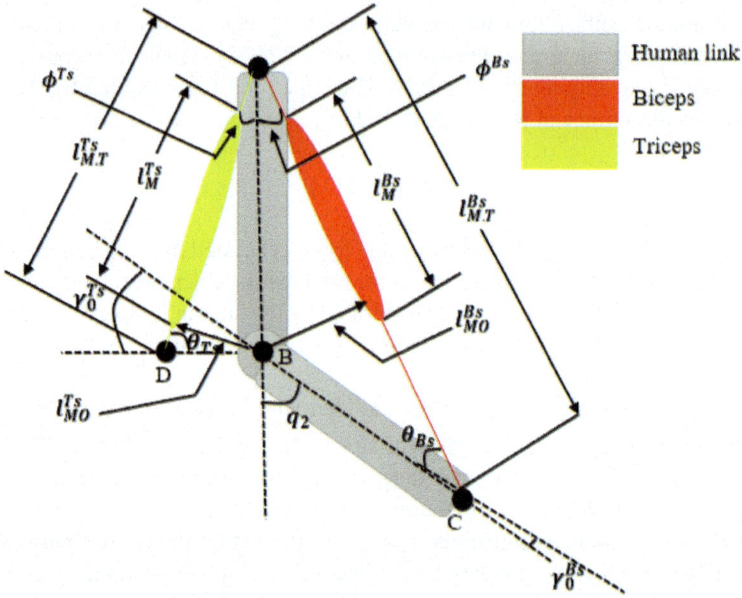

Fig. 2 Kinematic model of the human upper limb

Table 1 The kinematic parameters of human upper limb [46]

Parameters	Symbol	Value
Length of human link	l_{AB}	0.24 m
Length of human link	l_{BC}	0.031 m
Length of human link	l_{BD}	0.016 m
Length of biceps muscle	l_M^{Bs}	0.094 m
Length of biceps tendon length	l_T^{Bs}	0.152 m
Pennation angle of biceps	α^{Bs}	0°
Offset angle of biceps	γ_0^{Bs}	10°
Length of triceps muscle	l_M^{Ts}	0.056 m
Length of triceps tendon	l_T^{Ts}	0.121 m
Pennation angle of triceps	α^{Ts}	12°
Offset angle of triceps	γ_0^{Ts}	62°

$$\theta_{Bs,Ts} = \left(q_2 + \gamma_0^{Bs,Ts} + \phi^{Bs,Ts}\right) \tag{7}$$

The angle $\phi^{Bs,Ts}$ is calculated using:

$$\phi^{Bs,Ts} = \arctan\left(\frac{l_{BC,CD}\sin\left(q_2 + \gamma_0^{Bs,Ts}\right)}{l_{BC,CD}\sin\left(q_2 + \gamma_0^{Bs,Ts}\right) + l_{AB}}\right) \tag{8}$$

Furthermore, we can determine the moment arm for the biceps and triceps with the following expression:

$$l_{MO}^{Bs,Ts} = l_{BC,CD} \sin\left(\theta_{Bs,Ts}\right) \tag{9}$$

Here, $l_{MO}^{Bs,Ts}$ is instrumental in calculating the elbow joint effort exerted by the biceps and triceps muscles.

3.2 *Analysis of Shoulder Mechanics and Elbow Joint Torque via Hill's Model*

The method employed to extract force data from recorded surface electromyography (sEMG) signals through Hill's model (refer to Fig. 3) involves a series of systematic steps. Initially, the sEMG signal undergoes correction, filtering, and normalization against its peak values, resulting in a processed signal denoted as $e_s(t)$. Next, it is essential to consider the activation dynamics of the muscle, given that the sEMG signal relates to its past values. This process yields a measure known as neural activation, represented by $u(t)$. Subsequently, the neural activation can be converted into muscle activation, denoted as $a(t)$, which serves as the input for Hill's model [47]. This allows for the determination of muscle forces, as illustrated in Fig. 3.

The dynamics of muscle activation can be articulated concerning neural activation $u(t)$ as follows:

$$a(t)^{Bs,Ts} = \frac{e^{Au(t)^{Bs,Ts}-1}}{e^{A}-1} \tag{10}$$

where $a(t)^{Bs,Ts}$ signifies the activation dynamics of the biceps and triceps. The parameter $-3 < A < 0$ represents a nonlinear component. Hill's model provides a comprehensive framework for estimating joint torque by characterizing the muscle-tendon function [48]. Typically, this model comprises three primary elements: a contractile component, a parallel elastic component, and a series elastic component (see Fig. 4). The force produced by each musculotendinous unit can be expressed as:

$$F_{M,T}^{Bs,Ts} = F_{M}^{Bs,Ts} \cos(\alpha^{Bs,Ts}) \tag{11}$$

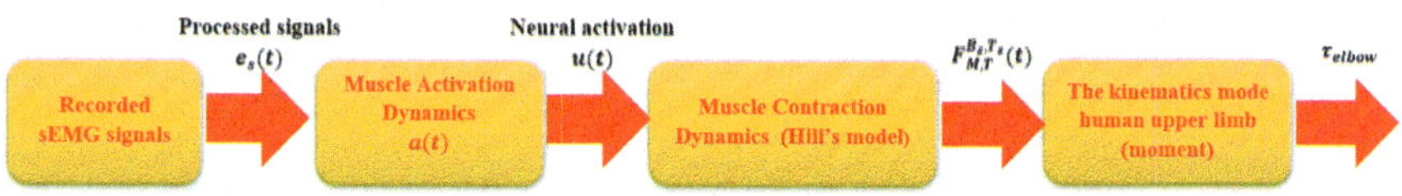

Fig. 3 Musculoskeletal model depicting the torque exerted by the elbow joint

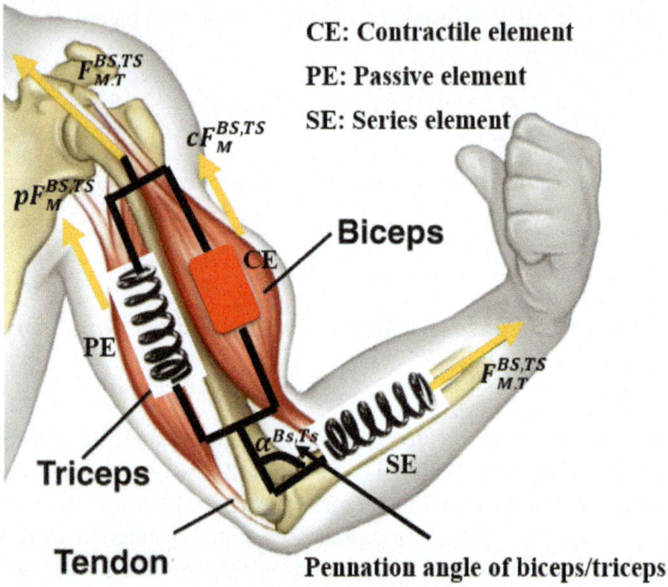

Fig. 4 Hill's model representation for the human upper limb

where $F_M^{Bs,Ts}$ represents the force generated by the biceps or triceps, and $\alpha^{Bs,Ts}$ indicates the pennation angle of the respective muscles. The generated force $F_M^{Bs,Ts}$ can be described as:

$$F_M^{Bs,Ts} = cF_M^{Bs,Ts} + pF_M^{Bs,Ts} \tag{12}$$

where $cF_M^{Bs,Ts}$ and $pF_M^{Bs,Ts}$ denote the contractile muscle force and the passive elastic normalized force-length relation for the biceps/triceps, respectively. The passive elastic normalized force-length relationship can be expressed as:

$$pF_M^{Bs,Ts} = \frac{e^{(l_M^{Bs,Ts})-1}}{e^5} \tag{13}$$

while the contractile muscle force is defined as:

$$cF_M^{Bs,Ts} = f(\vartheta^{BS,TS}) f(l_M^{Bs,Ts}) a(t)^{Bs,Ts} f_{M0}^{Bs,Ts} \tag{14}$$

where $f_{M0}^{Bs,Ts}$ represents the maximum isometric force of the biceps/triceps. Here, $f(l_M^{Bs,Ts})$ illustrates the force-length relationship of the contractile element (for biceps/triceps), and $f(\vartheta^{BS,TS})$ denotes the force-contraction velocity relationship, given by:

$$f(\vartheta^{BS,TS}) = \frac{f_{M0}^{Bs,Ts} b - a\vartheta^{BS,TS}}{b + \vartheta^{BS,TS}} \tag{15}$$

where $a \approx 2.5 f_{MO}^{Bs,Ts}$, $b = \frac{a}{f_{MO}^{Bs,Ts}} \vartheta_0^{Bs,Ts}$, $\vartheta^{BS,TS}$ represents the contractile velocity of the muscle for the biceps/triceps, and $\vartheta_0^{Bs,Ts}$ indicates the maximum contraction velocity, estimated as $10 l_{MO}^{Bs,Ts}$ [47], with $l_{MO}^{Bs,Ts}$ denoting the maximum muscle length (biceps/triceps) [49]. Furthermore, $f(l_M^{Bs,Ts})$ in Eq. (14) is defined as:

$$f(l_M^{Bs,Ts}) = -\delta_1 (\tilde{l}_M^{Bs,Ts})^2 + \delta_2 (\tilde{l}_M^{Bs,Ts})^2 + \delta_3 \tag{16}$$

where $\delta_1 = 6.25$, $\delta_2 = 12.5$, and $\delta_3 = 5.25$ are constant factors [47]. Here, $\tilde{l}_M^{Bs,Ts} = \frac{l_M^{Bs,Ts}}{l_{MO}^{Bs,Ts}}$ [49].

By integrating the passive force of the muscles from Eq. (13) and the contractile muscle forces from Eq. (14) for the biceps/triceps muscles into Eq. (11) and utilizing Eq. (12), we derive:

$$F_{M,T}^{Bs,Ts} = (cF_M^{Bs,Ts} + pF_M^{Bs,Ts}) \cos(\alpha^{Bs,Ts}) \tag{17}$$

where $F_{M,T}^{Bs,Ts}$ represents the combined force generated by the shoulder, incorporating both biceps and triceps contributions. From Eq. (17), we can ascertain the torque of the elbow joint using the kinematic model of the human upper limb, specifically by calculating the muscle force (biceps/triceps) and multiplying it by the moment arm for these muscles as follows:

$$\tau_{elbow} = F_M^{Bs} l_{MO}^{Bs} + F_M^{Ts} l_{MO}^{Ts} \tag{18}$$

where l_{MO}^{Bs} and l_{MO}^{Ts} are defined in Eq. (9). The elbow joint torque, computed from biological EMG signals, can be related to the moment arm through the following expression:

$$F_{elbow} = l_{fa} \tau_{elbow} \tag{19}$$

where l_{fa} denotes the length of the robotic forearm. Thus, the forces produced by the shoulder and elbow can be consolidated as follows:

$$f_{EMG}^{Bs,Ts} = \begin{bmatrix} F_{M,T}^{Bs,Ts} \\ F_{elbow} \end{bmatrix} \tag{20}$$

3.3 Refining the Calibration of Hill's Model

This section details the calibration procedures employed for optimizing Hill's model (Eqs. 10–17). An experiment was conducted with a healthy male subject (95 kg, 177 cm, 35 years old). The subject was positioned upright in front of the setup, grasping the wrist interface of a specially designed robotic apparatus that generates torque corresponding to tip of the wrist joint (see Fig. 1). This joint was outfitted with a 6-axis force-torque sensor to accurately capture the applied forces. Guided by visual feedback, the subject was instructed to exert and sustain forces of 1N, 2N, 4N, 8 N, 12 N, and 16 N in four cardinal directions: $[\pm X, \pm Y]$ and along the $[\pm Z]$ axes. The subject repeated these tasks five times to facilitate a robust estimation of Hill's model parameters. Throughout the trials, surface electromyography (sEMG) data were recorded from four distinct muscles using surface electrodes. The sampling rate was set to 1024 Hz, and the signals underwent bandpass filtering in the 10–500 Hz range, along with notch filtering at 50 Hz. All filters implemented were Butterworth filters, configured within the software environment. To align the extracted signals with actual muscle activations, the maximum voluntary contraction (MVC) values were employed for normalization. The endpoint forces and the muscle activation levels were meticulously documented during the experimental sessions, with parameter estimation carried out via a least-squares error approach, utilizing data from 62 trials. During the experiments, the subject's limb remained tense between 5 to 20 s, while in the remaining intervals, it was allowed to relax. Notable variations in the endpoint forces can be observed, illustrated by the thick blue line, which represents the average stiffness across the X, Y, and Z directions throughout the five tasks.

4 Estimation of DMI

4.1 Dynamic Modeling of the Human Upper Limb

The precise assessment of the DMI is crucial for ensuring that the exoskeleton adapts seamlessly to its user, thereby facilitating effective active rehabilitation. The dynamic representation of the human upper limb can be articulated as a second-order system, specifically a mass-damper-spring configuration [37]:

$$\mathcal{M}_h\ddot{\theta}_h + \mathcal{C}_h\dot{\theta}_h + \mathcal{G}_h\theta_h = \mathcal{J}^T f_h \tag{21}$$

In this formulation, $\mathcal{M}_h$, $\mathcal{C}_h$, and $\mathcal{G}_h \in \mathbb{R}^{7\times 7}$ represent the inertia, damping, and spring matrices of the human end-effector, respectively. Here, $\theta_h \in \mathbb{R}^7$ denotes the user's motion intention.

In contrast to conventional approaches [37, 50, 51] that simplify the dynamic model to merely a damping-spring system, our analysis considers the comprehensive dynamics of the user's upper limb as expressed in Eq. (21). Without compromising generality, the user's dynamic model matrices can be represented as follows:

$$\begin{cases} \mathcal{M}_h = \mathcal{M}_{h0} + \Delta\mathcal{M}_h \\ \mathcal{C}_h \; = \mathcal{C}_{h0} + \Delta\mathcal{C}_h \end{cases} \tag{22}$$

where $\mathcal{M}_{h0}$ and $\mathcal{C}_{h0}$ denote the known components of the mass and damping matrices, while $\Delta\mathcal{M}_h$ and $\Delta\mathcal{C}_h$ account for the uncertainties. Introducing new variables, we define $z_1 = \theta_h$ and $z_2 = \dot{\theta}_h$. Consequently, the system described in Eq. (22) can be restructured as follows:

$$\begin{cases} \dot{z}_1 = z_2 \\ \dot{z}_2 = \mathcal{U}(t) + f_0(t) + \omega(t) \end{cases} \tag{23}$$

where $\mathcal{U} = \mathcal{M}_{h0}^{-1}[\mathcal{J}^T f_h]$, $\omega(t) = \mathcal{M}_{h0}^{-1}[-\Delta\mathcal{M}_h\ddot{\theta}_h - \Delta C_h\dot{\theta}_h - \mathcal{G}_h\theta_h]$, and $f_0(t) = \mathcal{M}_{h0}^{-1}[-C_{h0}\dot{\theta}_h]$.

To estimate the state z_2, we design the following estimator:

$$\dot{\hat{z}}_2 = \mathcal{U}(t) + f_0(t) + \hat{\mathcal{W}}^T \mathcal{S}(Z) + \mathcal{E} - \lambda_1 \, |\tilde{z}_2| \, \text{sign}(\tilde{z}_2) \tag{24}$$

where $\tilde{z}_2 = \hat{z}_2 - z_2$ represents the estimation error, and $\lambda_1 = diag(\lambda_{1ii}) > 0$, $i \in 1, \ldots, 7$. Given that $\omega(t)$ represents the uncertain part of the user's dynamic model, we employ a Radial Basis Function Neural Network (RBFNN) to approximate this unknown component as follows:

$$\begin{cases} \dot{\hat{z}}_2 = \mathcal{U}(t) + f_0(t) + \hat{\mathcal{W}}^T \mathcal{S}(Z) + \mathcal{E} - \lambda_1 \, |\tilde{z}_2| \, \text{sign}(\tilde{z}_2) \\ \dot{\hat{\mathcal{W}}} = \Gamma\left(\mathcal{S}(Z)\tilde{z}_2 + \varkappa\hat{\mathcal{W}}\right) \end{cases} \tag{25}$$

where $\Gamma = diag(\Gamma_{1ii}) > 0, i \in 1, \ldots, 7$, and $\mathcal{E}$ denotes the bounded neural network approximation error. The term $\hat{\mathcal{W}}^T\mathcal{S}(Z)$ approximates $\mathcal{W}^T\mathcal{S}(Z)$, with $\mathcal{W}$ being the optimal weights and $\mathcal{S}(Z)$ the basis functions of the neural network. When the upper limb is coupled to the exoskeleton, we can define $Z = \left[\theta^T, \dot{\theta}^T, \ddot{\theta}^T\right]$. The parameter $\varkappa$ is determined as follows: $\lim_{t\to\infty} \varkappa = 0$ and $\int_0^t \varkappa dt = Q_i < \infty$. In practical scenarios, we choose $\varkappa$ as: $\varkappa = \frac{1}{1+t^2}$, where t represents the execution time.

Next, we define the error dynamics of the estimator as follows:

$$\begin{cases} \dot{\tilde{z}}_2 = -\lambda_1 \, |\tilde{z}_2| \, \text{sign}(\tilde{z}_2) + \tilde{\mathcal{W}}^T \mathcal{S}(Z) + \mathcal{E} \\ \dot{\hat{\mathcal{W}}} = \Gamma\left(\mathcal{S}(Z)\tilde{z}_2 + \varkappa\hat{\mathcal{W}}\right) \end{cases} \tag{26}$$

where $\tilde{\mathcal{W}} = \hat{\mathcal{W}} - \mathcal{W}$. To evaluate the stability of the estimator, we propose the following Lyapunov function:

$$\mathcal{V}_e = \frac{1}{2}\tilde{z}_2^T\tilde{z}_2 + \frac{1}{2}\sum_{i=1}^{7}\tilde{\mathcal{W}}_i^T\Gamma_i^{-1}\tilde{\mathcal{W}}_i \tag{27}$$

Calculating the time derivative of $\mathcal{V}_e$ and utilizing Eq. (26), we arrive at:

$$\dot{\mathcal{V}}_e = -\tilde{z}_2^T \lambda_1 |\tilde{z}_2| + \tilde{z}_2^T \tilde{\mathcal{W}}^T \mathcal{S}(Z) + \tilde{z}_2^T \mathcal{E} - \sum_{i=1}^{7} \tilde{\mathcal{W}}_i^T \mathcal{S}_i(Z)\tilde{z}_2 - \sum_{i=1}^{7} \tilde{\mathcal{W}}_i^T \varkappa_i \hat{\mathcal{W}}_i \tag{28}$$

Utilizing the inequality : $-\tilde{\mathcal{W}}_i^T \hat{\mathcal{W}}_i = -\tilde{\mathcal{W}}_i^T (\mathcal{W}_i + \tilde{\mathcal{W}}_i) = -\tilde{\mathcal{W}}_i^T \mathcal{W}_i - \tilde{\mathcal{W}}_i^T \tilde{\mathcal{W}}_i$ and $-\tilde{\mathcal{W}}_i^T \mathcal{W}_i \leq \frac{1}{2}(\tilde{\mathcal{W}}_i^T \tilde{\mathcal{W}}_i + \mathcal{W}_i^T \mathcal{W}_i)$. we have: $-\tilde{\mathcal{W}}_i^T \hat{\mathcal{W}}_i = -\frac{1}{2}\tilde{\mathcal{W}}_i^T \tilde{\mathcal{W}}_i + \frac{1}{2}\mathcal{W}_i^T \mathcal{W}_i$ and $\tilde{z}_2^T \mathcal{E} = \frac{\tilde{z}_2^T z_2}{2} + \frac{\|\mathcal{E}\|^2}{2}$. Taking into consideration this fact, we obtain:

$$\dot{\mathcal{V}}_e \leq -\tilde{z}_2^T (\lambda_1 - \frac{1}{2} I_{7\times 7}) |\tilde{z}_2| - \frac{1}{2}\sum_{i=1}^{7} \varkappa_i \tilde{\mathcal{W}}_i^T \tilde{\mathcal{W}}_i + \frac{1}{2}\sum_{i=1}^{7} \varkappa_i \mathcal{W}_i^T \mathcal{W}_i + \frac{\|\mathcal{E}\|^2}{2}$$

$$\dot{\mathcal{V}}_e \leq -\alpha_1 \mathcal{V}_e + \alpha_2 \tag{29}$$

In this context, we define

$$\alpha_1 = \min\left(\gamma_{\min}\left(\lambda_1 - \frac{1}{2} I_{7\times 7}\right), \frac{\varkappa_i}{2}\right)$$

and

$$\alpha_2 = \frac{\|\mathcal{E}\|^2}{2},$$

where $\gamma_{\min}$ denotes the smallest eigenvalue of the matrix $\left(\lambda_1 - \frac{1}{2} I_{7\times 7}\right)$. Additionally, the term

$$\left(-\frac{1}{2}\sum_{i=1}^{7} \varkappa_i \tilde{\mathcal{W}}_i^T \tilde{\mathcal{W}}_i + \frac{1}{2}\sum_{i=1}^{7} \varkappa_i \mathcal{W}_i^T \mathcal{W}_i\right) \to 0$$

follows from the established definition of $\varkappa$ in Eq. (25).

To confirm stability, we need to check the condition

$$\left(\lambda_1 - \frac{1}{2} I_{7\times 7}\right) > 0.$$

This condition indicates that $\dot{\mathcal{V}}_e \leq 0$ holds true whenever

$$\mathcal{V}_e \geq \frac{\alpha_2}{\alpha_1}.$$

According to Lemma 1, this guarantees that the estimation error remains bounded, ensuring that the estimated DMI (θ_h) is also constrained. The proof is thereby concluded. ■

We can now establish the desired trajectory for the reference model, denoted as $\hat{\theta}_h$, which represents the user's DMI. This can be expressed as follows:

$$\theta_d = \theta + \hat{\theta}_h \tag{30}$$

In this expression, θ represents the current position of the exoskeleton robot, whereas $\hat{\theta}_h$ corresponds to the user's desired motion, estimated using the model described in Eq. (25). When $\hat{\theta}_h \rightarrow 0$, it indicates that the wearer of the exoskeleton has ceased to exert any forces, resulting in a gradual reduction in the robot's motion. Consequently, as θ_d approaches θ, the exoskeleton will come to rest at its last recorded position. This operational mode allows the exoskeleton to effectively align with the user's DMI.

Next, we will implement a reference model-based adaptive impedance control strategy. This approach aims to ensure that the dynamics of the exoskeleton arm, as described in Eq. (1), converge toward the specified reference impedance model outlined in Eq. (2).

5 Dynamic Adaptive Impedance Control Strategy

Following on the approximation of the user's DMI outlined in the preceding section, the updated planned trajectory is formulated as shown in Eq. (30). To achieve this, we introduce an new adaptive impedance control strategy leveraging a Modified Function Approximation Technique (FAT). This method is designed to guarantee precise alignment of the exoskeleton robot with the target reference impedance model, promoting optimal interaction and coordination between the user and the exoskeleton. The proposed control framework is depicted in Fig. 5.

To begin with, we must ascertain that the dynamics of the exoskeleton robot can align with the dynamics of the desired reference model. This process involves calculating the discrepancy between the system dynamics and the reference impedance model, commonly referred to as the matching error [52, 53]. From the reference model in Eq. (2), we obtain:

$$\mathbb{P}(t) = \mathcal{M}_d \ddot{e} + C_d \dot{e} + \mathcal{G}_d e - \mathcal{J}^T f_h \tag{31}$$

where $e = \theta - \theta_d$. The objective is to achieve $\lim_{t \to \infty} \mathbb{P}(t) = 0$, indicating that the exoskeleton's dynamics as described in Eq. (1) perfectly match the reference impedance model outlined in Eq. (2). To facilitate the analysis, we define the augmented matching error as follows:

$$\bar{\mathbb{P}}(t) = \mathcal{K}_f \mathbb{P}(t) = \ddot{e} + \mathcal{K}_d \dot{e} + \mathcal{K}_p e - \mathcal{K}_f f_H \tag{32}$$

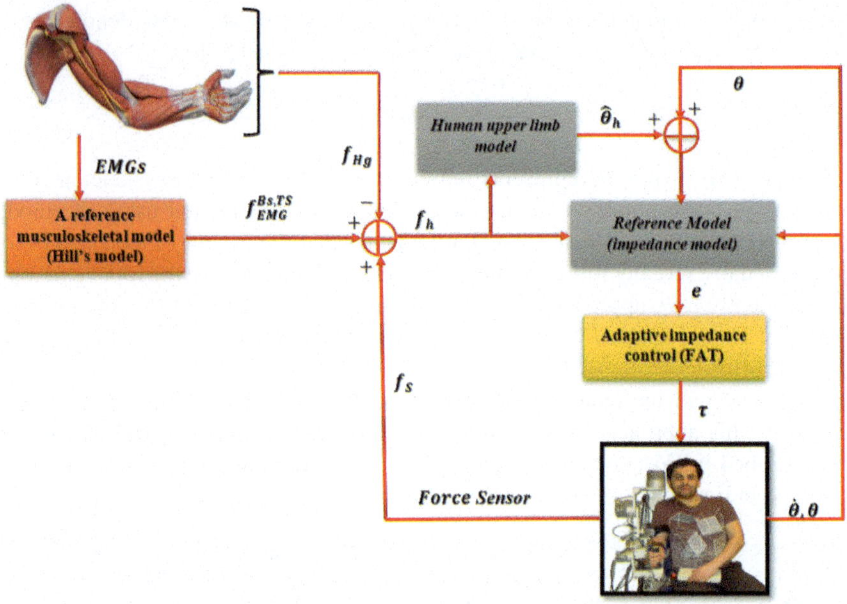

Fig. 5 Overall control strategy diagram

Here, $f_H = \mathcal{J}^T f_h, \mathcal{K}_f = \mathcal{M}_d^{-1}, \mathcal{K}_d = \mathcal{M}_d^{-1} C_d$, and $\mathcal{K}_p = \mathcal{M}_d^{-1} \mathcal{G}_d$. We will select two positive-definite matrices, Λ and Υ, that satisfy $\Lambda + \Upsilon = \mathcal{K}_d$ and $\Lambda\Upsilon + \dot{\Lambda} = \mathcal{K}_p$. The augmented matching error can then be reformulated as:

$$\bar{\mathbb{P}}(t) = \ddot{e} + (\Lambda + \Upsilon)\dot{e} + \left(\Lambda\Upsilon + \dot{\Lambda}\right)e - \dot{\mathcal{F}} - \Upsilon\mathcal{F} \tag{33}$$

where $\mathcal{F}$ satisfies $\mathcal{K}_f f_H = \dot{\mathcal{F}} + \Upsilon\mathcal{F}$. We can now introduce an auxiliary variable, $\mathbb{S}$, defined as the filtered matching error:

$$\mathbb{S} = \dot{e} + \Lambda e - \mathcal{F} \tag{34}$$

Using Eq. (34), we can express the augmented error in Eq. (33) as follows:

$$\bar{\mathbb{P}}(t) = \dot{\mathbb{S}} + \Upsilon\mathbb{S} \tag{35}$$

It is evident from Eq. (35) that $\lim_{t\to\infty} \mathbb{S}(t) = 0$ implies $\lim_{t\to\infty} \dot{\mathbb{S}}(t) = 0$, assuming that the limit $\lim_{t\to\infty} \dot{\mathbb{S}}(t) = 0$ exists. Considering Eqs. (32) and (33), we find that $\lim_{t\to\infty} \mathbb{P}(t) = 0$ if $\lim_{t\to\infty} \mathbb{S}(t) = 0$. Thus, the control objective is achieved by ensuring:

$$\lim_{t\to\infty} \mathbb{S}(t) = 0 \tag{36}$$

To achieve this objective, the error states of the system can be defined as follows:

$$\mathbb{S} = \dot{\theta} - \psi \tag{37}$$
$$\psi = \dot{\theta}_d + \Lambda e - F \tag{38}$$
$$\sigma = \dot{\psi} \tag{39}$$
$$e = \theta - \theta_d \tag{40}$$

where $\theta_d \in \mathbb{R}^7$ represents the desired joint trajectory, and $\theta \in \mathbb{R}^7$ is the measured joint position. Based on Eqs. (34–40), we can reformulate the dynamic equation of the exoskeleton robot, originally described in Eq. (1), as follows:

$$\mathcal{M}\dot{\mathbb{S}} + \mathcal{C}\mathbb{S} + \mathcal{M}\sigma + \mathcal{C}\psi + \mathcal{G} = \tau + f_H \tag{41}$$

where $f_H = \mathcal{J}^T f_h$. Considering that the system matrices $\mathcal{M}$, $\mathcal{C}$, and $\mathcal{G}$ are not explicitly known, a control law can be formulated to guarantee the stability of the robot system as follows:

$$\tau = \hat{\mathcal{M}}\sigma + \hat{\mathcal{C}}\psi + \hat{\mathcal{G}} - K_1\mathbb{S} - f_H \tag{42}$$

where $K_1 = diag(K_{1ii}) > 0$ and $i \in i = 1, \ldots, 7$, and $\hat{\mathcal{M}}$, $\hat{\mathcal{C}}$, and $\hat{\mathcal{G}}$ are the estimates of $\mathcal{M}$, $\mathcal{C}$, and $\mathcal{G}$, respectively. Rather than utilizing the traditional FAT approach to approximate the dynamic model matrices [39, 40], we eliminate the use of basis functions by defining the following expressions:

$$\begin{cases} \mathcal{M}_e = \mathcal{M} - \mathcal{E}_{\mathcal{M}} \\ \mathcal{C}_e \ = \mathcal{C} - \mathcal{E}_{\mathcal{C}} \\ \mathcal{G}_e \ = \mathcal{G} - \mathcal{E}_{\mathcal{G}} \end{cases} \tag{43}$$

where $\mathcal{M}_e \in \mathbb{R}^{7\times7}$, $\mathcal{C}_e \in \mathbb{R}^{7\times7}$, and $\mathcal{G}_e \in \mathbb{R}^7$ represent the optimal approximations of the inertia matrix, the Coriolis and centrifugal matrix, and the gravitational vector, respectively. The terms $\mathcal{E}_{\mathcal{M}} \in \mathbb{R}^{7\times7}$, $\mathcal{E}_{\mathcal{C}} \in \mathbb{R}^{7\times7}$, and $\mathcal{E}_{\mathcal{G}} \in \mathbb{R}^7$ correspond to the bounded errors associated with these approximations. The control input defined in Eq. (42) can be reformulated as:

$$\tau = \hat{\mathcal{M}}_e\sigma + \hat{\mathcal{C}}_e\psi + \hat{\mathcal{G}}_e - K_1\mathbb{S} - f_H \tag{44}$$

By substituting this expression into Eq. (41), we derive the closed-loop dynamics of the exoskeleton system as follows:

$$\mathcal{M}\dot{\mathbb{S}} + \mathcal{C}\mathbb{S} + K_1\mathbb{S} = -\tilde{\mathcal{M}}_e\sigma - \tilde{\mathcal{C}}_e\psi - \tilde{\mathcal{G}}_e + \mathcal{E}_D \tag{45}$$

Here, $\tilde{\mathcal{M}}_e = \mathcal{M}_e - \hat{\mathcal{M}}_e$, $\tilde{\mathcal{C}}_e = \mathcal{C}_e - \hat{\mathcal{C}}_e$, and $\tilde{\mathcal{G}}_e = \mathcal{G}_e - \hat{\mathcal{G}}_e$ represent the estimation errors for $\mathcal{M}_e$, $\mathcal{C}_e$, and $\mathcal{G}_e$, respectively. Additionally, $\mathcal{E}_D = \mathcal{E}(\mathcal{E}_{\mathcal{M}}, \mathcal{E}_{\mathcal{C}}, \mathcal{E}_{\mathcal{G}}) \in \mathbb{R}^7$ denotes the combined approximation error. Assuming an appropriate adaptation law for $\tilde{\mathcal{M}}_e$, $\tilde{\mathcal{C}}_e$, and $\tilde{\mathcal{G}}_e$, we can establish:

$$\tilde{\mathcal{M}}_e \to 0, \quad \tilde{\mathcal{C}}_e \to 0, \quad \tilde{\mathcal{G}}_e \to 0$$

As a result, the control input defined in Eq. (44) guarantees both the system's stability and the achievement of the desired performance.

Remark 2 It is important to emphasize that $\mathcal{E}_{\mathcal{M}}$, $\mathcal{E}_{\mathcal{C}}$, and $\mathcal{E}_{\mathcal{G}}$ cannot be overlooked. Nevertheless, their fluctuations remain within defined limits, encapsulated by the overall approximation error $\mathcal{E}D$, which adheres to the following constraint:

$$\|\mathcal{E}_D\| \leq \gamma_c$$

where γ_c is a small positive constant.

To demonstrate the stability of the exoskeleton system, we define a Lyapunov candidate function as:

$$\mathcal{V} = \frac{1}{2}\mathbb{S}^T \mathcal{M}\mathbb{S} + \frac{1}{2}\,\mathrm{Tr}\left(\tilde{\mathcal{M}}_e^T \Lambda_{\mathcal{M}_e} \tilde{\mathcal{M}}_e + \tilde{\mathcal{C}}_e^T \Lambda_{\mathcal{C}_e} \tilde{\mathcal{C}}_e + \tilde{\mathcal{G}}_e^T \Lambda_{\mathcal{G}_e} \tilde{\mathcal{G}}_e\right) \tag{46}$$

where $\Lambda_{\mathcal{M}_e} \in \mathbb{R}^{7\times7}$, $\Lambda_{\mathcal{C}_e} \in \mathbb{R}^{7\times7}$, and $\Lambda_{\mathcal{G}_e} \in \mathbb{R}^{7\times7}$ are diagonal positive-definite matrices, and $\mathrm{Tr}(\cdot)$ denotes the matrix trace operation. The temporal derivative of Eq. (46) is expressed as:

$$\dot{\mathcal{V}} = \mathbb{S}^T \mathcal{M}\dot{\mathbb{S}} + \frac{1}{2}\mathbb{S}^T \dot{\mathcal{M}}\mathbb{S} + \mathrm{Tr}\left(\dot{\tilde{\mathcal{M}}}_e^T \Lambda_{\mathcal{M}_e} \tilde{\mathcal{M}}_e + \dot{\tilde{\mathcal{C}}}_e^T \Lambda_{\mathcal{C}_e} \tilde{\mathcal{C}}_e + \dot{\tilde{\mathcal{G}}}_e^T \Lambda_{\mathcal{G}_e} \tilde{\mathcal{G}}_e\right) \tag{47}$$

Substituting $\mathcal{M}\dot{\mathbb{S}}$ from Eq. (45), and applying the control input from Eqs. (44, 47) is transformed into:

$$\begin{aligned}\dot{\mathcal{V}} = \mathbb{S}^T \left(\frac{1}{2}\dot{\mathcal{M}} - \mathcal{C}\right)\mathbb{S} - \mathbb{S}^T K_1 \mathbb{S} - \mathbb{S}^T \tilde{\mathcal{M}}_e \sigma - \mathbb{S}^T \tilde{\mathcal{C}}_e \psi - \mathbb{S}^T \tilde{\mathcal{G}}_e + \mathbb{S}^T \mathcal{E}_D \\ + Tr\left(\dot{\tilde{\mathcal{M}}}_e^T \Lambda_{\mathcal{M}_e} \tilde{\mathcal{M}}_e + \dot{\tilde{\mathcal{C}}}_e^T \Lambda_{\mathcal{C}_e} \tilde{\mathcal{C}}_e + \dot{\tilde{\mathcal{G}}}_e^T \Lambda_{\mathcal{G}_e} \tilde{\mathcal{G}}_e\right)\end{aligned} \tag{48}$$

Using Property 2, the above equation simplifies to:

$$\begin{aligned}\dot{\mathcal{V}} = -\mathbb{S}^T K_1 \mathbb{S} + \mathbb{S}^T \mathcal{E}_D + \mathrm{Tr}\Big[\left(\dot{\tilde{\mathcal{M}}}_e^T \Lambda_{\mathcal{M}_e} - \sigma \mathbb{S}^T\right)\tilde{\mathcal{M}}_e + \left(\dot{\tilde{\mathcal{C}}}_e^T \Lambda_{\mathcal{C}_e} - \psi \mathbb{S}^T\right)\tilde{\mathcal{C}}_e \\ + \left(\dot{\tilde{\mathcal{G}}}_e^T \Lambda_{\mathcal{G}_e} - \mathbb{S}^T\right)\tilde{\mathcal{G}}_e\Big]\end{aligned} \tag{49}$$

Considering the relationship $\dot{\tilde{(\cdot)}}e = (\dot{.})e - \dot{\hat{(\cdot)}}_e$, Eq. (49) can be rewritten as:

$$\dot{\mathcal{V}} = -\mathbb{S}^T K_1 \mathbb{S} + \mathbb{S}^T \mathcal{E}_D + Tr\left(\left(\dot{\hat{\mathcal{M}}}_e^T \Lambda_{\mathcal{M}_e} - \sigma \mathbb{S}^T\right) \tilde{\mathcal{M}}_e\right) + Tr\left(\left(\dot{\hat{C}}_e^T \Lambda_{C_e} - \psi^T\right) \tilde{C}_e\right)$$
$$+Tr\left(\left(\dot{\hat{\mathcal{G}}}_e^T \Lambda_{\mathcal{G}_e} - s^T\right) \tilde{\mathcal{G}}_e\right) \tag{50}$$

The adaptation laws are formulated as follows to guarantee the stability of the exoskeleton system:

$$\begin{cases} \dot{\hat{\mathcal{M}}}_e^T = -\left(\sigma \mathbb{S}^T + \kappa_{\mathcal{M}_e} \hat{\mathcal{M}}_e^T\right) \Lambda_{\mathcal{M}_e}^{-1}, \\ \dot{\hat{C}}_e^T = -\left(\psi \mathbb{S}^T + \kappa_{C_e} \hat{C}_e^T\right) \Lambda_{C_e}^{-1}, \\ \dot{\hat{\mathcal{G}}}_e^T = -\left(\mathbb{S}^T + \kappa_{\mathcal{G}_e} \hat{\mathcal{G}}_e^T\right) \Lambda_{\mathcal{G}_e}^{-1}. \end{cases} \tag{51}$$

Here, $\kappa_{(\cdot)_e}$ is chosen such that:$\lim_{t\to\infty} \kappa_{(\cdot)_e} = 0$ and $\int_0^t \kappa_{(\cdot)_e}\, dt = Q_c < \infty$. A practical choice is $\kappa_{(\cdot)_e}$ as: $\kappa_{(\cdot)_e} = \dfrac{1}{1+t^2}$. By substituting Eq. (51) into Eq. (50), we obtain:

$$\dot{\mathcal{V}} = -\mathbb{S}^T K_1 \mathbb{S} + \mathbb{S}^T \mathcal{E}_D + \kappa_{\mathcal{M}_e} \operatorname{Tr}\left(\hat{\mathcal{M}}_e^T \tilde{\mathcal{M}}_e\right) + \kappa_{C_e} \operatorname{Tr}\left(\hat{C}_e^T \tilde{C}_e\right) + \kappa_{\mathcal{G}_e} \operatorname{Tr}\left(\hat{\mathcal{G}}_e^T \tilde{\mathcal{G}}_e\right) \tag{52}$$

Applying Remark 2, we use the inequality:

$$\mathbb{S}^T \mathcal{E}_D \leq \frac{\mathbb{S}^T \mathbb{S}}{2} + \frac{\gamma_c^2}{2}$$

$$Tr\left(\left(\hat{\cdot}_e\right)^T \left(\tilde{\cdot}_e\right)\right) \leq \frac{1}{2} Tr\left((\cdot_e)^T (\cdot_e)\right) - \frac{1}{2} Tr\left(\left(\tilde{\cdot}_e\right)^T \left(\tilde{\cdot}_e\right)\right)$$

With these relations, Eq. (52) is refined into:

$$\begin{aligned} \dot{\mathcal{V}} \leq -\mathbb{S}^T \left(K_1 - \frac{1}{2} I_{7\times7}\right) \mathbb{S} - \frac{\kappa_{\mathcal{M}_e}}{2} \operatorname{Tr}\left(\tilde{\mathcal{M}}_e^T \tilde{\mathcal{M}}_e\right) - \frac{\kappa_{C_e}}{2} \operatorname{Tr}\left(\tilde{C}_e^T \tilde{C}_e\right) \\ - \frac{\kappa_{\mathcal{G}_e}}{2} \operatorname{Tr}\left(\tilde{\mathcal{G}}_e^T \tilde{\mathcal{G}}_e\right) + \frac{\kappa_{\mathcal{M}_e}}{2} \operatorname{Tr}\left(\mathcal{M}_e^T \mathcal{M}_e\right) \\ + \frac{\kappa_{C_e}}{2} \operatorname{Tr}\left(C_e^T C_e\right) + \frac{\kappa_{\mathcal{G}_e}}{2} \operatorname{Tr}\left(\mathcal{G}_e^T \mathcal{G}_e\right) + \frac{\gamma_c^2}{2} \\ \dot{\mathcal{V}} \leq -\alpha_1 + \alpha_2 \end{aligned} \tag{53}$$

where

$$\alpha_1 = \min\left(\gamma_{\min}\left(K_1 - \frac{1}{2} I_{7\times7}\right), \frac{\kappa_{\mathcal{M}_e}}{2}, \frac{\kappa_{C_e}}{2}, \frac{\kappa_{\mathcal{G}_e}}{2}\right)$$

and

$$\alpha_2 = \frac{\gamma_c^2}{2},$$

where $\gamma_{\min}$ represents the smallest eigenvalue of the matrix $\left(K_1 - \frac{1}{2} I_{7\times 7}\right)$. Furthermore, the following condition holds:

$$\frac{\kappa_{\mathcal{M}_e}}{2} \operatorname{Tr}\left(\mathcal{M}_e^T \mathcal{M}_e\right) + \frac{\kappa_{\mathcal{C}_e}}{2} \operatorname{Tr}\left(\mathcal{C}_e^T \mathcal{C}_e\right) + \frac{\kappa_{\mathcal{G}_e}}{2} \operatorname{Tr}\left(\mathcal{G}_e^T \mathcal{G}_e\right) \to 0$$

which results directly from the definition of $\kappa_{(\cdot)_e}$ in Eq. (51). By imposing the following positive definiteness condition:

$$\left(K_1 - \frac{1}{2} I_{7\times 7}\right) > 0,$$

we can ensure that:

$$\dot{\mathcal{V}} \leq 0 \quad \text{whenever} \quad \mathcal{V} \geq \frac{\alpha_2}{\alpha_1}.$$

Thus, based on Lemma 1, all error signals within the closed-loop system are guaranteed to be bounded. This completes the proof. ■

6 Experimental Validation

6.1 Experimental Procedures

To initiate this investigation, the subject engages in the calibration of impedance parameters. As illustrated in Fig. 3, the processed surface electromyography (sEMG) signals from each muscle, denoted as $u(t)^{Bs,Ts}$, are utilized to derive muscle activations $a(t)^{Bs,Ts}$. Subsequently, these activations yield the muscle-tendon forces $f_{EMG}^{Bs,Ts}$, which contribute to the net shoulder force and elbow joint torque. We focus on two antagonistic muscle pairs, specifically the primary surface muscles influencing the shoulder and elbow joints, to develop the musculoskeletal model. Electrodes are strategically placed on the biceps long head, triceps medial head, pectoralis major (clavicular part), and deltoid (scapular part) muscles. The raw sEMG signals undergo several processing steps, including high-pass filtering, full-wave rectification, low-pass filtering, and normalization against the maximum voluntary contraction (MVC) levels of the respective muscles. The resulting exerted forces f_h interact with the dynamic model of the human upper limb, as described in Eq. (21), and a Radial Basis Function Neural Network (RBFNN) is employed to estimate the DMI of the subject, denoted as θ_h.

With the impedance model established, we utilize the shoulder and elbow joints of the robot to experiment, enabling the upper limb exoskeleton to track the estimated DMI of the subject. The controller parameters are manually selected as follows: $k_1 = 150 \cdot I_{7\times7}, \kappa_{(\cdot)_e} = \varkappa = \frac{1}{1+t^2}, \lambda_1 = 0.75 \cdot I_{7\times7}$, and $\Lambda_{\mathcal{M}_e} = \Lambda_{C_e} = \Lambda_{\mathcal{G}_e} = 10 \cdot I_{7\times7}$. The initial values are set to $(\hat{\mathcal{M}}_{eii} = 1)$, $(\hat{C}_{eii} = 1)$, and $(\hat{\mathcal{G}}_{ei} = 1)$. A total of $N = 2^8$ nodes are employed for the RBFNN $\hat{\mathcal{W}}^T \mathcal{S}(Z)$, with centers uniformly distributed within the range: $[-0.5, 0.5] \times [-0.5, 0.5] \times [-0.5, 0.5] \times [-0.5, 0.5] \times [-0.5, 0.5] \times [-0.5, 0.5] \times [-0.5, 0.5] \times [-0.5, 0.5]$, where each parameter can take on values of either -0.5 or 0.5. The parameters for updating the neural network are defined as $\Gamma = 0.001 \cdot I_{7\times7}$. The known parameters of the human upper limb are chosen artificially as follows: $\mathcal{M}_{h0} = 0.1 \cdot I_{7\times7}$ and $C_{h0} = 0.02 \cdot I_{7\times7}$

Remark 3 Although the controller operates within the joint space, the results are presented in task space through the use of the Jacobian and homogeneous transformation matrices derived from the exoskeleton robot's direct kinematics.

6.2 Experimenta Results

Scenario 1: In the first scenario, subject A (male, 95 kg, 177 cm, aged 35 years) is securely fastened to the exoskeleton robot with the elbow joint flexed at 90°. The subject is instructed to generate a straightforward line. The proposed control scheme aims to estimate his DMI and guide the robot to accurately follow the trajectory produced by the subject while ensuring the safety and stability of the human-robot system. The experimental results are presented in Figs. 6, 7 and 8.

Figures 6, 7 and 8 represent the experiment results of the proposed control approach to perform a simple trajectory. Figure 6 compares the desired (DMI—blue solid line) and measured (red dashed line) trajectory of a robotic end effector, displayed from four different viewpoints: 3D perspective, Up View (Y-Z plane), Lateral Right View (X-Z plane), and Front View (Y-Z plane). The measured trajectory generally follows the desired path but exhibits notable deviations, particularly in the Y-direction where there are more oscillations and wider movements compared to the desired smooth path. Mostly, the Z-direction tracking showing better accuracy than the Y-direction. Figure 7 shows the position tracking errors of a the exoskeleton robot system over 25 s across six dimensions: three translational (x, y, z in centimeters) and three rotational (γ, β, and α in degrees). The x-position error remains mostly stable with a notable step change to (1 cm) around (15 s), while the y-position shows consistent oscillatory behavior within ± 0.5 cm. The z-position error exhibits stability initially, followed by oscillations and a sharp dip at (20 s). For rotational errors, γ shows excellent tracking except for sharp spikes at 15 s reaching $\pm 2°$, β displays a gradual drift between $-1.6°$ and $-1.7°$ before recovering, and α maintains the smallest errors within $\pm 0.02°$. Figure 8 presents the control input torques for a 7-joint robotic system, where each joint exhibits distinct characteristics: Joints 1 and 3 show similar oscillatory patterns within ± 1 Nm and ± 0.4 Nm respectively; Joint 2 demonstrates a significant torque transition from -2 to $+2$ Nm; Joint 4 displays initial high-frequency oscillations followed by stabilization at a slight negative torque; Joint 5 shows a steady increase from 0 to 0.3 Nm; Joint 6 exhibits moder-

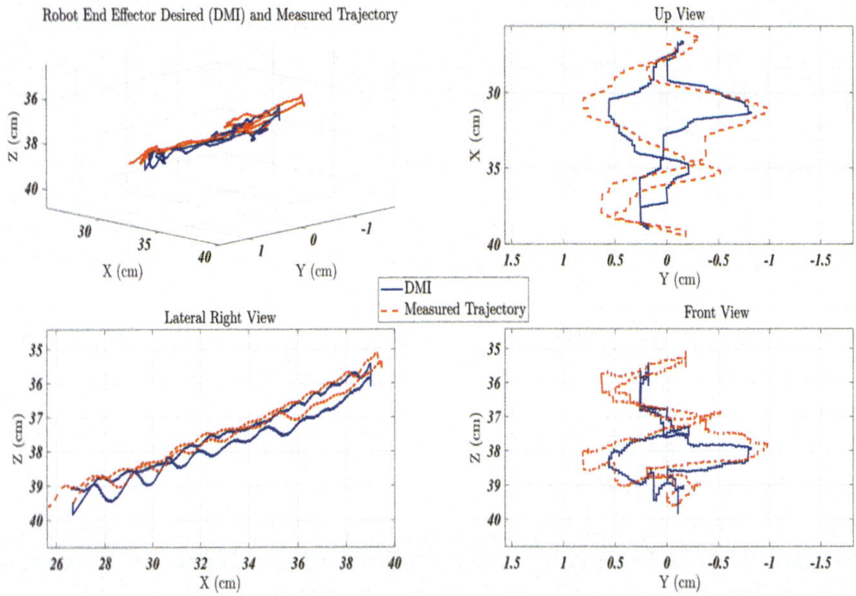

Fig. 6 End-effector tracking performance in task space under scenario 1

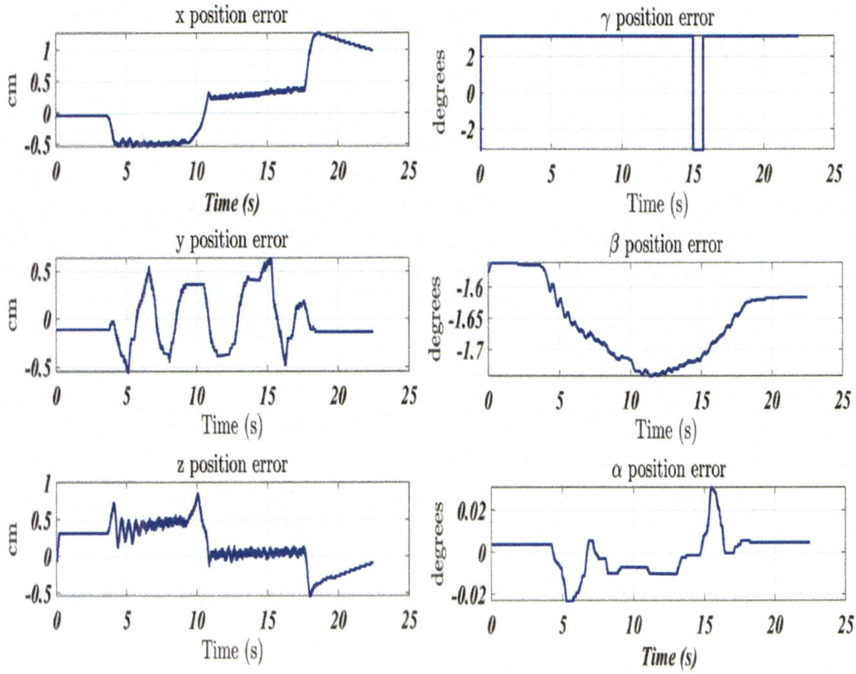

Fig. 7 Cartesian tracking errors under scenario 1

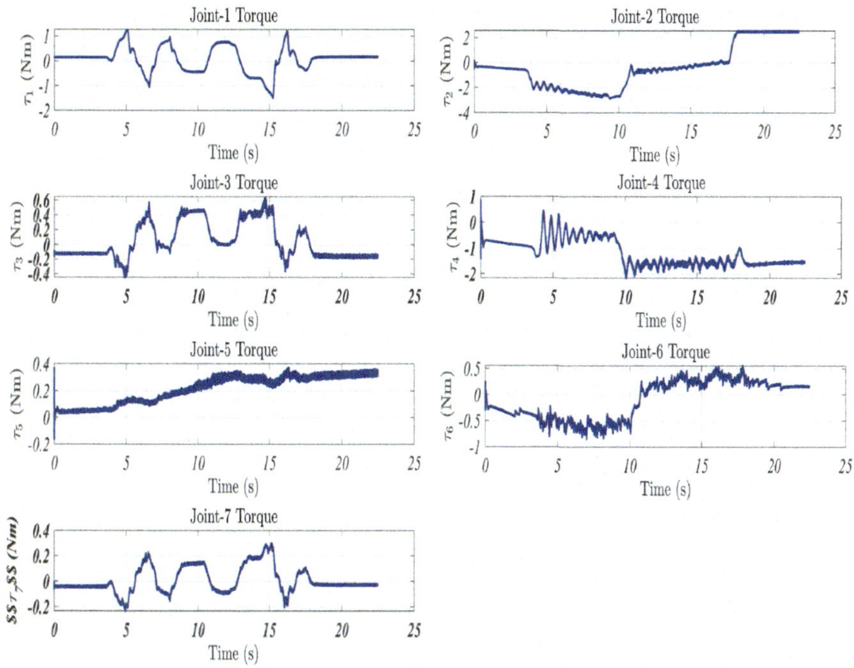

Fig. 8 Torque input under scenario 1

ate oscillations within ±0.5 Nm; and Joint 7 follows controlled oscillations within ±0.2 Nm. These varying torque profiles reflect the complex dynamic requirements of the robot's motion control during task execution. The proposed control approach demonstrates good position tracking with most errors remaining small, though some periodic oscillations and specific time-point deviations are evident, particularly in the Y-direction due to a delay in estimation of DMI caused by the nonlinear estimator, the training quality data for training the radial basis function neural network (RBFNN) and the calibration of the Hill's model.

Scenario 2: In the second scenario, subject A is instructed to create a complex square-shaped trajectory in space. The goal is to evaluate the effectiveness of the estimated DMI and assess the robustness of the proposed control system. The results obtained from this experiment are illustrated in Figs. 9, 10 and 11.

Figures 9, 10 and 11 illustrate the experimental outcomes of the proposed control strategy for executing a complex trajectory in three-dimensional space. Figure 9 offers a comprehensive visualization that effectively showcases the tracking performance and accuracy of the control system. It highlights deviations of the actual movement from the desired motion intention (DMI) from four distinct viewpoints: a 3D perspective, a top view (Y-Z plane), a lateral right view (X-Z plane), and a front view (Y-Z plane). Overall, the measured trajectory closely follows the intended path, demonstrating improved accuracy compared to the robot's initial performance while

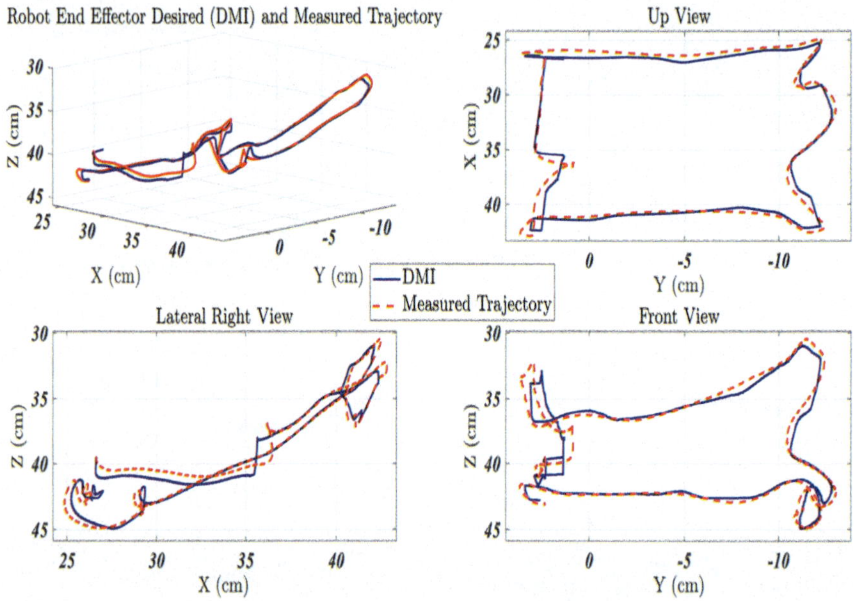

Fig. 9 End-effector tracking performance in task space under scenario 2

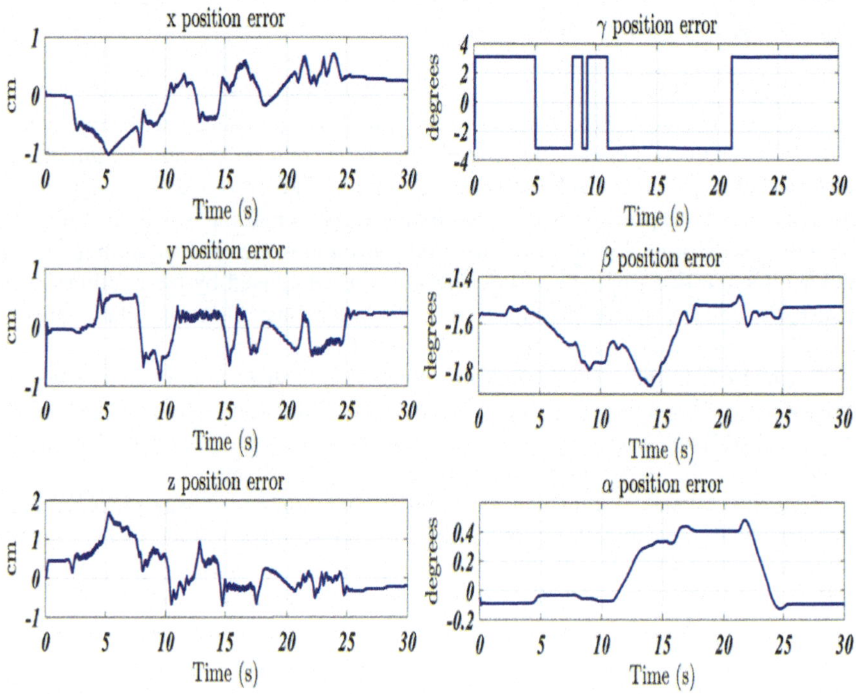

Fig. 10 Cartesian tracking errors under scenario 2

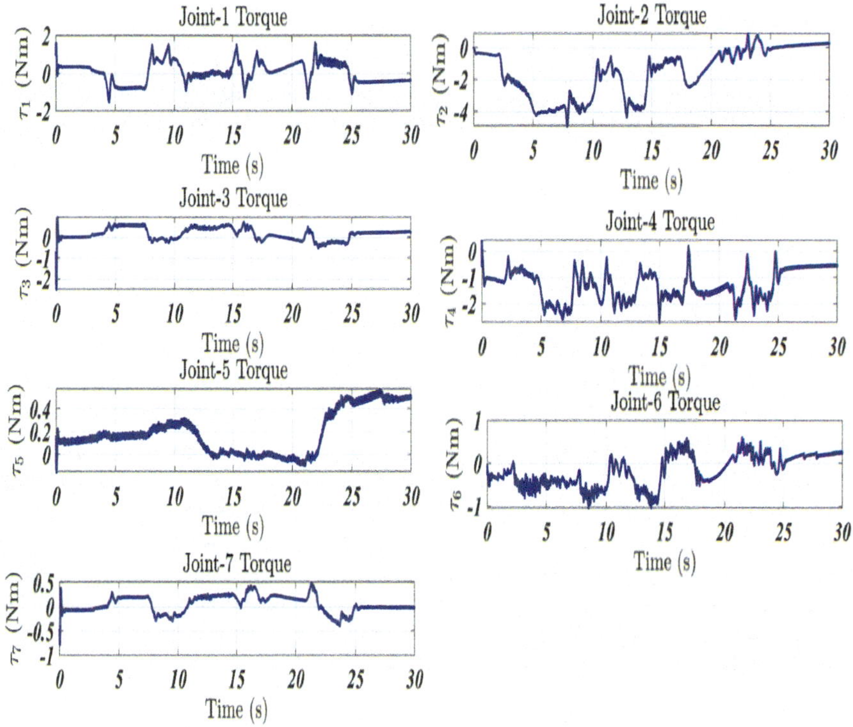

Fig. 11 Torque input under scenario 2

tracking a linear trajectory, as shown in Fig. 6. This enhancement can be attributed to the training data derived from trial-and-error methods. Figure 10 displays the position tracking errors of the exoskeleton robot system over a 30-s duration across six dimensions: three translational (x, y, z in centimeters) and three rotational (γ, β, and α in degrees). Notably, the position errors remain below 1 cm, indicating effective tracking. Similarly, the rotational errors also demonstrate excellent tracking capabilities. In Fig. 11, the control input torques for the seven-joint robotic system are depicted. Most control inputs exhibit smooth variations, except for joints 5 and 6, which display relatively high-frequency fluctuations. These varying torque profiles underscore the complex dynamic demands of the robot's motion control during task execution and reflect the robustness of the control inputs in continuously mitigating the effects of unknown dynamics. The proposed control approach demonstrates effective position tracking, with most errors remaining minimal throughout the rehabilitation task. These results highlight the effectiveness of the proposed scheme in real-time applications, capable of online estimation of the desired motion intention (DMI) while accommodating unknown dynamics and external forces.

7 Conclusion

This chapter introduces an adaptive impedance control strategy for a wearable exoskeleton robot with unknown dynamic parameters, developed to facilitate active rehabilitation processes. The main challenge of this approach was ensuring the exoskeleton accurately interprets the wearer's motion intention to provide precise assistance. Due to human movements' nonlinear and time-varying nature, estimating the user's desired motion intention (DMI) is complex. To fix this dilemma, the model's human forces are modeled by using the human arm gravity, force sensors input, and sEMG signals through the Hill model. A Radial Basis Function Neural Network (RBFNN) is used to estimate the real-time DMI. This information is then added to the adaptive impedance control, which lets the robot follow a target impedance. Experimental results from two scenarios validate the effectiveness of the proposed control scheme. In the first scenario, the robot successfully tracked a simple linear trajectory, showing stable control despite minor deviations in certain directions. The second scenario tested the system's ability to track a complex 3D trajectory, demonstrating improved accuracy and robustness, with errors remaining minimal and control inputs generally smooth. The proposed control strategy ensures the exoskeleton can collaborate effectively with the user by predicting and adapting to their motion intentions, providing a comfortable and responsive rehabilitation experience. These results confirm the potential of this approach for real-time, adaptive assistance in rehabilitation contexts.

Future work will focus on enhancing the accuracy of DMI estimation by incorporating advanced machine-learning models and refining the calibration process of the Hill model. Additionally, testing the proposed control scheme with a broader range of subjects, including individuals with different levels of motor impairment, will help generalize the approach's effectiveness. Expanding the experimental scenarios to include dynamic tasks and real-world rehabilitation activities can further validate the system's adaptability and robustness. Finally, exploring the integration of this control strategy with other assistive technologies, such as virtual reality for immersive feedback, may open new avenues for personalized and engaging rehabilitation experiences.

Ethical Approval

The control approach described in this study was applied to healthy subjects (chapter authors). According to ETS and the Province of Quebec policies, no ethical approval is required for experiments involving healthy adult subjects when the research poses minimal risk and the subjects are the researchers themselves.

References

1. Narayanan, K.L., Ahmed, T., Khan, M.M.R., Kawser, T., Fareh, R., Wang, I., Brahmi, B., Rahman, M.H.: Robotic devices for upper limb rehabilitation: a review. Med. Healthc. Robot. 123–156 (2023)
2. Hughes, A.-M., et al.: Evaluation of upper extremity neurorehabilitation using technology: a European Delphi consensus study within the EU cost action network on robotics for neurorehabilitation. J. Neuroeng. Rehabil. **13**(1), 86 (2016)
3. Keller, U., van Hedel, H.J., Klamroth-Marganska, V., Riener, R.: Charmin: the first actuated exoskeleton robot for pediatric arm rehabilitation. IEEE/ASME Trans. Mechatron. **21**, 2201–2213 (2016)
4. Lu, R., et al.: Development and learning control of a human limb with a rehabilitation exoskeleton. IEEE Trans. Ind. Electron. **61**(7), 3776–3785 (2014)
5. Brahmi, B., Rahman, M.H., Saad, M.: Impedance learning adaptive super–twisting control of a robotic exoskeleton for physical humanâŁłrobot interaction. IET CyberâŁłSyst. Robot. **5**(1), e12077 (2023)
6. Li, Z., Huang, Z., He, W., et al.: Adaptive impedance control for an upper limb robotic exoskeleton using biological signals. IEEE Trans. Ind. Electron. **64**(2), 1664–1674 (2017)
7. Lakshminarayanan, K., Shah, R., Ramu, V., Madathil, D., Yao, Y., Wang, I., Brahmi, B., Rahman, M.H.: Author spotlight: enhancing neurorehabilitation through EEG, motor imagery, and virtual reality. JoVE (J. Vis. Exp.) **207**, e66859 (2024)
8. Xiong, D., Zhang, D., Zhao, X., Zhao, Y.: Deep learning for EMG-based human-machine interaction: a review. IEEE/CAA J. Autom. Sin. **8**(3), 512–533 (2021)
9. Khan, A.M., Yun, D.-w., Zuhaib, K.M., et al.: Estimation of desired motion intention and compliance control for upper limb assist exoskeleton. Int. J. Control Autom. Syst. **15**(2), 802–814 (2017)
10. Brahmi, B., El Bojairami, I., Laraki, M.-H., El-Bayeh, C.Z., Saad, M.: Impedance learning control for physical human-robot cooperative interaction. Math. Comput. Simul. **190**, 1224–1242 (2021)
11. Lalitharatne, T.D., et al.: EEG-based evaluation for perception-assist in upper-limb power-assist exoskeletons. In: IEEE World Automation Congress (WAC) (2014)
12. Witkowski, M., et al.: Enhancing brain-machine interface (BMI) control of a hand exoskeleton using electrooculography (EOG). J. Neuroeng. Rehabil. **11**(1), 1 (2014)
13. Khokhar, Z.O., Xiao, Z.G., Menon, C.: Surface EMG pattern recognition for real-time control of a wrist exoskeleton. Biomed. Eng. Online **9**(1), 41 (2010)
14. Huang, J., et al.: Control of upper-limb power-assist exoskeleton using a human-robot interface based on motion intention recognition. IEEE Trans. Autom. Sci. Eng. **12**(4), 1257–1270 (2015)
15. Li, Z., et al.: Human-robot coordination control of robotic exoskeletons by skill transfers. IEEE Trans. Ind. Electron. **64**(6), 5171–5181 (2017)
16. Carmichael, M., Liu, D.: Estimating physical assistance need using a musculoskeletal model. IEEE Trans. Biomed. Eng. **60**(7), 1912–1919 (2013)
17. Jlassi, S., Tliba, S., Chitour, Y.: An event-controlled online trajectory generator based on the human-robot interaction quality for upper limb rehabilitation. Int. J. Adv. Robot. Syst. **11**(4), 51 (2014)
18. Kim, W., Lee, H., Lim, D., Han, J., Shin, K., Han, C.: Development of a muscle circumference sensor to estimate torque of the human elbow joint. Sens. Actuators A Phys. **208**, 95–103 (2014)
19. Khan, A.M., Yun, D.W., Ali, M.A., Zuhaib, K.M., Yuan, C., Iqbal, J., et al.: Passivity based adaptive control for upper extremity assist exoskeleton. Int. J. Control Autom. Syst. **14**(1), 291–300 (2016)
20. Khan, A.M., Usman, M., Ali, A., Khan, F., Yaqub, S., Han, C.: Muscle circumference sensor and model reference-based adaptive impedance control for upper limb assist exoskeleton robot. Adv. Robot. **30**(24), 1515–1529 (2016)

21. Lee, H.-D., Lee, B.-K., Kim, W.-S., Han, J.-S., Shin, K.-S., Han, C.-S.: Human-robot cooperation control based on a dynamic model of an upper limb exoskeleton for human power amplification. Mechatronics **24**(2), 168–176 (2014)
22. Ye, H., et al.: Investigation of human intention recognition method based on biosignals for human-robot cooperation systems. Robot. Comput.-Integr. Manuf. **30**(1), 47–54 (2014)
23. Bayon, C., Raya, R., Lerma, C., Ramirez, O., Serrano, J.I., Rocon, E., Pons, J.L.: Robotics in neurorehabilitation: a review on current and future trends. NeuroRehabilitation **41**, 345–368 (2017)
24. Chen, Y., et al.: EMG-based hands-free control for a power assist exoskeleton robot. Int. J. Adv. Robot. Syst. **10**(5), 1 (2013)
25. Alonso-Vargas, G., et al.: Control strategies for upper limb exoskeletons in rehabilitation therapies. Robot. Auton. Syst. **93**, 70–83 (2017)
26. Gopura, R.A.R.C., Kiguchi, K., Li, Y.: Sueful-7: a 7dof upper-limb exoskeleton robot with muscle-model-oriented EMG-based control. IEEE/ASME Trans. Mechatron. **16**, 109–117 (2011)
27. Xiong, B., Zeng, N., Li, H., Yang, Y., Li, Y., Huang, M., Shi, W., Min, D., Zhang, Y.: Intelligent prediction of human lower extremity joint moment: an artificial neural network approach. IEEE Access **7**, 29973–29980 (2019)
28. Iqbal, J., Khan, A.M., Khalid, U., Ullah, A., Majeed, M.: Developing wearable robotic exoskeletons for the disabled. IEEE Potent. **35**(3), 16–22 (2016)
29. Zheng, J., Li, Z., Xu, T., Song, R.: An electromyographic-based control using gaussian mixture model on an upper-limb cable-driven rehabilitation robot. Adv. Intell. Syst. 2400505 (2024)
30. Zeng, Y., Yang, J.T., Peng, C., Yin, Y.: Evolving gaussian process autoregression based learning of human motion intent using improved energy kernel method of EMG. IEEE Trans. Biomed. Eng. **66**(9), 2556–2565 (2019)
31. Louis, N., Gorce, P.: Upper limb muscle forces during a simple reach-to-grasp movement: a comparative study. Med. Biol. Eng. Comput. **47**, 1173–1179 (2009)
32. DeVore, R., Hanin, B., Petrova, G.: Neural network approximation. Acta Numer. **30**, 327–444 (2021)
33. Rocon, E., Raya, R., Lerma, C., Pons, J.L.: Neurorehabilitation using wearable exoskeletons: challenges and future directions. NeuroRehabilitation **41**(4), 375–382 (2017)
34. Mann, G., Reinkensmeyer, D.J., Scott, S.H.: Robotic neurorehabilitation technologies: transferring research into the clinics. Curr. Opin. Biomed. Eng. **14**, 1–11 (2019)
35. Bortole, M., Venkatakrishnan, A., Zhu, F., Moreno, J.C., Francisco, G.E., Pons, J.L., Contreras-Vidal, J.L.: The H2 robotic exoskeleton for gait rehabilitation after stroke: early findings from a clinical study. J. NeuroEng. Rehabil. **12**, 54 (2015)
36. Riener, R., Nef, T., Colombo, G.: Robot-aided neurorehabilitation of the upper extremities. Med. Biol. Eng. Comput. **43**, 2–10 (2005)
37. Fasoli, S.E., Krebs, H.I., Stein, J., Frontera, W.R., Hughes, R., Hogan, N.: Robotic therapy for chronic motor impairments after stroke: follow-up results. Arch. Phys. Med. Rehabil. **84**(4), 477–482 (2003)
38. Hogan, N., Krebs, H.I., Rohrer, B., Palazzolo, J.J., Ternman, L., Lum, P.S., Hogan, N.: Motions or muscles? Some behavioral factors underlying robotic assistance of motor recovery. J. Rehabil. Res. Dev. **43**(5), 605–618 (2006)
39. Tejima, T., Tadano, S., Suzuki, Y., Kawashima, K., Maeda, Y.: Development of an upper-limb assistive device driven by pneumatic artificial muscles. J. Robot. Mechatron. **29**(2), 319–328 (2017)
40. Mehrholz, J., Hädrich, A., Platz, T., Kugler, J., Pohl, M.: Electromechanical and robot-assisted arm training for improving arm function and activities of daily living after stroke. Cochrane Database Syst. Rev. **6**, CD006876 (2012)
41. Brahmi, B., Saad, M.: Adaptive control of an electrically driven exoskeleton robot (theory and experiments). J. Vib. Eng. Technol. **11**(7), 3399–3412 (2023)
42. Rahman, M.H., Saad, M., Kenné, J.P., Archambault, P.S.: Nonlinear sliding mode control implementation of an upper limb exoskeleton robot to provide passive rehabilitation therapy.

In: Intelligent Robotics and Applications: 5th International Conference, ICIRA 2012, Montreal, Canada, October 3-5, 2012, Proceedings, Part II 5, pp. 52–62. Springer (2012)
43. Reinkensmeyer, D.J., Emken, J.L., Cramer, S.C.: Robotics, motor learning, and neurorehabilitation. Neurorehabil. Neural Repair **18**(3), 240–252 (2004)
44. Nef, T., Riener, R.: Armin—design of a novel arm rehabilitation robot. In: 6th International Workshop on Virtual Rehabilitation, pp. 1–6 (2008)
45. Nef, T., Guidali, M., Riener, R.: Armin III—arm therapy exoskeleton with an ergonomic shoulder actuation. Appl. Bionics Biomech. **6**, 127–142 (2009)
46. Masia, L., Krebs, H.I., Cappa, P., Hogan, N.: Design and characterization of hand module for whole-arm rehabilitation following stroke. IEEE/ASME Trans. Mechatron. **12**(4), 399–407 (2007)
47. Colombo, R., Pisano, F., Mazzone, A., Delconte, C., Micera, S., Carrozza, M.C., Dario, P., Minuco, G.: Design strategies to improve patient motivation during robot-aided rehabilitation. J. NeuroEng. Rehabil. **4**, 3 (2007)
48. Cempini, M., Cortese, M., Vitiello, N.: A powered elbow exoskeleton for post-stroke early neurorehabilitation: preliminary tests on a hemiparetic patient. In: IEEE International Conference on Rehabilitation Robotics, pp. 1–6 (2013)
49. Bai, S., Song, A., Xu, B.: A novel EEG-driven lower-limb rehabilitation robot for active training. IEEE Trans. Ind. Electron. **61**(7), 3776–3785 (2014)
50. Marchal-Crespo, L., Reinkensmeyer, D.J.: Review of control strategies for robotic movement training after neurologic injury. J. NeuroEng. Rehabil. **6**, 20 (2009)
51. Vitiello, N., et al.: Neuroexos: a powered elbow exoskeleton for subjects with upper limb impairment. IEEE Trans. Robot. **29**(1), 220–235 (2013)
52. Brahmi, B., Driscoll, M., El Bojairami, I.K., Saad, M., Brahmi, A.: Novel adaptive impedance control for exoskeleton robot for rehabilitation using a nonlinear time-delay disturbance observer. ISA Trans. **108**, 381–392 (2021)
53. Lo, H.S., Xie, S.Q.: Exoskeleton robots for upper-limb rehabilitation: state of the art and future prospects. Med. Eng. Phys. **34**(3), 261–268 (2012)

Computer-Aided Design and Control of an Integrated Exoskeleton-Upper Limb System

Sana Bembli, Nahla Khraief Haddad, and Safya Belghith

Abstract This research outlines the various stages involved in constructing Computer-Aided Design (CAD) simulation models using the combined environments of SolidWorks and Matlab/Simulink. The study specifically focuses on developing an exoskeleton model designed for rehabilitation purposes, coupled with a human upper limb model. These simulation models are essential for visualizing the system's ability to track desired trajectories during movement, providing a virtual testbed for evaluating control strategies and mechanical behavior. The first part of the research is dedicated to the design and analysis of the exoskeleton-upper limb system's CAD model. This stage involved creating a detailed 3D mechanical representation of the exoskeleton and human arm in SolidWorks, where joint configurations, mechanical constraints, and movement capabilities were defined. The next phase of the research focused on extracting the Electromyography (EMG) signals generated during elbow flexion and extension movements. To control the exoskeleton-upper limb system, a Model-Free Adaptive Terminal Sliding Mode Controller with Gravity Compensation was implemented. This advanced control algorithm was chosen for its robustness in handling the complex, nonlinear dynamics of the exoskeleton interacted with the human. The effectiveness and performance of both the models and the control algorithm were validated through simulation results.

Keywords Exoskeleton · Human upper-limb · Computer aided design (CAD) · Model free adaptive terminal sliding mode with gravity compensation controller · EMG signal

S. Bembli (✉) · N. Khraief Haddad · S. Belghith
RISC Laboratory, National Engineering School of Tunis, University of Tunis El-Manar, Tunis, Tunisia
e-mail: bembli.sana@hotmail.fr

J. Narayan and H. Gritli (eds.), *Design and Control of Rehabilitation Robots*, Studies in Systems, Decision and Control 585, https://doi.org/10.1007/978-3-031-86977-8_13

1 Introduction

The increasing prevalence of musculoskeletal disorders and neurological impairments necessitates the development of advanced assistive technologies to enhance the quality of life and autonomy of affected individuals. Traditional rehabilitation methods often lack the precision, adaptability, and customization required for optimal patient outcomes [1–3].

Recent technological advancements have resulted in a growing demand for well-designed products. The superior value and functionality of more offerings have led to diverse market situations and heightened competition. This environment necessitates that engineers create machines with significant flexibility and features tailored for various applications. However, due to constraints related to cost and time, it is often unfeasible to develop prototypes for every type of experimental research device [4, 5].

Current exoskeleton systems face limitations in terms of ergonomics, user comfort, and real-time adaptability. Additionally, the control mechanisms often do not account for the variability in user needs and the dynamic nature of human limb movements [6]. There is a critical need for a computer-aided design approach that can optimize the mechanical structure of the exoskeleton while incorporating intelligent control systems to provide real-time, adaptive assistance and rehabilitation.

The advent of computer-aided design (CAD) software facilitates not only the modeling of geometric objects but also the kinematic, dynamic, and force analysis of these models within professional CAD systems. Nevertheless, integrating control systems into these applications can be complex, if not impossible. Consequently, control systems are typically modeled in the Matlab environment using Simulink blocks [5, 7].

Simulation models are generally constructed in stages; a design model is first created in a CAD program (such as SolidWorks or Catia). These applications ideally interface with the Matlab/Simulink environment. In simulation studies, a set of components can be utilized, as this allows for the analysis of interactions between different parts within the control system during motion. Therefore, each simulation model consists of multiple individual components or subsets.

This chapter focuses on modeling an exoskeleton-upper extremity system. An exoskeleton is an artificially enhanced device attached to the arm, designed to assist with manual tasks and dynamically support natural movements. It is engineered to carry the weight of the arm and any objects being held while maintaining full freedom of motion [8].

Exoskeletons aim to eliminate the probability of hurt from repetitive strain and burnout, which is a significant concern for social security systems. They connect from the shoulder to the wrist through a series of articulations, allowing for a full range of arm motion. A motorized mechanism enables the locking of the arm's position and can assist in moving heavy objects.

To achieve this, designers are developing various mechanisms to accurately replicate the complex kinematics of human joints while accommodating important morphological variations among users. Moreover, the intricate dynamics of the exoskeleton system have driven researchers to formulate various control laws to ensure adherence to the desired trajectory. Among these, Model Predictive Control (MPC) [9] is an advanced algorithm that leverages a mathematical model of the exoskeleton to forecast its future behavior. This capability allows for real-time optimization of control signals to achieve intended movement trajectories. Additionally, Proportional-Integral-Derivative (PID) control [10–12] employs sensor feedback to modulate the exoskeleton's movements in relation to the user's actions, making it suitable for systems requiring precise control and stability. Furthermore, researchers have explored Impedance control [13, 14], which adjusts the exoskeleton's resistance to user movements, thereby providing varying levels of assistance based on the task. Moreover, sliding mode variable structure control demonstrates strong performance in trajectory tracking [15–17]. It is also recognized for its robustness against parametric uncertainties, modeling errors, and external disturbances. Additionally, Terminal sliding mode control (TSMC) has been proposed as an alternative to linear sliding variables, utilizing a nonlinear sliding variable instead [18, 19]. This approach introduces a nonlinear fractional power element to the sliding variable during the sliding phase, offering advantages such as finite-time convergence of state variables, improved tracking accuracy, and faster response.

The challenge lies in designing and controlling an integrated exoskeleton-upper limb system that can seamlessly assist and rehabilitate users by mimicking natural limb movements. This involves complex biomechanical and physiological considerations, as well as the integration of advanced control algorithms to ensure synchronization with the user's intent and adaptive responses to their movements. This project seeks to tackle these challenges by creating a comprehensive framework for the computer-aided design and control of a fully integrated exoskeleton-upper limb system. The proposed solution will leverage advanced computational tools and algorithms to enhance the design process and implement sophisticated control strategies for improved user interaction and rehabilitation outcomes. Therefore, the first contribution of this chapter is the development of a Simulink model for the exoskeleton-upper limb system, facilitated by communication between Matlab and SolidWorks. The objective is to control the system and enable 3D simulation. The second contribution involves the integration of adaptive control and gravity compensation approaches with terminal sliding mode and model-free controllers to enhance trajectory tracking. This integration results in a robust Model Free Adaptive Terminal Sliding Mode with Gravity Compensation (MFATSMGC) control law.

The structure of the chapter is as follows: Sect. 2 describes the Computer-Aided Design model of the exoskeleton-upper limb system. Section 3 outlines the extraction of EMG signals. Section 4 discusses the control of the system using the MFTSMGC algorithm. Section 5 describes simulation results and discussions, while Sect. 6 concludes the chapter and outlines future work.

2 Computer Aided Design Model of the Exoskeleton-Upper Limb System

Computer-aided design (CAD) software is utilized to develop digital representations of physical objects or systems. These digital models serve various functions, such as design, engineering analysis, and manufacturing. Typically, CAD software provides tools for both 2D and 3D geometry creation and editing, along with features for applying materials and textures, managing lighting and shading, and producing animations and simulations [20].

2.1 MATLAB/SolidWorks Communication

In CAD software, kinematic and dynamic models are generally developed using specialized tools that enable users to specify constraints, forces, and other parameters governing the behavior of the modeled system. These models serve various purposes, including simulating system behavior under different conditions, optimizing design parameters, and diagnosing issues that may arise during the design process [21].

The integration of SolidWorks and MATLAB enhances model design and control through effective communication, conceptualization, and 3D simulation. By leveraging the capabilities of both programs, designers can create more complex and optimized models featuring advanced control systems and detailed 3D simulations. The data exchange between SolidWorks and MATLAB facilitates the utilization of each program's strengths [22].

In this context, we will demonstrate how to establish communication between SolidWorks and MATLAB to create a Simulink model of an exoskeleton-upper limb system, allowing for monitoring with various control signals. The following section outlines the steps necessary to establish this communication.

To integrate a SolidWorks model into MATLAB/Simulink, the process starts with installing the required files to establish communication between the two platforms. This software is then set up within MATLAB. Next, the CAD model is saved in SimMechanics Link format, which generates the required XML file. The final step involves importing this XML file into the MATLAB/Simulink environment to build the simulation model [23].

2.2 Exoskeleton-Upper Limb System Model

This work aims to model an exoskeleton in conjunction with a human upper limb to regulate their interaction. The human arm features 3 degrees of freedom (D.O.F) at the shoulder and 1 D.O.F at the elbow. The system under study includes an exoskeleton interacting with the human upper limb, offering 2 D.O.F for flexion and extension movements at the shoulder and elbow (see Fig. 1) [24–26].

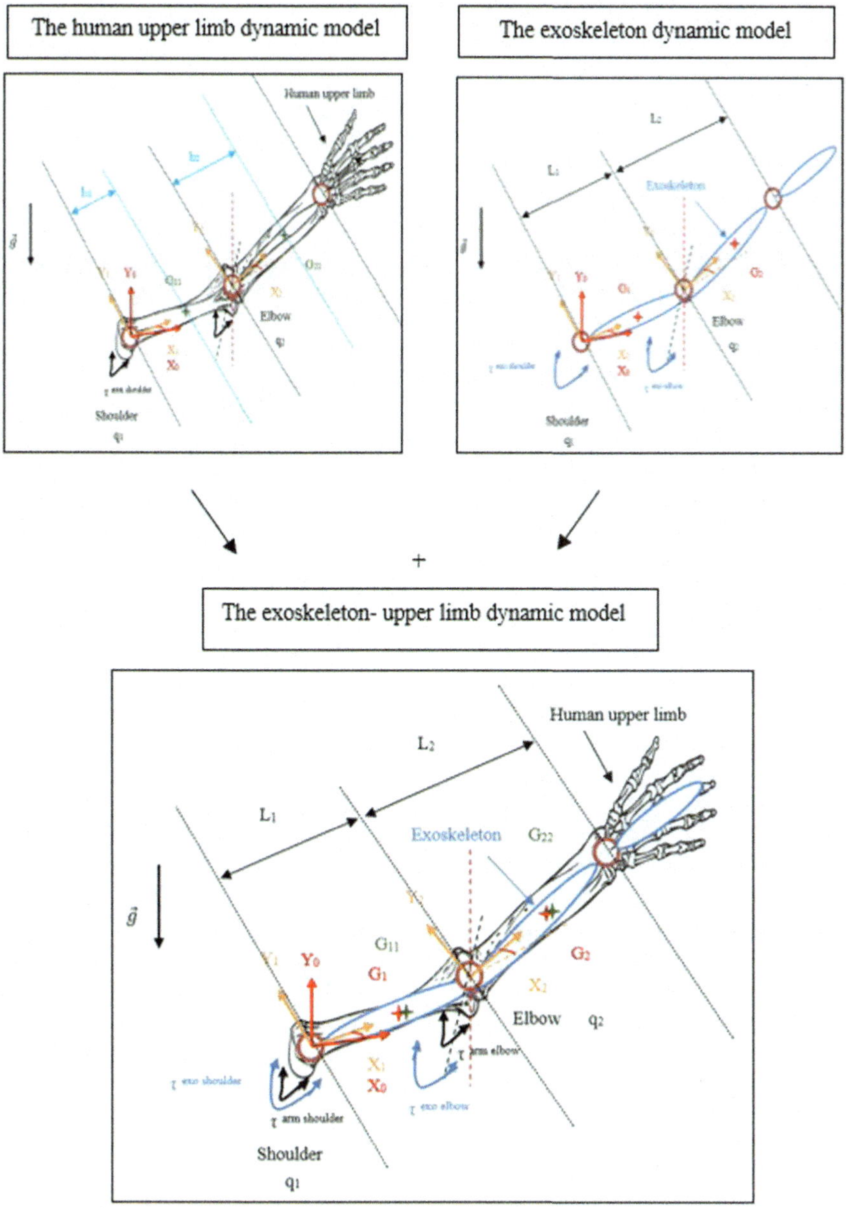

Fig. 1 The treated system dynamic model

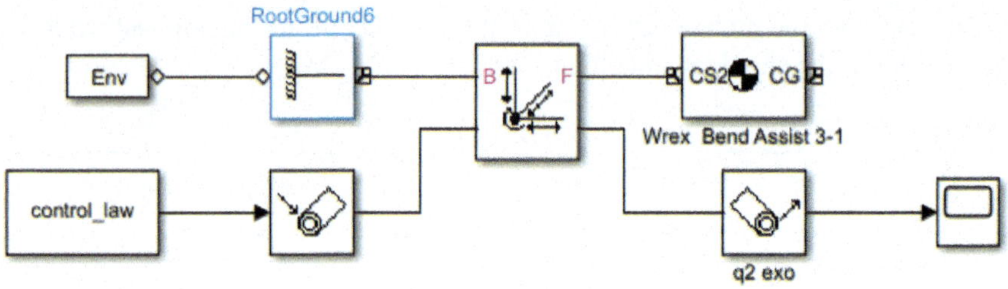

Fig. 2 The exoskeleton elbow joint SimMechanics block scheme

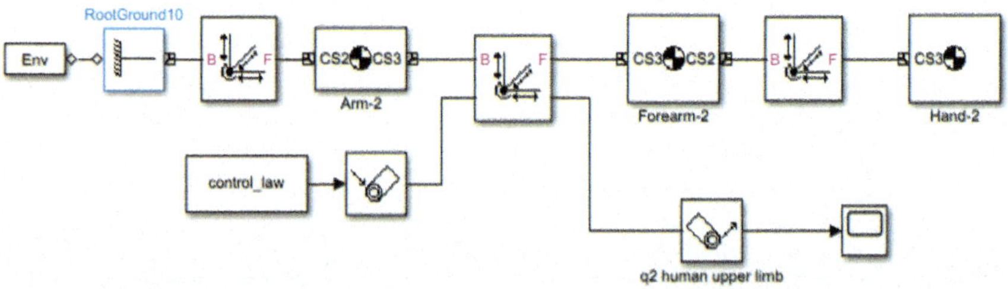

Fig. 3 The human upper limb elbow joint SimMechanics block scheme

In this section, we will design models of both the exoskeleton and the upper limb using SolidWorks, followed by converting these designs into the MATLAB/Simulink environment. The focus of the subsequent work will be on controlling the elbow joint. Figures 2 and 3 illustrate the SimMechanics block diagrams for the exoskeleton and the human upper limb elbow joint, respectively. Additionally, Fig. 4 presents the 3D models of both the elbow joint of the treated system.

3 EMG Signal Extraction

This paragraph focuses on extracting the EMG signal during the flexion/extension movement of the elbow. The exoskeleton-upper limb system will then be monitored using the "Model-Free Adaptive Terminal Sliding Mode with Gravity Compensation" algorithm, where the EMG signals will serve as the desired inputs.

To achieve this, we began by analyzing the signals (from the skin surface) that the brain sends to the muscles in order to replicate the same joint movements. The muscle signal is first acquired and amplified, then filtered to eliminate the DC component. Finally, it undergoes a third stage to remove high-frequency noise [27]. Figure 5 illustrates the different stages of muscle signal extraction with A1 represents the amplification gain of the first stage and A2 represents the amplification gain of the second stage.

The main steps for EMG signal extraction are as follows:

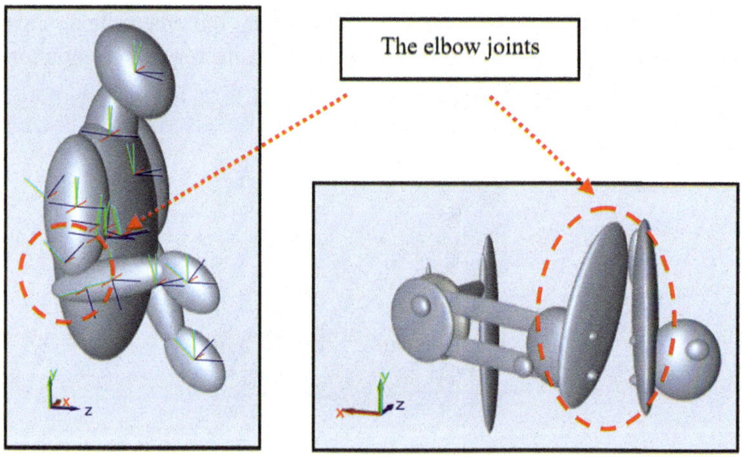

Fig. 4 The human upper limb and the exoskeleton elbow joint 3D models

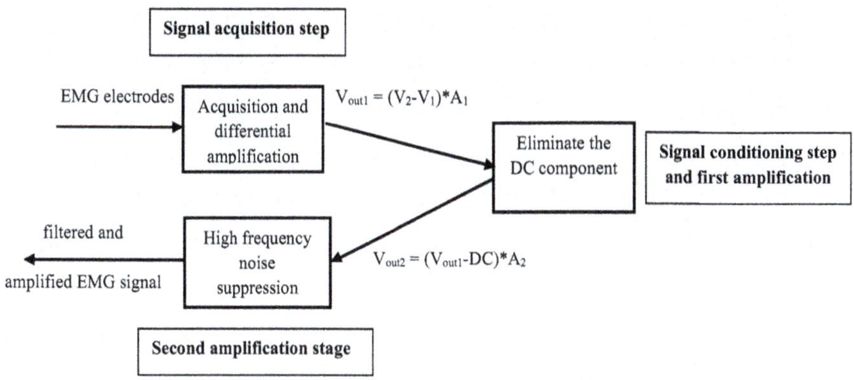

Fig. 5 Different steps of muscle signal extraction

- **Signal Acquisition and Differential Muscle Signal Amplification**: This phase involves measuring electrical impulses from the nervous system. It utilizes two electrodes placed on the skin of the arm, and a reference electrode. The instrumentation amplifier amplifies the difference between the two electrodes, while canceling out noise.
- **Signal Conditioning: Removal of DC Component**: In this stage, a first-order active high-pass filter is used to remove the DC offset and low-frequency noise from the low-amplitude signal produced in the first stage.
- **Signal Enhancement and High-Frequency Noise Removal**: The final step uses an active low-pass filter to eliminate high-frequency noise, yielding a clean signal for the microcontroller.

Figure 6 shows the EMG signal extraction during the elbow flexion/extension movement. This signal will be utilized as required trajectory when monitoring the exoskeleton and the human upper limb.

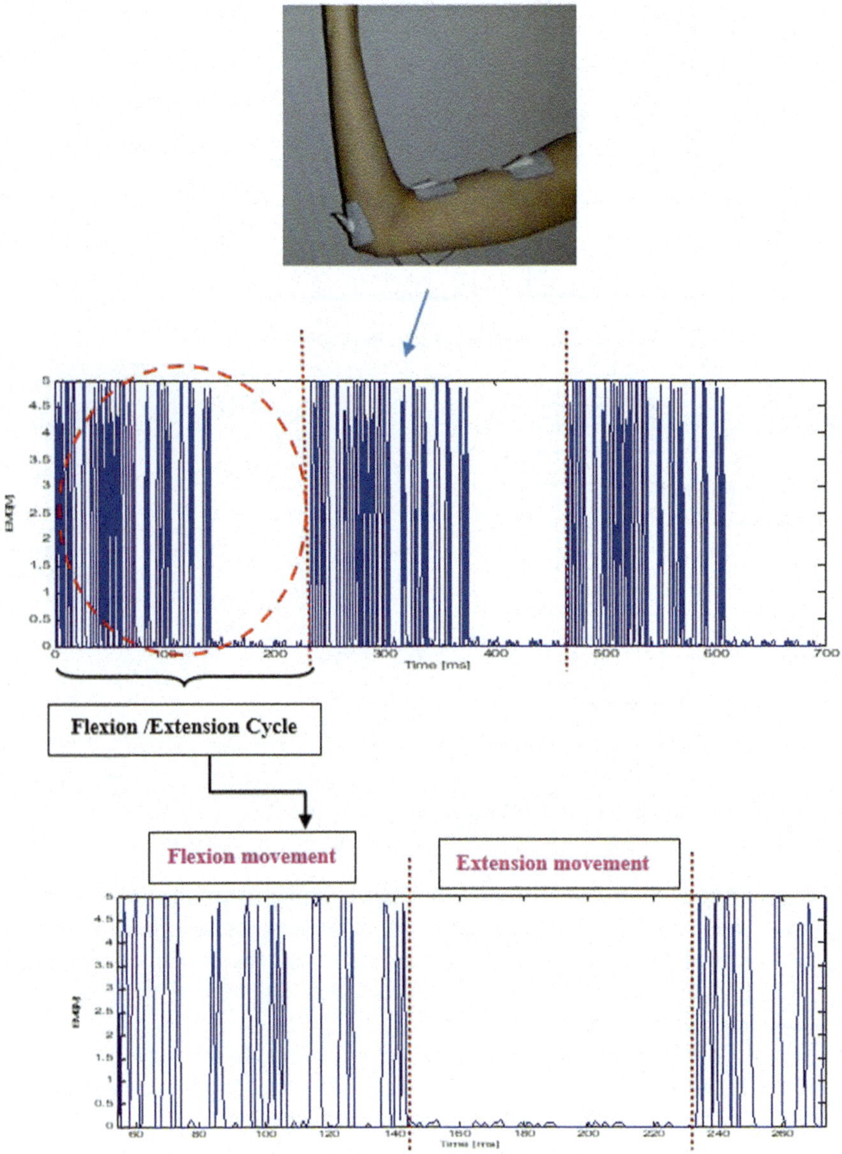

Fig. 6 EMG signal acquired through the elbow flexion/extension movement

4 Exoskeleton-Upper Limb System Control

To control the exoskeleton-upper limb system, we developed a novel Model-Free Adaptive Terminal Sliding Mode with Gravity Compensation (MFATSMGC) algorithm. This algorithm integrates multiple control strategies (Fig. 7) to address the challenges associated with non-linear dynamics and external disturbances, such as gravitational forces, that affect the system.

The MFATSMGC algorithm is the result of combining adaptive control, gravity compensation, model-free control, and terminal sliding mode control into a unified approach. Each component is essential in improving the control system's overall performance and robustness. The Adaptive control allows the system to dynamically adjust its parameters based on changes in the environment or the user's condition, such as varying load or muscle fatigue. This ensures that the exoskeleton continues to operate effectively even when faced with unpredictable conditions. Gravity compensation is essential for systems like exoskeletons, which must work against the gravity force to support or assist human movement. By compensating for gravitational forces, the controller enables smoother and more accurate motion, particularly in rehabilitation scenarios where precision is critical. The Model-free control eliminates the need for an exact mathematical model of the system, which is particularly useful when dealing with complex systems like the human body and its interaction with the exoskeleton. Instead, it relies on real-time data and feedback to make control decisions, reducing the complexity of the control design. Finally, the Terminal sliding mode control adds robustness to the system by ensuring finite-time convergence to the desired trajectory, even in the presence of disturbances or uncertainties. This feature is particularly important for ensuring the system can precisely follow the user's movements as indicated by EMG signals.

By integrating these approaches, the MFATSMGC algorithm provides a highly adaptable and efficient control strategy that ensures smooth, accurate, and reliable control of the exoskeleton-upper limb system, even in dynamic or uncertain environments.

This combination gives the following MFTASMGC:

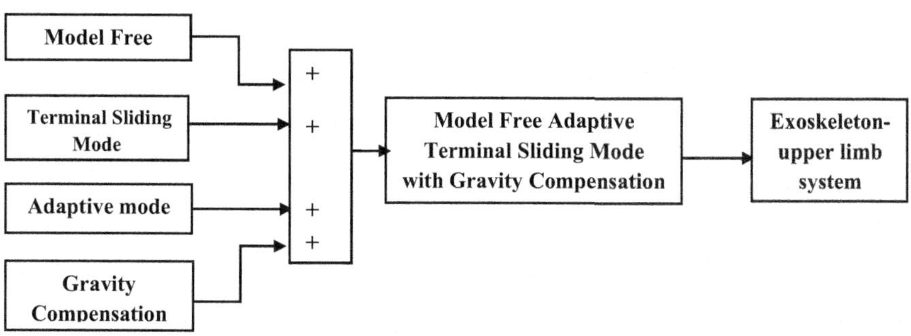

Fig. 7 The proposed controller (MFATSMGC)

$$U_{MFATSMGC} = u_{MF} + u_{TSM} + u_{AM} + u_{GC} \tag{1}$$

where

- u_{TSM} describes the Terminal Sliding Mode control law.
- u_{MF} describes the Model Free controller.
- u_{AM} presents the adaptive mode control law.
- u_{GC} presents the gravity compensation approach.

The model free approach is given by [28, 29]:

$$\ddot{q} = F(t) + \alpha u_{MF}(t) \tag{2}$$

With:

- q is the system output.
- $\ddot{q}$ presents the output acceleration.
- α is a positive gain.
- F is an unknown nonlinear term including unmodeled dynamics and uncertainties.
- u_{MF} is the corresponding input signal.

We close the loop by an intelligent proportional-derivative controller (iPD), we get:

$$u_{MF} = -\frac{\hat{F} - \ddot{q}_d + u_C}{\alpha} \tag{3}$$

where u_c is the feedback controller to follow the required output signal; q_d is a required output trajectory; $\dot{q}_d$ is the required velocity and $\hat{F}$ is an estimate of F.

u_c is defined as the classic PD controller with:

$$u_c = K_D\dot{e} + K_p e \tag{4}$$

and

$$e = q - q_d \tag{5}$$

with e presents the error of trajectory following; K_d is the derivative gain matrix and K_p is the integral gain matrix.

From Eqs. (3) and (4) we get:

$$u_{MF} = -\frac{\hat{F} - \ddot{q}_d + K_D\dot{e} + K_p e}{\alpha} \tag{6}$$

By combining Eqs. (4) and (6), we get:

$$\ddot{e} + u_c = F - \hat{F} \tag{7}$$

The estimation of F is expressed as:

$$F(t) = \hat{F}(t) = F(t - \varepsilon) = \ddot{q}(t - \varepsilon) - \alpha u_{MF}(t - \varepsilon) \quad (8)$$

where ε is a small-time delay.

Based on the Eqs. (7) and (4), we have:

$$\ddot{e} + K_D\dot{e} + K_p e = e_{est} \quad (9)$$

with: $e_{est} = F - \hat{F}$.

We start by combining the Adaptive Terminal Sliding mode with the Model Free Control, so we get:

$$U_{MFATSM} = -\frac{\hat{F} - \ddot{q}_d + K_D\dot{e} + K_p e}{\alpha} + u_{ATSM} \quad (10)$$

We get from (9) and (7):

$$\ddot{e} + K_D\dot{e} + K_p e = e_{est} + \alpha u_{ATSM} \quad (11)$$

with $e_{est} = F - \hat{F}$.

We considered the system expressed as:

$$\begin{cases} x_1 = e \\ x_2 = \dot{e} \end{cases} \quad (12)$$

We get:

$$\begin{cases} \dot{x}_1 = x_2 \\ \dot{x}_2 = -K_D\dot{e} - K_p e + e_{est} + \alpha u_{ATSM} \end{cases} \quad (13)$$

The Adaptive Terminal Sliding mode in expressed as [30]:

$$u_{ATSM} = u_{eq} + u_n \quad (14)$$

with $u_n = -\frac{\hat{K}}{\alpha}\text{sign}(S_t)$.

The sliding surface, defined in Eq. 14, is designed to ensure fast tracking of the desired trajectory while avoiding the singularity issue [31–33]:

$$S_t = x_1 + \frac{1}{\beta}x_2^{r/s} \quad (15)$$

with:

- $\beta > 0$;

– r and s are positive odd integers ensuring the condition 1< r/s < 2.

$\hat{K}$ is given by the following adaptation law:

$$\dot{\hat{K}} = \|\beta\| \frac{r}{s} \left\| x_{-I}^{r/s-1} S_t \right\| \tag{16}$$

u_{eq} is obtained when the condition $\dot{S}_t = 0$ is satisfied. So, we estimate $\dot{S}_t$:

$$\dot{S}_t = \dot{x}_1 + \frac{1}{\beta}\frac{r}{s} x_2^{\left(r/s\right)-1} \dot{x}_2 \tag{17}$$

$$\dot{S}_t = \dot{x}_1 + \frac{1}{\beta}\frac{r}{s} x_2^{\left(r/s\right)-1} \left(-K_D\dot{e} - K_p e + e_{est} + \alpha u_{ATSM}\right) \tag{18}$$

We get:

$$u_{eq} = -\frac{\beta s}{\alpha r}\dot{e}^{2-r/s} + \frac{K_D}{\alpha}\dot{e} + \frac{K_p}{\alpha}e - \frac{e_{est}}{\alpha} \tag{19}$$

We have:

$$u_{ATSM} = -\frac{\beta s}{\alpha r}\dot{e}^{2-r/s} + \frac{K_D}{\alpha}\dot{e} + \frac{K_p}{\alpha}e - \frac{\hat{K}}{\alpha}sign(S_t) - \frac{e_{est}}{\alpha} \tag{20}$$

From (10) and (20), we obtain the MFATSM Controller given by:

$$U_{MFATSM} = -\frac{\beta s}{\alpha r}\dot{e}^{2-r/s} - \frac{F - \ddot{q}_d}{\alpha} - \frac{\hat{K}}{\alpha}sign(S_t) \tag{21}$$

With the inclusion of the gravity compensation approach, we ultimately arrive at the global Model-Free Adaptive Terminal Sliding Mode with gravity compensation (MFATSMGC), represented by the following expression (Fig. 8):

$$U_{MFATSMGC} = -\frac{\beta s}{\alpha r}\dot{e}^{2-r/s} - \frac{F - \ddot{q}_d}{\alpha} - \frac{\hat{K}}{\alpha}sign(S_t) + G \tag{22}$$

The mathematical model of exoskeleton-upper limb system dynamics is given in [23]. Therefore, to prove the stability of the system under consideration, we utilized the following Lyapunov function:

$$V = \frac{1}{2}\dot{q}^T M(q)\dot{q} + \frac{1}{2}S_t^T S_t \tag{23}$$

$$\dot{V} = \dot{q}^T M(q)\ddot{q} + \frac{1}{2}\dot{q}^T \dot{M}(q)\dot{q} + S_t^T \dot{S}_t \tag{24}$$

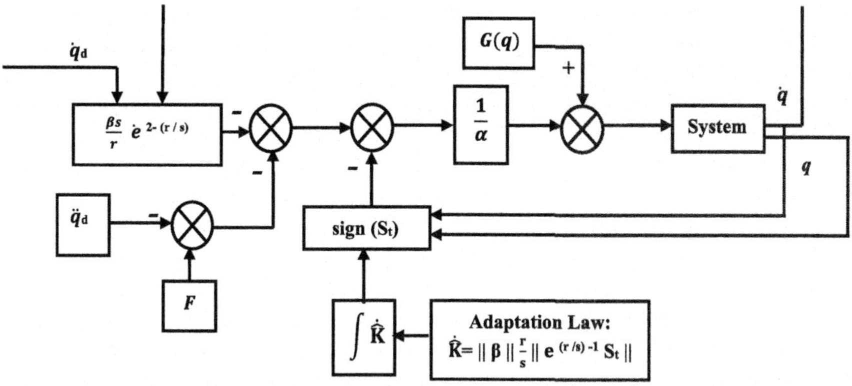

Fig. 8 Block diagram of the model free adaptive terminal sliding mode with gravity compensation controller (MFATSMGC)

By replacing $\dot{S}_t$ by its expression, we have:

$$\dot{V}=\dot{q}^T\left[-\frac{\beta s}{\alpha r}\dot{e}^{2-{r}/{s}}-\frac{F-\ddot{q}_d}{\alpha}-\frac{\hat{K}}{\alpha}sign(S_t)-C(q,\dot{q})\dot{q}\right]+\frac{1}{2}\dot{q}^T\dot{M}(q)\dot{q}$$
$$+S_t^T\left[\frac{1}{\beta}\frac{r}{s}x_2^{\left({r}/{s}\right)-1}\left(e_{est}-\hat{K}sign(S_t)\right)\right] \quad (25)$$

$$\dot{V}=\dot{q}^T\left[-\frac{\beta s}{\alpha r}\dot{e}^{2-{r}/{s}}-\frac{F-\ddot{q}_d}{\alpha}-\frac{\hat{K}}{\alpha}sign(S_t)\right]+\frac{1}{2}\dot{q}^T\left[\dot{M}(q)\dot{q}-2C(q,\dot{q})\dot{q}\right]$$
$$+\left[S_t\frac{1}{\beta}\frac{r}{s}x_2^{\left({r}/{s}\right)-1}\left(e_{est}-\hat{K}sign(S_t)\right)\right] \quad (26)$$

The inertia and Coriolis matrices are asymmetric and adhere to the following relationship:

$$\frac{1}{2}\dot{q}^T\left[\dot{M}(q)\dot{q}-2C(q,\dot{q})\dot{q}\right]=0 \quad (27)$$

Based on (27), we get:

$$\dot{V}=-\dot{q}^T\left[\frac{\beta s}{\alpha r}\dot{e}^{2-{r}/{s}}+\frac{F-\ddot{q}_d}{\alpha}-\frac{\hat{K}}{\alpha}sign(S_t)\right]+\frac{1}{\beta}\frac{r}{s}x_2^{\left({r}/{s}\right)-1}\left(S_te_{est}-S_t\hat{K}sign(S_t)\right)] \quad (28)$$

$$\dot{V}<-|\dot{q}|^T\frac{\beta s}{\alpha r}\dot{e}^{2-{r}/{s}}-\frac{\hat{K}}{\alpha}|\dot{q}|^T-\frac{1}{\beta}\frac{r}{s}\hat{K}x_2^{\left({r}/{s}\right)-1}S_t\hat{K}|S_t| \quad (29)$$

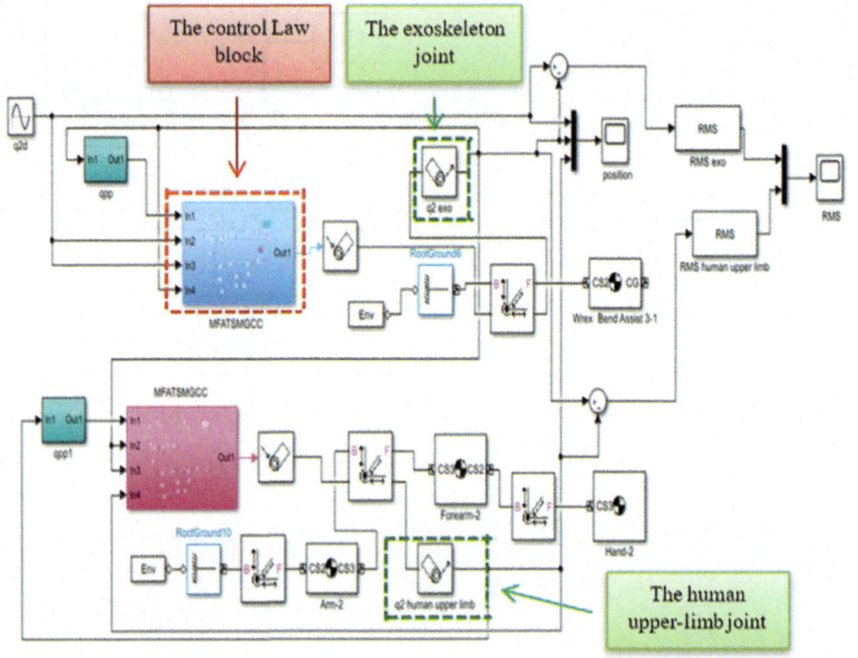

Fig. 9 SimMechanics block scheme of the system using the model free adaptive terminal sliding mode with gravity compensation controller

As r and s satisfied the following condition: $1 < r/s < 2$, so $x_2^{(r/s)-1} > 0$ for all $x_2 \neq 0$ and $\dot{e}^{2-r/s} > 0$ and $\hat{K} > 0$, $\lambda > 0$ and $\alpha > 0$, then, for all $x_2 \neq 0$, we have $\dot{V} < 0$, The system is capable of reaching the sliding surface $S_t = 0$ in finite time.

In the following step, we will implement the control law in Simulink. Figure 9 illustrates the MFATSMGC command applied to both the elbow articulation of the exoskeleton model and the human arm.

The two Simulink models are linked, with the exoskeleton controlled to follow a specified trajectory. The output from this block serves as the input for the upper limb block.

5 Simulation and Results

The simulation results are presented to validate the performance of the proposed dynamic model and control law. In the first scenario, a sinusoidal signal is used as the desired trajectory, and the system's response is studied. The desired trajectory is defined as $q_{2d} = \sin(2\pi t)$; The control law parameters are set to: $r = 7$, $s = 5$, α

= 1, λ = 60, and K = 20. The exoskeleton and the human upper-limb simulation parameters are given by Table 1.

The simulation results for this system are shown in Fig. 10, which displays both the desired trajectory and the actual trajectory followed by the exoskeleton and the human upper limb.

Table 1 Exoskeleton and human upper-limb simulation parameters

Joint	1	2
Masses of the exoskeleton (kg)	4.3	3.5
Lengths of the exoskeleton (m)	0.35	0.20
Solid friction coefficient of the exoskeleton (N m)	0.031	0.028
Masses of the human upper limb (kg)	1.95	1.04
Lengths of the human upper limb (m)	0.329	0.253
Solid friction coefficient of the human upper limb (N m)	0.002	0.006

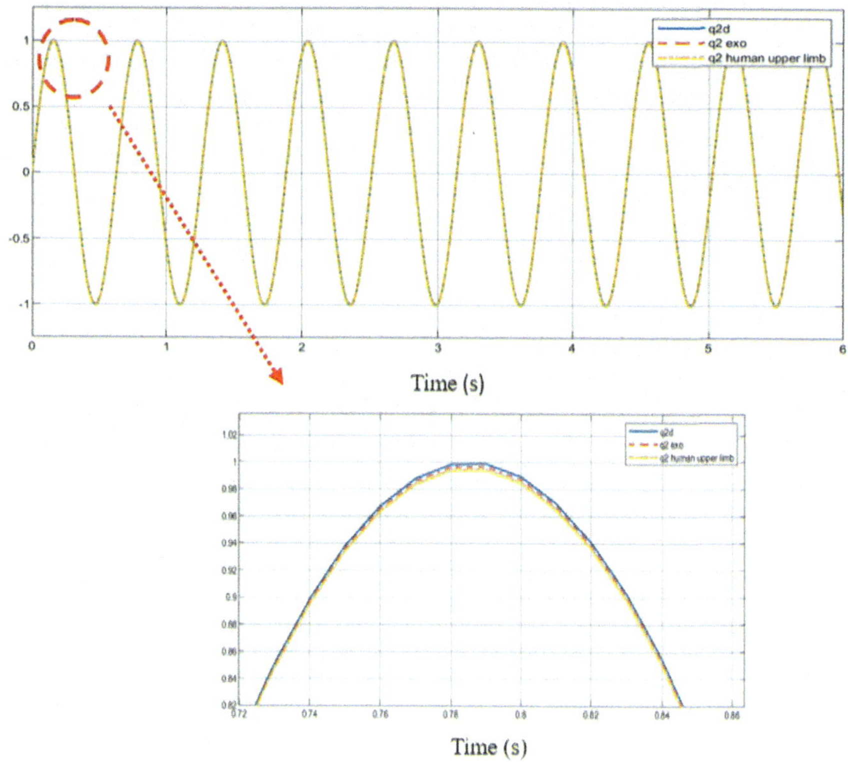

Fig. 10 Simulation results of the required trajectories following by the elbow joint of the system using the MFTSMGC controller

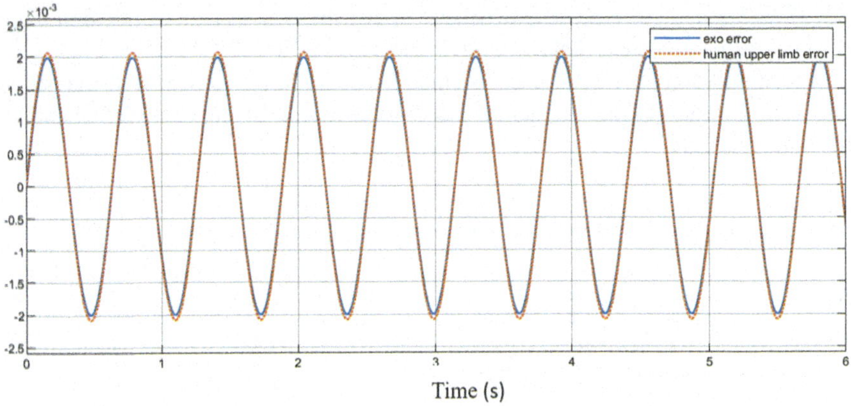

Fig. 11 The error calculation when tracking the sinusoidal signal by the exoskeleton and the upper limb controlled by the MFATSMGCC

Figure 11 illustrates the position tracking errors for both the exoskeleton and the upper limb, showing an oscillating error of approximately $\pm 2 \times 10^{-3}$ radians.

In the second scenario, the EMG signal is applied as the desired trajectory, and the system's response during elbow flexion and extension is evaluated.

Figures 12 and 14 present the measured and desired trajectories during these movements for both the exoskeleton and the upper limb, demonstrating accurate position tracking.

Figures 13 and 15 show the tracking error of the EMG signal for elbow flexion and extension, highlighting that the MFTSMGC performs well in following the required trajectories for both the exoskeleton and the upper limb.

To further demonstrate the control law's efficiency, a statistical analysis is conducted, including the calculation of RMS error, mean error, and standard deviation error (std) when following the desired position trajectories. Table 2 provides the e-RMS, e-Mean, and e-Std values for the considered system control. The results indicate that the combination of adaptive control, gravity compensation, and the model-free terminal sliding mode controller improves the system's ability to accurately follow the desired trajectories, with RMS errors on the order of 10^{-4}.

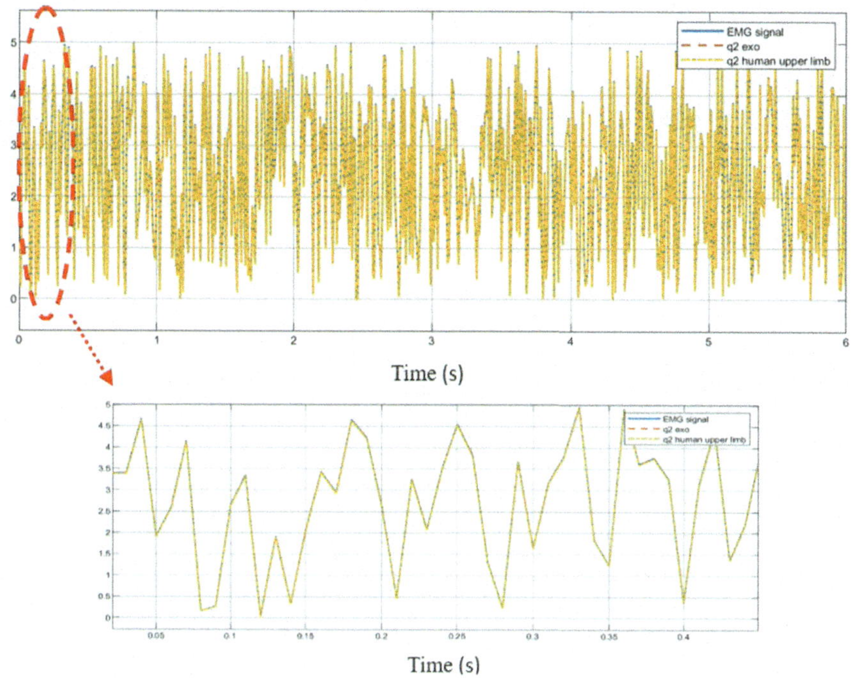

Fig. 12 Simulation results of the EMG signal tracking during the flexion movement of the elbow by the considered system using the MFATSMGCC

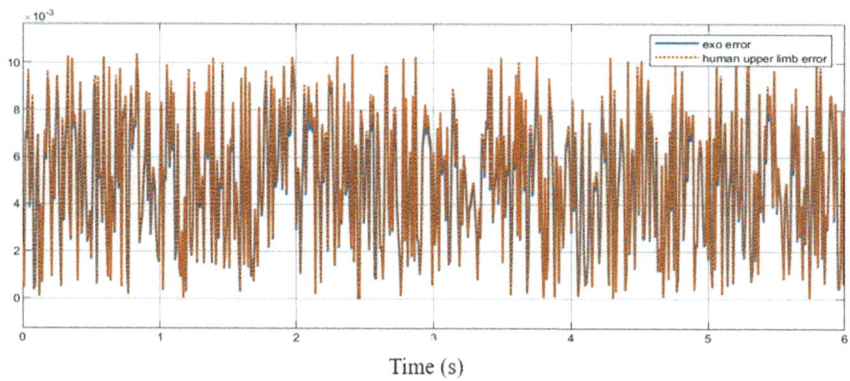

Fig. 13 The error calculation when tracking the EMG signal during the flexion movement using the MFATSMGCC

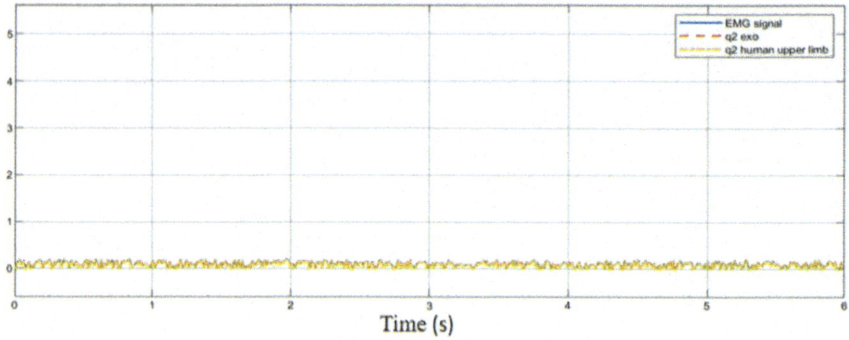

Fig. 14 Simulation results of the EMG signal following during the extension movement of the elbow by the treated system using the MFATSMGCC

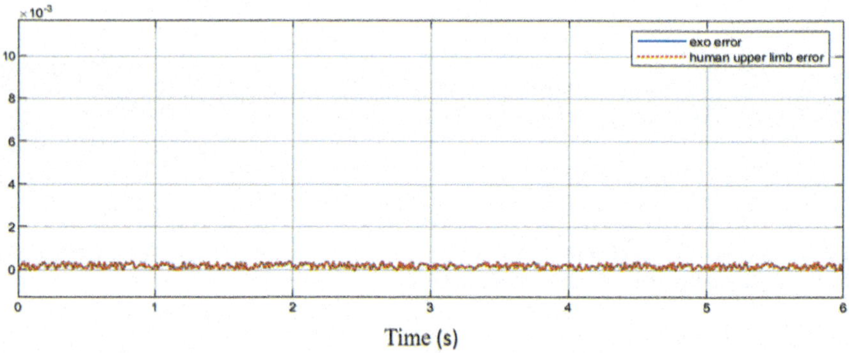

Fig. 15 The error calculation when tracking the EMG signal during the extension movement using the MFATSMGCC

Table 2 Calculation of the e-RMS, the e-mean and the e-std of the elbow articulation

	Model free adaptive terminal sliding mode with gravity compensation			
	Sine wave signal		EMG signal	
	Exoskeleton control	Upper limb control	Exoskeleton control	Upper limb control
e-RMS 10^{-3} (rad)	0.60	0.63	5.8	6.1
e-mean 10^{-3} (rad)	0.54	0.59	5.1	5.3
e-std 10^{-3} (rad)	0.65	0.71	6.2	6.7

6 Conclusion

This chapter focuses on the development of both an exoskeleton model and a human upper limb model, as well as the control of this system using EMG signals. The work involved several key stages, beginning with the exploration of communication between Matlab/Simulink and SolidWorks environments. This step was crucial for ensuring seamless interaction between the control algorithms designed in Matlab and the 3D mechanical models created in SolidWorks. The integration of these two platforms allowed us to simulate the movements of the exoskeleton and the human arm within a unified Simulink system, enabling 3D simulations that follow Matlab-based control commands. The next stage of this research involved implementing the exoskeleton and human arm models into Simulink, creating a comprehensive system model. This model was used to simulate the interaction between the human arm and the exoskeleton in real-time, particularly in response to EMG signals, which were extracted from the user's muscles. Detailed steps for EMG signal extraction were provided, including signal acquisition, amplification, filtering, and noise removal, ensuring a clean input for the control system. A Model-Free Adaptive Terminal Sliding Mode Controller with Gravity Compensation was then developed to govern the movement of the exoskeleton and human upper limb system. This control strategy was specifically designed to handle the non-linear dynamics of the system and to provide robust performance in tracking the desired trajectories derived from the EMG signals. To validate the proposed control approach, simulation results were analyzed, and a statistical study was conducted. This study included metrics such as RMS error, mean error, and standard deviation, which demonstrated the controller's efficiency and performance in accurately tracking the desired movement trajectories. The results showed that the proposed control law could deliver precise control with minimal tracking errors. As future work, the objective is to move from simulation to a physical model of the upper limb exoskeleton, designed with two DoF to control both the shoulder and elbow articulations. The findings from this chapter will be applied to the prototype, further enhancing the integration of the human upper limb with the exoskeleton system. This next step will involve real-world testing, using EMG signals to monitor the exoskeleton's movement, and expanding the application of the developed control strategies to this more complex mechanical structure.

References

1. Suminski, A.J., Zimbelman, J.L., Scheidt, R.A.: Design and validation of a MR-compatible pneumatic manipulandum. J. Neurosci. Methods **163**, 255–266 (2007)
2. Jawad, Y., Alessio, P., Matteo Lavit, N., Tito, D., Hermes, G., Matteo, M.: A planar parallel device for neurorehabilitation. Robotics **9**, 104 (2020)
3. Fong, J., Crocher, V., Tan, Y., Oetomo, D., Mareels, I.: EMU: a transparent 3D robotic manipulandum for upper-limb rehabilitation. In: 2017 International Conference on Rehabilitation Robotics (ICORR) QEII Centre, London, UK, July 17–20, 2017 (2017)

4. Cekus, D., Posiadała, B., Warys, P.: Integration of modeling in solidworks and Matlab/Simulink environments. Mech. Eng. **61**(1) (2014)
5. Gran, R.J.: Numerical Computing with Simulink. Volume I, Creating Simulation. Society for Industrial and Applied Mathematics, Philadelphia (2007)
6. Schiele, A., van der Helm, F.C.T.: Kinematic design to improve ergonomics in human machine interaction. IEEE Trans. Neural Syst. Rehabil. Eng. **14**, 456–469 (2006)
7. Posiadała, B., Warys, P., Cekus, D., Tomala, M.: The dynamics of the forest crane during the working cycle. Int. J. Struct. Stabil. Dyn. **13**(7) (2013)
8. Jarrassé, N., Roby-Brami, A., Saint-Bauzel, L., Morel, G.: Les Exosquelettes Robotisés: Réalités Et Perspectives. Institut Des Systèmes Intelligents Et De Robotique – Isir – Équipe Agathe (Assistance Au Geste Et Applications Thérapeutiques), Inserm U1150, Le Magazine De La Chirurgie Francophone (2014)
9. Syed Yarooq, R., Syed Faiz, A., Athar, A., Kushsairy A.K., Kamran, J., Sheroz, K., Zuriati, J.: Model predictive control for upper limb rehabilitation robotic system under noisy condition. In: 2018 IEEE 5th International Conference on Smart Instrumentation, Measurement and Application (ICSIMA), November 2018 (2018). https://doi.org/10.1109/ICSIMA.2018.8688747
10. Kittel Ouimet, T.: Commande d'un bras exosquelette robotique à sept degrés de liberté. Montréal 1–5 (2012)
11. Wen, Y., Jacob, R., Xiaoou, L.: PID admittance control for an upper limb exoskeleton. In: Proceedings of the 2011 American Control Conference, San Francisco, CA, USA, 29 June 2011–01 July 2011 (2011). https://doi.org/10.1109/ACC.2011.5991147
12. Wen, Y., Jacob, R.: A novel linear PID controller for an upper limb exoskeleton. In: 49th IEEE Conference on Decision and Control (CDC), January 2011 (2011). https://doi.org/10.1109/CDC.2010.5716985
13. Abdul Manan, K., Deok-won, Y., Mian Ashfaq, A., Jungsoo, H., Kyoosik, S., Changsoo, H.: Adaptive impedance control for upper limb assist exoskeleton. In: 2015 IEEE International Conference on Robotics and Automation (ICRA), 26–30 May 2015, Seattle, WA, USA (2015). https://doi.org/10.1109/ICRA.2015.7139801
14. Quan, L., Yang, L., Yi, L., Chang, Z., Wei, M., Qingsong, A., Sheng Q.X.: Path planning and impedance control of a soft modular exoskeleton for coordinated upper limb rehabilitation. Front. Neurorobot. **15** (2021). https://doi.org/10.3389/fnbot.2021.745531
15. Feng, Y., Yu, X., Man, Z.: Non-singular terminal sliding mode control of rigid manipulators. Automatica **38**, 2159–2167 (2002)
16. Bouteraa, Y., Ben Abdallah, I.: Estimated model-based sliding mode controller for an active exoskeleton robot. Chapter 10: Applications of Sliding Mode Control, Decision and Control, pp. 175–189 (2016)
17. Sira-Ramirez, H.: On the dynamical sliding mode control of nonlinear systems. Int. J. Control 1039–1061 (1993)
18. Yong, F., Yu, X., Man, Z.: Non-singular terminal sliding mode control of rigid manipulators. Automatica **38**, 2159–2167 (2002)
19. Wu, Y., Yu, X., Man, Z.: Terminal sliding mode control design for uncertain dynamic systems. Syst. Control Lett. **34**, 281–287 (1998)
20. Bonsa Regassa, H., Abraham Debebe, W.: Future prospects of computer-aided design (CAD)—a review from the perspective of artificial intelligence (AI), extended reality, and 3D printing. Results Eng. **14** (2022). https://doi.org/10.1016/j.rineng.2022.100478
21. Arumugavel, S.: Computer aided design in the field of mechanical fabrication. Iconic Res. Eng. J. (IRE J.) **2**(4) (2018). ISSN: 2456-8880
22. Bembli, S., Khraief Haddad, N., Belghith, S.: Computer aided decision model to control an exoskeleton-upper limb system. In: The 3rd International Conference on Advanced Systems and Electrical Technologies (IC_ASET), 19–22 March 2019, Hammamet, Tunisia (2019). https://doi.org/10.1109/ASET.2019.8871010
23. Poyi, G.T., Wu, M.H., Bousbaine, A., Wiggins, B.: Validation of a quad-rotor helicopter Matlab/Simulink and solidworks models. In: Control and Automation 2013, Uniting Problems and Solutions, September 2013 (2013)

24. Bembli, S., Khraief Haddad, N., Belghith, S.: Robust control of an exoskeleton-upper limb system: a comparative study. Int. J. Mechatron. Autom. (IJMA) **8**(4), 217–228 (2021). https://doi.org/10.1504/IJMA.2021.120376
25. Bembli, S., Khraief Haddad, N., Belghith, S.: Modelling, control and robustness analysis of a 2-DoF exoskeleton-upper limb system. Int. J. Intell. Eng. Inform. (IJIEI) **9**(6), 578–603 (2021). https://doi.org/10.1504/IJIEI.2021.122438
26. Bembli, S., Khraief Haddad, N., Belghith, S.: An integrated exoskeleton-upper limb system: modeling, control, stability study and robustness analysis. In: Advances in Robust Control and Applications, pp. 1–24. Springer International Publishing (2023). https://doi.org/10.1007/978-981-99-3463-8_1
27. Bembli, S., Khraief Haddad, N., Belghith, S.: An exoskeleton—upper limb system control using a robust model free terminal sliding mode with EMG signal. In: The 5th International Conference on Control, Automation and Diagnosis (ICCAD'21), November 3–5, 2021, Grenoble, France (2021)
28. Fliess, M., Join, C., Mboup, M., SiraRamirez, H.: Vers une commande multivariable sans modèle. In: Conférence internationale francophone d'automatique (CIFA 2006), Bordeaux, France (2006)
29. Fliess, M., Join, C.: Model-free control and intelligent PID controllers: towards a possible trivialization of nonlinear control. In: 15th IFAC Symposium on System Identification (SYSID 2009), Saint-Malo, France (2009). https://doi.org/10.3182/20090706-3-FR-2004.00256
30. Bembli, S., Khraief Haddad, N., Belghith, S.: Human-in-the-loop dynamic modeling and control of an upper-limb exoskeleton. In: The International Conference on Information Technology & Smart Industrial Systems (ITSIS 2022), July 15–17, 2022, Paris, France (2022)
31. Bembli, S., Khraief Haddad, N., Belghith, S.: A terminal sliding mode control using EMG signal: application to an exoskeleton-upper limb system. In: 16th International Conference on Informatics in Control, Automation and Robotics (ICINCO-2019), Prague, 29–31 July 2019 (2019)
32. Bembli, S., Khraief Haddad, N., Belghith, S.: An exoskeleton—upper limb system control using a robust model free terminal sliding mode. In: The 4th International Conference on Advanced Systems and Electrical Technologies (IC_ASET), 15–18 December 2020, Hammamet, Tunisia (2020)
33. Bembli, S., Khraief Haddad, N., Belghith, S.: A robust model free terminal sliding mode with gravity compensation control of a 2 DoF exoskeleton-upper limb system. J. Control Autom. Electr. Syst. **32**(3), 632–641 (2021). https://doi.org/10.1007/s40313-021-00687-z

Robust Stabilization of a 1-DoF Knee Rehabilitation Exoskeleton Using an Enhanced LMI-Based Composite Nonlinear Controller

Sahar Jenhani, Hassène Gritli, and Jyotindra Narayan

Abstract This chapter presents a control strategy aimed at robustly stabilizing a one-degree-of-freedom (1-DoF) knee rehabilitation exoskeleton robot. The proposed approach combines a linear state-feedback controller with a nonlinear control law to effectively manage the nonlinear dynamics characteristic of knee-joint exoskeleton systems, particularly in the presence of parameter uncertainties and disturbances. To simplify the model, we approximate the nonlinear function in the robot dynamics. Primarily, we assume that such a nonlinear function is bounded by a constant value, thus leading to the first design approach for the development of the Linear Matrix Inequality (LMI) condition. Following this, we assume that the nonlinear function remains bounded within the constraint of a linear equation. Under this assumption, we develop two distinct methods for the design of the LMI condition. To ensure stability and robustness, a quadratic Lyapunov function is used to derive LMI conditions for the computation of the feedback gain of the control law. In these three design approaches, the development of the LMI conditions is achieved through various technical lemmas such as Young's inequality, the S-procedure, the Schur complement, and the matrix inversion lemma. The robustness of the controller is validated through simulations that introduce external disturbances. Additionally, a comparative analysis of the proposed control strategies is conducted. Numerical simulations confirm the effectiveness of our control approach in stabilizing the knee-joint exoskeleton robot, even in the presence of external disturbances and parametric uncertainties.

S. Jenhani · H. Gritli (✉)
Laboratory of Robotics, Informatics and Complex Systems (RISC Lab - LR16ES07), National Engineering School of Tunis, University of Tunis El Manar, Tunis, Tunisia
e-mail: grhass@yahoo.fr

H. Gritli
Higher Institute of Information and Communication Technologies, University of Carthage, Technopole of Borj Cédria, Ben Arous, Tunisia

J. Narayan
Department of Mechanical Engineering, Indian Institute of Technology Patna, Patna, India

Department of Computing, Imperial College London, London, UK

J. Narayan and H. Gritli (eds.), *Design and Control of Rehabilitation Robots*, Studies in Systems, Decision and Control 585, https://doi.org/10.1007/978-3-031-86977-8_14

Keywords Knee rehabilitation exoskeleton · Robust control strategy · Nonlinear control law · Nonlinear dynamics · Linear matrix inequality · External disturbances

1 Introduction

The growing elderly population worldwide is driving an increasing need for innovative rehabilitation solutions to maintain mobility and enhance the quality of life for elderly individuals. According to data from the United Nations, the proportion of individuals aged 60 and over has significantly increased over the past two decades and is expected to reach 21.1% by 2050. This demographic trend highlights the necessity of developing assistive technologies that can meet the needs for rehabilitation and mobility support [1]. Among these technologies, exoskeletons, particularly those designed for lower limb rehabilitation, stand out due to their ability to provide motor support and facilitate rehabilitation exercises [2–8].

Rehabilitation exoskeletons are wearable devices that allow users to perform specific movements with mechanical and/or motorized assistance, thus helping individuals with mobility impairments to recover or improve their locomotor functions [9, 10]. These systems are particularly useful for patients suffering from neurological disorders, spinal cord injuries, strokes, or degenerative diseases that affect gait and balance. Moreover, in the context of an aging population, these devices can be used to maintain the mobility of elderly individuals, thereby reducing the risk of falls and enhancing daily autonomy. However, the design and control of rehabilitation exoskeletons, especially for complex joints like the knee, pose numerous challenges. The knee, as a central load-bearing joint, plays a crucial role in locomotion and body support, and exoskeletons must not only replicate the natural movements of this joint but also adapt to the specific needs of each user [11–15]. The nonlinear dynamics of the knee joint, combined with parametric uncertainties and external disturbances such as changes in muscle resistance or unforeseen external forces, further complicate the task of designing robust and efficient controllers [8, 16–22].

In recent years, considerable research has been directed toward developing advanced control systems for knee exoskeletons, particularly for rehabilitation of individuals with spinal cord injuries. These systems must overcome significant challenges, such as managing nonlinear dynamics, parametric uncertainties, and external disturbances like friction, which are inherent in the motion of the knee joint. Various control strategies have emerged to address these complexities, including affine Proportional-Derivative (PD) controllers, which have shown promising results in enhancing the robustness and stability of knee exoskeletons [17]. Additionally, state-feedback controllers have been extensively investigated for their ability to regulate knee joint movements effectively while accounting for factors like viscous and solid frictions [23–26]. To further improve control performance in the face of nonlinearities, researchers have also explored adaptive control methods, such as Model Reference Adaptive Control (MRAC), which dynamically adjusts to changes in system parameters, and Fuzzy PID control, which offers flexibility in handling varying operational conditions [13, 27–30]. Another significant advancement is the development of composite control strategies, which combine linear state-feedback with nonlinear control laws to achieve robust stabilization of the knee exoskeleton under different types of disturbances [31–33]. In addition to these control methods, more advanced adaptive controllers have been introduced, such as the RISE controller, designed to handle the high levels of nonlinearity present in exoskeleton systems [34]. Comparative studies have also been conducted to evaluate different control approaches for knee exoskeletons, assessing their kinematic performance during repetitive tasks such as squatting motions [35]. However, despite the effectiveness of traditional control techniques, they often face limitations when dealing with the full complexity of nonlinear and uncertain exoskeleton systems. To overcome these limitations, recent research has increasingly focused on control strategies based on LMI. LMI-based techniques provide a robust mathematical framework for deriving stability and performance conditions, which can be solved algorithmically. This makes them particularly well-suited for complex systems like knee exoskeletons, where nonlinearities, parameter uncertainties, and external disturbances must be accounted for [36–38]. The use of quadratic Lyapunov functions within the LMI framework allows for robust control design that guarantees system stability even in the presence of these challenges [14, 39].

By integrating linear state-feedback controllers with nonlinear control laws within the LMI framework, it is possible to effectively model and mitigate the effects of system nonlinearities while maintaining robustness across a wide range of operating conditions [40]. This flexibility not only enhances the stability and adaptability of the control system but also allows for optimization of controller performance, making LMI-based methods particularly valuable in rehabilitation settings. Overall, these recent advances in control design represent a significant step forward in the development of knee exoskeletons, providing more effective and adaptable solutions that contribute to improved rehabilitation outcomes for patients. Compared to other studies that use composite control strategies combining linear state-feedback with nonlinear control laws to achieve robust position control of the knee exoskeleton [31, 32], this work adopts a different formulation for the nonlinear term in the composite control strategy. This approach enables us to develop a more versatile control strategy. Additionally, previous works such as [31] focused their analysis on the nonlinear dynamic model without considering uncertainties. In contrast, this research takes into account the uncertainties present in the nonlinear dynamic model.

In this chapter, we present a robust control strategy for a 1-DoF knee rehabilitation exoskeleton [14, 23, 24, 31]. This device is designed to stabilize the knee joint in response to different dynamic conditions encountered during rehabilitation exercises. Our approach combines a composite nonlinear controller with LMI techniques to effectively handle the system's nonlinearities, parameter uncertainties, and external disturbances. Using mathematical tools such as Lyapunov stability theory, the S-procedure, the matrix inversion lemma, the Young's inequality and the Schur complement [41–45], we derive LMI conditions to ensure system stability and robustness. We develop three different design methods to formulate the LMI conditions, taking into account different scenarios of modeling the nonlinearities. In the first case, we approximate the nonlinear function in the robot dynamics by assuming it is constrained by a constant value. This simplification allows us to develop a first design method for the LMI criteria. Next, we consider the nonlinear function remains constrained within the limits of a linear equation, which leads to the development of two other distinct design methods for the LMI condition. To verify the effectiveness of the adopted control approach, we present numerical simulations that illustrate the controller's ability to stabilize the knee exoskeleton even in the presence of disturbances and parametric uncertainties. Additionally, a comparison of the developed control strategies is conducted to identify the most effective method in various scenarios.

The key contributions of this chapter include the following:

- Development of a control strategy that combines a linear state-feedback control law with a nonlinear control law to stabilize a 1-DoF knee rehabilitation exoskeleton robot. This approach is designed to effectively manage the nonlinear dynamics of knee-joint exoskeleton systems, particularly in the presence of disturbances and parameter uncertainties.
- Formulation of LMI conditions for the controller's feedback gain using a quadratic Lyapunov function to ensure stability and robustness. The development of these LMI conditions is supported by a theoretical framework that incorporates technical lemmas including the matrix inversion lemma, the Schur complement, the S-procedure and the Young's inequality.
- The efficacy of the proposed control strategy is validated through numerical simulations that introduce external disturbances. These simulations demonstrate the robustness of the controller in stabilizing the exoskeleton robotic system, even in the presence of parametric uncertainties and disturbances.
- A comparative evaluation of the proposed control strategies is conducted to evaluate their effectiveness in various scenarios.

This chapter is structured like so: Sect. 2 outlines the dynamic model of the knee rehabilitation exoskeleton robot. In Sect. 3, we explore the development of the robust control methodology for the knee exoskeleton. Section 4 details the formulation of the stability conditions using LMI approach. Section 5 presents simulation results assessing the controller's performance. Finally, Sect. 6 offers concluding remarks and future research directions

2 Dynamic Modeling of the Knee Rehabilitation Exoskeleton

2.1 *Structure of the Knee Exoskeleton System*

This research centers on controlling wearable exoskeletons, especially those featuring one to three DoF in the sagittal plane. These exoskeletons may include passive, active, or fixed joints, and are designed to aid in rehabilitation or enhance mobility.

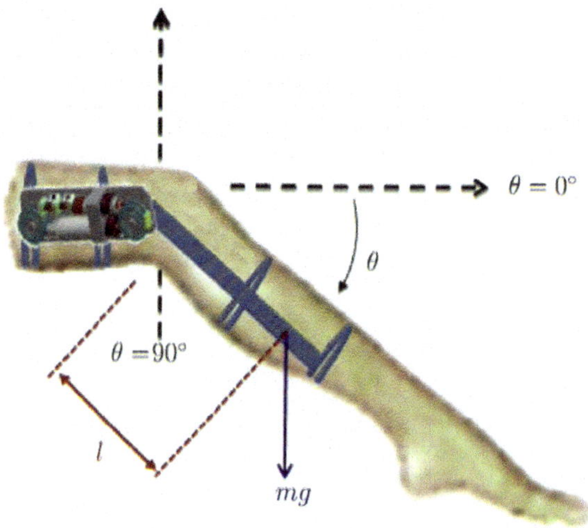

Fig. 1 Illustration of the knee exoskeleton robotic system [23, 47]

Our primary focus is on a knee exoskeleton with a single degree of freedom, specifically tailored to assist and rehabilitate the knee joint. The 1-DoF knee exoskeleton, shown in Fig. 1 [18, 25, 26, 46, 47], was selected for its simplicity and ease of use, which is particularly beneficial for individuals with limited knee mobility [47].

Figure 1 illustrates the knee exoskeleton used in this study, featuring a streamlined design to facilitate easy application and removal, making it well-suited for users with knee joint restrictions [18, 23, 25, 26, 47]. The choice of this specific orthosis is driven by its straightforward and user-centric design, aligning perfectly with the objectives of our research [18].

2.2 *Nonlinear Dynamic Model of the Adopted Knee Exoskeleton*

This part explores the nonlinear dynamic model of the lower-limb rehabilitation exoskeleton, specifically focusing on its application to the knee joint (see Fig. 1). The presence of uncertainties from parameter variations and unmodeled dynamics makes it challenging to accurately determine the dynamic model parameters. To address these issues, a detailed nonlinear dynamic model is developed for the exoskeleton, which accounts for external disturbances such as solid friction, viscous friction, and parameter uncertainties. The governing equation for the nonlinear dynamic model of the knee exoskeleton, as shown in Fig. 1, is expressed as:

$$\vartheta\ddot{\theta} = u - f_v\dot{\theta} + a\cos(\theta) - f_s\text{sgn}(\dot{\theta}) + \delta_t \tag{1}$$

In this equation, ϑ represents the inertia of the exoskeleton, f_v is the viscous friction coefficient, f_s denotes the solid friction coefficient and δ_t denotes the load torque. Additionally, $a = mgl$, where m is the mass of the exoskeleton, l is its length, and g is the acceleration due to gravity.

In [31], previous studies base their analysis on a nonlinear dynamic model without accounting for uncertainties. In this work, we separate the dynamics into a nominal component and an uncertain component, allowing us to better capture the variability in system behavior. The system's parameters are presented as follows:

$$\begin{aligned} a &= a_\star + \Delta a & \vartheta &= \vartheta_\star + \Delta\vartheta \\ f_v &= f_v^\star + \Delta f_v & f_s &= f_s^\star + \Delta f_s \end{aligned} \tag{2}$$

By substituting the parameters from (2) into the Eq. (1), the following dynamic model is obtained:

$$\vartheta_\star\ddot{\theta} = a_\star \cos(\theta) - f_v^\star\dot{\theta} - f_s^\star \mathrm{sgn}(\dot{\theta}) + u + \Delta_t \tag{3}$$

In this context, Δ_t denotes the total or aggregated disturbance, which can be defined as:

$$\Delta_t = \delta_t - \Delta\vartheta\ddot{\theta} + \Delta a \cos(\theta) - \Delta f_s \mathrm{sgn}(\dot{\theta}) - \Delta f_v\dot{\theta} \tag{4}$$

In the following analysis, we assume that Δ_t adheres to the following constraint:

$$|\Delta_t| \leq \bar{\delta}, \qquad \bar{\delta} > 0 \tag{5}$$

By introducing the variable $\phi = \theta - \theta_d$, where θ_d denotes the desired position for controlling the knee exoskeleton, and noting that $\dot{\theta} = \dot{\phi}$ and $\ddot{\theta} = \ddot{\phi}$, we can express the dynamic model (1) like so:

$$\vartheta_\star\ddot{\phi} = a_\star \cos(\phi + \theta_d) - f_v^\star\dot{\phi} - f_s^\star \mathrm{sgn}(\dot{\phi}) + u + \Delta_t \tag{6}$$

2.3 *Required Control Effort at the Desired Equilibrium Position*

The objective of this section is to quantify the control effort required to maintain the knee exoskeleton robot at the desired equilibrium position θ_d. Once the robot reaches this position, continuous control is needed to hold it steady ($\dot{\theta}_d = 0$).

The required control effort u_d is given by:

$$u_d = -a_\star \cos(\theta_d) \tag{7}$$

where $a_\star$ represents system stiffness. This effort ensures the robot remains at the target angle, compensating for disturbances or internal dynamics. As the position stabilizes at θ_d, the control effort adjusts to maintain equilibrium, counteracting any forces that could cause deviations.

3 Robust Control Methodology for the Knee Exoskeleton

3.1 Problem Formulation

This research focuses on the control strategies for a lower-limb exoskeleton designed to support individuals with disabilities, focusing especially on the knee joint. Although exoskeleton robots have various applications, the primary objective of this study is to enhance mobility for individuals facing mobility challenges. Consequently, such wearable technologies, including exoskeletons and orthotics, are frequently utilized to assist older adults and people with limb impairments in improving their physical mobility.

Interest in controlling lower-limb exoskeletons has surged among robotics researchers due to their growing range of applications. This study aims to develop control methods for a knee-joint rehabilitation exoskeleton robot. The focus is on a knee exoskeleton robot with 1-DoF, as depicted in Fig. 1.

The objective is to ensure the efficient regulation and stabilization of the exoskeleton robotic system shown in Fig. 1, which operates based on the nonlinear dynamics described in Eq. (1). To achieve this, we use the LMI approach to formulate conditions that guarantee the robustness of the control system for the knee rehabilitation robot. These LMI conditions are developed through three distinct design methods, each accounting for a specific scenario in modeling the nonlinearities. In the first method, we approximate the nonlinear function in the robot dynamics by assuming it is bounded by a constant value, which leads to the formulation of the initial LMI design. Next, we consider the nonlinear function as constrained within a linear equation, resulting in the development of two additional design methods. The LMI framework employs several key mathematical tools, including Young's inequality, the S-procedure, the Schur complement, and the matrix inversion lemma [42–44], which will be discussed in detail in the following section.

3.2 State Space Formulation

Let the state vector be defined as $\mathcal{X} = \begin{bmatrix} \phi & \dot{\phi} \end{bmatrix}^{\mathrm{T}}$. The motion equation (6) can then be expressed as follows:

$$\dot{\mathcal{X}} = \mathcal{A}\mathcal{X} + \mathcal{B}u + \mathcal{D}_1\Phi_1(\mathcal{X}) + \mathcal{D}_2\Phi_2(\mathcal{X}) + \mathcal{B}\Delta_t \tag{8}$$

Here, $\Phi_1(\mathcal{X})$ and $\Phi_2(\mathcal{X})$ are defined as:

$$\Phi_1(\mathcal{X}) = \text{sgn}(\dot{\phi}), \qquad \text{and} \qquad \Phi_2(\mathcal{X}) = \cos(\phi + \theta_d) \tag{9}$$

The matrices $\mathcal{A}$, $\mathcal{B}$, $\mathcal{D}_1$ and $\mathcal{D}_2$ are given by:

$$\mathcal{A} = \begin{bmatrix} 0 & 1 \\ 0 & -\frac{f_v^\star}{\vartheta_\star} \end{bmatrix}, \quad \mathcal{B} = \begin{bmatrix} 0 \\ \frac{1}{\vartheta_\star} \end{bmatrix}, \quad \mathcal{D}_1 = \begin{bmatrix} 0 \\ -\frac{f_s^\star}{\vartheta_\star} \end{bmatrix}, \quad \text{and} \quad \mathcal{D}_2 = \begin{bmatrix} 0 \\ \frac{a_\star}{\vartheta_\star} \end{bmatrix} \tag{10}$$

3.3 Composite Control Approach

This study aims to design a control strategy that ensures stabilization of the knee exoskeleton robot at the target state θ_d. To accomplish this, a composite control strategy is proposed, given by:

$$u = u_{nl} + u_l \tag{11}$$

Here, u_l is an affine control law, expressed as follows:

$$u_l = u_d + \Delta u_l \tag{12}$$

Here, u_d is defined in (7), and Δu_l is a state feedback controller, specified as:

$$\Delta u_l = \mathcal{K}\mathcal{X} \tag{13}$$

In this expression, $\mathcal{K}$ is the gains matrix that will be designed in later stages.

The term u_{nl} in (11) represents the nonlinear control component, which will be designed subsequently. This additional control term is intended to address the issue of the lumped disturbance Δ_t.

Remark 1 In this study, we propose a composite control strategy (11) that combines a linear component u_l and a nonlinear component u_{nl}, each addressing distinct control objectives. The linear term, defined by (12) with Δu_l specified in (13), provides basic stabilization toward the target state θ_d by using state feedback to correct small deviations. The nonlinear term u_{nl} is designed to counteract disturbances Δ_t arising from nonlinear dynamics and external forces that could disrupt stability. Together, u_l and u_{nl} ensure robust stabilization and precision in the exoskeleton's position control.

3.4 Dynamics Under the Composite Control Law

When the composite control strategy given in (11) is applied to the nonlinear dynamics described in (8), the resulting closed-loop dynamic model is formulated as follows:

$$\dot{\mathcal{X}} = (\mathcal{A} + \mathcal{B}\mathcal{K})\,\mathcal{X} + \mathcal{B}\,u_{nl} + \mathcal{D}_1\Phi_1(\mathcal{X}) + \mathcal{D}_2\Phi_3(\mathcal{X}) + \mathcal{B}\Delta_t \tag{14}$$

In this expression, $\Phi_3(\mathcal{X})$ is defined as:

$$\Phi_3(\mathcal{X}) = \cos(\phi + \theta_d) - \cos(\theta_d) \tag{15}$$

4 Development of the Nonlinear Controller and LMI Conditions

This section describes three distinct methods for designing the nonlinear controller u_{nl} using the Lyapunov approach. Following this, we will design the linear state feedback controller Δu_l. Additionally, we will utilize the LMI framework to derive stability conditions necessary for computing the gain of the linear control component. These LMI conditions are crucial for ensuring the stability of the exoskeleton robotic system.

To achieve our objective, positive-definite candidate Lyapunov function will be utilized, defined as:

$$\mathcal{V}(\mathcal{X}) = \mathcal{X}^{\mathrm{T}}\mathcal{P}\mathcal{X} > 0 \tag{16}$$

where $\mathcal{P} = \mathcal{P}^{\mathrm{T}} > 0$.

The time derivative of this Lyapunov function is expressed as:

$$\dot{\mathcal{V}}(\mathcal{X}) = 2\mathcal{X}^{\mathrm{T}}\mathcal{P}\dot{\mathcal{X}} \tag{17}$$

To enhance control performance and ensure exponential stability, we introduce the following condition:

$$\dot{\mathcal{V}}(\mathcal{X}) + \eta\mathcal{V}(\mathcal{X}) < 0 \tag{18}$$

where $\eta > 0$ is a positive scalar representing the decay rate of the Lyapunov function.

By substituting into Eq. (17) with (14), we can rewrite it as follows:

$$\begin{aligned}\dot{\mathcal{V}}(\mathcal{X}) + \eta\mathcal{V}(\mathcal{X}) = {} & 2\mathcal{X}^{\mathrm{T}}\mathcal{P}\,(\mathcal{A} + \mathcal{B}\mathcal{K})\,\mathcal{X} + 2\mathcal{X}^{\mathrm{T}}\mathcal{P}\mathcal{B}u_{nl} + 2\mathcal{X}^{\mathrm{T}}\mathcal{P}\mathcal{D}_1\Phi_1(\mathcal{X}) \\ & + 2\mathcal{X}^{\mathrm{T}}\mathcal{P}\mathcal{D}_2\Phi_3(\mathcal{X}) + 2\mathcal{X}^{\mathrm{T}}\mathcal{P}\mathcal{B}\Delta_t + \eta\mathcal{X}^{\mathrm{T}}\mathcal{P}\mathcal{X}\end{aligned} \tag{19}$$

To ensure the stabilization of the controlled exoskeleton robot, we will rely on the developed expression (19). This requires the formulation of three distinct LMI criteria for the gain matrix $\mathcal{K}$, which will be detailed in the subsequent analysis.

4.1 First LMI Design Strategy

The primary objective is to achieve the stabilization of the controlled robot using the expanded form of Eq. (19). Following this, we will establish a specific LMI constraint to determine the gain $\mathcal{K}$. In this initial method, we suppose that there is a constant upper bound on the function $\Phi_3(\mathcal{X})$.

Given that $\Phi_1(\mathcal{X}) = \text{sgn}(\dot{\phi}) = \pm 1$, it leads to $\Phi_1(\mathcal{X})^2 = 1$. By applying the Young inequality [43, 44], we obtain:

$$2\mathcal{X}^{\mathrm{T}}\mathcal{P}\mathcal{D}_1\Phi_1(\mathcal{X}) \leq \mathcal{X}^{\mathrm{T}}\mathcal{P}\mathcal{M}_1\mathcal{P}\mathcal{X} + \mathcal{D}_1^{\mathrm{T}}\mathcal{M}_1^{-1}\mathcal{D}_1 \tag{20}$$

here $\mathcal{M}_1 = \mathcal{M}_1^{\mathrm{T}} > 0$.

Given that $\Phi_3(\mathcal{X})$ is a scalar function defined in Eq. (15), applying the Young inequality results in:

$$2\mathcal{X}^{\mathrm{T}}\mathcal{P}\mathcal{D}_2\Phi_3(\mathcal{X}) \leq \mathcal{X}^{\mathrm{T}}\mathcal{P}\mathcal{M}_2\mathcal{P}\mathcal{X} + \mathcal{D}_2^{\mathrm{T}}\mathcal{M}_2^{-1}\mathcal{D}_2\Phi_3(\mathcal{X})^2 \tag{21}$$

where $\mathcal{M}_2 = \mathcal{M}_2^{\mathrm{T}} > 0$.

Considering $0^\circ \leq \theta \leq 90^\circ$, the following inequalities hold:

$$\cos^2(\theta) \leq 1, \qquad \text{and} \qquad -2\cos(\theta) \leq 0 \tag{22}$$

Utilizing (15), it follows that $\Phi_3(\mathcal{X})^2 = \cos^2(\theta) - 2\cos(\theta_d)\cos(\theta) + \cos^2(\theta_d)$. Furthermore, from inequality (22), we derive:

$$\Phi_3(\mathcal{X})^2 \leq \mu_{11}, \qquad \text{where} \qquad \mu_{11} = 1 + \cos^2(\theta_d) > 0 \tag{23}$$

Therefore, expression (21) can be rewritten as:

$$2\mathcal{X}^{\mathrm{T}}\mathcal{P}\mathcal{D}_2\Phi_3(\mathcal{X}) \leq \mathcal{X}^{\mathrm{T}}\mathcal{P}\mathcal{M}_2\mathcal{P}\mathcal{X} + \mu_{11}\mathcal{D}_2^{\mathrm{T}}\mathcal{M}_2^{-1}\mathcal{D}_2 \tag{24}$$

Additionally, considering that Δ_t is a scalar function and based on assumption (5), we can apply the Young inequality to obtain:

$$2\mathcal{X}^{\mathrm{T}}\mathcal{P}\mathcal{B}\Delta_t \leq m_3\bar{\delta}^2\mathcal{X}^{\mathrm{T}}\mathcal{P}\mathcal{B}\mathcal{B}^{\mathrm{T}}\mathcal{P}\mathcal{X} + m_3^{-1}\left(\frac{\Delta_t}{\bar{\delta}}\right)^2 \tag{25}$$

where $m_3 = m_3^{\mathrm{T}} > 0$.

Assuming that the total disturbance term Δ_t adheres to the following condition:

$$\left|\frac{\Delta_t}{\bar{\delta}}\right| \leq 1 \tag{26}$$

it follows from relation (26) that $\left(\frac{\Delta_t}{\bar{\delta}}\right)^2 \leq 1$. Consequently, we can establish the following expression:

$$2\mathcal{X}^{\mathrm{T}}\mathcal{P}\mathcal{B}\Delta_t \leq m_3\bar{\delta}^2\mathcal{X}^{\mathrm{T}}\mathcal{P}\mathcal{B}\mathcal{B}^{\mathrm{T}}\mathcal{P}\mathcal{X} + m_3^{-1} \tag{27}$$

By substituting Eqs. (20), (24), and (27) into expression (19) and considering Eq. (18), we derive the subsequent expression:

$$\begin{aligned}\dot{\mathcal{V}}(\mathcal{X}) + \eta\mathcal{V}(\mathcal{X}) = \mathcal{X}^{\mathrm{T}} \big(&(\mathcal{P}\mathcal{A} + \mathcal{P}\mathcal{B}\mathcal{K}) + (\mathcal{P}\mathcal{A} + \mathcal{P}\mathcal{B}\mathcal{K})^{\mathrm{T}} \\ &+\mathcal{P}\mathcal{M}_1\mathcal{P} + \mathcal{P}\mathcal{M}_2\mathcal{P} + m_3\bar{\delta}^2\mathcal{P}\mathcal{B}\mathcal{B}^{\mathrm{T}}\mathcal{P} + \eta\mathcal{P}\big)\mathcal{X} + 2\mathcal{X}^{\mathrm{T}}\mathcal{P}\mathcal{B}u_{nl} \\ &+ \mathcal{D}_1^{\mathrm{T}}\mathcal{M}_1^{-1}\mathcal{D}_1 + \mu_{11}\mathcal{D}_2^{\mathrm{T}}\mathcal{M}_2^{-1}\mathcal{D}_2 + m_3^{-1}\end{aligned} \tag{28}$$

Let us define:

$$\lambda = \mathcal{D}_1^{\mathrm{T}}\mathcal{M}_1^{-1}\mathcal{D}_1 + \mu_{11}\mathcal{D}_2^{\mathrm{T}}\mathcal{M}_2^{-1}\mathcal{D}_2 + m_3^{-1} \tag{29}$$

Therefore, we specify the auxiliary nonlinear controller u_{nl} as:

$$u_{nl} = \begin{cases} -\frac{1}{2}\frac{\mathcal{B}^{\mathrm{T}}\mathcal{P}\mathcal{X}}{\mathcal{X}^{\mathrm{T}}\mathcal{P}\mathcal{B}\mathcal{B}^{\mathrm{T}}\mathcal{P}\mathcal{X}}\lambda & \text{if } \|\mathcal{B}^{\mathrm{T}}\mathcal{P}\mathcal{X}\|^2 > \kappa \\ -\frac{1}{2}\frac{\mathcal{B}^{\mathrm{T}}\mathcal{P}\mathcal{X}}{\kappa}\lambda & \text{if } \|\mathcal{B}^{\mathrm{T}}\mathcal{P}\mathcal{X}\|^2 \leq \kappa \end{cases} \tag{30}$$

where $\kappa > 0$.

When $\|\mathcal{B}^{\mathrm{T}}\mathcal{P}\mathcal{X}\|^2 > \kappa$, by substituting the expression for the nonlinear controller u_{nl} from Eq. (30) into inequality (28), we simplify it to the following form:

$$\begin{aligned}\dot{\mathcal{V}}(\mathcal{X}) + \eta\mathcal{V}(\mathcal{X}) \leq \mathcal{X}^{\mathrm{T}} \big(&(\mathcal{P}\mathcal{A} + \mathcal{P}\mathcal{B}\mathcal{K}) + (\mathcal{P}\mathcal{A} + \mathcal{P}\mathcal{B}\mathcal{K})^{\mathrm{T}} \\ &+\mathcal{P}\mathcal{M}_1\mathcal{P} + \mathcal{P}\mathcal{M}_2\mathcal{P} + m_3\bar{\delta}^2\mathcal{P}\mathcal{B}\mathcal{B}^{\mathrm{T}}\mathcal{P} + \eta\mathcal{P}\big)\mathcal{X}\end{aligned} \tag{31}$$

Alternatively, in the case where $\|\mathcal{B}^{\mathrm{T}}\mathcal{P}\mathcal{X}\|^2 \leq \kappa$, using the definition of u_{nl} from Eq. (30) leads to the following form of inequality (28):

$$\begin{aligned}\dot{\mathcal{V}}(\mathcal{X}) + \eta\mathcal{V}(\mathcal{X}) \leq \mathcal{X}^{\mathrm{T}} \big(&(\mathcal{P}\mathcal{A} + \mathcal{P}\mathcal{B}\mathcal{K}) + (\mathcal{P}\mathcal{A} + \mathcal{P}\mathcal{B}\mathcal{K})^{\mathrm{T}} + \mathcal{P}\mathcal{M}_1\mathcal{P} \\ &+\mathcal{P}\mathcal{M}_2\mathcal{P} + m_3\bar{\delta}^2\mathcal{P}\mathcal{B}\mathcal{B}^{\mathrm{T}}\mathcal{P} + \eta\mathcal{P}\big)\mathcal{X} + \lambda\left(1 - \frac{\|\mathcal{B}^{\mathrm{T}}\mathcal{P}\mathcal{X}\|^2}{\kappa}\right)\end{aligned} \tag{32}$$

Considering that $\frac{\|\mathcal{B}^{\mathrm{T}}\mathcal{P}\mathcal{X}\|^2}{\kappa} \leq 1$ and $\lambda > 0$, it can be concluded that:

$$\lambda\left(1 - \frac{\|\mathcal{B}^{\mathrm{T}}\mathcal{P}\mathcal{X}\|^2}{\kappa}\right) \geq 0 \tag{33}$$

Therefore, it is evident that the defined $\mathcal{V}(\mathcal{X})$ in (16) is constrained. As a result, if $\|\mathcal{B}^{\mathrm{T}}\mathcal{P}\mathcal{X}\|^2 \leq \kappa$, the trajectory of the closed-loop system remains restricted. Thus, the problem reduces to verifying the condition given in (31).

To derive the LMI conditions, let's first define the function $\mathcal{Z}(\mathcal{X})$ as follows: $\mathcal{Z}(\mathcal{X}) = 2\mathcal{X}^{\mathrm{T}}\mathcal{P}(\mathcal{A}+\mathcal{B}\mathcal{K})\mathcal{X} + \mathcal{X}^{\mathrm{T}}\mathcal{P}\mathcal{M}_1\mathcal{P}\mathcal{X} + \mathcal{X}^{\mathrm{T}}\mathcal{P}\mathcal{M}_2\mathcal{P}\mathcal{X} + m_3\bar{\delta}^2\mathcal{X}^{\mathrm{T}}\mathcal{P}\mathcal{B}\mathcal{B}^{\mathrm{T}}$ $\mathcal{P}\mathcal{X} + \eta\mathcal{X}^{\mathrm{T}}\mathcal{P}\mathcal{X}$. From this expression, we can derive the following inequality:

$$\mathcal{Z}(\mathcal{X}) = \begin{bmatrix}\mathcal{X}\\ 1\end{bmatrix}^{\mathrm{T}}\begin{bmatrix}\Upsilon_{11} & \mathcal{O}\\ \star & \mathcal{O}\end{bmatrix}\begin{bmatrix}\mathcal{X}\\ 1\end{bmatrix} < 0 \tag{34}$$

with $\Upsilon_{11} = (\mathcal{P}\mathcal{A}+\mathcal{P}\mathcal{B}\mathcal{K}) + (\mathcal{P}\mathcal{A}+\mathcal{P}\mathcal{B}\mathcal{K})^{\mathrm{T}} + \mathcal{P}\mathcal{M}_1\mathcal{P} + \mathcal{P}\mathcal{M}_2\mathcal{P} + m_3\bar{\delta}^2\mathcal{P}\mathcal{B}\mathcal{B}^{\mathrm{T}}$ $\mathcal{P} + \eta\mathcal{P}$.

In (34) the notation $\star$ signifies the term derived by taking the transpose of the corresponding element. Additionally, $\mathcal{O}$ and $\mathcal{I}$ represent the zero and identity matrices, respectively, of suitable dimensions.

The conditions governing the nonlinear controller u_{nl}, as introduced in expression (30), may be reformulated like so:

$$\begin{bmatrix}\mathcal{X}\\ 1\end{bmatrix}^{\mathrm{T}}\begin{bmatrix}\mathcal{P}\mathcal{B}\mathcal{B}^{\mathrm{T}}\mathcal{P} & \mathcal{O}\\ \star & -\kappa\end{bmatrix}\begin{bmatrix}\mathcal{X}\\ 1\end{bmatrix} > 0 \tag{35}$$

By applying the S-procedure [43, 44], and considering expressions (34) and (35), the following condition can be derived:

$$\begin{bmatrix}\Upsilon_{11} + \beta_1\mathcal{P}\mathcal{B}\mathcal{B}^{\mathrm{T}}\mathcal{P} & \mathcal{O}\\ \star & -\kappa\beta_1\end{bmatrix} < 0 \tag{36}$$

where $\beta_1 > 0$.

To proceed, we multiply condition (36) from both sides by $\mathcal{Q} = \mathrm{diag}(\mathcal{S}, \mathcal{I})$, where $\mathcal{S} = \mathcal{P}^{-1}$. This results in the subsequent LMI:

$$\begin{bmatrix}\Theta_{11} & \mathcal{O}\\ \star & -\kappa\beta_1\end{bmatrix} < 0 \tag{37}$$

where $\Theta_{11} = (\mathcal{A}\mathcal{S} + \mathcal{B}\mathcal{R}) + (\mathcal{A}\mathcal{S} + \mathcal{B}\mathcal{R})^{\mathrm{T}} + \mathcal{M}_1 + \mathcal{M}_2 + m_3\bar{\delta}^2\mathcal{B}\mathcal{B}^{\mathrm{T}} + \eta\mathcal{S} + \beta_1$ $\mathcal{B}\mathcal{B}^{\mathrm{T}}$ and $\mathcal{R} = \mathcal{K}\mathcal{S}$.

Thus, we have formulated the LMI condition (37), which is defined with respect to the unknown matrix variables $\mathcal{S}$, $\mathcal{R}$, $\mathcal{M}_1$, and $\mathcal{M}_2$, as well as the unknown scalar variables β_1 and m_3.

Theorem 1 *Let us assume the presence of positive constants* $\bar{\delta}$ *and* κ, *as well as matrices* $\mathcal{S}$, $\mathcal{R}$, $\mathcal{M}_1$, *and* $\mathcal{M}_2$, *alongside scalar variables* β_1 *and* m_3. *For a fixed scalar* $\eta \geq 0$ *the following condition:*

$$\begin{bmatrix} \left\{ \begin{array}{c} (\mathcal{AS}+\mathcal{BR})+(\mathcal{AS}+\mathcal{BR})^{\mathrm{T}}+\mathcal{M}_1+\mathcal{M}_2 \\ +m_3\bar{\delta}^2\mathcal{BB}^{\mathrm{T}}+\eta\mathcal{S}+\beta_1\mathcal{BB}^{\mathrm{T}} \end{array} \right\} & \mathcal{O} \\ \star & -\kappa\beta_1 \end{bmatrix} < 0 \tag{38}$$

is satisfied, then the composite nonlinear control law described by (11), (12), and (30) will effectively stabilize the nonlinear dynamics of the knee exoskeleton robot given by (1). The feedback control of the linear part is given like so:

$$\mathcal{K} = \mathcal{RS}^{-1} \tag{39}$$

4.2 Second and Third LMI Design Strategy

In the subsequent development of two distinct methods for establishing LMI conditions, we presume that the function $\Phi_3(\mathcal{X})$ is consistently constrained by a linear expression. As previously outlined, we continue to use the same formulation (19) to establish two distinct approaches for determining the LMI criteria related to the feedback gain matrix $\mathcal{K}$.

As $\Phi_1(\mathcal{X}) = \mathrm{sgn}(\dot{\phi}) = \pm 1$, applying Young's inequality yields expression (20).

Considering $\Phi_3(\mathcal{X})$ as a scalar function as outlined in (9), we can utilize Young's inequality in expression (19) to derive the following equation:

$$2\mathcal{X}^{\mathrm{T}}\mathcal{P}\mathcal{D}_2\Phi_3(\mathcal{X}) \leq m_2\mathcal{X}^{\mathrm{T}}\mathcal{P}\mathcal{D}_2\mathcal{D}_2^{\mathrm{T}}\mathcal{P}\mathcal{X} + m_2^{-1}\Phi_3(\mathcal{X})^2 \tag{40}$$

where $m_2 = m_2^{\mathrm{T}} > 0$.

Using the expression from (9), it follows that:

$$\Phi_3(\mathcal{X})^2 = \cos^2(\theta) + \cos^2(\theta_d) - 2\cos(\theta_d)\cos(\theta) \tag{41}$$

The two inequalities presented below can be proven:

$$\cos^2(\theta) \leq 1, \quad \text{and} \quad -2\cos(\theta) \leq a_2\theta + b_2, \quad \text{where} \quad a_2 = \frac{4}{\pi}, \quad \text{and} \quad b_2 = -2 \tag{42}$$

Thus, we can derive the following condition:

$$-2\cos(\theta_d)\cos(\theta) \leq a_2\cos(\theta_d)\theta_d + a_2\cos(\theta_d)\phi + b_2\cos(\theta_d) \tag{43}$$

Then, we get:

$$\Phi_3(\mathcal{X})^2 \leq 2\mu_{21}\phi + \mu_{22} \tag{44}$$

For all desired angles θ_d in the range $0 \leq \theta_d \leq 90°$, the parameters satisfy the following conditions:

$$\mu_{21} = \frac{a_2}{2}\cos(\theta_d) > 0 \tag{45}$$

$$\mu_{22} = 1 + a_2\cos(\theta_d)\theta_d + b_2\cos(\theta_d) + \cos^2(\theta_d) > 0 \tag{46}$$

Noting that the quantity μ_{22} can be rewritten like so

$$\mu_{22} = a_2\cos(\theta_d)\theta_d + (1 - \cos(\theta_d))^2 \tag{47}$$

Since $a_2 > 0$ ad $\cos(\theta_d) \geq 0$, if follows that $\mu_{22} > 0$, as noted previously in expression (46).

Assuming $\phi = \mathcal{C}_1\mathcal{X}$, where $\mathcal{C}_1 = \begin{bmatrix} 1 & 0 \end{bmatrix}$, expression (44) can be rewritten as:

$$\Phi_3(\mathcal{X})^2 \leq 2\mu_{21}\mathcal{C}_1\mathcal{X} + \mu_{22} \tag{48}$$

As a result, Eq. (40) becomes:

$$2\mathcal{X}^{\mathrm{T}}\mathcal{P}\mathcal{D}_2\Phi_3(\mathcal{X}) \leq m_2\mathcal{X}^{\mathrm{T}}\mathcal{P}\mathcal{D}_2\mathcal{D}_2^{\mathrm{T}}\mathcal{P}\mathcal{X} + 2\mu_{21}m_2^{-1}\mathcal{C}_1\mathcal{X} + \mu_{22}m_2^{-1} \tag{49}$$

Additionally, considering Δ_t as a scalar function, we can apply the Young inequality lemma to the term $2\mathcal{X}^{\mathrm{T}}\mathcal{P}\mathcal{B}\Delta_t$ in Eq. (19). Assuming that Δ_t adheres to the constraint provided in (26), we can then derive the inequalities presented in (27).

By substituting (20), (49), and (27) into the inequality (19) and incorporating the constraint from (18), we derive the subsequent inequality:

$$\begin{aligned}\dot{\mathcal{V}}(\mathcal{X}) + \eta\mathcal{V}(\mathcal{X}) = \mathcal{X}^{\mathrm{T}}\big((&\mathcal{P}\mathcal{A} + \mathcal{P}\mathcal{B}\mathcal{K}) + (\mathcal{P}\mathcal{A} + \mathcal{P}\mathcal{B}\mathcal{K})^{\mathrm{T}} + \mathcal{P}\mathcal{M}_1\mathcal{P} \\ &+ m_2\mathcal{P}\mathcal{D}_2\mathcal{D}_2^{\mathrm{T}}\mathcal{P} + m_3\bar{\delta}^2\mathcal{P}\mathcal{B}\mathcal{B}^{\mathrm{T}}\mathcal{P} + \eta\mathcal{P}\big)\mathcal{X} + 2\mathcal{X}^{\mathrm{T}}\mathcal{P}\mathcal{B}u_{nl} \\ &+ \mathcal{D}_1^{\mathrm{T}}\mathcal{M}_1^{-1}\mathcal{D}_1 + 2\mu_{21}m_2^{-1}\mathcal{C}_1\mathcal{X} + \mu_{22}m_2^{-1} + m_3^{-1}\end{aligned} \tag{50}$$

In the subsequent analysis, we will explore two distinct formulations of the parameter λ, leading to the development of two separate methodologies for deriving the LMI conditions associated with the feedback gain matrix. These approaches will be based on the framework established in expression (50).

4.2.1 Second LMI Design Strategy

In this approach, we introduce the following specific formulation for λ:

$$\lambda = \mathcal{D}_1^{\mathrm{T}}\mathcal{M}_1^{-1}\mathcal{D}_1 + 2\mu_{21}m_2^{-1}\mathcal{C}_1\mathcal{X} + \mu_{22}m_2^{-1} + m_3^{-1} \tag{51}$$

Additionally, we consider the same auxiliary nonlinear controller u_{nl} defined in Eq. (30).

When $\|\mathcal{B}^{\mathrm{T}}\mathcal{P}\mathcal{X}\|^2 > \kappa$, substituting the expression for u_{nl} into inequality (50) simplifies it to:

$$\begin{aligned}\dot{\mathcal{V}}(\mathcal{X}) + \eta\mathcal{V}(\mathcal{X}) \leq \mathcal{X}^{\mathrm{T}}\big((\mathcal{P}\mathcal{A} + \mathcal{P}\mathcal{B}\mathcal{K}) + (\mathcal{P}\mathcal{A} + \mathcal{P}\mathcal{B}\mathcal{K})^{\mathrm{T}} \\ + \mathcal{P}\mathcal{M}_1\mathcal{P} + m_2\mathcal{P}\mathcal{D}_2\mathcal{D}_2^{\mathrm{T}}\mathcal{P} + m_3\bar{\delta}^2\mathcal{P}\mathcal{B}\mathcal{B}^{\mathrm{T}}\mathcal{P} + \eta\mathcal{P}\big)\mathcal{X}\end{aligned} \tag{52}$$

Alternatively, when $\|\mathcal{B}^{\mathrm{T}}\mathcal{P}\mathcal{X}\|^2 \leq \kappa$, inequality (50) transforms into:

$$\begin{aligned}\dot{\mathcal{V}}(\mathcal{X}) + \eta\mathcal{V}(\mathcal{X}) \leq \mathcal{X}^{\mathrm{T}}\big((\mathcal{P}\mathcal{A} + \mathcal{P}\mathcal{B}\mathcal{K}) + (\mathcal{P}\mathcal{A} + \mathcal{P}\mathcal{B}\mathcal{K})^{\mathrm{T}} \\ + \mathcal{P}\mathcal{M}_1\mathcal{P} + m_2\mathcal{P}\mathcal{D}_2\mathcal{D}_2^{\mathrm{T}}\mathcal{P} + m_3\bar{\delta}^2\mathcal{P}\mathcal{B}\mathcal{B}^{\mathrm{T}}\mathcal{P} + \eta\mathcal{P}\big)\mathcal{X} \\ + \lambda\left(1 - \frac{\|\mathcal{B}^{\mathrm{T}}\mathcal{P}\mathcal{X}\|^2}{\kappa}\right)\end{aligned} \tag{53}$$

Considering expression (33), $\mathcal{V}(\mathcal{X})$ is restricted, ensuring that the closed-loop system's trajectory remains stable when $\|\mathcal{B}^{\mathrm{T}}\mathcal{P}\mathcal{X}\|^2 \leq \kappa$. The problem thus reduces to verifying the condition in (52).

To derive the LMI conditions, we first define the function $\mathcal{Z}(\mathcal{X})$ as follows: $\mathcal{Z}(\mathcal{X}) = 2\mathcal{X}^{\mathrm{T}}\mathcal{P}(\mathcal{A} + \mathcal{B}\mathcal{K})\mathcal{X} + \mathcal{X}^{\mathrm{T}}\mathcal{P}\mathcal{M}_1\mathcal{P}\mathcal{X} + m_2\mathcal{X}^{\mathrm{T}}\mathcal{P}\mathcal{D}_2\mathcal{D}_2^{\mathrm{T}}\mathcal{P}\mathcal{X} + m_3\bar{\delta}^2\mathcal{X}^{\mathrm{T}}\mathcal{P}$ $\mathcal{B}\mathcal{B}^{\mathrm{T}}\mathcal{P}\mathcal{X} + \eta\mathcal{X}^{\mathrm{T}}\mathcal{P}\mathcal{X}$. We can represent this as an inequality:

$$\mathcal{Z}(\mathcal{X}) = \begin{bmatrix}\mathcal{X} \\ 1\end{bmatrix}^{\mathrm{T}}\begin{bmatrix}\Upsilon_{12} & \mathcal{O} \\ \star & \mathcal{O}\end{bmatrix}\begin{bmatrix}\mathcal{X} \\ 1\end{bmatrix} < 0 \tag{54}$$

where Υ_{12} is given by:

$$\Upsilon_{12} = (\mathcal{P}\mathcal{A} + \mathcal{P}\mathcal{B}\mathcal{K}) + (\mathcal{P}\mathcal{A} + \mathcal{P}\mathcal{B}\mathcal{K})^{\mathrm{T}} + \mathcal{P}\mathcal{M}_1\mathcal{P} + m_2\mathcal{P}\mathcal{D}_2\mathcal{D}_2^{\mathrm{T}}\mathcal{P} + m_3\bar{\delta}^2\mathcal{P}\mathcal{B}\mathcal{B}^{\mathrm{T}}\mathcal{P} + \eta\mathcal{P}$$

Applying the S-procedure [43, 44] to expressions (54) and (35), we derive the following condition:

$$\begin{bmatrix}\Upsilon_{12} + \beta_1\mathcal{P}\mathcal{B}\mathcal{B}^{\mathrm{T}}\mathcal{P} & \mathcal{O} \\ \star & -\kappa\beta_1\end{bmatrix} < 0 \tag{55}$$

where $\beta_1 > 0$.

Next, we multiply the condition (55) from both sides by $\mathcal{Q} = \text{diag}(\mathcal{S}, \mathcal{I})$, where $\mathcal{S} = \mathcal{P}^{-1}$. This yields the subsequent LMI:

$$\begin{bmatrix} \Theta_{12} & \mathcal{O} \\ \star & -\kappa\beta_1 \end{bmatrix} < 0 \tag{56}$$

where $\Theta_{12} = (\mathcal{AS} + \mathcal{BR}) + (\mathcal{AS} + \mathcal{BR})^{\mathrm{T}} + \mathcal{M}_1 + m_2\mathcal{D}_2\mathcal{D}_2^{\mathrm{T}} + m_3\bar{\delta}^2\mathcal{BB}^{\mathrm{T}} + \eta\mathcal{S} + \beta_1\mathcal{BB}^{\mathrm{T}}$ and $\mathcal{R} = \mathcal{KS}$.

We have established LMI condition, specifically (56), which is characterized by the unknown matrix variables $\mathcal{S}, \mathcal{R}$, and $\mathcal{M}_1$, as well as the unknown scalar variables β_1, m_2, and m_3.

Theorem 2 *Given the existence of positive constants $\bar{\delta}$ and κ, as well as matrices $\mathcal{S}$, $\mathcal{R}$, and $\mathcal{M}_1$, and scalar variables β_1, m_2, and m_3. For a fixed scalar $\eta \geq 0$, the following condition:*

$$\begin{bmatrix} \left\{ \begin{matrix} (\mathcal{AS} + \mathcal{BR}) + (\mathcal{AS} + \mathcal{BR})^{\mathrm{T}} + \mathcal{M}_1 + \\ m_2\mathcal{D}_2\mathcal{D}_2^{\mathrm{T}} + m_3\bar{\delta}^2\mathcal{BB}^{\mathrm{T}} + \eta\mathcal{S} + \beta_1\mathcal{BB}^{\mathrm{T}} \end{matrix} \right\} & \mathcal{O} \\ \star & -\kappa\beta_1 \end{bmatrix} < 0 \tag{57}$$

is satisfied, therefore the control strategy defined by (11), (12), and (30) will successfully stabilize the dynamics of the knee exoskeleton robot as represented by (1). Moreover, the feedback gain $\mathcal{K}$ of the linear controller is defined by expression (39).

4.2.2 Third LMI Design Strategy

In this method, we propose a distinct formulation for λ as follows:

$$\lambda = \mathcal{D}_1^{\mathrm{T}}\mathcal{M}_1^{-1}\mathcal{D}_1 + m_3^{-1} \tag{58}$$

We also incorporate the same auxiliary nonlinear controller u_{nl} as defined in (30).

If $\|\mathcal{B}^{\mathrm{T}}\mathcal{P}\mathcal{X}\|^2 > \kappa$, substituting the nonlinear controller u_{nl} from Eq. (30) into inequality (50) simplifies it to:

$$\begin{aligned} \dot{\mathcal{V}}(\mathcal{X}) + \eta\mathcal{V}(\mathcal{X}) \leq \mathcal{X}^{\mathrm{T}} \big(&(\mathcal{PA} + \mathcal{PBK}) + (\mathcal{PA} + \mathcal{PBK})^{\mathrm{T}} + \mathcal{PM}_1\mathcal{P} \\ &+ m_2\mathcal{PD}_2\mathcal{D}_2^{\mathrm{T}}\mathcal{P} + m_3\bar{\delta}^2\mathcal{PBB}^{\mathrm{T}}\mathcal{P} + \eta\mathcal{P} \big) \mathcal{X} + 2\mu_{21}m_2^{-1}\mathcal{C}_1\mathcal{X} + \mu_{22}m_2^{-1} \end{aligned} \tag{59}$$

When $\|\mathcal{B}^{\mathrm{T}}\mathcal{P}\mathcal{X}\|^2 \leq \kappa$, using the same definition for u_{nl}, inequality (50) transforms to:

$$\begin{aligned}\dot{\mathcal{V}}(\mathcal{X}) + \eta\mathcal{V}(\mathcal{X}) \leq \mathcal{X}^{\mathrm{T}} \big((\mathcal{P}\mathcal{A} + \mathcal{P}\mathcal{B}\mathcal{K}) + (\mathcal{P}\mathcal{A} + \mathcal{P}\mathcal{B}\mathcal{K})^{\mathrm{T}} \\ + \mathcal{P}\mathcal{M}_1\mathcal{P} + m_2\mathcal{P}\mathcal{D}_2\mathcal{D}_2^{\mathrm{T}}\mathcal{P} + m_3\bar{\delta}^2\mathcal{P}\mathcal{B}\mathcal{B}^{\mathrm{T}}\mathcal{P} + \eta\mathcal{P}\big)\mathcal{X} \\ + 2\mu_{21}m_2^{-1}\mathcal{C}_1\mathcal{X} + \mu_{22}m_2^{-1} + \lambda\left(1 - \frac{\|\mathcal{B}^{\mathrm{T}}\mathcal{P}\mathcal{X}\|^2}{\kappa}\right)\end{aligned} \tag{60}$$

Taking into account expression (33), $\mathcal{V}(\mathcal{X})$ remains bounded. Consequently, the trajectory of the closed-loop system remains constrained provided that $\|\mathcal{B}^{\mathrm{T}}\mathcal{P}\mathcal{X}\|^2 \leq \kappa$. This reduces the problem to verifying the condition given in (59).

To establish the LMI conditions, we begin by defining the function $\mathcal{Z}(\mathcal{X})$ like so: $\mathcal{Z}(\mathcal{X}) = 2\mathcal{X}^{\mathrm{T}}\mathcal{P}(\mathcal{A}+\mathcal{B}\mathcal{K})\mathcal{X} + \mathcal{X}^{\mathrm{T}}\mathcal{P}\mathcal{M}_1\mathcal{P}\mathcal{X} + m_2\mathcal{X}^{\mathrm{T}}\mathcal{P}\mathcal{D}_2\mathcal{D}_2^{\mathrm{T}}\mathcal{P}\mathcal{X} + m_3\bar{\delta}^2\mathcal{X}^{\mathrm{T}}\mathcal{P}$ $\mathcal{B}\mathcal{B}^{\mathrm{T}}\mathcal{P}\mathcal{X} + \eta\mathcal{X}^{\mathrm{T}}\mathcal{P}\mathcal{X} + 2\mu_{21}m_2^{-1}\mathcal{C}_1\mathcal{X} + \mu_{22}m_2^{-1}$, where $\mathcal{Z}(\mathcal{X})$ can be expressed in matrix notation as:

$$\mathcal{Z}(\mathcal{X}) = \begin{bmatrix}\mathcal{X}\\ 1\end{bmatrix}^{\mathrm{T}} \begin{bmatrix}\Upsilon_{13} & \mu_{21}m_2^{-1}\mathcal{C}_1^{\mathrm{T}}\\ \star & \mu_{22}m_2^{-1}\end{bmatrix} \begin{bmatrix}\mathcal{X}\\ 1\end{bmatrix} < 0 \tag{61}$$

where Υ_{13} is given by $\Upsilon_{13} = (\mathcal{P}\mathcal{A}+\mathcal{P}\mathcal{B}\mathcal{K}) + (\mathcal{P}\mathcal{A}+\mathcal{P}\mathcal{B}\mathcal{K})^{\mathrm{T}} + \mathcal{P}\mathcal{M}_1\mathcal{P} + m_2$ $\mathcal{P}\mathcal{D}_2\mathcal{D}_2^{\mathrm{T}}\mathcal{P} + m_3\bar{\delta}^2\mathcal{P}\mathcal{B}\mathcal{B}^{\mathrm{T}}\mathcal{P} + \eta\mathcal{P}$.

Applying the S-procedure [43, 44] and using expressions (61) and (35), we derive the following condition:

$$\begin{bmatrix}\Upsilon_{13} + \beta_1\mathcal{P}\mathcal{B}\mathcal{B}^{\mathrm{T}}\mathcal{P} & \mu_{21}m_2^{-1}\mathcal{C}_1^{\mathrm{T}}\\ \star & \mu_{22}m_2^{-1} - \kappa\beta_1\end{bmatrix} < 0 \tag{62}$$

where $\beta_1 > 0$.

Next, multiplying both sides of inequality (62) by $\mathcal{Q} = \mathrm{diag}(\mathcal{S}, \mathcal{I})$, with $\mathcal{S} = \mathcal{P}^{-1}$, we obtain:

$$\begin{bmatrix}\Gamma_{11} & \mu_{21}m_2^{-1}\mathcal{S}\mathcal{C}_1^{\mathrm{T}}\\ \star & \mu_{22}m_2^{-1} - \xi^{-1}\end{bmatrix} < 0 \tag{63}$$

where $\Gamma_{11} = (\mathcal{A}\mathcal{S} + \mathcal{B}\mathcal{R}) + (\mathcal{A}\mathcal{S} + \mathcal{B}\mathcal{R})^{\mathrm{T}} + \mathcal{M}_1 + m_2\mathcal{D}_2\mathcal{D}_2^{\mathrm{T}} + m_3\bar{\delta}^2\mathcal{B}\mathcal{B}^{\mathrm{T}} + \eta\mathcal{S} + \beta_1\mathcal{B}\mathcal{B}^{\mathrm{T}}$, with $\mathcal{R} = \mathcal{K}\mathcal{S}$, and $\xi^{-1} = \kappa\beta_1$.

By applying the Schur complement lemma to expression (63), we obtain:

$$\mu_{22}m_2^{-1} - \xi^{-1} < 0 \tag{64a}$$

$$\Gamma_{11} + \left(\mu_{21}m_2^{-1}\mathcal{S}\mathcal{C}_1^{\mathrm{T}}\right)\left(-\mu_{22}m_2^{-1} + \xi^{-1}\right)^{-1}\left(\mu_{21}m_2^{-1}\mathcal{S}\mathcal{C}_1^{\mathrm{T}}\right)^{\mathrm{T}} < 0 \tag{64b}$$

Let's define Π as:

$$\Pi = \left(\mu_{21}m_2^{-1}\mathcal{S}\mathcal{C}_1^{\mathrm{T}}\right)\left(-\mu_{22}m_2^{-1} + \xi^{-1}\right)^{-1}\left(\mu_{21}m_2^{-1}\mathcal{S}\mathcal{C}_1^{\mathrm{T}}\right)^{\mathrm{T}} \tag{65}$$

Utilizing the matrix inversion lemma [43, 44], we derive:

$$\left(\xi^{-1} - \mu_{22} m_2^{-1}\right)^{-1} = \xi - \xi^2 \tau^{-1} \tag{66}$$

where

$$\tau = \xi - \frac{1}{\mu_{22}} m_2 \tag{67}$$

Subsequently, referring to Eq. (66), we deduce that the expression (65) can be expressed as:

$$\Pi = \left(\mu_{21} m_2^{-1} \mathcal{S} C_1^{\mathrm{T}}\right)\left(\xi - \xi^2 \tau^{-1}\right)\left(\mu_{21} m_2^{-1} \mathcal{S} C_1^{\mathrm{T}}\right)^{\mathrm{T}} \tag{68}$$

Expanding expression (68), we obtain:

$$\Pi = \xi \left(\mu_{21} m_2^{-1}\right)^2 \mathcal{S} C_1^{\mathrm{T}} C_1 \mathcal{S} - \xi^2 \left(\mu_{21} m_2^{-1}\right)^2 \tau^{-1} \mathcal{S} C_1^{\mathrm{T}} C_1 \mathcal{S} \tag{69}$$

Relying on expression (67), we can further illustrate:

$$\xi \left(\mu_{21} m_2^{-1}\right)^2 \mathcal{S} C_1^{\mathrm{T}} C_1 \mathcal{S} = \left(\mu_{21} m_2^{-1}\right)^2 \tau \mathcal{S} C_1^{\mathrm{T}} C_1 \mathcal{S} + \frac{\mu_{21}^2}{\mu_{22}} m_2^{-1} \mathcal{S} C_1^{\mathrm{T}} C_1 \mathcal{S} \tag{70a}$$

$$\xi^2 \left(\mu_{21} m_2^{-1}\right)^2 \tau^{-1} \mathcal{S} C_1^{\mathrm{T}} C_1 \mathcal{S} = \left(\mu_{21} m_2^{-1}\right)^2 \tau \mathcal{S} C_1^{\mathrm{T}} C_1 \mathcal{S} + 2\frac{\mu_{21}^2}{\mu_{22}} m_2^{-1} \mathcal{S} C_1^{\mathrm{T}} C_1 \mathcal{S}$$

$$+ \frac{\mu_{21}^2}{\mu_{22}^2} \tau^{-1} \mathcal{S} C_1^{\mathrm{T}} C_1 \mathcal{S} \tag{70b}$$

By utilizing (70), we can simplify Eq. (69) like so:

$$\Pi = -\frac{\mu_{21}^2}{\mu_{22}} m_2^{-1} \mathcal{S} C_1^{\mathrm{T}} C_1 \mathcal{S} - \frac{\mu_{21}^2}{\mu_{22}^2} \tau^{-1} \mathcal{S} C_1^{\mathrm{T}} C_1 \mathcal{S} \tag{71}$$

Subsequently, by substituting the previous equation (71) into expression (64b), we obtain:

$$\Gamma_{11} - \frac{\mu_{21}^2}{\mu_{22}} m_2^{-1} \mathcal{S} C_1^{\mathrm{T}} C_1 \mathcal{S} - \frac{\mu_{21}^2}{\mu_{22}^2} \tau^{-1} \mathcal{S} C_1^{\mathrm{T}} C_1 \mathcal{S} < 0 \tag{72}$$

By examining the augmented matrix $\bar{C}_1 = \begin{bmatrix} C_1 \\ \mathcal{O} \end{bmatrix}$, we can easily show that:

$$-\frac{\mu_{21}^2}{\mu_{22}} m_2^{-1} \mathcal{S} C_1^{\mathrm{T}} C_1 \mathcal{S} = \mu_3^{-1} m_2^{-1} \mathcal{S} \bar{C}_1^{\mathrm{T}} \bar{C}_1 \mathcal{S} \tag{73}$$

with $\mu_3^{-1} = -\frac{\mu_{21}^2}{\mu_{22}}$.

Therefore, by using (73), we can express the inequality (72) in the following form:

$$\Gamma_{11} + \mu_3^{-1} m_2^{-1} \mathcal{S}\bar{\mathcal{C}}_1^{\mathrm{T}} \bar{\mathcal{C}}_1 \mathcal{S} - \frac{\mu_{21}^2}{\mu_{22}^2} \tau^{-1} \mathcal{S}\mathcal{C}_1^{\mathrm{T}} \mathcal{C}_1 \mathcal{S} < 0 \tag{74}$$

Since $\mu_3^{-1} < 0$ and $m_2^{-1} > 0$, then, applying Young's inequality leads to the following result:

$$\mu_3^{-1} m_2^{-1} \mathcal{S}\bar{\mathcal{C}}_1^{\mathrm{T}} \bar{\mathcal{C}}_1 \mathcal{S} \leq -\mu_3 m_2 \mathcal{I} - \bar{\mathcal{C}}_1 \mathcal{S} - \mathcal{S}\bar{\mathcal{C}}_1^{\mathrm{T}} \tag{75}$$

By substituting (75) into Eq. (74), we can rewrite it as follows:

$$\Gamma_{11} - \mu_3 m_2 \mathcal{I} - \bar{\mathcal{C}}_1 \mathcal{S} - \mathcal{S}\bar{\mathcal{C}}_1^{\mathrm{T}} - \frac{\mu_{21}^2}{\mu_{22}^2} \tau^{-1} \mathcal{S}\mathcal{C}_1^{\mathrm{T}} \mathcal{C}_1 \mathcal{S} < 0 \tag{76}$$

Since $\mu_{22} > 0$, $\xi > 0$ and $m_2 > 0$, it follows that $\tau^{-1} < 0$. Consequently, using the Schur complement we can reformulate expression (76) in the following manner:

$$\begin{bmatrix} \Gamma_{11} - \mu_3 m_2 \mathcal{I} - \bar{\mathcal{C}}_1 \mathcal{S} - \mathcal{S}\bar{\mathcal{C}}_1^{\mathrm{T}} & \mathcal{S}\mathcal{C}_1^{\mathrm{T}} \\ \star & \frac{\mu_{22}^2}{\mu_{21}^2} \tau \mathcal{I} \end{bmatrix} < 0 \tag{77}$$

By setting $\alpha_1 = \beta_1^{-1}$ and modifying the expressions for Γ_{11} and τ in Eq. (77), using the schur complement we obtain the following LMI:

$$\begin{bmatrix} \Xi_{11} & \mathcal{S}\mathcal{C}_1^{\mathrm{T}} & \mathcal{B} \\ \star & \frac{\mu_{22}^2}{\mu_{21}^2} \frac{\alpha_1}{\kappa} - \frac{\mu_{22}}{\mu_{21}^2} m_2 & \mathcal{O} \\ \star & \star & -\alpha_1 \mathcal{I} \end{bmatrix} < 0 \tag{78}$$

where $\Xi_{11} = (\mathcal{A}\mathcal{S} + \mathcal{B}\mathcal{R}) + (\mathcal{A}\mathcal{S} + \mathcal{B}\mathcal{R})^{\mathrm{T}} + \mathcal{M}_1 + m_2 \mathcal{D}_2 \mathcal{D}_2^{\mathrm{T}} + m_3 \bar{\delta}^2 \mathcal{B}\mathcal{B}^{\mathrm{T}} + \eta \mathcal{S} + \frac{\mu_{22}}{\mu_{21}^2} m_2 \mathcal{I} - \bar{\mathcal{C}}_1 \mathcal{S} - \mathcal{S}\bar{\mathcal{C}}_1^{\mathrm{T}}$.

Furthermore, by applying (64a) and multiplying by ξ^2, we can obtain the following expression:

$$\xi^2 \mu_{22} m_2^{-1} - \xi < 0 \tag{79}$$

Since $\mu_{22} m_2^{-1} > 0$ and $\xi > 0$, we can apply the Schur complement to inequality (79), leading to the following expression:

$$\begin{bmatrix} -\xi & \xi \\ \star & -\frac{1}{\mu_{22}} m_2 \end{bmatrix} < 0 \tag{80}$$

We obtain the following second LMI:

$$\begin{bmatrix} -\frac{\alpha_1}{\kappa} & \frac{\alpha_1}{\kappa} \\ \star & -\frac{1}{\mu_{22}} m_2 \end{bmatrix} < 0 \tag{81}$$

In summary, we have established two LMIs (78) and (81), involving the matrix variables $\mathcal{S}$, $\mathcal{R}$ and $\mathcal{M}_1$, and the scalar variables α_1, m_2 and m_3.

Theorem 3 *Let there exist positive constants $\bar{\delta}$ and κ, as well as matrices $\mathcal{S}$, $\mathcal{R}$, and $\mathcal{M}_1$, and scalar variables α_1, m_2 and m_3. For a fixed scalar $\eta \geq 0$, the subsequent conditions are satisfied:*

$$\begin{bmatrix} \left\{ \begin{array}{c} (\mathcal{AS}+\mathcal{BR})+(\mathcal{AS}+\mathcal{BR})^{\mathrm{T}}+\mathcal{M}_1 \\ +m_2\mathcal{D}_2\mathcal{D}_2^{\mathrm{T}}+m_3\bar{\delta}^2\mathcal{BB}^{\mathrm{T}}+\eta\mathcal{S} \\ +\frac{\mu_{22}}{\mu_{21}^2}m_2\mathcal{I}-\bar{\mathcal{C}}_1\mathcal{S}-\mathcal{S}\bar{\mathcal{C}}_1^{\mathrm{T}} \end{array} \right\} & \mathcal{S}\mathcal{C}_1^{\mathrm{T}} & \mathcal{B} \\ \star & \frac{\mu_{22}^2}{\mu_{21}^2}\frac{\alpha_1}{\kappa}-\frac{\mu_{22}}{\mu_{21}^2}m_2 & \mathcal{O} \\ \star & \star & -\alpha_1\mathcal{I} \end{bmatrix} < 0 \tag{82a}$$

$$\begin{bmatrix} -\frac{\alpha_1}{\kappa} & \frac{\alpha_1}{\kappa} \\ \star & -\frac{1}{\mu_{22}} m_2 \end{bmatrix} < 0 \tag{82b}$$

therefore the composite nonlinear control law specified in Eqs. (11), (12), and (30) will ensure the stability of the nonlinear dynamics of the robotic system as described in (1). Moreover, the feedback gain $\mathcal{K}$ of the linear controller of the composite nonlinear control law is defined by expression (39).

Remark 2 Different expressions for λ have been used in previous research. For example, in [31], λ is defined as

$$\lambda = \mathcal{D}_1^{\mathrm{T}}\mathcal{M}_1^{-1}\mathcal{D}_1 + \mu_1\mathcal{D}_2^{\mathrm{T}}\mathcal{M}_2^{-1}\mathcal{D}_2 + \bar{\delta}^2\mathcal{B}^{\mathrm{T}}\mathcal{M}_3^{-1}\mathcal{B}$$

while in [32], a similar parameterization of λ is adopted. The modification of the parameter λ enables the development of a distinct control strategy, providing specific variations to enhance the controller's adaptability.

In this study, we employ two distinct expressions for λ, given in Eqs. (51) and (58), to address different needs within the control design process. The first expression of λ (51) incorporates additional terms to account for both the system dynamics and disturbances, offering a more comprehensive approach and ensuring robust control in the presence of significant disturbances. In contrast, the second expression of λ (58) is a simplified version, omitting certain terms and suited to situations where a less complex control design suffices, especially when disturbances have a minor effect.

By utilizing these two formulations, we achieve a balance between control performance and computational efficiency. The first expression provides enhanced robustness, although at the cost of increased complexity, while the second expression is computationally more efficient, making it suitable for simpler cases.

Remark 3 Notice that, in the first design method, we approximated the nonlinear term $\Phi_3(\mathcal{X})$ using a constant constraint (23), which is a simplification that provides an upper bound for the nonlinear term under the assumption that $0^\circ \leq \theta \leq 90^\circ$. This approach ensures that the constraints remain easy to handle while guaranteeing stability for the exoskeleton's position control. For the second and third design methods, we used a first-order linear constraint for $\Phi_3(\mathcal{X})$, which is more conservative. The first-order linear constraint introduces terms involving θ_d and ϕ (see Eq. (48)), which allow for better handling of nonlinearities but are more conservative than the constant constraint used in the first method.

While it is possible to use a higher-degree or nonlinear constraint, we opted for the first-order linear constraint as it provides a balance between modeling accuracy and computational simplicity. The linear approach allows for more tractable stability conditions, particularly in the context of the LMI-based design, while still capturing the essential nonlinear dynamics that influence the system's behavior.

5 Simulation Results

This section showcases the simulation outcomes for the three control methods implemented in this study. The primary objective is to demonstrate the efficacy of the developed LMI-based criteria in determining the controller gain $\mathcal{K}$, ensuring the stability of the knee exoskeleton, as shown in Fig. 1. The various parameters of the robot, as outlined in Eq. (1), are detailed in Table 1.

Our focus here is on providing numerical simulation outcomes for the different control strategies employed. A comparative analysis of these methods will follow to highlight their respective performances.

Let us set the intended angular position θ_d to a constant value, specifically $\theta_d = 30^\circ$. Additionally, the external disturbance Δ_t will be modeled as a random signal, active only during the time duration between 2 and 7 s. Outside this range, Δ_t will be set to zero. The expression for Δ_t is given by:

Table 1 The adopted values of the different parameters of the 1-DoF knee rehabilitation exoskeleton robot

Parameter	Value	Nominal parameter	Nominal value	Unit
ϑ	0.323 + 0.1	$\vartheta_\star$	0.323	Kg m^2/rad
m	0.1 + 0.01	$m_\star$	0.1	Kg
l	0.25 + 0.05	$l_\star$	0.25	m
f_v	0.35 + 0.1	$f_v^\star$	0.35	N m s/rad
f_s	0.059 + 0.05	$f_s^\star$	0.059	N m/A
g	9.8			

$$\Delta_t = \bar{\delta} \times (2 \times \text{rand} - 1) \tag{83}$$

where the function rand represents a random signal ranging from 0 to 1.

5.1 Results Obtained Using the First Design Methodology

In this section, we use the LMI stability condition as defined by Theorem 1. The control gain $\mathcal{K}$ for the controller given by (11) (or equivalently by (12) and (13)) is determined through the matrices $\mathcal{S}$ and $\mathcal{R}$, which are the solutions of the LMI problem presented in Theorem 1.

By applying the fixed values $\alpha = 2$, $\kappa = 1$, and $\bar{\delta} = 1$ to the LMI problem derived from the theorem, we obtain the following results:

$$\mathcal{S} = \begin{bmatrix} 7.2169 & -14.3125 \\ -14.3125 & 35.8420 \end{bmatrix} \qquad \mathcal{R} = \begin{bmatrix} -7.3404 & -63.7134 \end{bmatrix}$$

$$\mathcal{M}_1 = \begin{bmatrix} 4.7304 & 0.0 \\ 0.0 & 18.0649 \end{bmatrix} \qquad \mathcal{M}_2 = \begin{bmatrix} 4.7304 & 0.0 \\ 0.0 & 18.0649 \end{bmatrix}$$

$$m_3 = 18.0649 \qquad \beta_1 = 18.0649$$

As a result, the feedback gain $\mathcal{K}$ is computed as follows:

$$\mathcal{K} = \begin{bmatrix} -21.8326 & -10.4959 \end{bmatrix} \tag{84}$$

It is important to highlight that the previous numerical results, including the computed feedback gain $\mathcal{K}$, were derived by addressing the LMI (38) as outlined in Theorem 1, utilizing predetermined values for κ and $\bar{\delta}$. To validate the feasibility of the LMI (38) numerically, it is crucial to explore various configurations of these two parameters. To simplify the computational challenges associated with solving the LMI problem, we select specific values for κ and $\bar{\delta}$, which will allow us to assess whether the LMI problem is feasible or infeasible.

The selected values for κ include: 10^2, 10^1, 1.0, 10^{-1}, and 10^{-2}. In contrast, the values chosen for $\bar{\delta}$ are: 10.0, 5.0, 2.0, 1.0, 0.8, 0.4, and 0.1. By varying these parameters, we will solve the LMI (38) from Theorem 1 for every combination of κ and $\bar{\delta}$, while fixing α at 2. The focus will be on presenting the resulting gains $\mathcal{K}$.

Table 2 displays the numerical outcomes of the gain $\mathcal{K}$ obtained by resolving the LMI (38) in Theorem 1 for the different combinations of κ and $\bar{\delta}$. Notably, the LMI consistently demonstrated feasibility across all scenarios examined. Furthermore, from the data in Table 2, we observe that as κ decreases and $\bar{\delta}$ increases, the feedback

Table 2 Numerical results, $\mathcal{K}^{\mathrm{T}}$, of the LMI (38) in Theorem 1 for different values of κ and $\bar{\delta}$

$\bar{\delta}\backslash\kappa$	100	10	1	10^{-1}	10^{-2}
10.0	$10^3\begin{bmatrix}-1.2590\\-0.5950\end{bmatrix}$	$\begin{bmatrix}-995.5565\\-483.4953\end{bmatrix}$	$\begin{bmatrix}-758.0836\\-381.7393\end{bmatrix}$	$10^3\begin{bmatrix}-1.0073\\-0.4892\end{bmatrix}$	$10^3\begin{bmatrix}-1.2718\\-0.6011\end{bmatrix}$
5.0	$\begin{bmatrix}-322.9781\\-152.2694\end{bmatrix}$	$\begin{bmatrix}-256.4667\\-124.1775\end{bmatrix}$	$\begin{bmatrix}-200.3177\\-100.4943\end{bmatrix}$	$\begin{bmatrix}-268.1673\\-129.8659\end{bmatrix}$	$\begin{bmatrix}-335.7466\\-158.3091\end{bmatrix}$
2.0	$\begin{bmatrix}-60.8881\\-28.2980\end{bmatrix}$	$\begin{bmatrix}-49.5215\\-23.5685\end{bmatrix}$	$\begin{bmatrix}-44.1433\\-21.7457\end{bmatrix}$	$\begin{bmatrix}-61.2221\\-29.2569\end{bmatrix}$	$\begin{bmatrix}-73.6566\\-34.3376\end{bmatrix}$
1.0	$\begin{bmatrix}-23.4467\\-10.5877\end{bmatrix}$	$\begin{bmatrix}-19.9579\\-9.1958\end{bmatrix}$	$\begin{bmatrix}-21.8326\\-10.4959\end{bmatrix}$	$\begin{bmatrix}-31.6585\\-14.8842\end{bmatrix}$	$\begin{bmatrix}-36.2152\\-16.6274\end{bmatrix}$
0.8	$\begin{bmatrix}-18.9537\\-8.4625\end{bmatrix}$	$\begin{bmatrix}-16.4103\\-7.4710\end{bmatrix}$	$\begin{bmatrix}-19.1554\\-9.1459\end{bmatrix}$	$\begin{bmatrix}-28.1109\\-13.1594\end{bmatrix}$	$\begin{bmatrix}-31.7222\\-14.5022\end{bmatrix}$
0.4	$\begin{bmatrix}-12.9631\\-5.6289\end{bmatrix}$	$\begin{bmatrix}-11.6801\\-5.1714\end{bmatrix}$	$\begin{bmatrix}-15.5857\\-7.3460\end{bmatrix}$	$\begin{bmatrix}-23.3807\\-10.8598\end{bmatrix}$	$\begin{bmatrix}-25.7316\\-11.6685\end{bmatrix}$
0.1	$\begin{bmatrix}-11.0910\\-4.7434\end{bmatrix}$	$\begin{bmatrix}-10.2019\\-4.4528\end{bmatrix}$	$\begin{bmatrix}-14.4701\\-6.7835\end{bmatrix}$	$\begin{bmatrix}-21.9026\\-10.1412\end{bmatrix}$	$\begin{bmatrix}-23.8595\\-10.7830\end{bmatrix}$

gain $\mathcal{K}$ tends to rise significantly. In particular, for the values $\bar{\delta} = 10$ and $\bar{\delta} = 5$, the gain $\mathcal{K}$ reaches impractically high levels, rendering it unsuitable for effective control of the joint exoskeleton robot. Consequently, these specific values of $\bar{\delta}$ should not be considered for achieving reliable performance in the control strategy.

We present the graphical results depicted in Fig. 2, which utilize the feedback gain $\mathcal{K}$ defined by Eq. (84) within the proposed controller (11). The results clearly illustrate that the exoskeleton robot successfully reached the target position of $\theta_d = 30°$. Additionally, once the knee exoskeleton robot stabilizes at the optimal position θ_d, the control effort u converge to u_d.

5.1.1 Results Obtained Using the Second Design Methodology

In this section, we utilize the LMI stability condition established by Theorem 2. The control gain $\mathcal{K}$ for the adopted controller (11) is determined through the matrices $\mathcal{S}$ and $\mathcal{R}$, which serve as solutions to the LMI problem defined in Theorem 2.

By applying fixed values of $\alpha = 2$, $\kappa = 1$, and $\bar{\delta} = 1$, we obtain the following results:

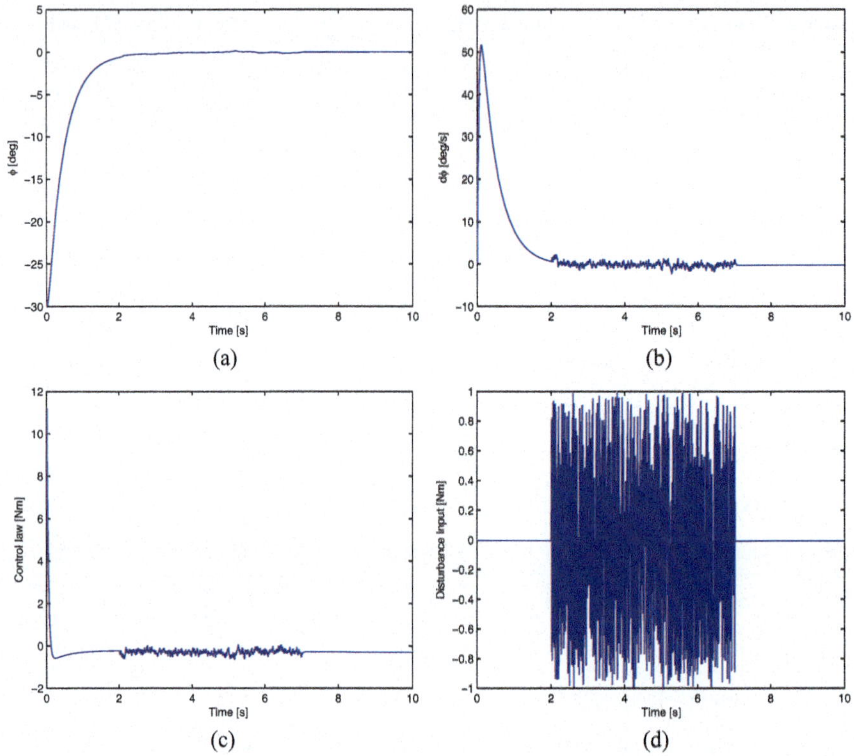

Fig. 2 Simulation results derived from the first design method. Temporal variations of: **a** the angular position error ϕ, **b** the angular velocity error $\dot{\phi}$, **c** the composite nonlinear controller u, and **d** the lumped disturbance Δ_t

$$\mathcal{S} = \begin{bmatrix} 19.4664 & -32.3115 \\ -32.3115 & 84.0894 \end{bmatrix} \qquad \mathcal{R} = \begin{bmatrix} -17.5967 & -133.2980 \end{bmatrix}$$

$$\mathcal{M}_1 = \begin{bmatrix} 12.8451 & 0 \\ 0 & 38.6027 \end{bmatrix} \quad m_2 = 38.6027 \quad m_3 = 38.6027 \quad \beta_1 = 38.6027$$

Thus, the feedback gain $\mathcal{K}$ is calculated as follows:

$$\mathcal{K} = \begin{bmatrix} -9.7603 & -5.3356 \end{bmatrix} \tag{85}$$

As in the previous section, these numerical outcomes and the corresponding feedback gain $\mathcal{K}$ were derived by solving the LMI (57) from Theorem 2 with predetermined values for κ and $\bar{\delta}$. To further investigate the feasibility of the LMI, it is crucial to vary these parameters. We select the same range of values for κ as before: 10^2, 10^1, 1.0, 10^{-1}, and 10^{-2}, while for $\bar{\delta}$, we consider: 10.0, 5.0, 2.0, 1.0, 0.8, 0.4, and

Table 3 Numerical results, $\mathcal{K}^{\mathrm{T}}$, of the LMI (57) in Theorem 2 for different values of κ and $\bar{\delta}$

$\bar{\delta}\backslash\kappa$	100	10	1	10^{-1}	10^{-2}
10.0	$\begin{bmatrix} -326.6099 \\ -196.2353 \end{bmatrix}$	$\begin{bmatrix} -359.8560 \\ -213.5153 \end{bmatrix}$	$\begin{bmatrix} -332.1684 \\ -199.5738 \end{bmatrix}$	$\begin{bmatrix} -364.4499 \\ -216.2479 \end{bmatrix}$	$\begin{bmatrix} -329.9745 \\ -198.2624 \end{bmatrix}$
5.0	$\begin{bmatrix} -84.1083 \\ -50.1300 \end{bmatrix}$	$\begin{bmatrix} -92.9086 \\ -54.7238 \end{bmatrix}$	$\begin{bmatrix} -87.9199 \\ -52.4237 \end{bmatrix}$	$\begin{bmatrix} -97.5025 \\ -57.4564 \end{bmatrix}$	$\begin{bmatrix} -87.4729 \\ -52.1571 \end{bmatrix}$
2.0	$\begin{bmatrix} -16.2078 \\ -9.2205 \end{bmatrix}$	$\begin{bmatrix} -18.1633 \\ -10.2621 \end{bmatrix}$	$\begin{bmatrix} -19.5303 \\ -11.2216 \end{bmatrix}$	$\begin{bmatrix} -22.7572 \\ -12.9948 \end{bmatrix}$	$\begin{bmatrix} -19.5724 \\ -11.2476 \end{bmatrix}$
1.0	$\begin{bmatrix} -6.5078 \\ -3.3763 \end{bmatrix}$	$\begin{bmatrix} -7.4854 \\ -3.9105 \end{bmatrix}$	$\begin{bmatrix} -9.7603 \\ -5.3356 \end{bmatrix}$	$\begin{bmatrix} -12.0793 \\ -6.6431 \end{bmatrix}$	$\begin{bmatrix} -9.8723 \\ -5.4034 \end{bmatrix}$
0.8	$\begin{bmatrix} -5.3438 \\ -2.6750 \end{bmatrix}$	$\begin{bmatrix} -6.2040 \\ -3.1483 \end{bmatrix}$	$\begin{bmatrix} -8.5879 \\ -4.6293 \end{bmatrix}$	$\begin{bmatrix} -10.7979 \\ -5.8809 \end{bmatrix}$	$\begin{bmatrix} -8.7083 \\ -4.7021 \end{bmatrix}$
0.4	$\begin{bmatrix} -3.7917 \\ -1.7399 \end{bmatrix}$	$\begin{bmatrix} -4.4956 \\ -2.1320 \end{bmatrix}$	$\begin{bmatrix} -7.0247 \\ -3.6875 \end{bmatrix}$	$\begin{bmatrix} -9.0895 \\ -4.8647 \end{bmatrix}$	$\begin{bmatrix} -7.1563 \\ -3.7670 \end{bmatrix}$
0.1	$\begin{bmatrix} -3.3067 \\ -1.4477 \end{bmatrix}$	$\begin{bmatrix} -3.9617 \\ -1.8144 \end{bmatrix}$	$\begin{bmatrix} -6.5362 \\ -3.3932 \end{bmatrix}$	$\begin{bmatrix} -8.5556 \\ -4.5471 \end{bmatrix}$	$\begin{bmatrix} -6.6713 \\ -3.4748 \end{bmatrix}$

0.1. The LMI (57) in Theorem 2 is solved for each combination of κ and $\bar{\delta}$, with α fixed at 2. The resulting gains $\mathcal{K}$ will be presented.

Table 3 summarizes the numerical results for the gain $\mathcal{K}$ obtained by resolving the LMI (57) in Theorem 2 for the specified parameter values. Notably, the LMI was consistently feasible. Furthermore, as seen in Table 3, a decrease in κ and an increase in $\bar{\delta}$ lead to a significant rise in the feedback gain $\mathcal{K}$. Particularly, for $\bar{\delta} = 10$ and $\bar{\delta} = 5$, the gain becomes excessively high, rendering it impractical for effective control of the joint exoskeleton robot.

The graphical results illustrated in Fig. 3 utilize the gain $\mathcal{K}$ from Eq. (85) within the proposed control law (11). The results confirm that the exoskeleton robot successfully reached the target position of $\theta_d = 30°$. Once stabilized at this optimal position, the control effort u gradually decreased to the targeted value -0.212.

In addition to the simulation results, we analyzed the time evolution of the variable λ (see Eq. 51), as shown in Fig. 4. The variation of λ reflects the controller's dynamic adjustments to ensure system stability. As the exoskeleton reaches the target position, λ stabilizes.

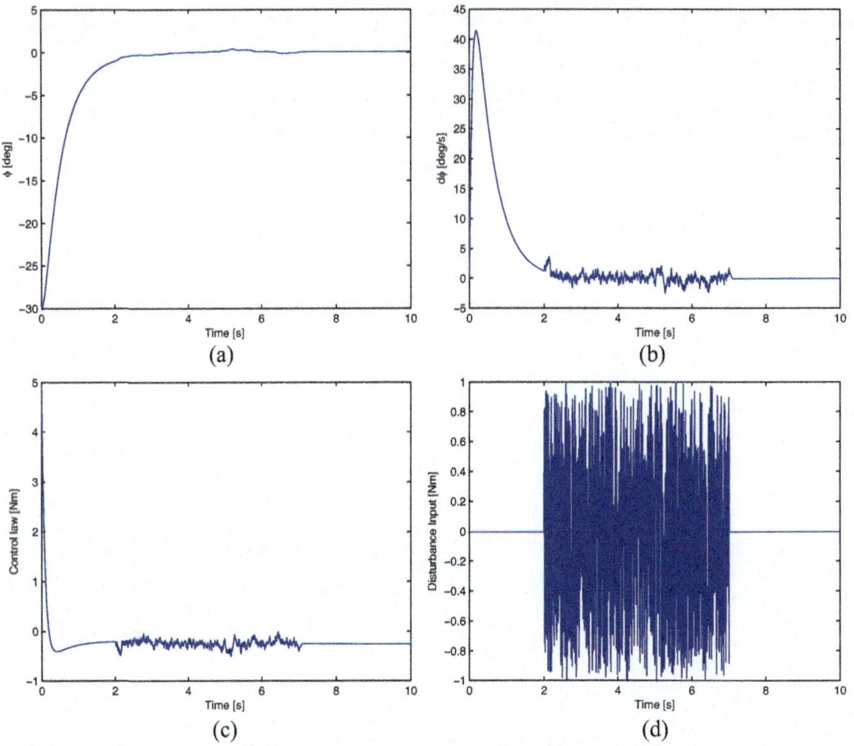

Fig. 3 Simulation results derived from the second design method. Temporal variations of: **a** the angular position error ϕ, **b** the angular velocity error $\dot{\phi}$, **c** the composite nonlinear controller u, and **d** the lumped disturbance Δ_t

5.1.2 Results Obtained Using the Third Design Methodology

In this section, we analyze the LMI stability condition as established by Theorem 3. The control gain $\mathcal{K}$ for the adopted controller (11) is derived from the matrices $\mathcal{S}$ and $\mathcal{R}$, which are the solutions to the LMI problem outlined in Theorem 3.

By applying fixed values of $\alpha = 2$, $\kappa = 1$, and $\bar{\delta} = 1$, we obtain the following results:

$$\mathcal{S} = \begin{bmatrix} 95.4609 & -184.6071 \\ -184.6071 & 518.0442 \end{bmatrix} \qquad \mathcal{R} = \begin{bmatrix} -132.6161 & -441.9774 \end{bmatrix}$$

$$\mathcal{M}_1 = \begin{bmatrix} 50.7463 & 0 \\ 0 & 197.7943 \end{bmatrix} \quad m_2 = 115.1720 \quad m_3 = 197.7943 \quad \beta_1 = 0.0136$$

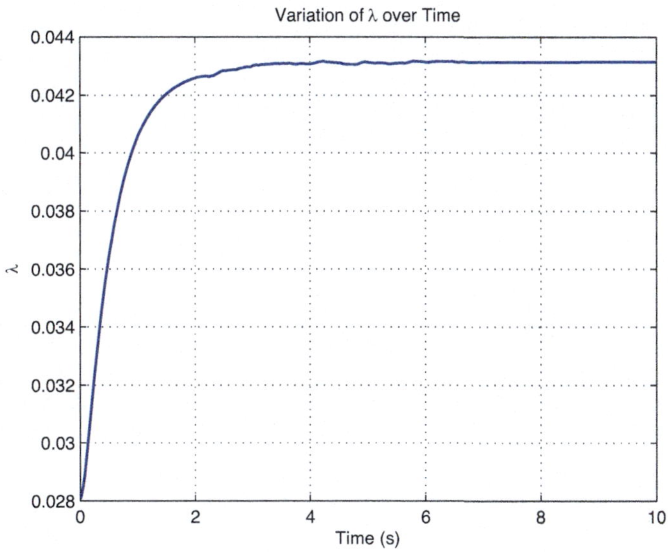

Fig. 4 Variation of the function λ over time

Thus, the feedback gain $\mathcal{K}$ is computed as:

$$\mathcal{K} = \begin{bmatrix} -9.7763 & -4.3370 \end{bmatrix} \tag{86}$$

Similar to the previous sections, these numerical results and the associated gain $\mathcal{K}$ were derived by solving the LMIs (82a) and (82b) from Theorem 3 with predetermined values for κ and $\bar{\delta}$. To further investigate the feasibility of the LMI, it is necessary to vary these parameters. The same values for κ as in prior sections are considered: 10^2, 10^1, 1.0, 10^{-1}, and 10^{-2}. For $\bar{\delta}$, we use: 10.0, 5.0, 2.0, 1.0, 0.8, 0.4, and 0.1. We then resolve the LMI problem in Theorem 3 for every combination of κ and $\bar{\delta}$, with α fixed at 2. The resulting gains $\mathcal{K}$ will be reported.

Table 4 presents the numerical outcomes for the gain $\mathcal{K}$ obtained by resolving the LMIs (82a) and (82b) in Theorem 3 for the specified parameter values. Notably, the LMI consistently proved to be feasible. Furthermore, Table 4 shows that a decrease in κ and an increase in $\bar{\delta}$ lead to a significant increase in the feedback gain $\mathcal{K}$. In particular, for $\bar{\delta} = 10$ and $\bar{\delta} = 5$, the gain becomes excessively high, making it impractical for effective control of the joint exoskeleton robot.

The graphical results displayed in Fig. 5 utilize the gain $\mathcal{K}$ from Eq. (86) within the proposed controller (11). The results indicate that the exoskeleton robot successfully reached the target position of $\theta_d = 30°$. the control effort u progressively aligned with the desired control effort u_d.

Table 4 Numerical results, $\mathcal{K}^{\mathrm{T}}$, of the LMIs (82a) and (82b) in Theorem 3 for various values of κ and $\bar{\delta}$

$\bar{\delta}\backslash\kappa$	100	10	1	10^{-1}	10^{-2}
10.0	$\begin{bmatrix} -926.9735 \\ -514.8173 \end{bmatrix}$	$10^3 \begin{bmatrix} -1.4402 \\ -0.7533 \end{bmatrix}$	$\begin{bmatrix} -373.7761 \\ -192.5624 \end{bmatrix}$	$\begin{bmatrix} -307.9554 \\ -154.9395 \end{bmatrix}$	$\begin{bmatrix} -321.7173 \\ -159.8111 \end{bmatrix}$
5.0	$\begin{bmatrix} -239.2392 \\ -132.3920 \end{bmatrix}$	$\begin{bmatrix} -371.4637 \\ -193.8289 \end{bmatrix}$	$\begin{bmatrix} -98.0187 \\ -49.9674 \end{bmatrix}$	$\begin{bmatrix} -81.5047 \\ -40.4195 \end{bmatrix}$	$\begin{bmatrix} -85.0467 \\ -41.6252 \end{bmatrix}$
2.0	$\begin{bmatrix} -46.6735 \\ -25.3129 \end{bmatrix}$	$\begin{bmatrix} -72.2303 \\ -37.1734 \end{bmatrix}$	$\begin{bmatrix} -20.8066 \\ -10.0408 \end{bmatrix}$	$\begin{bmatrix} -18.0985 \\ -8.3539 \end{bmatrix}$	$\begin{bmatrix} -18.7790 \\ -8.5331 \end{bmatrix}$
1.0	$\begin{bmatrix} -19.1642 \\ -10.0159 \end{bmatrix}$	$\begin{bmatrix} -29.4826 \\ -14.7940 \end{bmatrix}$	$\begin{bmatrix} -9.7763 \\ -4.3370 \end{bmatrix}$	$\begin{bmatrix} -9.0405 \\ -3.7731 \end{bmatrix}$	$\begin{bmatrix} -9.3122 \\ -3.8057 \end{bmatrix}$
0.8	$\begin{bmatrix} -15.8630 \\ -8.1803 \end{bmatrix}$	$\begin{bmatrix} -24.3529 \\ -12.1085 \end{bmatrix}$	$\begin{bmatrix} -8.4527 \\ -3.6525 \end{bmatrix}$	$\begin{bmatrix} -7.9535 \\ -3.2234 \end{bmatrix}$	$\begin{bmatrix} -8.1761 \\ -3.2384 \end{bmatrix}$
0.4	$\begin{bmatrix} -11.4615 \\ -5.7327 \end{bmatrix}$	$\begin{bmatrix} -17.5133 \\ -8.5278 \end{bmatrix}$	$\begin{bmatrix} -6.6878 \\ -2.7399 \end{bmatrix}$	$\begin{bmatrix} -6.5042 \\ -2.4905 \end{bmatrix}$	$\begin{bmatrix} -6.6614 \\ -2.4820 \end{bmatrix}$
0.1	$\begin{bmatrix} -10.0861 \\ -4.9679 \end{bmatrix}$	$\begin{bmatrix} -15.3759 \\ -7.4088 \end{bmatrix}$	$\begin{bmatrix} -6.1363 \\ -2.4547 \end{bmatrix}$	$\begin{bmatrix} -6.0513 \\ -2.2615 \end{bmatrix}$	$\begin{bmatrix} -6.1881 \\ -2.2457 \end{bmatrix}$

5.2 *Discussion*

The proposed controller successfully tackles the challenge of regulating the position of the knee exoskeleton robot illustrated in Fig. 1. This achievement is realized through the implementation of three distinct methodologies for calculating the feedback gain. As a result, the robot is accurately reach the desired target position, denoted as θ_d. Consequently, the angular position error ϕ diminishes to zero, as shown in Figs. 2a, 3a, and 5a. Similarly, the angular velocity error $\dot{\phi}$ also approaches zero, as depicted in Figs. 2a, 3b, and 5b. While some fluctuations are observed due to the impact of a disturbing torque Δ_t, these results illustrate the controller's resilience and effectiveness in counteracting external disturbances.

Additionally, visual representations of the external disturbance affecting the knee exoskeleton are presented in Figs. 2d, 3d, and 5d. This disturbance is represented by the torque load Δ_t, the lumped disturbance is defined as a randomly varying signal within the range of $-\bar{\delta}$ and $\bar{\delta}$, applied exclusively within the time interval of 2–7 s. The subsequent simulation results influenced by this lumped disturbing torque Δ_t are illustrated in Figs. 2, 3, and 5. The knee exoskeleton system clearly demonstrates

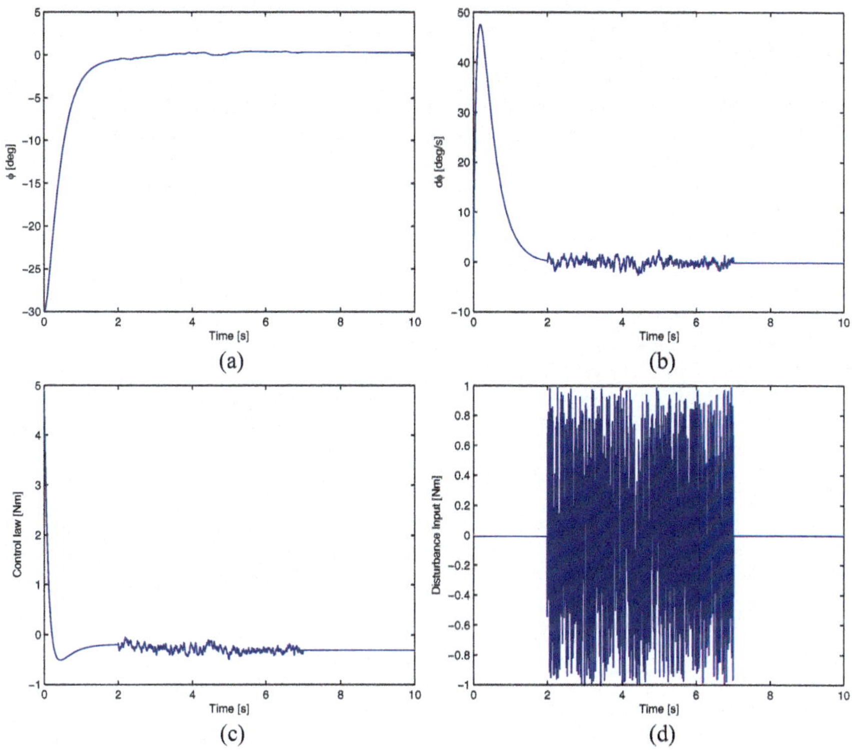

Fig. 5 Simulation results derived from the third design method. Temporal variations of: **a** the angular position error ϕ, **b** the angular velocity error $\dot{\phi}$, **c** the composite nonlinear controller u, and **d** the lumped disturbance Δ_t

oscillatory behavior around its equilibrium position. With the diminishing influence of the disturbing torque, the regulated robotic system progressively reaches stability and returns to its desired position.

Furthermore, the temporal development of the applied control law u is illustrated in Fig. 2c, 3c and 5c. It is clear that the control input u converges towards the intended control u_d after the knee exoskeleton robot stabilizes under the impact of the torque load. Moreover, as shown in Fig. 6, which illustrates the temporal variations of the composite nonlinear controller u during the time interval between 2 and 7 s, there are noticeable rapid fluctuations in the control signal around zero. These fluctuations are primarily due to the nonlinear control component, u_{nl}, which contributes to the dynamic behavior of the system.

Additionally, it is essential to highlight that the control effort u applied to the joint across all three developed approaches remains notably low, as shown in Figs. 2c, 3c and 5c. However, the third design methodology consistently yields a lower initial control effort compared to the first and second methods.

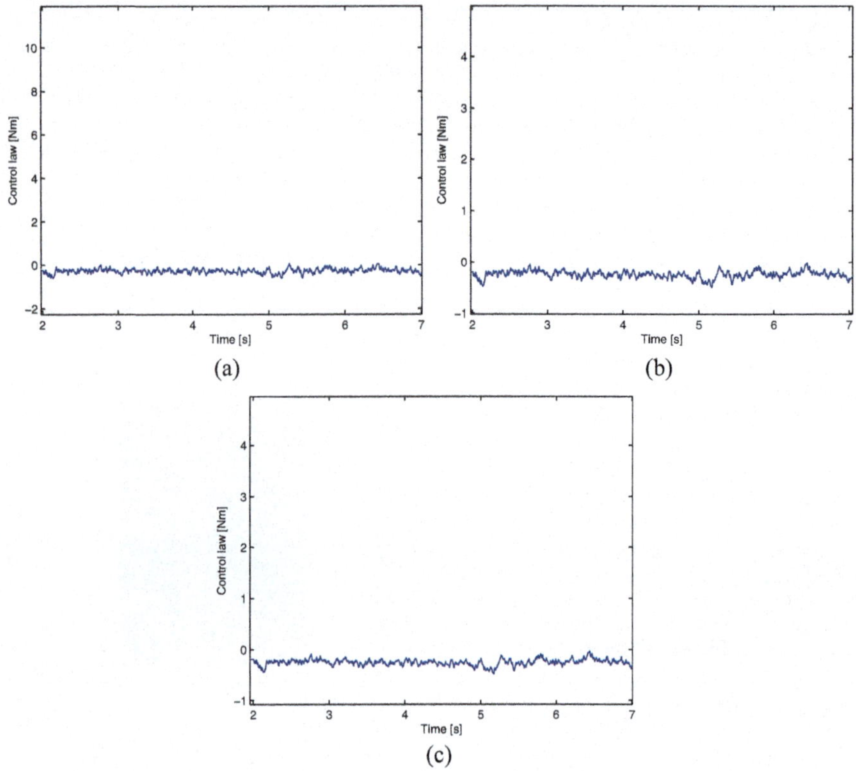

Fig. 6 Temporal variations of the the composite nonlinear controller u, during the time duration between 2 and 7 s: **a** using the first design method, **b** using the second design method, **c** using the third design method

Overall, the results from the three design methodologies demonstrate the successful regulation of the exoskeleton system, achieving the desired position $\theta_d = 30°$. Once the system was stabilized at this equilibrium, the primary objective was to maintain the position with minimal control effort. As the system approached the target position, the control effort u decreased and ultimately stabilized at the target value of -0.212. This reduction in control effort indicates that the robot reached a stable equilibrium at θ_d, where the force required to counteract disturbances and maintain the position is minimized. The graphs Figs. 2c, 3c and 5c clearly illustrate this gradual decrease in control effort, emphasizing the effectiveness and efficiency of the stabilization process.

Furthermore, The effectiveness of the proposed controller was evaluated through several metrics across the three design methodologies. The Mean Absolute Error (MAE) for the joint angle (ϕ) was calculated to be 0.0274 for the first approach, 0.0332 for the second approach, and 0.0318 for the third approach. In terms of joint velocity ($\dot{\phi}$), the MAE was consistently measured at 0.0560 for the first approach and

0.0565 for the second approach. The third approach yielded a slightly lower MAE of 0.0571.

These findings highlight the controller's effectiveness in achieving accurate and robust control of knee exoskeleton systems.

5.3 Influence of κ on the System Response

To study the influence of the parameter κ on the system's response, we will fix the values of α and $\bar{\delta}$ while varying κ across two different values. The third design method will be employed for this analysis. Specifically, we will illustrate the temporal variations of the applied control law u to examine the effect of κ on the system's behavior.

We analyze the LMI stability conditions established in Theorem 3 by selecting $\alpha = 2$ and $\bar{\delta} = 1$, and we will evaluate two different values of κ: 0.01 and 0.0001, to observe the impact of κ on the system's response.

The graphical results in Fig. 7a correspond to the third design method with $\kappa = 0.01$, while the results in Fig. 7b display the system's response with $\kappa = 0.0001$. As shown in Fig. 7b, the control input exhibits significant and rapid fluctuations, a phenomenon known as chattering. This behavior is particularly pronounced for smaller values of κ. The chattering effect, clearly visible in the initial seconds of the response, is characterized by high-frequency oscillations around the desired control value. In this case, reducing κ intensifies the chattering, as seen in the figure, where the control signal fluctuates significantly before stabilizing. However, as time progresses, the control signal stabilizes and successfully reaches the desired state with reduced oscillations.

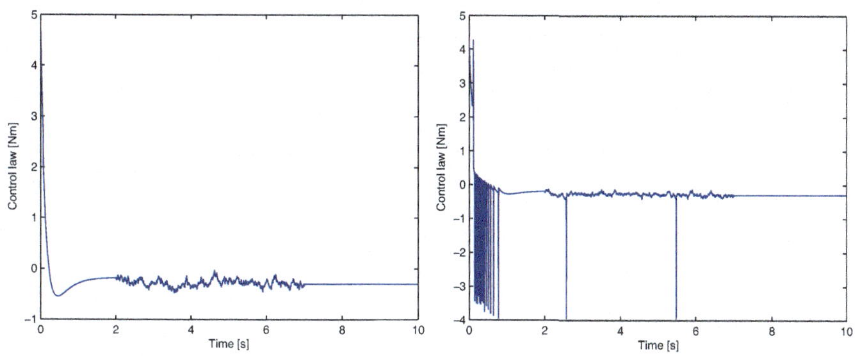

(a) (b)

Fig. 7 Temporal variations of the composite nonlinear controller u: **a** for $\kappa = 0.01$, **b** for $\kappa = 0.0001$

However, despite these fluctuations, the results indicate that the control effort u progressively converges towards the desired control effort u_d. Consequently, the exoskeleton robot successfully reaches the target position of $\theta_d = 30°$.

6 Conclusion and Future Works

This study presents a robust control strategy specifically designed for a 1-DoF knee rehabilitation exoskeleton robot [14, 25, 31]. Our focus was on achieving precise position control while addressing challenges such as parameter uncertainties, and external disturbances. To formulate the LMI stability criteria for the gains of the composite nonlinear control law, we employed three distinct design methodologies. Our approach leverages a quadratic Lyapunov function and incorporates advanced mathematical tools, including matrix inversion lemma, the Schur complement, Young's inequality and the S-procedure lemma. The resulting LMIs guarantee robust stabilization of the knee exoskeleton, even when faced with uncertainties and external disruptions. We conducted simulation results to verify the efficacy of the developed LMI stability conditions. The outcomes revealed the controller's capability to maintain stable knee exoskeleton position control, offering valuable insights for advancing rehabilitation strategies in exoskeleton control.

Looking ahead, our future work will focus on implementing a filtering mechanism to further mitigate duration or the amplitude of the oscillations detected in the regulated system. Additionally, we aim to carry out extensive experimental tests to validate our proposed LMI approach in real-world scenarios, assessing its robustness in various contexts while addressing challenges present in existing lower limb exoskeletons [48]. Moreover, the idea is to design a static output feedback controller based on the work achieved in [49] and also some nonlinear controllers [40, 50].

References

1. Gerber, A.M., Botes, R., Mostert, A., Vorster, A., Buskens, E.: A cohort study of elderly people in Bloemfontein, South Africa, to determine health-related quality of life and functional abilities. S. Afr. Med. J. **106**(3), 298–301 (2016)
2. Narayan, J., Dwivedy, S.K.: Preliminary design and development of a low-cost lower-limb exoskeleton system for paediatric rehabilitation. Proc. Inst. Mech. Eng. Part H J. Eng. Med. **235**(5), 530–545 (2021)
3. Baragash, R.S., Aldowah, H., Ghazal, S.: Virtual and augmented reality applications to improve older adults' quality of life: a systematic mapping review and future directions. Digit. Health **8**, 20552076221132099 (2022)
4. Clemson, L., Singh, M.A.F., Bundy, A., Cumming, R.G., Manollaras, K., O'Loughlin, P., Black, D.: Integration of balance and strength training into daily life activity to reduce rate of falls in older people (the life study): randomised parallel trial. BMJ **345** (2012)

5. Narayan, J., Dwivedy, S.K.: Lower limb exoskeletons for pediatric gait rehabilitation: a brief review of design, actuation, and control schemes. In: 2023 IEEE International Conference on Advanced Systems and Emergent Technologies (IC_ASET), pp. 1–6 (2023)
6. Bergmann, L., Voss, D., Leonhardt, S., Ngo, C.: Lower-limb exoskeleton with compliant actuators: human cooperative control. IEEE Trans. Med. Robot. Bionics **5**(3), 717–729 (2023)
7. Fareh, R., Elsabe, A., Baziyad, M., Kawser, T., Brahmi, B., Rahman, M.H.: Will your next therapist be a robot?—A review of the advancements in robotic upper extremity rehabilitation. Sensors **23**(11) (2023)
8. Narayan, J., Kalita, B., Dwivedy, S.K.: Adaptive backstepping human-cooperative control of a pediatric gait exoskeleton system with high- and low-level admittance. Proc. Inst. Mech. Eng. Part I J. Syst. Control Eng. **238**(3), 545–562 (2024)
9. Esquenazi, A., Talaty, M., Jayaraman, A.: Powered exoskeletons for walking assistance in persons with central nervous system injuries: a narrative review. PM&R **9**(1), 46–62 (2017)
10. Van der Loos, H.M., Reinkensmeyer, D.J., Guglielmelli, E.: Rehabilitation and Health Care Robotics. Springer Handbook of Robotics, pp. 1685–1728 (2016)
11. Jansen, C.M., Windau, J.E., Bonutti, P.M., Brillhart, M.V.: Treatment of a knee contracture using a knee orthosis incorporating stress-relaxation techniques. Phys. Ther. **76**(2), 182–186 (1996)
12. Lysholm, J., Nordin, M., Ekstrand, J., Gillquist, J.: The effect of a patella brace on performance in a knee extension strength test in patients with patellar pain. Am. J. Sports Med. **12**(2), 110–112 (1984)
13. Aljuboury, A.S., Hameed, A.H., Ajel, A.R., Humaidi, A.J., Alkhayyat, A., Mhdawi, A.K.A.: Robust adaptive control of knee exoskeleton-assistant system based on nonlinear disturbance observer. Actuators **11**(3) (2022)
14. Jenhani, S., Gritli, H.: Enhancing knee rehabilitation exoskeleton control via a robust LMI-based strategy and by considering state constraints. In: 2023 IEEE 11th International Conference on Systems and Control (ICSC), pp. 905–911 (2023)
15. Narayan, J., Auepanwiriyakul, C., Jhunjhunwala, S., Abbas, M., Dwivedy, S.K.: Hierarchical classification of subject-cooperative control strategies for lower limb exoskeletons in gait rehabilitation: a systematic review. Machines **11**(7), 764 (2023)
16. Wang, Z., Yang, C., Zhang, S., Zhang, S., Yi, C., Ding, Z., Wei, B., Jiang, F.: Knee flexion-assisted method for human-exoskeleton system. IEEE Trans. Neural Syst. Rehabil. Eng. **31**, 2800–2808 (2023)
17. Jenhani, S., Gritli, H.: LMI-based design of an affine PD controller for the robust stabilization of the knee joint of a lower-limb rehabilitation exoskeleton. In: 2023 IEEE International Conference on Advanced Systems and Emergent Technologies (IC_ASET), pp. 1–7. IEEE (2023)
18. Rifai, H., Mohammed, S., Daachi, B., Amirat, Y.: Adaptive control of a human-driven knee joint orthosis. In: 2012 IEEE International Conference on Robotics and Automation, pp. 2486–2491 (2012)
19. Narayan, J., Dwivedy, S.K.: Towards neuro-fuzzy compensated PID control of lower extremity exoskeleton system for passive gait rehabilitation. IETE J. Res. **69**(2), 778–795 (2023)
20. Gritli, H.: LMI-based design of an affine PD controller for the position control of robotic systems with application to human upper-limb rehabilitation via an exoskeleton robot. In: Intelligent Cyber-Physical Systems for Healthcare Solutions: From Theory to Practice, 1st edn, chap. 14, pp. 305–344. Springer Nature, Singapore (2024)
21. Jenhani, S., Gritli, H.: Robust position control of an exoskeleton robot for knee rehabilitation utilizing an LMI-based affine state-feedback controller. In: Intelligent Cyber-Physical Systems for Healthcare Solutions: From Theory to Practice, 1 edn, chap. 13, pp. 271–303. Springer Nature, Singapore (2024)
22. Jenhani, S., Gritli, H.: LMI-based robust control strategy for a knee rehabilitation exoskeleton: addressing state constraints, parameter uncertainties and external disturbances. In: Intelligent Cyber-Physical Systems for Healthcare Solutions: From Theory to Practice, 1 edn, chap. 12, pp. 237–270. Springer Nature, Singapore (2024)

23. Jenhani, S., Gritli, H.: An LMI-based robust state-feedback controller design for the position control of a knee rehabilitation exoskeleton robot: comparative analysis. Meas. Control **57**(9), 1326–1346 (2024)
24. Jenhani, S., Gritli, H.: Robust LMI-based state-feedback control of a knee exoskeleton robotic system. In: 2023 20th International Multi-Conference on Systems, Signals & Devices (SSD), pp. 121–128 (2023)
25. Gritli, H., Jenhani, S.: Position control of robotic systems via linear controllers with application to a lower limb rehabilitation exoskeleton robot: design and comparative analysis. In: Campos-Cantón, E., Huerta-Cuellar, G., Zambrano-Serrano, E., Tlelo-Cuautle, E. (eds.) Complex Syst. Appl., pp. 123–151. Springer Nature Switzerland, Cham (2024)
26. Jenhani, S., Gritli, H.: LMI-based design of a robust affine control law for the position control of a knee exoskeleton robot: comparative analysis of stability conditions. In: Study of Complex Systems and their Applications Conference, pp. 95–122. Springer (2023)
27. Daachi, M.E., Madani, T., Daachi, B., Djouani, K.: A radial basis function neural network adaptive controller to drive a powered lower limb knee joint orthosis. Appl. Soft Comput. **34**, 324–336 (2015)
28. Madani, T., Daachi, B., Djouani, K.: Adaptive controller based on uncertainty parametric estimation using backstepping and sliding mode techniques: application to an active orthosis. In: 2014 European Control Conference (ECC), pp. 1869–1874 (2014)
29. Al Rezage, G., Tokhi, M.O.: Fuzzy PID control of lower limb exoskeleton for elderly mobility. In: 2016 IEEE International Conference on Automation, Quality and Testing, Robotics (AQTR), pp. 1–6. IEEE (2016)
30. Narayan, J., Gritli, H., Dwivedy, S.K.: Fast terminal sliding mode control with rapid reaching law for a pediatric gait exoskeleton system. Int. J. Intell. Robot. Appl. **8**, 76–95 (2024)
31. Jenhani, S., Gritli, H.: A composite robust position control design for an actuated knee rehabilitation exoskeleton. In: 2023 20th International Multi-conference on Systems, Signals & Devices (SSD), pp. 129–134 (2023)
32. Jenhani, S., Gritli, H., Narayan, J.: An LMI-based composite nonlinear controller design for robust stabilization of a knee rehabilitation exoskeleton robot. In: 2024 10th International Conference on Control, Decision and Information Technologies (CoDIT), pp. 1430–1435. IEEE (2024)
33. Jenhani, S., Gritli, H., Narayan, J.: Robust stabilization of a knee exoskeleton robot using an LMI-based composite controller. In: 2024 Tenth Indian Control Conference (ICC), pp. 256–261. IEEE (2024)
34. Sherwani, K.I., Kumar, N., Chemori, A., Khan, M., Mohammed, S.: Rise-based adaptive control for EICoSI exoskeleton to assist knee joint mobility. Robot. Auton. Syst. **124**, 103354 (2020)
35. Petrič, T., Gams, A., Debevec, T., Žlajpah, L., Babič, J.: Control approaches for robotic knee exoskeleton and their effects on human motion. Adv. Robot. **27**(13), 993–1002 (2013)
36. Boyd, S., El-Ghaoui, L., Feron, E., Balakrishnan, V.: Linear matrix inequalities in system and control theory. In: Studies in Applied and Numerical Mathematics, vol. 15, 1 edn. Society for Industrial and Applied Mathematics (SIAM), Philadelphia, PA (1994)
37. VanAntwerp, J.G., Braatz, R.D.: A tutorial on linear and bilinear matrix inequalities. J. Process Control **10**(4), 363–385 (2000)
38. Jenhani, S., Gritli, H., Narayan, J.: Robust position control of a knee-joint rehabilitation exoskeleton using a linear matrix inequalities-based design approach. Appl. Sci. **15**(1), 404 (2025)
39. Narayan, J., Abbas, M., Dwivedy, S.K.: Robust adaptive backstepping control for a lower-limb exoskeleton system with model uncertainties and external disturbances. Automatika: časopis za automatiku, mjerenje, elektroniku, računarstvo i komunikacije **64**(1), 145–161 (2023)
40. Gritli, H., Jenhani, S., Narayan, J.: Motion control of manipulator robots via two PD-based nonlinear controllers: design and comparative analysis. Autom. Control Comput. Sci. **58**(6), 603–616 (2024)
41. Gritli, H.: LMI-based robust stabilization of a class of input-constrained uncertain nonlinear systems with application to a helicopter model. Complexity **2020**, 7025761 (2020)

42. Turki, F., Gritli, H., Belghith, S.: An LMI-based design of a robust state-feedback control for the master-slave tracking of an impact mechanical oscillator with double-side rigid constraints and subject to bounded-parametric uncertainty. Commun. Nonlinear Sci. Numer. Simul. **82**, 105020 (2020)
43. Gritli, H., Belghith, S.: Robust feedback control of the underactuated inertia wheel inverted pendulum under parametric uncertainties and subject to external disturbances: LMI formulation. J. Franklin Inst. **355**(18), 9150–9191 (2018)
44. Gritli, H.: Robust master-slave synchronization of chaos in a one-sided 1-DoF impact mechanical oscillator subject to parametric uncertainties and disturbances. Mech. Mach. Theory **142**, 103610 (2019)
45. Gritli, H., Jenhani, S.: LMI-based state feedback control of the underactuated inertia wheel inverted pendulum to the unstable upright position. J. Vib. Test. Syst. Dyn. **8**(2), 207–234 (2024)
46. Jenhani, S., Gritli, H.: A robust LMI-based approach for the position control of a 1-DoF knee rehabilitation exoskeleton robot considering state constraints. In: 2024 IEEE International Conference on Interdisciplinary Approaches in Technology and Management for Social Innovation (IATMSI), vol. 2, pp. 1–6. IEEE (2024)
47. Rifaï, H., Mohammed, S., Hassani, W., Amirat, Y.: Nested saturation based control of an actuated knee joint orthosis. Mechatronics **23**(8), 1141–1149 (2013)
48. Gritli, H., Belghith, S.: LMI-based synthesis of a robust saturated controller for an underactuated mechanical system subject to motion constraints. Eur. J. Control **57**, 179–193 (2021)
49. Gritli, H., Zemouche, A., Belghith, S.: On LMI conditions to design robust static output feedback controller for continuous-time linear systems subject to norm-bounded uncertainties. Int. J. Syst. Sci. **52**(1), 12–46 (2021)
50. Jenhani, S., Gritli, H., Carbone, G.: Comparison between some nonlinear controllers for the position control of Lagrangian-type robotic systems. Chaos Theory Appl. **4**(4), 179–196 (2022)

Experimental Investigation of Robotic Ankle Foot Orthosis with Different Actuators for Foot Drop Rehabilitation

Gowrishankar Govindaraj and Arockia Selvakumar Arockia Doss

Abstract Patients with foot drop frequently utilize Ankle Foot Orthosis (AFO) devices to restore normal gait; however, conventional AFOs exhibit limitations in providing adequate ankle range of motion (ROM). This study aims to develop and integrate two distinct actuators, the Series Elastic Actuator (SEA) and Scissor Actuator (SA), into a robotic ankle-foot orthosis (RAFO). The process involved designing the actuators using SolidWorks, constructing prototypes, and integrating a fuzzy logic controller in MATLAB Simulink to control ankle movement, specifically plantarflexion (PF) and dorsiflexion (DF) motions. Experimental testing was conducted utilizing the bench test method to compare the performance of the two actuators using an Arduino Mega microcontroller 2560, rotary angle sensor, and ultrasonic sensor. The results indicated that SA provided a 14% greater total ROM than SEA. However, SEA demonstrated a significant advantage in energy efficiency, with 31% lower input power consumption than that of SA. Furthermore, in terms of torque generation, the SA outperformed the SEA by 32%. These findings suggest that developing a RAFO with a hybrid actuator that combines both spring and scissor mechanisms could potentially reduce power consumption while improving the ROM and torque generation. This hybrid approach may offer a more effective solution for individuals with foot drop, enhancing mobility and quality of life.

Keywords Series elastic actuator · Scissor actuator · Fuzzy logic controller · Robotic ankle foot orthosis

1 Introduction

Human mobility significantly influences the quality of life, and the ankle joint plays a crucial role in maintaining balance, stability, and flexibility during ambulation. With two degrees of freedom (DOF) in the sagittal plane, the ankle facilitates plantar flexion (PF) and dorsiflexion (DF) motions, which are essential for the natural gait

G. Govindaraj · A. S. A. Doss (✉)
Vellore Institute of Technology, Chennai, Tamil Nadu, India
e-mail: arockia.selvakumar@vit.ac.in

J. Narayan and H. Gritli (eds.), *Design and Control of Rehabilitation Robots*, Studies in Systems, Decision and Control 585, https://doi.org/10.1007/978-3-031-86977-8_15

patterns illustrated in Fig. 1 [1]. However, ankle dysfunction resulting from stroke, cerebral palsy, multiple sclerosis, or motor neuron disorders can severely impair locomotion and produce abnormal gait patterns, leading to a condition known as foot drop. This condition, characterized by a lack of active DF, results in an increased risk of tripping and falling during ambulation, significantly affecting activities of daily living and overall quality of life [2].

Traditionally, orthotic devices have been employed to treat foot drop, with passive AFOs being widely utilized due to their simplicity, lightweight construction, and cost-effectiveness. Passive AFOs function essentially as springs, maintaining the ankle in a neutral position to ensure toe clearance during the swing phase. Despite their advantages, passive AFOs have limitations as they restrict PF, which is crucial for the push-off phase, thereby impacting the natural movement and energy efficiency of the gait [3].

Functional Electrical Stimulation (FES) has been introduced as an alternative approach to address these limitations. FES devices stimulate the peroneal nerve to activate the dorsiflexor muscles, thus facilitating DF during the swing phase and permitting PF during push-off [4]. However, FES is only viable for patients with residual functional muscles and nerves, and some users report discomfort or inconsistent stimulation, which can limit the effectiveness and acceptance of FES-based devices [5].

In the past decade, powered AFOs have emerged as a promising solution to foot drop, overcoming the limitations of both passive and FES systems. Powered AFOs utilize actuators (such as electric motors or pneumatic systems) to actively assist ankle movement, thereby enabling both DF and PF. Early models primarily aimed to support DF for improved toe clearance and reduced foot slap, minimizing resistance during push-off [6]. Although these powered devices offer enhanced functionality, they face challenges, such as increased weight, size, noise, and limited operational time due to battery constraints. These factors often hinder their practical application in clinical or rehabilitation settings [7].

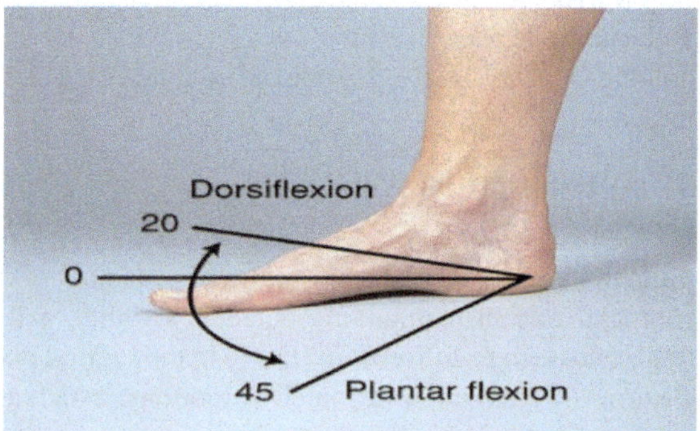

Fig. 1 Movement of ankle joint [38]

To address these issues, researchers are developing semi-active and hybrid AFOs that combine passive and active components to optimize energy efficiency and functional adaptability. Recent advances in actuator technology, such as series elastic actuators (SEAs) and soft pneumatic actuators, have enabled the creation of lightweight, low-power designs that provide effective gait assistance while conserving energy. For instance, hybrid systems integrate passive energy storage with powered actuation, achieving a balance between energy conservation and active support [8]. In particular, SEAs offer adaptable stiffness and precise force control, rendering them suitable for adjusting to the dynamic forces encountered during gait [9]. These devices not only enhance user mobility but also address the challenge of extended wearability, which is crucial for real-world application.

Current research on robotic gait assistance has further classified these devices into passive, active, and semi-active systems, based on their control strategies and energy sources. Passive systems store and release energy without external power input, whereas active systems generate additional energy to drive motion, resulting in higher responsiveness and greater power consumption [10]. Semi-active systems represent a compromise as they combine energy-efficient passive elements with minimal active control, allowing adaptable support that can respond to changes in walking speed, terrain, and user requirements [11].

Furthermore, RAFO equipped with intelligent control systems, such as fuzzy logic controllers, are being investigated to enhance gait adaptability. Unlike traditional PID controllers, fuzzy-logic controllers can adapt to individual patient variability and provide smooth transitions between gait phases, thereby offering a more naturalistic walking experience [12]. Studies indicate that integrating intelligent control into RAFO can enable adaptive assistance in real time, further addressing the gap between current limitations and the need for responsive, user-oriented devices.

In recent years, clinical studies on powered mobility aids have demonstrated promising outcomes for individuals with foot drop and other gait impairments. Participants reported improvements in walking stability, speed, and comfort. User feedback has frequently emphasized how powered AFOs create a sense of naturalistic walking, which is psychologically and functionally beneficial for the user. These advancements underscore the significance of further research and development to refine AFO technologies that enhance both mobility and quality of life for individuals with mobility impairments [13].

2 Literature Review

2.1 Semi-active Robots

Semi-active robotic devices facilitate ankle movement by regulating the braking mechanism of an AFO. Previous iterations exhibited limitations, including unintended impedance and insufficient toe clearance. The Smart AFO, developed by

Ohba et al., utilizes a magnetorheological fluid to regulate an elastic link, thereby storing and releasing energy for resistive torque [14]. Hassan et al. improved the braking performance of a Smart AFO. Semi-active AFOs provide enhanced ankle joint movement, decreased knee flexion, and accelerated recovery compared to passive AFOs [15].

2.2 *Active Rigid Actuators*

Active rigid actuators driven by motors provide direct assistance to impaired limbs, thereby overcoming the limitations of passive devices. In an effort to reduce costs, Shah et al. developed an economically viable robotic device utilizing a motor, a leaf spring AFO, and four-bar linkage systems to address foot drop [16]. Aboamer et al. introduced a cost-effective electromechanical orthosis incorporating a stepper motor and controller for foot drop [17]. Font-Llagunes et al. designed an adaptable robotic device comprising two robotic orthoses for the knee and ankle joints controlled by a motor-driven harmonic drive system. This device demonstrates improvements in walking speed, ankle motion, and power production [18].

2.3 *Pneumatic Active Actuators*

Traditional AFOs restrict ankle movement, whereas electric AFOs disrupt gait patterns. Pneumatically actuated AFOs (PAFO) provide cost-effective, user-friendly gait assistance. Shorter et al. developed the PAFO utilizing a rotational pneumatic actuator for both PF/DF motion [19]. Ulkir et al. designed a pneumatic AFO incorporating a pneumatic artificial muscle and a rotary actuator to address ankle joint weakness [20]. Hong et al. designed a compact AFO with a heel rocker mechanism and artificial muscles for increased dorsiflexion torque [21]. Kim et al. developed a portable pneumatic AFO powered by a wearable air compressor. Subsequent studies have enhanced the efficiency of AFOs using customized compressors and compact designs [22]. Salmeron et al. designed a lightweight soft robotic AFO with flexible materials and air-powered activation, enhancing push-off power and maintaining natural ankle movement [23]. Murphy et al. created a pneumatic robot for four functional movements of the ankle joint [24]. Park et al. developed a robotic device that replicates the muscle-tendon-ligament-skin system using pneumatic artificial muscles, offering ergonomic and adaptable assistive support [25].

2.4 Cable-Driven Active Actuators

Pneumatic artificial muscles and tendon-driven actuators typically necessitate antagonistic pairs to generate plantar flexion and dorsiflexion forces. Kwon et al. developed a robotic ankle-foot orthosis (AFO) incorporating flexible material and a single motor-driven cable-pulling mechanism [26]. Xia et al. investigated a portable ankle exoskeleton with a bidirectional cable-driven actuation system for push-off and stabilization [27]. T-Flex, a bio-inspired AFO, emulates antagonist muscles by modulating tendon stiffness. Cable-driven robotic devices demonstrate significant improvements in functional outcomes, gait parameters, and muscle strength among individuals with lower-limb impairments. These devices provide adjustable assistance and facilitate the implementation of personalized rehabilitation programs [28].

2.5 Active Series Elastic Actuators

Compliant actuators, such as Series Elastic Actuators (SEA), have been incorporated into drop foot robotics to enhance the safety of human-robot interaction by decoupling the actuator and transmission system inertia and friction from the load, thereby facilitating patient participation in therapy. Polinkovsky et al., Rossi et al., Moltedo et al., Kirtas's team, and Chen et al. have developed various active ankle-foot and knee-ankle orthoses utilizing SEAs, adaptive controllers, lever mechanisms, and specialized footwear, demonstrating improvements in walking performance and reduced energy expenditure for individuals with lower extremity disabilities [29–33].

2.6 Control System to Adapt Various Phases of Gait Cycle

Various control strategies incorporating sensors and PID controllers have been implemented to enhance force, stiffness, and response timing, which are crucial for accommodating different gait phases [34]. Modern devices integrate sensors to detect gait changes in real time, thereby improving user adaptability during activities such as ascending or descending stairs. This control mechanism supports individuals with foot drop during the swing phase, while facilitating unrestricted movement in the stance phase by alternating between modes based on gait stages [35]. Conventional PID control, which relies on mathematical modeling, demonstrates limitations in adapting to variable movements of the gait cycle. Fuzzy logic systems can emulate human decision-making without requiring mathematical models. To address this limitation, Kanthi et al. developed a control design utilizing a fuzzy controller to achieve a symmetrical gait in patients with foot drop [36]. Similarly, Adiputra et al. designed a passive-control AFO that employs a magnetorheological brake actuator governed by fuzzy logic control [37]. Despite significant advancements in developing RAFOs for addressing foot drop, a substantial gap remains in the research on the comparative evaluation of different actuation mechanisms within these devices. Specifically, limited research has systematically compared the performance of distinct actuators, such as SEA and SA, within RAFOs for foot drop patients.

Addressing this research gap is essential for several reasons. A comparative evaluation of different actuators can provide valuable insights into their respective capabilities for PF/DF motion, torque generation, energy efficiency, and adaptability to varying walking conditions. Such insights are crucial for informing the design and optimization of RAFOs tailored to the specific needs of patients with foot drop. This experimental study aimed to evaluate the performance of robotic ankle-foot orthoses (RAFOs) with two different actuators, SEA and SA, through bench test methods.

3 Materials and Methods

This section delineates the methodological approach employed in the development of the RAFO. The study investigates the implementation of various actuator configurations, including SEA and SA, to enhance AFO performance. The primary objective of this investigation was to assess the efficacy of SEA and SA actuators within an AFO framework through benchtop testing, with a focus on individuals experiencing foot drop. The methodological approach encompasses multiple phases: AFO adoption, actuator selection and integration, and device performance evaluation through benchtop testing. The RAFO comprised a polypropylene ankle-foot orthosis with a hinged ankle joint, which incorporated a sensor unit, controller, and actuator unit. The sensor unit consisted of an ultrasonic sensor and a rotary angle sensor. The SEA modulates the ankle joint based on signals from the controller. The total mass of the RAFO, including the SEA, was 3.4 kg.

3.1 Ankle Foot Orthosis (AFO)

A commercially available AFO with shank length 33 cm, foot length 25.5 cm, and thickness 0.5 cm was obtained from the Indian market, as illustrated in Fig. 2a. The dimensions of the AFO were utilized to develop a three-dimensional model of the AFO and integrate the SEA and SA using SolidWorks software. The AFO subsequently underwent a re-engineering process to facilitate PF/DF motion in the sagittal plane [38].

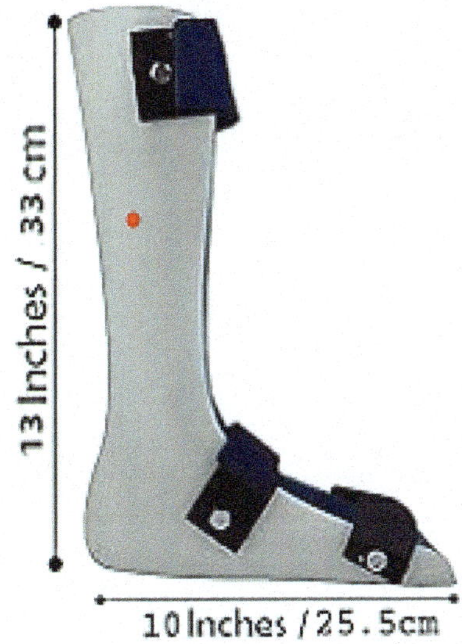

Fig. 2 Ankle foot orthosis

3.2 *Kinematics of AFO*

Figure 2 shows the ankle-foot system and the kinematic model of the AFO is presented in Fig. 3. The model incorporates shank length (a), foot length (b), and actuator linear displacement (x). The relationship among these parameters and the ankle angle (θ) is defined by Eq. (1). The objective was to regulate the ankle angle (θ) through the linear displacement of the actuator (x) to facilitate rehabilitation. The dynamic relationship between the motion of the actuator and ankle joint can be derived from these equations.

$$X^2 = a^2 + b^2 - 2ab\cos\theta \tag{1}$$

$$\theta = f(x) \tag{2}$$

3.3 *Construction of RAFO with SEA and SA*

The dynamic model of SEA with a simple mass-spring-damper system, comprising an equivalent motor mass m_m, damper coefficient b_m, and elastic constant k, is represented by

$$m_m x_m + b_m x_m = F_m - F_l \tag{3}$$

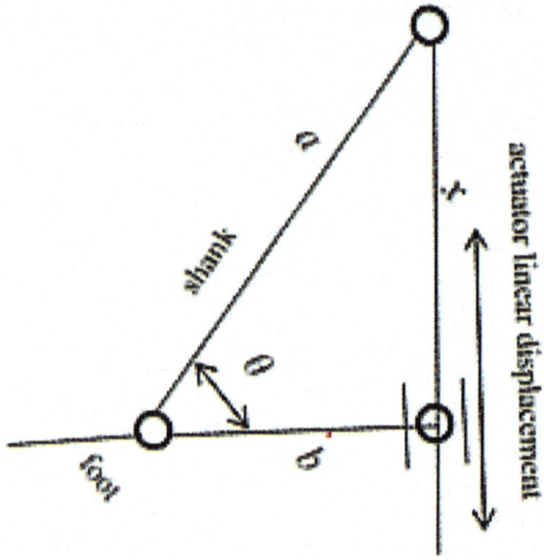

Fig. 3 Kinematics of ankle motion

With

$$F_l = k(x_m - x_l) \tag{4}$$

where x_m represents the linear position of the lead screw nut, x_l denotes the load position, and F_m signifies the force generated by the motor with F_l representing the output force. The schematic diagram is illustrated in Fig. 4.

A linear SEA was constructed by connecting an elastic element spring in series with a motor and ball screw. According to Hooke's law, a force is implied, and a sensor monitors the displacement of the elastic element. The force control operation was enhanced by incorporating a spring in series with the output of an electric motor. A 12 V DC motor (10 kg cm, 1000 RPM, Invento, India) was integrated with the SEA to manage the PF/DF motion. The SEA comprises a coupling, two fixed aluminum plates, two movable aluminum plates, eight compression springs, eight bushes, a ball screw, a ball nut, an aluminum mounting, two MS rods, and a lever arm connecting the device to the orthosis. The ball screw (BD 8 × 2.5 R) and ball nut converted the motor's rotational movements into translational motions. The movement of the ankle joint was regulated by adjustments to the SEA length, were driven by controlled

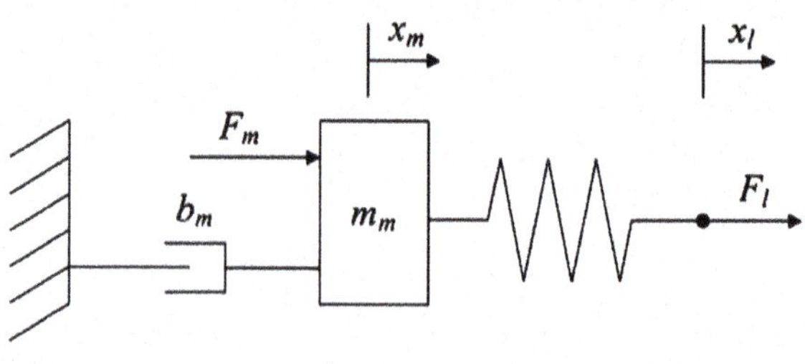

Fig. 4 Schematic diagram of SEA with spring, mass and damper

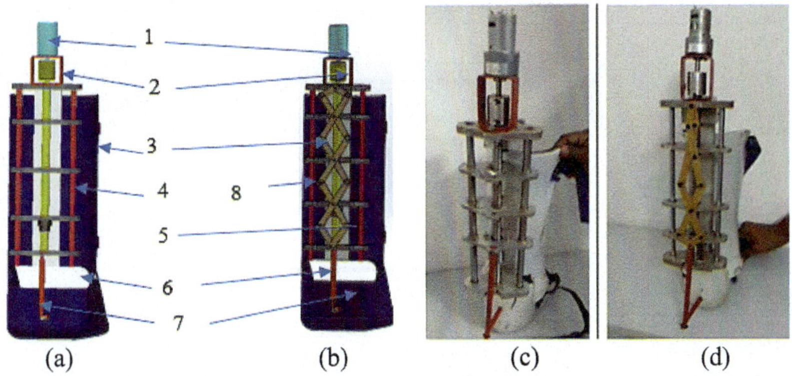

1. DC Motor, 2. Connector, 3. Shank, 4. Spring, 5. Ball Screw, 6. Foot, 7. Lever, 8. Scissor

Fig. 5 3D model (**a**, **b**) and prototype of RAFO with SEA and SA (**c**, **d**)

motor rotations. The eight compression springs within the SEA facilitated smooth adjustment of the length. A Scissor Actuator was developed by substituting scissors for springs. The SA consisted of 10 components employed in two sizes: six 3-hole flat plates and four 2-hole flat plates. Figure 5a, b depicts the 3D models [38] and Fig. 5c, d depicts the prototype of RAFO with SEA and SA.

3.4 *Electronic Module*

The electronic module incorporated an MCU-103 rotation angle sensor for precise ankle angle measurement, which necessitated the alignment of the potentiometer's rotational axis with the rotational axis of the AFO joint. An HC-SR04 ultrasonic sensor, comprising a transmitter, receiver, and control circuitry, was utilized to measure the displacement of the actuator.

3.5 *Controller*

An Arduino Mega 2560 microcontroller was utilized as the central processing unit to facilitate communication and control between the sensors and the actuator of the prototype. This microcontroller manages data acquisition from the sensors and provides control signals to the actuator. The L293D motor driver serves as an interface between the Arduino and DC motors, enabling precise motor control. Furthermore, a fuzzy logic controller, designed with 36 rule-based input and output membership functions and simulated in MATLAB/Simulink, regulates the motion of the system and collects sensor data. Figure 6 illustrates the fuzzy-logic controller with input

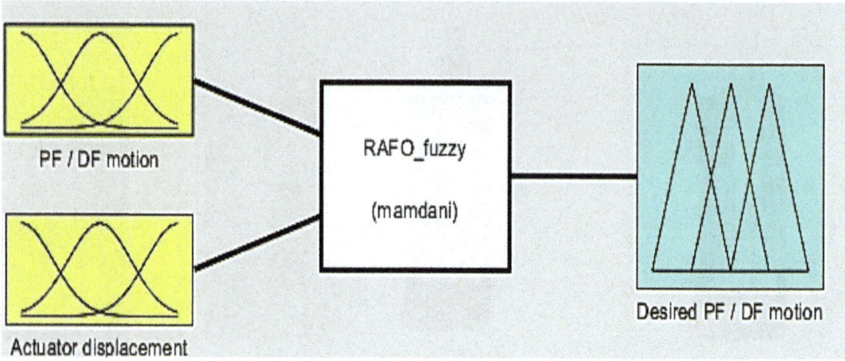

Fig. 6 Fuzzy logic controller with input and output membership functions

and output membership functions. Inputs: A set of 36 fuzzy logic rules has been developed for controlling a RAFO to achieve PF/DF motion within the range of −20° PF to 20° DF. The rules are based on two input variables: PF/DF motion (angle) and actuator displacement, and one output variable: Actuator Control Signal (U) [35].

The membership functions for the fuzzy reference PF/DF motion (input variable −1) range was established from −20° to +20° to facilitate a normal walking pattern. It is divided into 12 membership functions, namely N6, N5, N4, N3, N2, N1, P1, P2, P3, P4, P5, and P6. The extremity membership functions N6 and P6 encompass a broader range than the central functions, as the majority of actual and real-time angle values fall within −20° to +20°. The membership functions are triangular in shape, allowing for a uniform distribution of the membership value (u) across the entire range, thereby minimizing the width of the quantization steps for the input reference fuzzy set of the PF/DF motion.

The membership functions of the fuzzy actuator displacement (input variable 2) range were set from 6 to 14 cm. It was divided into three membership functions: small (S), medium (M), and large (L). The membership functions are triangular in shape, enabling a uniform distribution of the membership value (u) throughout the range, which reduces the width of the quantization steps for the input fuzzy set of actuator displacement.

The membership functions of the fuzzy variable actuation (output) are defined from −20° to +20° to support a typical walking pattern. The positive portion is related to the DF phase of the gait cycle, whereas the negative portion pertains to the PF phase. This is divided into six membership functions: A1 and A2. A3, A4, A5 and A6.

3.6 Triangular Membership Function

Let a, b, and c represent the x coordinates of the three vertices of μA (x) in a fuzzy set A (a: lower boundary and c: upper boundary where membership degree is zero, b: center where membership degree is 1) is represented in Fig. 7.

Rule Base for the Fuzzy Logic Controller (FLC) to achieve desired PF/DF motion, using the Ankle PF/DF motion and actuator displacement inputs to determine the corrective actuator output.

These rules account for the fine granularity of the PF/DF motion (12 membership functions) and three levels of actuator displacement (3 membership functions). The goal is to create a control action that assists or opposes motion based on the conditions of the defective foot, thereby mimicking the normal gait cycle. Figure 8 shows a surface view of the relationship between the input and output variables. Table 1 represents the 36 rules to control the PF/DF motion of RAFO.

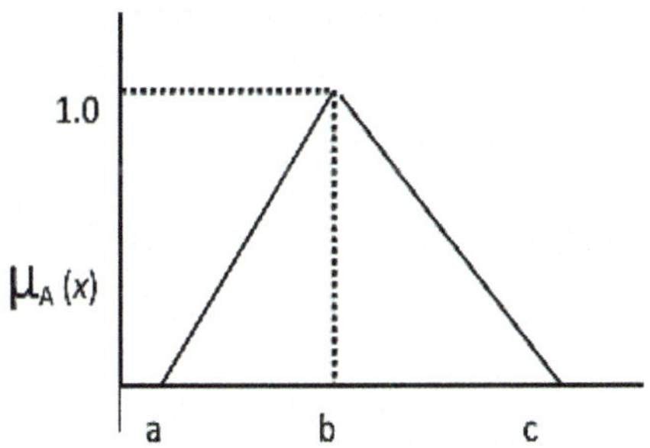

$$\mu_A(x) = \begin{cases} 0 & \text{if } x \le a \\ \dfrac{x-a}{b-a} & \text{if } a \le x \le b \\ \dfrac{c-x}{c-b} & \text{if } b \le x \le c \\ 0 & \text{if } x \ge c \end{cases}$$

Fig. 7 Schematic diagram of triangular membership function

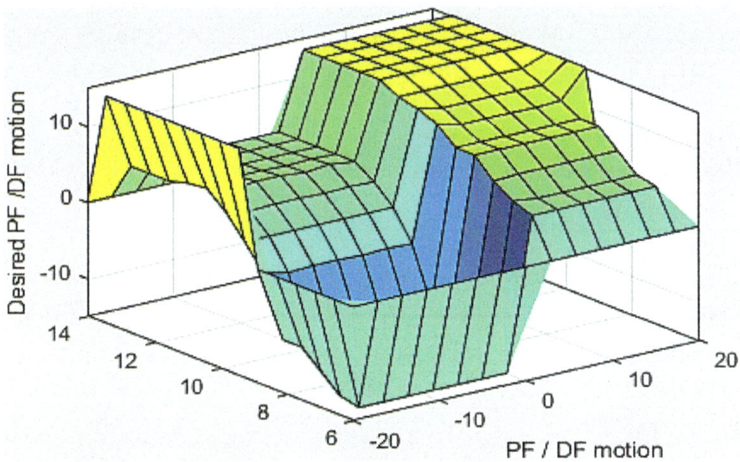

Fig. 8 Surface view of input and output variables relationship

Table 1 Rules to control the PF/DF motion of RAFO

S. No	Reference PF/DF motion (deg)	Actuator displacement (cm)	Desired PF/DF motion (deg)
	(Input variable 1)	(Input variable 2)	(Output variable)
1	**N6** (−20, −15, −10)	**S** (6, 7, 8)	A1 (−20, −15, −10)
2	**N6** (−20, −15, −10)	**M** (7, 9, 11)	A2 (−15, −10, −5)
3	**N6** (−20, −15, −10)	**L** (9, 11, 14)	A3 (−10, −5, 0)
4	**N5** (−15, −10, −5)	**S** (6, 7, 8)	A1 (−20, −15, −10)
5	**N5** (−15, −10, −5)	**M** (7, 9, 11)	A2 (−15, −10, −5)
6	**N5** (−15, −10, −5)	**L** (9, 11, 14)	A3 (−10, −5, 0)
7	**N4** (−10, −5, −2)	**S** (6, 7, 8)	A1 (−20, −15, −10)
8	**N4** (−10, −5, −2)	**M** (7, 9, 11)	A2 (−15, −10, −5)
9	**N4** (−10, −5, −2)	**L** (9, 11, 14)	A3 (−10, −5, 0)
10	**N3** (−5, −2, −1)	**S** (6, 7, 8)	A1 (−20, −15, −10)
11	**N3** (−5, −2, −1)	**M** (7, 9, 11)	A2 (−15, −10, −5)
12	**N3** (−5, −2, −1)	**L** (9, 11, 14)	A3 (−10, −5, 0)
13	**N2** (−2, −1, 0)	**S** (6, 7, 8)	A1 (−20, −15, −10)
14	**N2** (−2, −1, 0)	**M** (7, 9, 11)	A2 (−15, −10, −5)
15	**N2** (−2, −1, 0)	**L** (9, 11, 14)	A3 (−10, −5, 0)
16	**N1** (−1, 0, 2)	**S** (6, 7, 8)	A1 (−20, −15, −10)
17	**N1** (−1, 0, 2)	**M** (7, 9, 11)	A2 (−15, −10, −5)
18	**N1** (−1, 0, 2)	**L** (9, 11, 14)	A3 (−10, −5, 0)
19	**P1** (0, 1, 2)	**S** (6, 7, 8)	A4 (0, 5, 10)
20	**P1** (0, 1, 2)	**M** (7, 9, 11)	A5 (5, 10, 15)
21	**P1** (0, 1, 2)	**L** (9, 11, 14)	A6 (10, 15, 20)
22	**P2** (2, 5 8)	**S** (6, 7, 8)	A4 (0, 5, 10)
23	**P2** (2, 5 8)	**M** (7, 9, 11)	A5 (5, 10, 15)
24	**P2** (2, 5 8)	**L** (9, 11, 14)	A6 (10, 15, 20)
25	**P3**(5,8, 10)	**S** (6, 7, 8)	A4 (0, 5, 10)
26	**P3**(5,8, 10)	**M** (7, 9, 11)	A5 (5, 10, 15)
27	**P3**(5,8, 10)	**L** (9, 11, 14)	A6 (10, 15, 20)
28	**P4** (8, 10, 12)	**S** (6, 7, 8)	A4 (0, 5, 10)
29	**P4** (8, 10, 12)	**M** (7, 9, 11)	A5 (5, 10, 15)
30	**P4** (8, 10, 12)	**L** (9, 11, 14)	A6 (10, 15, 20)
31	**P5** (10, 12, 15)	**S** (6, 7, 8)	A4 (0, 5, 10)
32	**P5** (10, 12, 15)	**M** (7, 9, 11)	A5 (5, 10, 15)
33	**P5** (10, 12, 15)	**L** (9, 11, 14)	A6 (10, 15, 20)
34	**P6** (12, 15, 20)	**S** (6, 7, 8)	A4 (0, 5, 10)
35	**P6** (12, 15, 20)	**M** (7, 9, 11)	A5 (5, 10, 15)
36	**P6** (12, 15, 20)	**L** (9, 11, 14)	A6 (10, 15, 20)

3.7 *Bench Testing of RAFO*

Bench testing was conducted to assess the performance and functionality of the RAFO prior to initiating the clinical trial. The model was mounted on a test rig and preliminary sensor adjustments were performed. The experiment involved simulating the PF/DF angle values using a fuzzy-logic controller. The PF/DF angle measurements were recorded using a rotary angle sensor, while an ultrasonic sensor was employed to track the actuator's movement. This experimental procedure was repeated thrice, and the values were recorded in a controlled environment. Figures 9 and 10 depicts the closed loop system with fuzzy logic controller and bench test setup of RAFO.

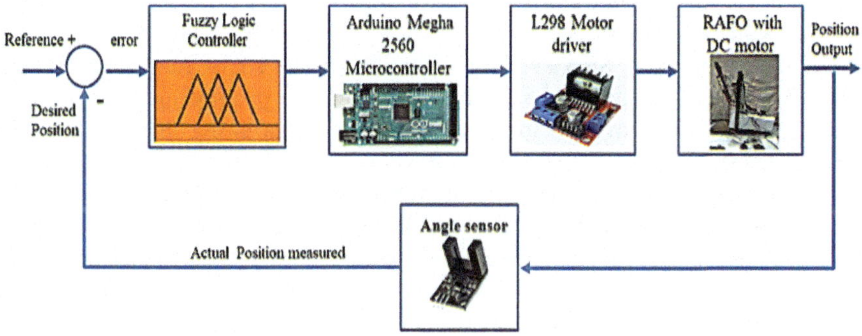

Fig. 9 Closed loop system with fuzzy logic controller

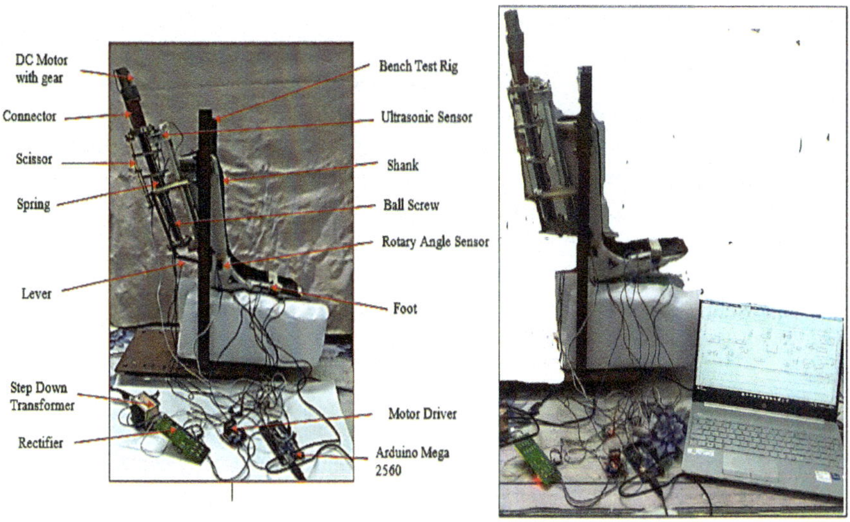

Fig. 10 Bench test setup of RAFO

4 Experimental Results of Bench Test

An experimental study of the RAFO utilizing two different actuators for foot drop patients produced several findings. This study compares the performance of a SEA and SA using a bench test method. The bench test results are presented in Table 2. The Fuzzy logic controller was adjusted to manage the PF and DF motions of the different actuators using MATLAB Simulink. The SEA-based actuators achieved a total range of motion of 24°, whereas the SA reached 27°.

The results indicate that the SA demonstrated a slightly larger total range of motion (27.4° ± 0.3°) than the SEA (24.1° ± 0.4°), suggesting that the SA might offer more flexibility in movement. This increased range of motion could potentially translate into improved gait mechanics for patients with foot drop, allowing for more natural ankle movement during walking. In terms of PF/DF motion, the SA again showed superior performance with a PF of −9.4° ± 0.60° and a DF of 18.7° ± 0.40°, compared to the SEA's PF of −8.7° ± 0.80° and DF of 15.4° ± 0.60°. This enhanced angular range could contribute to better foot positioning during the gait cycle, potentially improving both foot clearance during the swing phase, and stability during the stance phase.

The torque generation capabilities of both actuators were examined, with the SA showing a higher torque range (2.54–7.85 Nm) than the SEA (0.940–5.97 Nm). This increased torque output of the SA could provide stronger assistance to patients, particularly those who require substantial support owing to severe muscle weakness. However, the SA's improved performance comes at the cost of a higher power consumption. The input power consumption for the SA (52.4 ± 0.6 Watts) was significantly higher than that of the SEA (39.9 ± 0.8 Watts). This difference in power efficiency is an important consideration, as it can affect the battery life and overall portability of the device.

Graphical representations of the PF/DF motion over a 60-s period are illustrated in Fig. 11, showing the cyclical nature of movement for both actuators, which mimics the natural motion of the ankle during walking.

The torque generation graphs visually confirm the superior torque output of the SA across its range of motion. Figure 12 shows that the torque produced by the SEA ranges from 0.940 to 5.97 Nm and the torque generated by the SA ranges from 2.54 to 7.85 Nm. Figure 13 shows that the input power consumption of the SEA was 15–40 Watts and that of the SA was 24–52 Watts, respectively.

Figure 14a–c illustrates the box plot graph for the PF/DF motion with SEA and SA, the torque generation with SEA and SA, and the input power consumption with

Table 2 Experimental results of SEA and SA actuators

Actuators	Total ROM (deg)	PF (deg)	DF (deg)	Input power consumption (Watts)	Torque (N-m)
SEA	24.1 ± 0.4	−8.7 ± 0.80	15.4. ± 0.6	39.9 ± 0.8	5.97 ± 0.7
SA	27.4 ± 0.3	−9.4 ± 0.60	18.7 ± 0.40	52.4 ± 0.6	7.85 ± 0.3

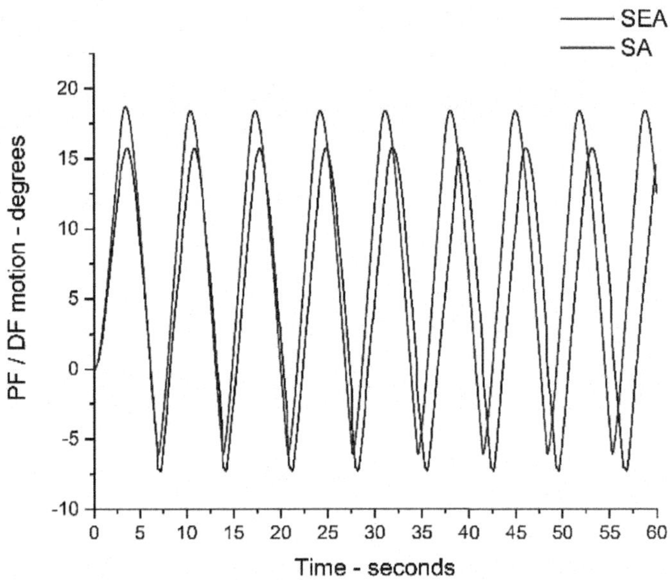

Fig. 11 PF/DF motion versus time of SEA and SA

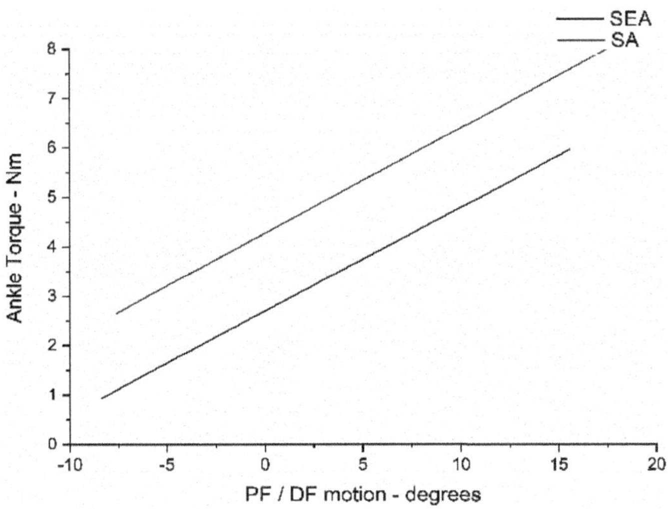

Fig. 12 Torque generation of RAFO with SEA and SA

SEA and SA. Although the SA demonstrated advantages in ROM, DF angle, and torque generation, the SEA exhibited better performance in PF motion and energy efficiency. The choice between these actuators for devices depends on the specific needs of the patient, balancing factors such as the required assistive force, range of

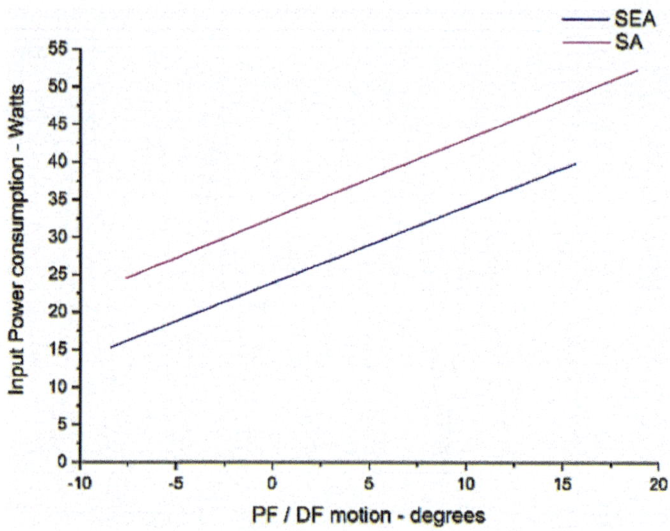

Fig. 13 Input power consumption of RAFO with SEA and SA

motion, and importance of power efficiency for prolonged use. These findings offer valuable insights into the development and customization of devices, potentially leading to improved mobility, independence, and quality of life in individuals with foot drop.

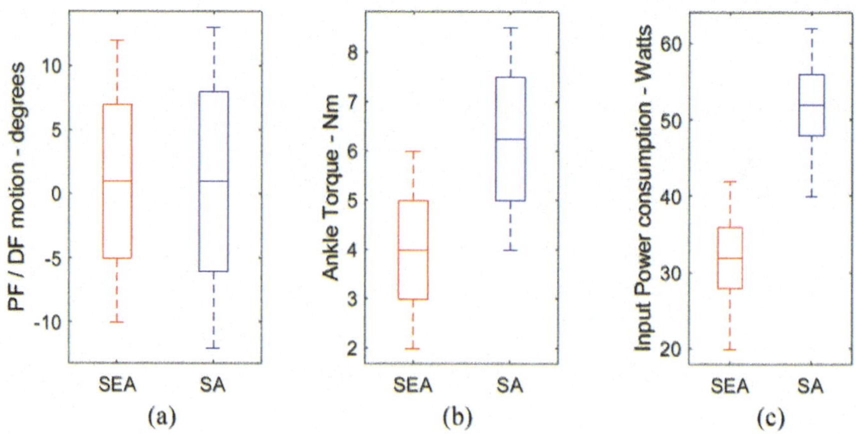

Fig. 14 **a** PF/DF motion of SEA and SA, **b** Torque generation of SEA and SA, and **c** Input power consumption of SEA and SA

5 Conclusion

Experiments were conducted to evaluate the performance of a RAFO equipped with SEA and SA for patients with foot drop, using a bench test method. The significant findings have implications for the development of assistive technology in rehabilitation engineering and mobility assistance. This research focuses on the key findings and assesses the strengths and limitations of SEA and SA actuators in terms of clinical benefits. It effectively simulates the human foot by regulating the PF/DF motion, thereby assisting in foot clearance during the swing phase and ensuring stability during the stance phase. These findings suggest that the PF/DF movement facilitated by RAFO has the potential to enhance the gait cycle and walking patterns of patients with foot drop. In addition, the bench test method proved useful for customizing to meet the specific needs of individual patients.

The assessment of performance based on joint angles and actuator force helps to identify stress points and areas of discomfort, which are critical for improving patient comfort. Although the proposed design offers adequate support and comfort in a controlled environment, it is crucial to validate these findings in real-time applications with patients. However, bench testing, despite being beneficial for initial evaluation, has limitations. It does not account for the dynamic interactions between various factors, such as patient variability, terrain, walking speed, and environmental conditions. Future research should focus on refining and optimizing the technology based on real-time feedback to better support patients with foot drop, ultimately enhancing their mobility, independence, and quality of life.

This study compared the performance of a RAFO using SEA and SA for foot drop patients. The key findings are.

- **Range of Motion (ROM)**

The SA demonstrated a 14% greater total range of motion (ROM) compared to the SEA. This increased ROM is crucial, as it directly contributes to a more natural gait cycle by allowing for a wider range of DF movements. However, SEA performs the PF movement more effectively than SA. A greater ROM can potentially enhance the ability of patients with foot drop to achieve smoother transitions between the swing and stance phases of walking.

- **Input Power Consumption**

A significant finding of this study is that the input power consumption of the SA is 31% higher than that of SEA. While SA offers superior mechanical performance, the higher power demand could present a challenge in terms of energy efficiency and battery life in portable applications. This trade-off between performance and energy consumption is critical, especially for long-term use in assistive devices, where power efficiency is as important as mechanical effectiveness.

- **Torque Generation**

The torque generated by the SA was 32% higher than that generated by the SEA. The torque at the ankle joint is vital for stabilizing the foot during walking, particularly during the push-off phase. The higher torque generated by the SA suggests that it may be more effective in providing necessary support and propulsion, which could lead to improved stability and balance in patients with foot drop.

6 Future Work

- Real-world Validation: Conduct clinical trials with foot drop patients to evaluate performance under actual walking conditions.
- Long-term studies have investigated the prolonged impact of use on gait improvement, muscle health, and mobility.
- Energy Efficiency: Optimize the power consumption of the Scissor Actuator (SA) while preserving the mechanical performance.
- Adaptive control: incorporating machine learning for adaptive control to address individual patient variability and environmental challenges.
- Enhanced Sensing: Integrate sensors to track parameters, such as fatigue and gait symmetry, enabling better feedback and personalized adjustments.
- Terrain Adaptability: Test and improve functionality across diverse terrains such as stairs or inclines.
- Comparative analysis: Comparison of SEA and SA actuators with existing devices for effectiveness and areas for refinement.
- Portability Improvements: Focus on miniaturization and weight reduction for greater user comfort and convenience.
- Battery Innovations: Explore advanced battery systems or energy-harvesting techniques to extend operational life.
- User-Friendly Interfaces: Design intuitive controls for patients and clinicians to easily fine-tune the settings.
- These efforts aim to enhance the usability, effectiveness, and adaptability of patients with foot drop, ultimately improving their quality of life.

Acknowledgements The authors would like to appreciate the efforts of the editors and reviewers. This research did not receive any specific grants from funding agencies in the public, commercial, or not-for-profit sectors.

Funding The authors received no financial support for the research, authorship, or publication of this study.

Declaration of Interest The author confirms that there are no conflicts of interest to declare.

References

1. Delp, S.L., Zajac, F.E.: Force and power output from the foot during toe-off in human walking. J. Biomech. **35**(2), 89–98 (2020)
2. Perry, J., Burnfield, J.M.: Gait Analysis: Normal and Pathological Function, 2nd edn. SLACK Incorporated (2019)
3. Bregman, D.J.J., et al.: The effects of ankle-foot orthoses on walking ability in stroke patients: a systematic review. Clin. Rehabil. **25**, 639–647 (2020)
4. Popović, D., Sinkjaer, T.: Functional electrical stimulation for foot drop: application in stroke. Phys. Ther. **95**(3), 251–259 (2020)
5. Knutson, J.S., et al.: Interfacing functional electrical stimulation and rehabilitation robotics: considerations for the future. J. Neuroeng. Rehabil. **13**(1), 72 (2021)
6. Agrawal, S., Banala, S.K., Fattah, A.: A powered ankle-foot orthosis for rehabilitation of foot drop. IEEE Trans. Neural Syst. Rehabil. Eng. **26**(4), 807–815 (2022)
7. Dollar, A.M., Herr, H.: Lower extremity exoskeletons and active orthoses: challenges and state-of-the-art. IEEE Trans. Robot. **24**(1), 144–158 (2022)
8. Angelo, A., et al.: Recent advancements in robotic lower limb orthoses for gait assistance. IEEE Rev. Biomed. Eng. **15**, 123–139 (2023)
9. Moltedo, M., et al.: Semi-active ankle-foot orthoses with integrated series elastic actuators: balancing energy efficiency and gait adaptability. Sensors **23**(9), 4020–4035 (2023)
10. Yuan, K., et al.: Classification of active and passive gait assistance systems: a review. J. Rehabil. Res. Dev. **60**, 334–344 (2022)
11. Asbeck, A.T., et al.: Soft robotic exosuits for assisting human locomotion. Int. J. Robot. Res. **36**(4), 500–513 (2021)
12. Kim, J., Park, S.: Innovative actuation methods in robotic ankle-foot orthoses for adaptive walking assistance. Front. Robot. AI **10**(2), 201–218 (2023)
13. Awad, L.N., et al.: A soft robotic exosuit improves walking in patients after stroke. Sci. Transl. Med. **9**(482) (2021)
14. Ohba, K., Sugiyama, T., Ueno, A., Takahashi, H.: Smart AFO design with magnetorheological fluid for resistive torque control. Int. J. Fluid Power **9**(3), 19–24 (2008). https://doi.org/10.1080/14399776.2008.10785984
15. Hassan, M., Yagi, K., Kadone, H., Ueno, T., Mochiyama, H., Suzuki, K.: Optimized design of a variable viscosity link for robotic AFO. In: Proc. 41st Annu. Int. Conf. IEEE Eng. Med. Biol. Soc. (EMBC), pp. 6220–6223 (2019)
16. Shah, B., McNally, D., Patel, K., Frone, S., Sutaria, S.: Design and fabrication of an intuitive leg-assist device to address lower extremity weakness. In: Proc. IEEE 37th Annu. Northeast Bioeng. Conf. (NEBEC), pp. 1–2 (2011)
17. Aboamer, M.A., Mohamed, N.A.R., Alrashide, B.: Low-cost electromechanical bionic model for foot drop gait disability. In: Proc. 2nd Int. Conf. Comput. Inf. Sci. (ICCIS), pp. 1–3 (2020)
18. Font-Llagunes, J.M., Lugrís, U., Clos, D., Alonso, F.J., Cuadrado, J.: Design, control, and pilot study of a lightweight and modular robotic exoskeleton for walking assistance after spinal cord injury. J. Mech. Robot. **12**(3), Art. no. 031008 (2020)
19. Shorter, K.A., Li, Y., Morris, E.A., Kogler, G.F., Hsiao-Wecksler, E.T.: Experimental evaluation of a portable powered ankle-foot orthosis. In: Proc. Annu. Int. Conf. IEEE Eng. Med. Biol. Soc. (EMBS), pp. 624–627 (2011)
20. Ulkir, O., Akgun, G., Kaplanoglu, E.: Mechanical design and analysis of a pneumatic ankle foot orthosis. In: Proc. Electr. Electron., Comput. Sci., Biomed. Eng. Meeting (EBBT), pp. 1–4 (2018)
21. Hong, J.-C., Suzuki, S., Fukushima, Y., Yasuda, K., Ohashi, H., Iwata, H.: Development of high-dorsiflexion assistive robotic technology for gait rehabilitation. In: Proc. IEEE Int. Conf. Syst., Man, Cybern. (SMC), pp. 3801–3806 (2018)
22. Kim, S.J., Na, Y., Lee, D.Y., Chang, H., Kim, J.: Pneumatic AFO powered by a miniature custom compressor for drop foot correction. IEEE Trans. Neural Syst. Rehabil. Eng. **28**(8), 1781–1789 (2020)

23. Salmeron, L.J., Juca, G.V., Mahadeo, S.M., Ma, J., Yu, S., Su, H.: An untethered electro-pneumatic exosuit for gait assistance of people with foot drop. Front. Biomed. Devices **83549**, Art. no. V001T09A009 (2020)
24. Murphy, P., Adolf, G., Daly, S., Bolton, M., Maurice, O., Bonia, T., Mavroidis, C., Yen, S.-C.: Test of a customized compliant ankle rehabilitation device in unpowered mode. In: Proc. 36th Annu. Int. Conf. IEEE Eng. Med. Biol. Soc. (EMBC), pp. 3057–3060 (2014)
25. Park, Y.L., Chen, B.R., Pérez-Arancibia, N.O., Young, D., Stirling, L., Wood, R.J., Goldfield, E.C., Nagpal, R.: Design and control of a bio-inspired soft wearable robotic device for ankle-foot rehabilitation. Bioinspir. Biomim. **9**(1), Art. no. 016007 (2014)
26. Kwon, J., Park, J.-H., Ku, S., Jeong, Y., Paik, N.-J., Park, Y.-L.: A soft wearable robotic ankle-foot-orthosis for post-stroke patients. IEEE Robot. Autom. Lett. **4**(3), 2547–2552 (2019)
27. Xia, H., Kwon, J., Pathak, P., Ahn, J., Shull, P.B., Park, Y.-L.: Design of a multifunctional soft ankle exoskeleton for foot drop prevention, propulsion assistance, and inversion/eversion stabilization. In: Proc. 8th IEEE RAS/EMBS Int. Conf. Biomed. Robot. Biomechatronics (BioRob), pp. 118–123 (2020)
28. Manchola, M., Serrano, D., Gómez, D., Ballen, F., Casas, D., Munera, M., Cifuentes, C.A.: T-FLEX: variable stiffness ankle-foot orthosis for gait assistance. In: Proc. Int. Symp. Wearable Robot Biosyst. Biorobotics, vol. 22, pp. 160–164 (2019)
29. Polinkovsky, A., Bachmann, R.J., Kern, N.I., Quinn, R.D.: Ankle-foot orthosis with insertion point eccentricity control. In: Proc. IEEE/RSJ Int. Conf. Intell. Robots Syst., pp. 1603–1608 (2012)
30. Rossi, S., Patané, F., Sette, D., Cappa, P.: WAKE-Up: a wearable ankle knee exoskeleton. In: Proc. 5th IEEE RAS/EMBS Int. Conf. Biomed. Robot. Bio Mechatronics, pp. 504–507 (2014)
31. Moltedo, M., Baček, T., Langlois, K., Junius, K., Vanderborght, B., Lefeber, D.: Design and experimental evaluation of a lightweight, high-torque, and compliant actuator for an active ankle-foot orthosis. In: Proc. Int. Conf. Rehabil. Robot. (ICORR), pp. 283–288 (2017)
32. Moltedo, M., Cavallo, G., Baček, T., Lataire, J., Vanderborght, B., Lefeber, D., Rodriguez-Guerrero, C.: Variable stiffness ankle actuator for use in robotic-assisted walking: control strategy and experimental characterization. Mech. Mach. Theory **134**, 604–624 (2019)
33. Chen, B., Zi, B., Wang, Z., Li, Y., Qian, J.: Development of robotic ankle-foot orthosis with series elastic actuator and magneto-rheological brake. J. Mech. Robot. **13**(1), Art. no. 011002 (2021)
34. Anam, K., Al-Jumaily, A.A.: Active exoskeleton control systems: state of the molted art. Proc. Eng. **41**, 988–994 (2012). https://www.sciencedirect.com/science/article/pii/S1877705812026732
35. Kanthi, M., Mruthyunjaya, H.S., George, V.I.: Fuzzy logic control for active ankle-foot orthosis. In: Proc. IEEE Int. Conf. Fuzzy Syst. (FUZZ-IEEE), pp. 1–6 (2013)
36. Martinez-Hernandez, U., Rubio-Solis, A., Cedeno-Campos, V., Dehghani-Sanij, A.A.: Towards an intelligent wearable ankle robot for assistance to foot drop. In: Proc. IEEE Int. Conf. Syst., Man Cybern. (SMC), pp. 3410–3415 (2019)
37. Arnez-Paniagua, V., Rifa, H., Amirat, Y., Ghedira, M., Gracies, J.M., Mohammed, S.: Adaptive control of an actuated ankle foot orthosis for paretic patients. Control. Eng. Pract. **90**, 207–220 (2019)
38. Gowrishankar, G., Arockia Selvakumar, A.: Evaluation of robotic ankle-foot orthosis with different actuators using Simscape multibody for foot-drop patients. Int. J. Online Biomed. Eng. (iJOE) **19**(10), 156–168 (2023). https://doi.org/10.3991/ijoe.v19i10.40375

Influence of Lazy Zone on Orthotropic Bone Adaptation in Human Femur: Pre- and Post-surgery Insights

Ankan Hazra, Souptick Chanda, and Debabrata Chakraborty

Abstract Orthotropic bone remodeling algorithm coupled with finite element models have been used extensively in bone adaptation studies in order to prognosticate the elastic material property distribution across human femur bone subjected to multiple routine activities. Peri-prosthetic bone resorption following a hip replacement surgery owing to post-operative bone remodeling in response to altered biomechanical environment around hip stem, may result in failure of prosthesis. Various numerical models have included a 'lazy zone' with no remodeling response to mechanical stimulus. Thus, for bio-mechanical simulations related to bone adaptation the selection of extent of lazy zone threshold is of crucial importance. This pre-clinical in silico study addresses the effect of size of lazy zone on orthotropic bone adaptation in an implanted femur post-surgery simulated with different physiological loading. In order to assess their impact on bone remodeling and material property distribution across implanted femur, this study incorporates eight distinct lazy zone thresholds (70, 60, 40, …, 10% and without lazy zone). It was observed that parametric variation in the size of lazy zone significantly affected bone remodeling stimulus. Decrease in lazy zone threshold was followed by elevated level of bone loss around hip stem, where an increase in threshold inhibited resorption. Regardless little noticeable agreement in elastic property distribution near cortex, the assessable differences in trabecular density patterns emphasizes the effects of the lazy zone on the orthotropic bone adaptation. A deeper understanding of these dynamics enables more precise predictions of bone behavior after surgery, helping to shape rehabilitation strategies.

A. Hazra (✉) · D. Chakraborty
Fracture Mechanics and Composite Structures Laboratory, Department of Mechanical Engineering, Indian Institute of Technology Guwahati, Guwahati, Assam, India
e-mail: ankan18@iitg.ac.in

D. Chakraborty
e-mail: chakra@iitg.ac.in

S. Chanda
Biomechanics and Simulations Laboratory, Department of Biosciences and Bioengineering, Indian Institute of Technology Guwahati, Guwahati, Assam, India
e-mail: csouptick@iitg.ac.in

J. Narayan and H. Gritli (eds.), *Design and Control of Rehabilitation Robots*, Studies in Systems, Decision and Control 585, https://doi.org/10.1007/978-3-031-86977-8_16

Keywords Lazy zone · Bone remodeling · Stimulus · Peri-prosthetic bone resorption · Biomechanical simulations · Implanted femur

1 Introduction

The human skeletal geometry, an extraordinary composition of 206 bones, fibro-cartilage and various connective tissues, caters a wide range of biomechanical functions. It offers structural support, facilitates movement, and protects essential organs, while also playing vital physiological roles such as acting as a mineral reservoir and supporting blood cell production [1]. Throughout the life, the skeletal architecture is skillfully adjusted in response to varying biomechanical forces, ensuring optimal strength and stability [2]. This inherent adaptability helps preventing fractures from heavy impacts and effectively maintaining bone health subjected to high stresses encountered during vigorous physical activities. The process of skeletal development is particularly pronounced during childhood and adolescence through two primary mechanisms: growth and modeling [3]. A crucial aspect of cellular activity in bone involves remodeling, a coordinated bone maintenance process that simultaneously removes old bone tissue and replaces it to preserve structural integrity and maintain mineral balance [4, 5].

Bone remodeling is an ongoing, dynamic process, persists throughout entire life and particularly prominent once bone reaches its peak mass (usually in the early 20s). Structural changes in the bone's microarchitecture arise from the concurrent activity of osteoclastic resorption and osteoblastic apposition, which occur in response to external loading, showcasing inherent adaptative nature of bone [6]. The remodeling process is generally classified into two phases: internal and external [7]. During internal remodeling, it is assumed that the geometry of the bone remains unchanged while the material properties evolve over time [8]. On the other hand, external remodeling presumes that the material properties, especially in cortical bone, are constant, with temporal changes occurring in the bone shape or geometry. In practical terms, both internal and external remodeling processes occur concurrently [9]. The adaptive remodeling theory suggests that bone tissue needs certain levels of biomechanical stimulation for its maintenance [10]; exceeding this threshold results in new bone formation (apposition), whereas inadequate stimulation leads to bone resorption. The threshold range of mechanical stimulus refers to a lazy or dead zone which is characterized by its remodeling inactivity as an adaptive response. If the remodeling stimulus fall within the range, the bone is presumed not to respond. Lazy zone plays a pivotal role in the remodeling dynamics of trabecular bone, as discussed in previous research by Huiskes et al. [11]. In reality, the lazy or dead zone appears to be more of a mathematical approximation than a strict mechanobiological principle. For instance, the behavior of real bones, as demonstrated by Beaupré et al. [12] shows a range of stimuli where the remodeling response is notably low, but not entirely absent. In their later work, Beaupré et al. [13] modeled this response by defining a dead zone where no net mass change occurs.

Over the past few deacdes, finite element (FE) methods combined with biomechanical models of adaptive bone remodeling have been widely employed to investigate the inherent responses (in form and function) of bones within the skeletal system, particularly in altered biomechanical environments [12–19]. However, the clinical relevance of these biomechanical simulations highlights the necessity for accurate modeling of biological structures and their surrounding physiological environments. Various numerical models and algorithms have been developed to understand and predict bone remodeling processes. Notably, orthotropic models have become instrumental in simulating the anisotropic nature of bone, which exhibits different mechanical properties depending on the direction of loading [17, 20, 21]. Lazy zone has been explored in several studies, with researchers demonstrating that varying the lazy zone threshold can significantly affect bone density and material property distribution. A previous study highlighted that an increased lazy zone threshold resulted in reduced bone resorption around implants, emphasizing the need for careful selection of this parameter in biomechanical simulations.

This pre-clinical in silico study aims to explore the impact of varying lazy zone thresholds on bone remodeling in post-surgical implanted femur model, subjected to different loading conditions. By systematically varying the lazy zone thresholds—from 70 to 10% and even considering scenarios without a lazy threshold—this research seeks to elucidate the relationship between lazy zone size, bone remodeling stimulus, and material property distribution. The rationale behind the selection of the lazy zone thresholds is grounded in a combination of both empirical findings and theoretical considerations, aimed at capturing a comprehensive range of potential remodeling behaviors. Our selection of these thresholds draws on previous research, where varying the lazy zone threshold was shown to significantly influence bone density and material property distribution, particularly in relation to implant stability. Thus, the thresholds are not arbitrary; they are chosen to reflect a broad spectrum of biologically relevant scenarios, addressing both theoretical predictions and clinical observations. This range allows us to observe how bone responds across a gradient of biomechanical loading conditions, offering insights into potential remodeling patterns that could guide implant design and rehabilitation strategies. We sought to explore how such variations influence the stimulus for bone remodeling and its effect on material property distribution.

In the orthopaedic simulation, a Multiple Load Case (MLC) scheme was employed to simulate musculoskeletal loads associated with various daily locomotor activities. These loads were applied successively at each iterative step to a consistent model to capture the activity-specific mechanical responses, particularly in the trabecular region. The load cases included in the MLC scheme, represent a broad range of typical physical activities to approximate the mechanical environments that bones are subjected to in real. However, variations in load magnitude and frequency, particularly in more extreme activity levels or higher-frequency loading, could indeed influence how the lazy zone interacts with bone adaptation. This approach is aimed at enhancing the reproducibility and relevance of the study by capturing a variety of outcomes that could arise from different mechanical environments and load-bearing conditions.

The overarching goal of this research is to investigate the relationship between lazy zone thresholds, bone remodeling stimuli, and material property distribution. By examining these factors, this study aims to contribute valuable insights for optimizing rehabilitation strategies, enhancing implant stability, and improving patient outcomes post-surgery. By using three-dimensional full-scale finite element modeling of a human femur, the study seeks to generate improving clinically relevant predictions that can inform both implant design and post-surgical rehabilitation. The findings are expected to deepen the understanding of the biomechanical environment surrounding hip implants, ultimately contributing to more effective rehabilitation protocols and improved patient outcomes following surgery. Furthermore, this work underscores the significance of tailored rehabilitation strategies in preventing prosthesis failure and enhancing long-term implant stability.

2 Materials and Methods

2.1 Finite Element Modeling

In this computational study, initially a right femur model was employed, which provides a clear distinction between porous trabecular and dense cortical bone. The 3D femur model encapsulates the complex trabecular network within a cortical envelope, allowing for a comprehensive analysis of the bone's structural characteristics and mechanical properties. The exterior cortical shell which is considerably stiffer than the trabecular bone, primarily serves a weight-bearing function and does not contribute to the adaptation process. Following that, FE modelling and analysis were conducted using ANSYS APDL 2019R5 [22]. A key aspect of modelling is the separation of the material model into distinct components. In this context, exterior cortical elements were assigned with constant properties while trabecular elements had variable properties for its complex adaptive nature. This approach not only improves the fidelity of the model but also ensures realistic simulations and more reliable predictions.

The solid model of the right femora (Fig. 1a), originally spans approximately 450 mm, was imported into SolidWorks CAD environment for the purpose of resection and virtual implantation. To optimize computational expensiveness, the distal condyles were removed from the intact model prior to this process. Subsequently, CAD model of a generic hip stem (Fig. 1b) was created in the same solid modelling environment based on specific dimensions and features as detailed in Table 1 [23], for a conventional hip prosthesis. In order to simulate an implanted hip following surgery, the solid model was further osteotomized at the neck above lesser trochanter and the stem was virtually implanted inside the femoral canal. Pre-operative planning parameters (resection level, angle of resection, canal preparation and stem positioning) were determined in accordance with the established protocols for hip reconstruction. In this manner, two distinct models were developed for pre- and post-operative bone remodeling simulations and then exported to ANSYS APDL for subsequent analysis. Both of the models were meshed (Fig. 1a, c) utilizing a 3D 10-noded tetrahedral structural element SOLID 187 (defined in ANSYS element reference). Higher-order tetrahedral elements with quadratic displacement behavior were chosen due to their improved convergence rate and superior simulation outcomes [24, 25]. Additionally, these elements are particularly efficacious at approximating complex morphological geometries, such as the femur, making them well-suited for this type of analysis [25, 26].

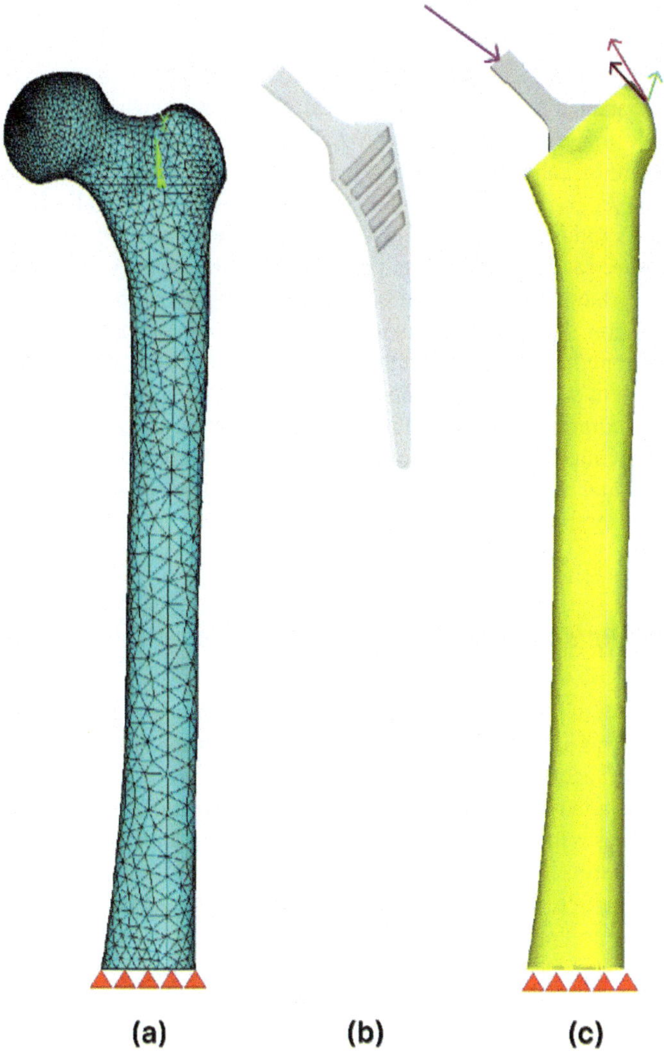

Fig. 1 **a** Intact femur model **b** hip stem **c** implanted femur post-THA

Table 1 Stem dimensions

Size: 10 standard offset	
Stem length (lateral)	140 mm
Stem length (medial)	118 mm
Offset with head	39.5 mm
Neck length with head	39 mm
Neck shaft angle	135°
Distal width	8 mm

2.2 *Material Properties*

In the subsequent FE analysis, the cortical bone was assigned constant isotropic linear elastic material properties with $E_{corticalbone} = 16.7\,\text{GPa}$ [25]. Initially, the inner structure of trabecular bone was modeled as homogeneous and isotropic with $E_{trabecular} = E_1 = E_2 = E_3 = 500\,\text{MPa}$ across all of the trabecular elements. The Poisson's ratios for both the isotropic cortical bone and the orthotropic trabecular bone were set at 0.3 [27]. The properties of Titanium 6AL4V (medical grade titanium alloy having $E_{stem} = 113.8\,\text{GPa}$), were assigned to the prosthesis for post-THA simulation. Selection of this alloy is attributed to its excellent biocompatibility and strength, making it suitable for use in surgical implants. In addition to that, the element-wise material properties for simulating an implanted femur immediately after surgery were straightaway borrowed from a converged intact femur model. Retrieving element-specific properties from the corresponding elements of a converged model provides a reliable basis for accurately simulating the immediate post-surgical condition.

2.3 *Loading and Boundary Conditions*

In both the cases of pre- and post-surgery orthopaedic simulation, MLC (multiple load case) scheme was used. Musculoskeletal loads representing different daily locomotor activities were successively applied in each iterative step one at a time to a consistent model with an aim to capture the activity-specific mechanical responses at trabecular region. Load cases in the multiple loading scheme were sourced from a priori study by Doblaré and García [22] and comprehensively outlined in Table 2. The orientations of these loads are referenced from standard anatomical planes. Muscle forces associated with femur, were uniformly applied across the set of exterior nodes of the greater trochanter corresponding to the areas of muscle attachment [28]. In the simulation of a post-THA hip joint, joint reaction force was applied at the centre of the femoral head. For both the pre- and post-surgical contexts, surgically resected femur was enforced with a zero-displacement constraint at distal diaphyseal end.

Table 2 Load cases applied to 3D model [17]

Load case	Hip joint force			Abductor muscle reaction		
	Magnitude	Orientation (in degrees)		Magnitude	Orientation (in degrees)	
	N	FP	SP	N	FP	SP
1	2317	24	6	703	28	15
2	1158	−15	35	351	−8	9
3	1548	56	−20	468	35	16

2.4 *Mathematical Modeling of Orthotropic Adaptation of Trabecular Bone*

In this computational study bone remodeling is represented as an iterative reorganisation process of trabecular bone. The biomechanical simulation involves two consecutive fundamental stages in each iteration: the realignment of local axis with changing stress directions and the revision of elastic moduli based on estimated directional stimuli. In each iteration, FEA was performed using MLC schemes and the load responses were obtained for consecutive load cases. Thereafter, the state of stress $[\sigma]_{(i,j)}$ was determined for each element across the entire domain subjected to successive loading, with specific elements identified by suffix i ($i = 1, 2, 3 \ldots i_{\max}$) and distinctive load cases denoted by another suffix j ($j = 1, 2, 3, \ldots j_{\max}$).

$$[\sigma]_{(i,j)} = \begin{bmatrix} \tau_{xx(i,j)} & \tau_{xy(i,j)} & \tau_{xz(i,j)} \\ \tau_{xy(i,j)} & \tau_{yy(i,j)} & \tau_{yz(i,j)} \\ \tau_{xz(i,j)} & \tau_{yz(i,j)} & \tau_{zz(i,j)} \end{bmatrix} \quad (1)$$

The principal stresses and their principal directions were derived for each element across different load cases. Considering that the elastic behavior of bone tissue subjected to compression and tension yield similar patterns [29], the largest magnitude of stress, regardless of its sign (tensile or compressive) is determined across all elements. Thereafter, under the successive action of MLC maximum absolute principal stress ($\sigma_{(i,p)}^{\max}$) is identified and the corresponding principal direction is denoted as the dominant load direction is designated by p that developed maximum stress magnitude follows:

$$\sigma_{(i,p)}^{\max \max\left(\left|\sigma_{1(i,j)}\right|,\left|\sigma_{2(i,j)}\right|,\left|\sigma_{3(i,j)}\right|\right)} \quad (2)$$

Under the premise of Wolff's Law, which posits that the bone architecture is dynamic in nature and capable of optimizing itself to withstand the maximum value of stress it experiences during varied physiological loading [30, 31], this functional adaptation includes realignment of the element material axes to the direction of the maximum principal stress determined from all loading scenario. Once the target orientations have been identified and local axes realigned accordingly, the next important step is to determine the orthotropic elastic moduli in three mutually perpendicular material directions. The second stage of the proposed remodeling algorithm incorporates the widely recognized mechanostat hypothesis. Its objective is to reinforce stiffness in the principal axis corresponding to predominant load case while reducing stiffness values in the other mutually perpendicular directions.

Instead of depending on a single scalar to represent mechanical stimulus, the approach involves assessing multi-directional stimuli for each element along the re-oriented material axes. This approach is capable of capturing the complexity of physiological loading environment, allowing for a more nuanced understanding of element-specific response and facilitating more precise adaptations based on the specific loading conditions encountered. The directional stimuli outlined here resembles the bone maintenance stimulus often found in isotropic bone remodelling [12]; guiding the trabecular bone material adaptation based according to MLC scheme. Traditional bone adaptation models might use an averaged strain energy density as a scalar variable to guide remodeling, but this approach acknowledges the complexity of stress distributions. By considering normal stresses in various material directions, it allows for a more accurate representation of how the material responds to different loading conditions. The formulation focuses on the mechanical stresses acting along transformed/modified orientation and also considers frequency of certain physiological activity and its contribution to the overall loading scheme. Elemental stresses in transformed local axes were calculated for all the loadings within the MLC scheme using stress transformation techniques follows as:

$$\sigma'_{(i,j)} = \lambda^T \sigma_{(i,j)} \lambda \tag{3}$$

Here, the term λ refers to DC (direction-cosine) matrix of pre-dominant principal stress directions, as outlined in Eq. 3. The formulation of directional bone remodelling stimulus $\varphi_{k(i)}$ in local material direction ($k = 1, 2, 3$), corresponding to the ith trabecular element undergoing successive loading during each iteration is expressed as:

$$\varphi_{k(i)} = \frac{1}{E_{k(i)}} \left[\sum_j N_j \left(\sigma'_{k(i,j)} \right)^{\Psi} \right]^{\frac{1}{\Psi}} \tag{4}$$

In this expression, E_k is Young's moduli and σ'_k designates the normal stress in transformed material direction (1–2–3). Meanwhile, the number of average daily cycles against each loading within MLC scheme is represented by N_j and the exponent Ψ is an empirical constant. Thus, daily mechanical stimuli for individual elements is defined in a way that accounts for both the magnitude of stress and frequency of each load cases. Furthermore, division by axial stiffness E_k adjusts the formulation closer a strain-based stimulus.

The study of bone adaptation in a 3D full-scale femur, particularly in the context of orthotropic materials necessitates detailed information on elastic constants along with axes orientation leading to greater computational complexity and expensiveness. An orthotropic material is defined by its unique properties in three mutually perpendicular directions, requires nine independent engineering constants $E_1, E_2, E_3, \nu_{12}, \nu_{23}, \nu_{31}, G_{12}, G_{23}, G_{13}$. Modifying elastic properties in orthotropic directions reflects the rate of apposition or resorption of bone, which is influenced by the local mechanical environment and various factors regulating bone remodeling. This process ensures that bone can adapt to meet both its structural requirements and metabolic functions, allowing it to maintain strength and integrity while responding to changes in load and stress. At the end of each iterative step, for every element orthotropic elastic modulus E_1, E_2, E_3 are modified in accordance with multi-linear remodelling rate equations which have introduced a lazy zone nonlinearity in the bone remodeling algorithm (refer Eq. 5).

$$\Delta E_{k(i)} = \begin{cases} c_f[\varphi_{k(i)} - \varphi_0(1+s)] \; if\, \varphi_{k(i)} > \varphi_0(1+s) \\ 0 \qquad\qquad\qquad\qquad\quad if\, \varphi_0(1-s) < \varphi_{k(i)} > \varphi_0(1+s) \\ c_r[\varphi_{k(i)} - \varphi_0(1-s)] \; if\, \varphi_{k(i)} < \varphi_0(1-s) \end{cases} \tag{5}$$

where c_f, c_r are the empirical constants that quantify rates of site-specific bone formation and resorption, respectively. φ_0 indicates a defined mechanical stimulus that is used as a benchmark or reference point for remodeling activities and guides the overall remodeling process based on mechanical demands whether bone formation or apposition should occur. Lazy zone or dead zone (s), a very common threshold concept exercised in many bone remodeling algorithms, represents a specific range of mechanical stimuli where bone does not exhibit any adaptive response (refer Fig. 2) [13, 32, 33] and when the remodeling stimulus exceeds the boundaries of the lazy zone, the bone tissue starts to adapt its intricate network to ensure the strain level remains within designated upper and lower limit [10, 34].

With the assumption that Poisson's ratios remain consistent throughout, the shear modulus has been estimated from the axial stiffness values in each iterative cycle as follows [35]:

$$G_{12} = \frac{(E_1 + E_2)}{4(1 + \nu_{12})} \tag{6}$$

Computationally emerged elastic moduli in due course of iteration might result in nonpositive strain energy for certain load conditions or yield impractical material properties, which is thermodynamically not admissible. In order to obviate this impermissible situation, a necessary condition was imposed to the algorithm to ensure positive definiteness of both the stiffness and compliance matrix [22, 36, 37]. Thus, it is ensured that the Poisson's ratios of all the ill-conditioned elements adhere to the following relationship:

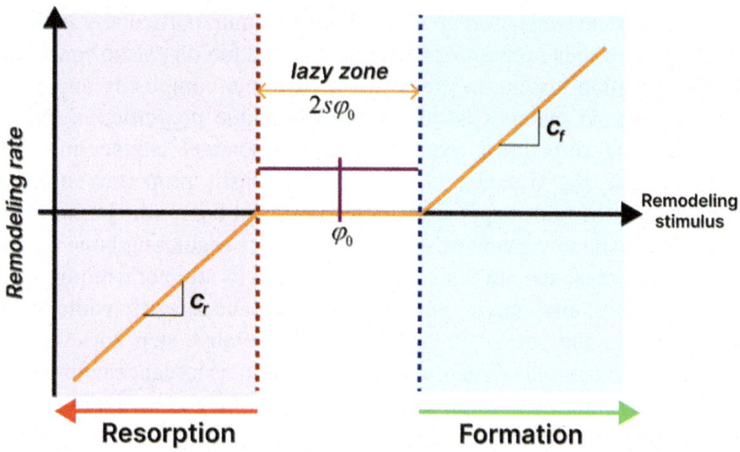

Fig. 2 Remodeling rate relationship with lazy zone used in the adaptive bone remodeling simulation

$$\frac{E_{1(i)}}{E_{2(i)}} > \nu_{12(i)}^2 \tag{7}$$

The estimated orthotropic elastic moduli after trabecular re-orientations, were constrained to a minimum of 20 MPa and an upper limit of 5 GPa (maximum permissible for trabecular bone), ensuring that the material behavior remains realistic and in line with known physiological limits [27]. The remodeling algorithm reached convergence when the average changes in orthotropic moduli were minimal (dropped below 1% of the overall averaged elastic moduli) across subsequent iterations. The mechanobiological simulation consisted of FE analysis using a MLC scheme, followed by implementation of bone remodeling algorithm. The remodeling subroutine was developed and executed within a computational framework using the MATLAB environment. At the conclusion of each simulation cycle, the updated elastic properties were estimated and the process was repeated until convergence was achieved.

3 Results

The primary objective of the study was to investigate the influence of the various lazy zones on the peri-prosthetic bone remodeling following a hip reconstruction surgery. To address this, different remodeling rate relationships incorporating varying sizes of the lazy zone have been explored. The study examines eight specific thresholds for lazy zones set at $s = 0.7, 0.6, 0.5, 0.4, 0.3, 0.2$, and 0.1, in addition to a condition with no lazy zone at all. This computational study specifically analyzes how changes in lazy zone size (different levels of remodeling inactivity) impact the estimation of bone material properties in the implanted femur. This comprehensive approach ultimately aims to improve understanding of bone adaptation around implants and enhance surgical outcomes. In our study, the driving stimuli for the remodeling algorithm were estimated at every iteration using constants from a well-established formulation, $\varphi_0 = 0.01, c_f = c_r = 5\,\text{GPa}, \Psi = 4$ as outlined in the prior work by Miller et al. [22]. This formulation has been widely validated in the literature, and the constants it uses have been extensively applied in similar biomechanical simulations of bone remodeling. We focused on these constants to ensure consistency with previous models and facilitate comparability with existing studies. A cross-sectional view of the femur was employed to study the internal distribution of stress, strain, and density resulting from the implantation.

Sectional views parallel to the femoral axis of the implanted femora were used to encompass the property distribution (elastic modulus, density etc.) in the trabecular and cortical regions to provide a clearer perspective on bone adaptation surrounding the hip prosthesis post-total hip arthroplasty. Figure 3 illustrates the spatial variation in axial stiffness across trabecular region within the cortical outline, derived from the remodeling study. The areas of increased stiffness are predominantly located in regions that experience higher physiological loading, such as the femoral neck and trochanteric region, while less-stressed areas show reduced stiffness values. The top left coronal view in Fig. 3 serves as a foundational baseline for comparison with the post-operative femur bone remodelling models, particularly concerning the varying sizes of the lazy zone. Marginal changes were observed in the spatial distribution of material properties with lazy zones set at $s = 0$, 0.1, 0.2 and 0.3 as compared to zones of s = 0.4, 0.5, 0.6, 0.7. It is evident that, with the increase of size of lazy zone appositional bone growth in lateral sides of the hip stem in the proximal metaphysis decreases. This reduction is attributed to a rise in cellular remodeling inactivity with the progressive increase in lazy zone. A gradual reduction in stiffness at distal femoral shaft was also evident. Aside from these, no substantial differences were observed. The coronal section was additionally subdivided into seven Gruen zones (GZ) to facilitate a detailed analysis of the spatial distribution, especially around the implant [38]. The gradual expansion of lazy zone led to significant redistribution of material properties, particularly observed in Gruen zones 1 and 7, indicating notable changes in modulus in these areas of proximal anchorage. Furthermore, substantial alterations were also evident around the distal tip, especially in GZ 4. In addition to these findings, no significant differences were observed in the remaining Gruen zones.

To evaluate the distribution of effective material properties around the implant, six transverse sectional views (A-F) were created based on the FEA results (see Fig. 4). The sections E and F, positioned at the distal end of the implant, exhibited significant similarities for all lazy zone sizes, with only slight variations near the stem tip for with lazy zone sizes $s = 0.7$, 0.6 and 0.5. The observed patterns in predicted modulus distribution across transverse section of proximal metaphyseal region (A-C) in the post-operative scenario, reported that the decrease in bone apposition with an increase in lazy zone, particularly on the medial and anterior aspects of the hip stem. Negligible changes were observed in section D, particularly in the mediolateral

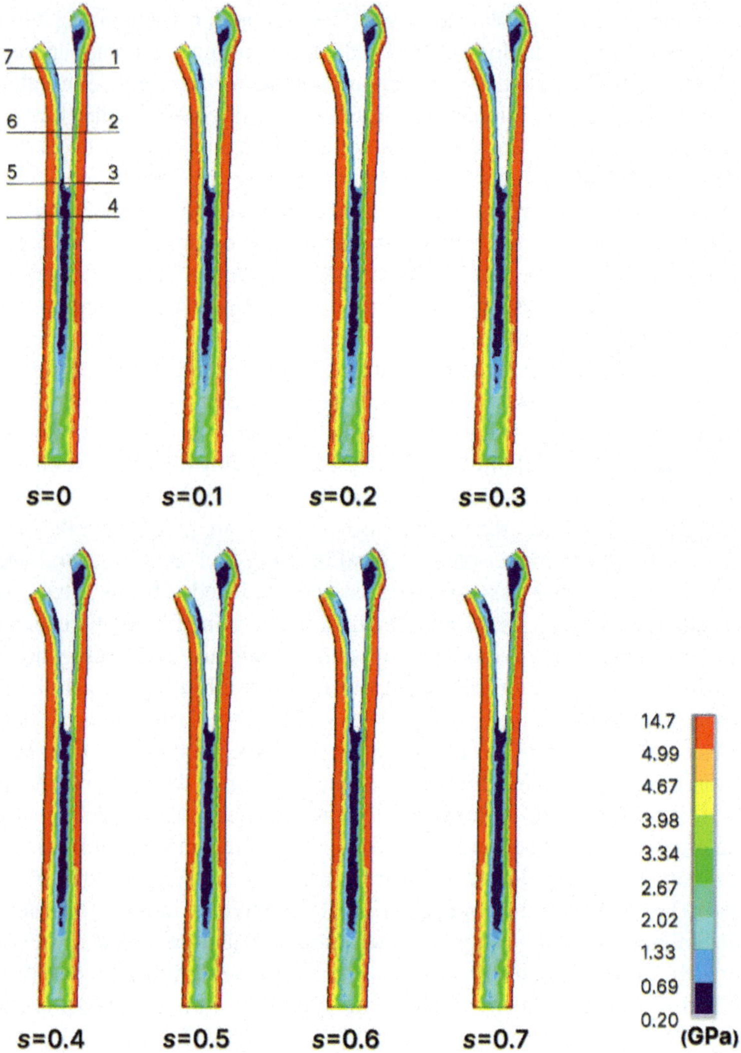

Fig. 3 Average axial young's modulus of post-operative femur at coronal section for eight different values of lazy zone

aspect. Distribution of material property in these transverse sections along the length of a hip prosthesis provides insights into bone strength and highlight potential areas of resorption or remodeling adjacent to the implant.

Subsequently, the apparent density of each trabecular element was estimated using the average of directional moduli and empirical power law as following $E_{rep} = 1904\rho^{1.64}$. The empirical relationship between Young's modulus and density was taken from a previous study [39]. Here, E_{rep} (representative Young's moduli) was

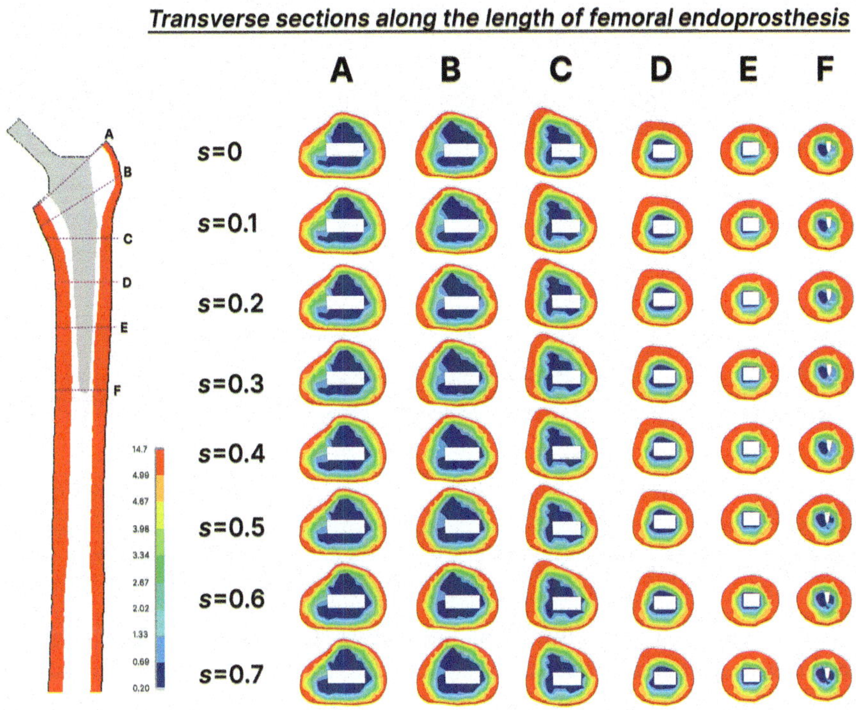

Fig. 4 Variations in young's modulus of the femur post-hip replacement across transverse sections

obtained as the arithmetic mean of the modulus values along orthogonal axes. Figure 5 illustrates the temporal evolution of averaged trabecular bone density for various *s* values.

4 Discussion

A crucial aspect of orthopaedic remodeling simulations is the inclusion of the lazy zone which has been highlighted in many recent computational works utilizing FE analysis. Trabecular elements remain inactive in the remodeling process within this specific boundary. A prior study by Huiskes et al. indicated that the effects of the lazy zone vary in the femur bone adaptation [11]. Our analysis focuses on the descriptive aspects of these variations rather than quantitative measurements, while Fig. 3 showing clear patterns in trabecular stiffness variations around the implant, suggesting a marginal influence on orthotropic bone remodeling. When comparing different size of lazy zones, differences in averaged density changes were found in the femoral trabecular bone due to remodeling. However, this study suggests that the effect of varying lazy zones on density changes in the femoral cancellous bone

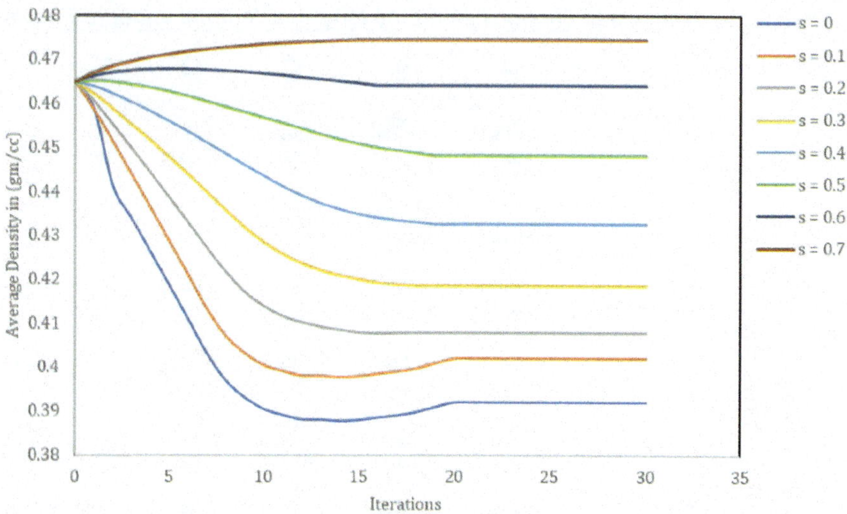

Fig. 5 Iteration-wise changes in average bone density for different values of lazy zone

is significant, albeit qualitative. While the current study provides a detailed examination of bone remodeling over iterative cycles, we acknowledge that translating these findings into patient-specific clinical insights, particularly for rehabilitation planning, is an important next step.

This study also offers significant potential for advancing the development and control of rehabilitation robots, especially in the context of post-surgical rehabilitation. By incorporating insights from bone remodeling dynamics and the effects of lazy zones, rehabilitation robots could be fine-tuned to provide more targeted and effective treatment. For instance, robots could adjust the intensity and nature of therapeutic movements based on the patient's bone remodeling progress and unique healing profile. This tailored approach could help prevent complications like excessive strain on under-remodeled bone areas, thereby supporting better recovery and promoting optimal bone healing. Additionally, rehabilitation robots that are responsive to the remodeling process could offer real-time feedback to both patients and healthcare providers, allowing for more adaptable and individualized rehabilitation plans. In the long run, this research could help pave the way for the creation of more sophisticated rehabilitation systems that are closely aligned with the biological healing process, leading to improved recovery outcomes for patients following implant surgery. In summary, while the current model provides valuable insights into bone remodeling dynamics, we recognize the importance of incorporating patient-specific variations and translating these results into long-term clinical outcomes. This remains a key area for future research, and we are committed to exploring these aspects in subsequent studies.

5 Conclusion

The current study inferred that the impact of increasing size of lazy zone on trabecular bone adaptation applied to post-THA human femur is significant. Despite that conducting analysis across diverse group of subjects considering patient factors such as age, medical history and sex, is crucial before drawing more precise conclusions regarding the effects of varying lazy zones on adaptive bone remodeling to ensure that findings are applicable to a broader population. The actual biomechanical response of bone tissue is highly complicated and requires additional elaboration beyond remodeling models. For a more comprehensive evaluation a combined study of mechano-regulation based tissue differentiation and osseointegration would be advantageous along with the current computational framework.

The implications of these findings for the design of in-vivo implants or prosthetics are considerable. Understanding the role of lazy zones in bone adaptation can help optimize implant designs that better promote osseointegration and minimize the risk of complications such as implant loosening or stress shielding. Future studies incorporating these factors could lead to more personalized and effective implant strategies, ultimately improving long-term outcomes for patients undergoing THA or similar surgical procedures.

6 Future Works

At this stage, our model primarily focuses on the biomechanical processes involved in remodeling around implants, with an emphasis on understanding how these processes evolve over time under various loading conditions. The current model does not yet account for the full range of patient-specific factors such as varying bone quality, patient activity levels, and biological healing rates, which can influence long-term outcomes. The "lazy zones" mentioned are important, and further investigations will be necessary to explore how these regions might affect implant stability and bone health in the long run. As a future direction, we plan to refine the model to incorporate patient-specific parameters such as bone density and remodeling rates, as well as the effects of different rehabilitation protocols. By doing so, we aim to provide a more personalized approach to patient care, which could better inform rehabilitation strategies.

Additionally, collaboration with clinical studies will help integrate real-world data on patient recovery and implant success, thus enabling the model to offer more practical insights for clinical application. Although the lack of direct comparison with such data is a limitation in this study, the current model still offers valuable insights into the key mechanical and material behaviours of femur-implant interactions. Furthermore, while we recognize the critical role that constants from prior studies play in the accuracy of bone adaptation predictions, our current research has focused more on examining broader biomechanical factors, such as loading conditions and lazy zone thresholds. In the future, a sensitivity analysis is to be included to assess how changes in these constants influence the remodeling outcomes.

References

1. Taichman, R.S.: Blood and bone: two tissues whose fates are intertwined to create the hematopoietic stem-cell niche. Blood **105**, 2631–2639 (2005). https://doi.org/10.1182/blood-2004-06-2480
2. Clarke, B.: Normal bone anatomy and physiology. Clin. J. Am. Soc. Nephrol. **3**(Suppl 3), S131–9 (2008). https://doi.org/10.2215/CJN.04151206
3. Seeman, E.: Periosteal bone formation—a neglected determinant of bone strength. N. Engl. J. Med. **349**, 320–323 (2003). https://doi.org/10.1056/NEJMp038101
4. Hauge, E.M., Qvesel, D., Eriksen, E.F., Mosekilde, L., Melsen, F.: Cancellous bone remodeling occurs in specialized compartments lined by cells expressing osteoblastic markers. J. Bone Miner. Res. Off. J. Am. Soc. Bone Miner. Res. **16**, 1575–1582 (2001). https://doi.org/10.1359/jbmr.2001.16.9.1575
5. Parfitt, A.M.: The bone remodeling compartment: a circulatory function for bone lining cells. J. Bone Miner. Res. Off. J. Am. Soc. Bone Miner. Res. **16**, 1583–1585 (2001). https://doi.org/10.1359/jbmr.2001.16.9.1583
6. Burr, D.B.: Targeted and nontargeted remodeling. Bone **30**, 2–4 (2002). https://doi.org/10.1016/s8756-3282(01)00619-6
7. Vaz, M.F., Canhão, H., Fonseca, J.E.: Bone: a composite natural material. In: Těšinova, P. (ed.) Advanced Composite Materials. IntechOpen, Rijeka (2011). https://doi.org/10.5772/17523
8. Wei, X., Cooper, D.M.L.: The various meanings and uses of bone "remodeling" in biological anthropology: a review. Am. J. Biol. Anthropol. **182**, 318–329 (2023). https://doi.org/10.1002/ajpa.24825
9. Goda, I., Ganghoffer, J.-F., Maurice, G.: Combined bone internal and external remodeling based on Eshelby stress. Int. J. Solids Struct. **94–95**, 138–157 (2016). https://doi.org/10.1016/j.ijsolstr.2016.04.036
10. Frost, H.M.: Bone's mechanostat: a 2003 update. Anat. Rec. Part A Discov. Mol. Cell. Evol. Biol. **275**, 1081–1101 (2003). https://doi.org/10.1002/ar.a.10119
11. Huiskes, R., Weinans, H., Van Rietbergen, B.: The relationship between stress shielding and bone resorption around total hip stems and the effects of flexible materials. Clin. Orthop. Relat. Res. 124–134 (1992). https://doi.org/10.1097/00003086-199201000-00014
12. Beaupre, G.S., Orr, T.E.: An approach for time-dependent bone modeling and remodeling—application: a preliminary remodeling simulation—Beaupré—2005. J. Orthopaed. Res. (Wiley Online Library) 662–670 (1990)
13. Beaupre, G.S., Orr, T.E.: An approach for time-dependent bone modeling and remodeling—theoretical development—Beaupré—2005. J. Orthopaed. Res. (Wiley Online Library) 651–661 (1990). http://onlinelibrary.wiley.com/doi/10.1002/jor.1100080506/abstract%5Cnpapers2://publication/uuid/945494C3-62EC-49A9-849A-91728212793D.
14. Jacobs, C.R., Levenston, M.E., Beaupré, G.S., Simo, J.C., Carter, D.R.: Numerical instabilities in bone remodeling simulations: the advantages of a node-based finite element approach. J. Biomech. **28** (1995). https://doi.org/10.1016/0021-9290(94)00087-K
15. Huiskes, R., Weinans, H., Grootenboer, H.J., Dalstra, M., Fudala, B., Slooff, T.J.: Adaptive bone-remodeling theory applied to prosthetic-design analysis. J. Biomech. **20**, 1135–1150 (1987). https://doi.org/10.1016/0021-9290(87)90030-3
16. Fernandes, P., Guedes, J.M., Rodrigues, H.: Topology optimization of three-dimensional linear elastic structures with a constraint on "perimeter". Comput. Struct. **73**, 583–594 (1999). https://doi.org/10.1016/S0045-7949(98)00312-5

17. Doblaré, M., García, J.M.: Application of an anisotropic bone-remodelling model based on a damage-repair theory to the analysis of the proximal femur before and after total hip replacement. J. Biomech. **34**, 1157–1170 (2001). https://doi.org/10.1016/S0021-9290(01)00069-0
18. Tsubota, K., Suzuki, Y., Yamada, T., Hojo, M., Makinouchi, A., Adachi, T.: Computer simulation of trabecular remodeling in human proximal femur using large-scale voxel FE models: approach to understanding Wolff's law. J. Biomech. **42**, 1088–1094 (2009). https://doi.org/10.1016/j.jbiomech.2009.02.030
19. Shefelbine, S.J., Augat, P., Claes, L., Simon, U.: Trabecular bone fracture healing simulation with finite element analysis and fuzzy logic. J. Biomech. **38**, 2440–2450 (2005). https://doi.org/10.1016/j.jbiomech.2004.10.019
20. Cowin, S.C.: Bone poroelasticity. In: Bone Mechanics Handbook, 2nd ed., vol. 32, pp. 23-1–23-31 (2001)
21. Guo, X.E., Jee, W.S.S.: Bone Mechanics Handbook, 2nd ed. CRC Press (2001)
22. Miller, Z., Fuchs, M.B., Arcan, M.: Trabecular bone adaptation with an orthotropic material model. J. Biomech. **35**, 247–256 (2002). https://doi.org/10.1016/S0021-9290(01)00192-0
23. Hazra, A., Ghosh, R., Chanda, S., Chakraborty, D.: In silico assessment of connective tissue formation on proximal surfaces of femoral stem with different implant materials. In: Kumari, P., Dwivedy, S.K. (eds.) Recent Advances in Mechanics of Functional Materials and Structure, pp. 575–586. Springer Nature Singapore, Singapore (2024)
24. Viceconti, M., Bellingeri, L., Cristofolini, L., Toni, A.: A comparative study on different methods of automatic mesh generation of human femurs. Med. Eng. Phys. **20**, 1–10 (1998). https://doi.org/10.1016/S1350-4533(97)00049-0
25. Ghosh, R., Hazra, A., Chanda, S., Chakraborty, D.: Computational assessment of growth of connective tissues around textured hip stem subjected to daily activities after THA. Med. Biol. Eng. Comput. **61**, 525–540 (2023). https://doi.org/10.1007/s11517-022-02729-3
26. Fraldi, M., Esposito, L., Perrella, G., Cutolo, A., Cowin, S.C.: Topological optimization in hip prosthesis design. Biomech. Model. Mechanobiol. **9**, 389–402 (2010). https://doi.org/10.1007/s10237-009-0183-0
27. Bazyar, P., Baumgart, A., Altenbach, H., Usbeck, A.: An overview of selected material properties in finite element modeling of the human femur. Biomechanics **3**, 124–135 (2023). https://doi.org/10.3390/biomechanics3010012
28. Duda, G.N., Brand, D., Freitag, S., Lierse, W., Schneider, E.: Variability of femoral muscle attachments. J. Biomech. **29**, 1185–1190 (1996). https://doi.org/10.1016/0021-9290(96)00025-5
29. Røhl, L., Larsen, E., Linde, F., Odgaard, A., Jørgensen, J.: Tensile and compressive properties of cancellous bone. J. Biomech. **24**, 1143–1149 (1991). https://doi.org/10.1016/0021-9290(91)90006-9
30. Hayes, W., Snyder, B.: Toward a quantitative formulation of Wolff's Law in trabecular bone. Am. Soc. Mech. Eng. Appl. Mech. Div. AMD **45** (1981)
31. Wolff, J.: The Law of Bone Remodelling (1986). https://doi.org/10.2106/00004623-198769030-00026
32. Weinans, H., Huiskes, R., Grootenboer, H.J.: The behavior of adaptive bone-remodeling simulation models. J. Biomech. **25**, 1425–1441 (1992). https://doi.org/10.1016/0021-9290(92)90056-7
33. Carter, D.R.: Mechanical loading histories and cortical bone remodeling. Calcif. Tissue Int. **36** (1984). https://doi.org/10.1007/BF02406129
34. Turner, C.H., Anne, V., Pidaparti, R.M.: A uniform strain criterion for trabecular bone adaptation: do continuum-level strain gradients drive adaptation? J. Biomech. **30**, 555–563 (1997). https://doi.org/10.1016/s0021-9290(97)84505-8
35. Ford, C.M., Keaveny, T.M.: The dependence of shear failure properties of trabecular bone on apparent density and trabecular orientation. J. Biomech. **29**, 1309–1317 (1996). https://doi.org/10.1016/0021-9290(96)00062-0

36. Lempriere, B.M.: Poisson's ratio in orthotropic materials. AIAA J. **6**, 2226–2227 (1968). https://doi.org/10.2514/3.4974
37. Cowin, S.C., Van Buskirk, W.C.: Thermodynamic restrictions on the elastic constants of bone. J. Biomech. **19**, 85–87 (1986). https://doi.org/10.1016/0021-9290(86)90112-0
38. Gruen, T.A., McNeice, G.M., Amstutz, H.C.: "Modes of failure" of cemented stem-type femoral components: a radiographic analysis of loosening. Clin. Orthop. Relat. Res. 17–27 (1979)
39. Wirtz, D.C., Schiffers, N., Forst, R., Pandorf, T., Weichert, D., Radermacher, K.: Critical evaluation of known bone material properties to realize anisotropic FE-simulation of the proximal femur. J. Biomech. **33**, 1325–1330 (2000). https://doi.org/10.1016/S0021-9290(00)00069-5

Ethical and Regulatory Considerations in Rehabilitation Robotics: Ensuring Safe and Equitable Care

Azamat Ali

Abstract The integration of robotics into rehabilitation therapies heralds promising advancements in patient care. Yet, it necessitates rigorous ethical scrutiny and adherence to regulatory frameworks to ensure safe and equitable outcomes. This chapter explores the multifaceted landscape of ethical and regulatory considerations in rehabilitation robotics. It emphasizes the imperative of ethical principles and compliance with global regulatory standards. Beginning with an overview of the ethical foundations underpinning healthcare technologies, it delves into the intricate web of regulatory requirements governing medical devices and robotics. This includes spotlighting key frameworks such as FDA guidelines and CE marking. Ethical issues intrinsic to robotic rehabilitation are also examined. These include patient autonomy, consent, and beneficence. Alongside these ethical considerations, robust safety standards and risk management protocols are discussed. These protocols are essential for mitigating potential harms. The chapter also addresses privacy concerns and strategies for safeguarding patient data. This is done in compliance with stringent privacy regulations like HIPAA and GDPR. Moreover, it underscores the importance of ensuring equitable access to rehabilitation robotics. This involves advocating for inclusive practices that transcend socioeconomic barriers. Through insightful case studies and best practices, this chapter illuminates pivotal ethical dilemmas and successful approaches. These are in navigating the evolving ethical and regulatory landscape of rehabilitation robotics. The chapter offers guidance for stakeholders committed to advancing safe and ethical care paradigms.

Keywords Rehabilitation robotics · Ethical principles · Regulatory compliance · Patient safety · Equitable access

---

A. Ali (✉)
School of Liberal and Creative Arts (Journalism and Mass Communication), Lovely Professional University, Phagwara, Punjab, India
e-mail: azamatcug@gmail.com

J. Narayan and H. Gritli (eds.), *Design and Control of Rehabilitation Robots*, Studies in Systems, Decision and Control 585, https://doi.org/10.1007/978-3-031-86977-8_17

1 Introduction

There are appropriate ethical and regulatory frameworks that govern the responsible development and use of healthcare technologies. As healthcare innovations incorporate AI, machine learning, and robotics into patient care, the frameworks avoid losing touch with the very foundation of ethical principles such as respect for persons, doing good, not doing harm, autonomy, and justice. They protect the rights of patients while encouraging positive health outcomes [1]. The technology that is used in health care should thus be beneficial, equitable, and secure. The technological advancement of rehabilitation robots, which is a game-changing device for recovery, has brought forth new issues, notably in terms of ethics and legislation. Ethics is at the heart of every health technology. Patient safety, autonomy, and equity are fundamental [2].

The development and application of all healthcare technology, including rehabilitation robots, ought to be guided by the ethical principles of beneficence, which means doing good, and non-maleficence, which means avoiding damage [3]. According to Beauchamp and Childress (2019), the technology that helps patients recover their mobility and autonomy after severe injury or illness may end up worsening the patient's case if the overall process adopted in developing them cannot help but do some patients more or greater harm than benefit [1]. In addition to ethical considerations, regulation provides that medical devices such as rehabilitation robots are at the highest standards of safety and effectiveness. Such regulatory bodies as the U.S. Food and Drug Administration (FDA) and the European Medicines Agency (EMA) are critical. They ensure that such technologies are safe for use with patients. These agencies ensure compliance with strict guidelines. These govern everything from the initial testing of devices to post-market surveillance [4]. Regulations are particularly important in rehabilitation robotics. These devices interact directly with patients, affecting their physical well-being, making it crucial to minimise risks.

Rehabilitation robots are defined as the robotic systems for the recovery of motor functions. They are used for patients with neurologic weaknesses, spinal cord injuries, or strokes [5]. These systems are being increasingly integrated into clinical physiotherapy. They offer the repetition and motion repetitions needed in the rehabilitation process from neurological diseases. Unlike conventional therapy, robotic systems administer homogenous and intensified treatment. Such can improve quicker recovery and sustainable positive outcomes over time [6]. For instance, gait trainers and robotic exoskeletons assist in reviving the lost ability of stroke survivors to walk. They allow helping the patient do movements that he cannot achieve without the support.

The inclusion of robotics into rehabilitation treatment offers medical benefits and ethical concerns. A major concern is patient autonomy and informed consent. Often, the robot imposes or prescribes specific movement patterns or exercises. The patients may feel a loss of control over the rehabilitation process [7]. The necessity should be communicated clearly to the patients on the functioning of the devices. Patients should also be given the autonomy to give their consent for it. Maintaining autonomy promotes integrity in the care of the patient.

Another ethical issue is the fairness of rehabilitation robotics distribution. Rehabilitation robots are quite expensive. This, in turn, limits access to patients who are part of lower socioeconomic groups or are residing in remote locations that do not have advanced healthcare facilities [8]. For the same reason, there are significant issues regarding health equality. These are patients who could use rehabilitation robotics the most. However, they cannot access them due to financial or geographical restrictions. That calls for policies of access to such cutting-edge technologies. This will therefore ensure that each patient of whatever origin and kind has access to the latest rehabilitation therapy. The regulatory frameworks also play a very significant role in that the FDA and the EMA require medical devices, including rehabilitation robots, to undergo stringent testing [9]. These tests demonstrate their safety and efficacy before they can be marketed and used in clinical settings.

Regulators have to ensure that unnecessary risks to the patient are not posed. They also have to ensure that well-documented therapeutic benefits are there. This is very important for public trust in healthcare technologies. Patients must be assured that the technological devices used in their care have the ultimate safety standards that may be desired. Moreover, regulatory frameworks are essential for protecting patient privacy [10]. Rehabilitation robots are increasingly integrated with data-driven technologies. Many of these systems collect and analyse patient data for optimising therapy. This does raise the question over the safety of such data and its privacy issues. Rules like the Health Insurance Portability and Accountability Act in America and the General Data Protection Regulation in Europe hold stringent definitions. The patient data is thus protected by these definitions [11]. These rules must be followed. It protects patient information and guarantees that healthcare technologies don't compromise the privacy of the patient.

In this backdrop, the chapter is attempting to analyse a comprehensive overview of the ethical and regulatory frameworks in rehabilitation robotics. The chapter is divided into nine sections. It begins with an introduction to the importance of ethics and regulations in healthcare, specifically for robotics used in patient care. Section 2 reviews global standards and regulatory bodies, such as the FDA and CE, which govern medical robotics. Section 3 covers ethical issues principles like autonomy, consent, and beneficence, highlighting unique ethical considerations for robotic therapies. Section 4 focuses on safety standards protocols and risk management, while Sect. 5 discusses privacy and data security compliance with the Health Insurance Portability and Accountability Act (HIPAA) and General Data Protection Regulation (GDPR) to protect patient data. Section 6 addresses equity and accessibility for fair access to robotics, aiming to reduce disparities across socioeconomic and geographic lines. Section 7 presents case studies that offer real-world examples of regulatory and ethical successes, and Sect. 8 explore future directions of upcoming challenges and regulatory needs as technology advances. Finally, the Sect. 9 provides a conclusion, summary and recommendations, ensuring responsible innovation in rehabilitation robotics. This structure builds a logical understanding of the regulatory and ethical dimensions essential to the field.

2 Regulatory Landscape for Rehabilitation Robotics

The regulatory environment for rehabilitation robotics is very tight and governed by global bodies. Such regulations enhance safety, effectiveness, and quality in medical devices. Given that rehabilitation robotics is a form of medical device, they can have many impacts on a patient's health and welfare, making it important to have adequate regulatory compliance from manufacturers in the marketplace. There are several major regulatory bodies and standards that govern medical devices and include rehabilitation robotics. These are a few of the major companies in which the U.S. Food and Drug Administration (FDA), the European Union's CE marking system, and others play their part in shaping the global regulatory framework for the sector (see Table 1).

2.1 U.S. Food and Drug Administration (FDA)

The FDA is one of the most significant regulatory authorities that oversee medical devices in the United States. Medical devices are classified into three categories: Class I, II, and III, based on the risk level of the medical device to patients. Rehabilitation robots fall under Class II typically. This classification requires demonstration of whether the device is substantially equivalent to a similar device that is already legally marketed in the United States. They do this through a 510(k) premarket notification process. Class III devices, which are considered to have the greatest risk, have to go through a more stringent premarket approval (PMA) procedure. It warrants their safety and efficacy based on clinical data [12].

FDA regulations are crucial for rehabilitation robotics because they enforce both pre-market and post-market requirements. These regulations ensure continuous monitoring of devices once they are in use. Post-market surveillance includes adverse event reporting and periodic updates on device performance. This helps

Table 1 Overview of regulatory framework for rehabilitation robotics

Regulatory agency	Key points
FDA (U.S.)	– Classifies devices into Class I, II, III based on risk – Class II uses 510(k) process; Class III requires stricter PMA – Monitors post-market use (e.g., ReWalk)
CE marking (EU)	– Regulated by MDR, replacing MDD in 2017 – Requires rigorous testing and clinical evidence – Emphasizes post-market surveillance (e.g., Ekso Bionics)
PMDA (Japan)	– Approves devices after strict safety, quality, and effectiveness checks – Promotes collaboration and post-market monitoring (e.g., HAL by Cyberdyne)
International standards	– ISO 13485 ensures global quality management – Harmonizes standards to simplify compliance and market entry

address any new issues promptly [13]. For example, the ReWalk exoskeleton, which enables individuals with spinal cord injuries to walk, complied with FDA standards to market its technology.

2.2 *CE Marking in the European Union*

The use of medical devices within the European Union is regulated by the Medical Device Regulation, MDR. This was adopted in 2017 and replaced the previous Medical Device Directive, MDD. CE marking is a certification for compliance with the essential requirements of the MDR, and therefore it is safe to be used in the EEA, European Economic Area. Rehabilitation robotics must be put under testing before CE marking, guaranteeing an assurance in regard to the safety and performance standards of the EU [14].

Another main characteristic of the MDR is its emphasis on post-market surveillance. This allows manufacturers to constantly monitor their devices and report any occurrences or risks. With rehabilitation robots, technical malfunctions can be unpredictable in the clinical environment. Devices, such as the Ekso Bionics exoskeleton, have successfully completed the CE marking process with demonstration of MDR requirements [15]. The MDD to the MDR increased the standard benchmark for medical device regulation in Europe. MDR implies that devices should have wider clinical evidence before approval. This shift makes it more difficult, but in the long run, it is safer for rehabilitation robots [16].

2.3 *Japan's Pharmaceuticals and Medical Devices Agency (PMDA)*

In Japan, the Pharmaceuticals and Medical Devices Agency governs medical devices. PMDA ascertains that the devices, for instance, the rehabilitation robots, meet specifications on safety, effectiveness, and quality standards before they enter the market. For example, Cyberdyne's Hybrid Assistive Limb HAL for enhancing mobility in patients with spinal cord injuries must meet the strict regulations of Japan before being approved [17]. The PMDA regulatory process emphasises both pre-market approval and post-market surveillance. Like the FDA and CE marking, the PMDA requires comprehensive clinical trials to demonstrate a device's safety and effectiveness. Additionally, the PMDA promotes collaboration between industry stakeholders and regulatory bodies, streamlining the approval process [18].

2.4 International Standards and Harmonization

International standards are essential for the harmonisation of global regulations on rehabilitation robotics. ISO and IEC are responsible for developing global standards for medical devices. Such a standard is ISO 13485 for quality management systems related to medical devices. This quality management system specifies requirements for manufacturers to ensure safety and quality criteria are met uniformly across different regions [19]. To make it an international product, their intention is to market these rehabilitation robots globally. Most manufacturers prefer standardising through the likes of ISO organizations. That way, they can work through regulatory issues in various countries with minimal divergences, reducing time and cost when penetrating different markets.

2.5 Challenges in Regulatory Compliance

In fact, navigating the regulatory landscape for rehabilitation robotics is challenging. Regimes are different across regions, and coupled with the complexities of robotic technologies, it poses immense hurdles for the entities in terms of regulation compliance. For instance, new robotic technologies perhaps may not altogether fit into the FDA's 510(k) process of proving equivalence to an existing device. Likewise, the shift from MDD to MDR in Europe has brought about more stringent regulations that have subsequently delayed approvals for devices [20]. The speed of technological advancement in rehabilitation robotics also does not make it easy going for regulators. Devices that employ artificial intelligence or machine learning may necessitate new regulatory frameworks to cover issues of safety and ethics. Regulation agencies, therefore, have to adjust in stride with the evolution of technology in this field [21].

3 Ethical Issues in Rehabilitation Robotics

This is an emerging field that combines these technologies with therapeutic interventions. Rehabilitation robotics raises several issues about the ethical concerns addressed, particularly in the context of patient care. Therefore, the ethics based upon the main principles governing healthcare practice include respect for autonomy, beneficence, non-maleficence, and justice, directed to the characteristics of robotic-assisted therapies. These principles help the health providers and the developers direct their efforts such that the rehabilitation robotics does not violate the dignity of the patients, do not violate their consent, and cause less harm and more good as far as therapy benefit is concerned.

3.1 Ethical Principles in Healthcare and Their Application to Robotics

Health care operates on four major ethical principles: autonomy, beneficence, non-maleficence, and justice. Autonomy emphasises the right of a patient to make an informed decision. The healthcare provider is expected to act in a way that would benefit the patient the most; that is, beneficence involves caring for the well-being of the patient. Non-maleficence calls upon the provider to avoid causing harm, and justice connects with equitable provision of resources to health care services [22]. These ethical principles apply to rehabilitation robotics, but with new dimensions. Robotics introduces complex dynamics where human control and machine autonomy overlap and require developing entirely new ethics. For example, the robotic exoskeleton needs patients to remain in control of their own bodies, maintaining a sense of autonomy. Ethical questions arise in terms of potential harm because malfunctions in robots pose a significant risk to patients' safety [23].

3.2 Patient Autonomy in Robotic Therapies

Autonomy is a fundamental right in healthcare and is more cherished in the new robotic therapies. In these therapies, patients employ machines to move or enhance functionality. Shared control happens in the interventions of the robotic-assisted therapy, where the control is passed between the patient and the robot. The question is the degree of autonomy the patient has in rehabilitation [24]. One challenge is educating patients about the capabilities and limitations of robotic systems. Such awareness would help them make informed decisions. However, advanced technologies like robotics have made informed consent really complicated. The system's workings and risks are unknown to patients. Healthcare providers have to ensure that clear and transparent information is given to patients on the potential benefits and risks of robotic therapies [25].

For example, robotic exoskeletons are useful in helping mobility-impaired patients. Immediately after a patient's mobility impairment is diagnosed, the balance between providing and providing-for must shift toward providing over for the patient [26]. Patients require control over their movements to avoid a dependent feeling or loss of agency. The systems therefore must be highly personalised so as to give primacy to patient input, which is based on a care philosophy informed by values of personhood that foreground the patient as a human being with some form of agency and autonmy.

3.3 Consent and Beneficence in Robotic Therapies

Informed consent in medical practice plays a crucial role in rehabilitation robotics. Most robots using advanced artificial intelligence and machine learning algorithms are, therefore, impossible for a patient to comprehend the whole idea of how they work. Health care providers should be clear in their communication with regards to risks, benefits, and alternatives to robotic therapies [27]. One of the major principles in robotics is beneficence, which is the principle of acting in the best interest of the patient. While robotic interventions may hasten rehabilitation, their use should thus be appraised to not cause harm to the patients. This principle is also tied to non-maleficence, with its main focus on not causing harm. While malfunctioning or ill programmed robots might harm the patients or otherwise delay their advancement. Hence, there is great importance in risk assessment and mitigation in ethical robotic care [28].

Specifically, for the purpose of serving as a medical assistant device in stroke rehabilitation, with the goal of assisting in the restoration of motor function in a patient [29]. The devices would need to be tailored to suit each patient's needs, and their use would not replace the human therapists but complement their efforts. The emphasis should be on ensuring the quality of the care and application with continuous monitoring and assessment to ensure that real benefit is delivered without patient safety being compromised.

3.4 Challenges and Ethical Dilemmas

Robotics in rehabilitation also raises bigger ethical concerns. These concerns include depersonalisation through a lack of human-to-human interaction. Robots can repeat certain tasks efficiently but are ineffective at empathetic and emotive support. Because of this, there is a possibility that treatment will become more formulaic rather than relational and maybe personal, which would reduce the quality of the relationship that exists between patients and carers [25]. The high development and deployment costs of robotic systems are likely to fuel healthcare inequality. Robotic therapies may not be affordable to many patients, mostly in low-income settings. These questions of justice and equitable access to care arise [27]. Questions of equitable and fair access to all are particularly crucial to be addressed in terms of policy and regulation with respect to the accessibility of robotic technologies to all patients, irrespective of their socioeconomic status [30].

4 Safety Standards and Risk Management

In the field of rehabilitation robotics, which is a field that brings together technical achievement with the benefit of greater health for patients, safety requirements and risk management measures are of the utmost importance. Rehabilitation robots are actually machines designed to help assist patients regain mobility and strength. However, there is a major concern regarding possible harms attributed to technical malfunctions, human error, or perhaps not having effective enough risk mitigation. This section discusses safety standards in rehabilitation robotics (Table 2), including some international regulations and guidelines. It provides different risk management approaches for ensuring safe and effective robotic-assisted therapy.

4.1 Safety Standards in Rehabilitation Robotics

The development of safety standards with regards to healthcare robotics is necessary, as it ensures that devices do not pose undue risks towards the patients. The development of such standards is accomplished by international organisations and regulatory bodies. Amongst these are the International Organisation for Standardisation (ISO), the International Electro technical Commission (IEC), the Food and Drug Administration in the United States, and CE marking for medical equipment by the European Commission. All these organisations present specific criteria with which rehabilitation robots must meet in order to achieve safety. Among the significant safety standards for rehabilitation robots is ISO 13482:2014. This standard highlights the safety requirements for personal care robots and rehabilitation, among others [31]. It focuses on the requirement for safety interactions between the robot and humans. It considers physical contact, movement caused by machinery, and electrical dangers. Rehabilitation movement-assisting robots ought to focus on these standards to avoid injury to patients.

Table 2 Safety standards and risk management in rehabilitation robotics

Aspect	Key points
Safety standards	– Adherence to international standards like ISO 13482:2014 (safe human–robot interaction) and ISO 14971:2019 (risk management) – Compliance with FDA and IEC 60601 for electrical safety – Regular post-market surveillance and modifications with risk analysis
Risk assessment	– Identification and mitigation of mechanical, software, and electrical risks – Examples: Redundant systems for mechanical failure, robust software validation, and adherence to IEC standards for electrical safety
Mitigation strategies	– Fail-safe mechanisms and sensors for hazard detection – Healthcare provider training for safe operations – Regular maintenance and software updates – Patient involvement through education and feedback

The rehabilitation robots also have to adhere to ISO 14971:2019 on the safety of requirements because this deals with risk management for medical devices [32]. This standard outlines how one is to identify, analyse, and mitigate risks. There must be risk monitoring throughout the whole product life cycle, from design right up to post-market surveillance. For example, any software or hardware modification should be performed only after adequate risk analysis has been done to allow new risks not resulting from the change. In the United States, medical devices, which the rehabilitation robot is under, are approved by the FDA. Class II devices, and this would include devices like robots for rehabilitation, require manufacturers to prove safety and effectiveness in use. Under these conditions, devices are mostly demonstrating substantial equivalence to existing or approved devices or undergoing strict clinical trials. The FDA's attention to safety translates to very stringent requirements met before rehabilitation robots are introduced into clinical practice [33].

4.2 *Risk Assessment in Robotic Rehabilitation Therapies*

The systematic process of risk assessment involves identification of potential hazards and the assessment of possible likelihood and severity of harm. It is vital for rehabilitation robots because robots frequently interact closely with patients. An effective risk assessment strategy ensures identifying hazards, estimating risks, determining acceptability, and then putting measures of mitigation in place [34]. The greatest risk is the mechanical failure of the robotic parts. Rehabilitation robots are used in assisting patients to walk or lift. A mechanical failure can result in a patient falling and getting injured. This can be stopped by having redundant systems. These ensure that in case a component fails, other mechanisms prevent accidents from occurring.

Software failure is a second significant risk that rehabilitation robots are exposed to. Rehabilitation robots rely significantly on sophisticated algorithms to facilitate movement and interaction. However, if the algorithms fail, the robot will begin moving erratically. To counter this, the software of the rehabilitation robot is thoroughly tested and validated to ensure minimal failure before deployment to hospitals for use. Additionally, consistent tracking and updating of the software ensure that the robot is not malfunctioning [35]. Electrical hazards, therefore, become another risk factor for rehabilitation robots. Since the operation of these robots depends on electricity, there is a risk of shock or burn if it malfunctions. This necessitates that these rehabilitation robots meet all the demands of IEC 60601, the medical standard for making sure that the medical devices are electrically safe [36]. This calls for proper insulation and safety systems for avoiding electrical shocks during its operation.

4.3 Mitigation Strategies in Rehabilitation Robotics

Reducing the possibility of damage is the concept of mitigation measures, and protecting the patient before, during, and after therapy treatment is a part of this. A fail-safe mechanism is one of the major strategies. Fail-safes ensure that the robot stops once it perceives probable danger, such as the obstacle in the path of the robot or a sudden change in the patient's status. In addition, some rehabilitation robots are equipped with sensors that can identify the movements of the patient and correlate to the necessary adjustments in the activity of the patient to prevent potential injuries [37]. Education and healthcare providers: Healthcare providers and education are also important in the mitigation of risk. Rehab robots are complex systems; thus, healthcare providers should be trained on safe operating and emergency responses. Training programs should include hands-on practice and simulation of emergency cases for minimal risks of accidents in a clinical setting.

Regular maintenance of rehabilitation robots is very crucial. The robots must be checked for mechanical wear and tear by regular inspections. In your software, update as soon as possible to patch up the vulnerabilities and bugs; otherwise, you're leaving yourself susceptible [38]. By proper maintenance, the robots will serve the desire for long. Another important point is the involvement of the patient in the risk mitigation process. The patient should be educated to know how to use the robot and what to do during an emergency. Patient feedback on the performance of the robot may lead to the determination of problems before their adverse manifestation in the health context [39].

5 Privacy and Data Security

With the sensitive patients' data collected and processed, privacy and data security are highly pertinent to rehabilitation robots. Robots increasingly become an indispensable part of rehabilitation services. Handling patient data behind secure walls requires handling patient data behind secure walls. Legally compliant data protection includes HIPAA in the U.S. and GDPR in the EU, which is indispensable for patients' trust to be not breaching the sensitive data within the health domain. This section provides insight into the question of rehabilitation robotics and privacy and defines strategies for safe data management and regulation.

5.1 Importance of Patient Data Privacy in Rehabilitation Robotics

Rehabilitation robots assist patients in recovery from injury or illness. Such robots collect vast amounts of information about the individual and their health status. For instance, it follows up on the progression of patients, records the movements of a patient, and detects physiological conditions. Advanced technologies are used to process large amounts of such data, such as AI and machine learning [40].

This data also contains medical histories and biometric details; thus, protection against unauthorised access is mandatory. A serious issue is that of a data breach. Unauthorised parties may get entry into the information of the patients to the potential opportunity of identity theft or discrimination. The interconnected systems and cloud storage also pose greater risks of cyberattacks. Hackers gain unauthorised entry into sensitive information through software vulnerabilities. Data privacy also ensures patient autonomy. Patients have a right to control how their data will be used. They ought to give informed consent before any data collection. This means they should understand how their information will be processed and why [41]. Rehabilitation robots should be designed with such concern.

5.2 Strategies for Secure Data Handling in Rehabilitation Robotics

Some technical, organisational, and regulatory strategies keep rehabilitation robotics handling data safe. Data encryption is one of the essential technical measures. Data encryption is a change in format that renders data unreadable to anyone who does not have access to it. Encrypting data at its collection point so that no third party may intercept or tamper with the information while being transported to healthcare providers does so [42]. Data encryption includes safe storage solutions, such as encrypted databases for inactive data.

Another important strategy is the use of strong authentication methods. Only authorised persons should be able to access the patient data. Multifactor authentication increases security. For instance, doctors would require both a password and biometric authentication before accessing the information. Another useful strategy is data anonymization. Removing personal identifiers, such as names or social security numbers, from patient data removes any information that might be used to identify them. This makes it possible for involved healthcare providers to use it for research without breaking the principle of privacy. Methods of anonymization need to be robust, though. Weak anonymization could potentially lead to the re-identification of patients [43].

5.3 Compliance with Privacy Regulations

Organisations that use rehabilitation robots have to comply with the privacy rules as outlined by the Health Insurance Portability and Accountability Act (HIPAA) and the General Data Protection Regulation (GDPR), since it requires particular protection requirements for patient data. HIPAA is a U.S. federal law that addresses the privacy and security of healthcare information. Consequently, under HIPAA, healthcare providers are mandated to include administrative, physical, and technical safeguards in place so that patient data are protected [44]. Moreover, patients have the right under HIPAA to view and obtain copies of their health information and to request corrections. In rehabilitation robotics, organisations must ensure that the robots store data securely and only allow access to the information by authorised personnel.

GDPR is one of the most elaborate privacy regulations globally. It applies to organisations operating in the EU or handling personal data relating to individuals in the EU [45]. GDPR has placed strictures on data collection, processing, and storage in an organization. This goal is to ensure absolute openness and garner consent from the user. Rehabilitation robots in the EU must be guided by this regulation. The patients should be kept informed that their data will be used, and they have the right to withdraw consent at any given time. Furthermore, under GDPR, very harsh consequences are authorised for unauthorised data breaches. This makes it imperative for such organisations to always focus on protecting data. By adopting these strategies and being compliant with laws, rehabilitation robotics can ensure that patient privacy and data security are maintained while performing effective care [46].

6 Equity and Accessibility in Rehabilitation Robotics

Equity and access must be achieved in rehabilitation robotics. These technologies will change patient care in terms of recovery assistance. Access for all types of patients, regardless of their geographic or socioeconomic location, can be quite demanding. There is a need to put healthcare equity and accessibility at the centre when developing, deploying, and distributing rehabilitation robotics. Without exploring the access gaps, the gains from these technologies might be a bit short. This will, therefore, exacerbate health inequities further.

6.1 Ensuring Equitable Access to Rehabilitation Robotics Technologies

Rehabilitation robotics is normally expensive. Their development entails the cost and the extent of infrastructure required. It entails specialised facilities and trained personnel. Overcoming these barriers involves the provision of making these technologies accessible to a wider population through cost reduction through having cost-reduction strategies. One of the ways rehabilitation robotics can be made available is by integrating them into public healthcare systems, which would then increase access [47]. In this regard, costs can be subsidised or through insurance. A country's healthcare policies come into play regarding these technologies for having access to them. Countries that have universal healthcare systems can make rehabilitation robotics regular care for all, but within privatised systems, access is only to those who are able to afford the costs or have good insurance. Government policies are needed through which rehabilitation robotics are incorporated in the public healthcare sector so that those people belonging to the lower-income group may avail themselves of it easily [48].

Devices also must accommodate various users. Users vary with different physical, cultural, and technological capacity levels. Customisable interfaces, multilingual support, and adaptive designs are key features. The features of such design ensure that rehabilitation robots meet patients with uncommon needs related to age, disability, or for any other reason [49]. Increasing user friendliness and the adaptability of rehabilitation robotics should be able to enhance access. It can also reduce disparities in usage.

6.2 Addressing Disparities in Access Based on Socioeconomic Factors

Socioeconomic factors have a great say in rehabilitation robotics access. In most cases, patients from rich families are usually in a position to afford the sophisticated treatments, such as rehabilitation robotics. The patients come from better economic backgrounds plus private health care. The weaker economic class of patients faces financial constraints. These constraints restrict most of them from seeking these services. It's a very painful reality to find that these patients tend to face higher disability rates in lower-income groups. This is the most constrained group that requires rehabilitation robotics [50].

It would reduce the disparities that are in place due to the policies. Costing rehabilitation robotics as low as possible and making them commonly available in public healthcare is a must. Subsidies, grants, and other incentives will be made for the purpose of promoting affordable use of these technologies. Rehabilitation robotics from the non-profit sectors and philanthropic efforts can be distributed to the disadvantaged or underprivileged communities. Cost-reduction programs with some grants offered to make it accessible to all income groups.

Education also determines accessibility. Patients and health professionals need to be aware of such technologies. They ought to appreciate their value and implement them into the treatment plans [27]. Education and training of patients and healthcare providers are required. Such courses can also improve knowledge on robotic therapies. Training, especially in developing and remote regions where advanced medical technologies are not readily available, should be provided.

6.3 *Addressing Geographic Disparities in Access*

The geographic location is another major aspect that affects rehabilitation robotics accessibility. Patients living in rural regions have a challenge with respect to advanced technologies pertaining to healthcare. It is characterised by a reduced number of specialised structures, fewer trained personnel, and infrastructure. In most cases, rehabilitation robots are only available in urban hospitals or research centres. This keeps the patients in rural regions at a disadvantage and cannot access the robotic therapies [51]. Solutions are found in telemedicine and remote rehabilitation services.

Telemedicine services can be merged with rehabilitation robotics. This eliminates the constraint of place and ensures home-based sessions of therapy. Access has also improved for rural patients. Moreover, with enhanced travel, rehabilitation can be more convenient and economical. There are indications that disparities based on geography could be addressed by increasing telemedicine services and integrating these services with rehabilitation robotics to make robotic therapies more accessible [52].

Another option is portable rehabilitation robots. Mobile units can be taken to different sites, offering temporary access to rehabilitation robotics. These can be used in mobile clinics to focus on rural or otherwise underserved communities. These clinics could be fitted with robotic devices and professional staff. They may be taken to locations where few people can access advanced health care technologies. In this sense, rehabilitation robotics goes directly to the patients who might not otherwise receive such therapies [52].

7 Case Studies and Best Practices in Rehabilitation Robotics

A very vibrant area in the technologies related to health care is rehabilitation robotics. The amount of potential it offers for improvements in patient outcomes is tremendous. At the same time, this advancement opens up various ethical dilemmas. Innovation must always be balanced with patient welfare, consent, and safety. In this section, we would look into real-world examples of ethical dilemmas in rehabilitation robotics and the successful implementations of ethical guidelines and regulatory compliance (Table 3). Through the complex context of rehabilitation robotics, the case studies demonstrate how many stakeholders, including clinicians and engineers, handle challenges in order to guarantee that these technologies are utilised in a responsible manner.

Table 3 Ethical dilemmas, successful implementations, and regulatory compliance in rehabilitation robotics

Category	Description
Ethical dilemmas	
Patient autonomy	Robotic systems may limit patient control over treatment
Bias	AI algorithms may cause unequal outcomes if not trained on diverse data
Data privacy	Sensitive data from robots must be securely encrypted to prevent breaches
Successful ethical implementations	
Ekso bionics exoskeleton	A wearable device that allows patient control, meeting FDA safety standards
RoboLaw project	EU project to establish ethical standards for healthcare robots, focusing on patient rights and data protection
Japan's approach	Japan ensures compliance with regulatory standards and ethical patient consent in robotic rehabilitation
Regulatory compliance	
FDA & CE marking	Robots are regulated by the FDA and CE standards for safety and efficacy

7.1 Ethical Dilemmas in Rehabilitation Robotics

One of the primary ethical questions with rehabilitation robots is the issue of a patient's autonomy. Even though numerous robotic systems are designed to operate independently, these can potentially constrain a patient's authority over his or her treatment. For instance, in stroke rehabilitation, robotic arms drive a patient's limb through certain motions. In truth, the use of this sort of recovery raises questions regarding the patient's retained autonomy. This report, by Fosch-Villaronga and Mahler (2021), unpacks the conflict between automated healthcare and patient control. Patients need to be well-informed and included in decision-making. It's through this that patients are not just recipients of care but also active participants in their treatment [27].

Bias is a third ethical issue that the robotic rehabilitation system may face. AI algorithms are liable to cause disproportionate treatment effects [27]. For example, an algorithm trained only on data from mainly affluent or white populations may not operate well for any other groups. An example is one in which a rehabilitation robot used in clinical trials poorly operated for patients with darker skin tones. This was due to the way the sensors of the robot were able to sense movement [53]. The incident brings out the ethical mandate that developers have been placed upon, such as the cross-population testing of the robotic systems to avoid healthcare disparities.

Another serious ethical issue is data privacy. Rehabilitation robots often collect sensitive information, such as motion tracking and biometric information. Such data is also vulnerable to breaches if not kept secure. Data security is therefore an important aspect of maintaining the trust of the patient and meeting EU requirements related to privacy under the GDPR and the Health Insurance Portability and Accountability Act. A case study revealing that poor encryption in a robotic rehabilitation system led to unauthorized access to patient data. This case stresses the need for robust cybersecurity measures in rehabilitation robotics [54].

7.2 *Successful Implementations of Ethical Guidelines*

While these are some ethical challenges, there are many institutions creating very effective frameworks. For example, Ekso Bionics developed an exoskeleton, a wearable robotic device to regain mobility in individuals with spinal cord injuries. The developers, therefore, ensured the device was very safe. It follows very strict testing procedures. Moreover, they acquired FDA approval, which mandates that one examine the safety and effectiveness of the device and associated risks. The Ekso Bionics exoskeleton contains controls that enable patients to decide on the level of robotic assistance they require. This aspect ensures maximum participation by the patient in the care process [55].

Another one is the RoboLaw project, which is cooperation between the European Commission and various robotic developers, a project that aimed to create an ethical standard of robotics in health care through legal channels. The project tended to place emphasis on transparency and accountability, as well as patient rights [28]. It provided recommendations on informed consent, patient autonomy, and data protection. These recommendations have been followed in many European rehabilitation clinics very successively. They ensure well-informed patients on their treatment options and how their data is handled during robotic therapy.

Rehabilitation robotics in Japan is an especially good example of compliance with regulatory models. Japan has one of the most advanced robotic industries in the world, and robots have been fully incorporated into routine rehabilitation care within its national health system [56]. The same testing and approval as other medical devices apply to rehabilitation robots under the regulation of the Ministry of Health, Labour, and Welfare. There are also ethical demands related to consent that are placed upon the Japanese hospitals. This has led to most health care sectors in Japan using the rehabilitation robots with exemplary integrity.

7.3 *Regulatory Compliance in Rehabilitation Robotics*

Another necessary factor in the ethical use of rehabilitation robots is regulatory compliance. This device is regulated by the FDA in the U.S., which means that its designing and usage are regulated. A lot of tests are conducted to verify safety and effectiveness for clinical uses before a rehabilitation robot is approved. This includes reviews of ethical considerations and potential risks and benefits [57]. Similarly, in the European Union, rehabilitation robots should be CE-marked. This implies that the product meets European standards for safety, health, and environment [58].

The developers engage agencies like the FDA early in the design process to ensure that their products meet the ethical and safety standards before they hit the market. A good practice that can make developers succeed in regulatory compliance is to show an early level of engagement with regulatory bodies. For instance, Hocoma, a Swiss company that produces the Lokomat robotic gait trainer, is an exemplary company that demonstrates this proactive approach [59]. Hocoma would go through the complex regulatory landscape by partnering with regulatory agencies when developing their product. They hence released their product in the market, ensuring it was in line with the required ethical and legal standards.

8 Future Directions and Challenges

As technologies like artificial intelligence, machine learning, and analytics of big data progress to maturity, the ethical and regulatory concerns become increasingly complex. On one hand, it offers many benefits but at the same time raises considerable ethical questions that have to be addressed in this regard. This section explores the emerging ethical issues, regulatory challenges, and advancements in the regulatory frameworks for future technologies.

8.1 Emerging Ethical Issues

Data privacy and consent are indeed the ethical concerns. They often remain unconscious about how their data is used, saved, and shared as organisations gather more personal data. This might lead to situations in which they provide neither informed nor meaningful consent. According to Acquisti et al. (2015), there should be transparency in data practices such that users ought to have comprehensive knowledge of the risks of their data being shared. Strong safeguards ensure that data misuse is not a danger, which would likely do serious harm to public trust [60].

Another ethical issue is algorithmic bias. Most of the machine learning algorithms work based on historical data, which can be biased. Such biases often get reflected in the automated systems of decision-making. For example, it has been established that the facial recognition system has a higher error rate when discussing people of colour and particularly women [61]. According to O'Neil, biased algorithms can even intensify social inequalities [62]. Therefore, there is a need for various data representations and monitoring outcomes.

The other is accountability in AI systems. As the authority of AI decisions becomes more and more autonomous, giving responsibility to people for making decisions from these systems becomes quite complex. Reed (2018) mention how difficult it is to allocate liability when AI malfunctions or gets its decisions wrong, particularly in such sensitive areas as healthcare or transportation. Lines of accountability are very important in developing the trust of the public and standards of ethics [63].

8.2 Regulatory Challenges

Technology regulation is faster than the advancements themselves. According to Zuboff (2019), digital technology issues have been found to be handily dealt with by weak modern regulation systems. With this, there exists the opportunity to develop less ethical practices. More adaptive regulatory frameworks are needed that can match the technological change [64]. The other critical regulatory barrier is that technology has become globally bound. Legal rules and ethical standards vary from country to country; hence, their respective regulatory frameworks also end up becoming fragmented in nature [65]. It makes compliance even more challenging for multinational companies and may raise moral dilemmas with respect to the practices that are considered acceptable in one country but objectionable in another country. There is an urgent need for international agreements on data protection and also on ethical standards to reduce such inconsistency in the regulatory framework.

A delicate balance has to be struck between innovation and regulation. Overregulation deters the progress of technology, while under regulation results in illegal activities. Thus, policymakers should balance this teeter-totter between innovation and public welfare to ensure that safety and security are safeguarded without thwarting innovation. Continuous communication between regulators, technologists, and the public has to be established in order to make certain that ethical matters are considered without strangling innovation.

8.3 Potential Advancements in Regulatory Frameworks and Ethical Guidelines

Innovations in adaptive regulatory models may address emergent challenges. Adaptive models have feedback loops where alterations are made in real-time as technology develops [66]. Thus, adaptable rules fitting the change in technology may be referred to as these ones. Regulatory sandboxes are another promising answer. There, companies get a chance to test new technologies under regulatory oversight within controlled environments. It opens up experimentation while being sure of ethics at all times [67]. What sandboxes do is encourage innovation by providing a safe space where developers can test their ideas without feeling the immediate backlash of regulatory constraints.

Ethical standards must also be made stronger. Ethical models that include all sides of the stakeholder's opinion should be taken into consideration in order to tackle the issues concerning fairness and equity. The ethical audits that will be initiated for high-tech industries will get together an organisational accountability system wherein every project that conforms to technology will follow the ethical standards applied to it. Therefore, it will be in vogue throughout the development of the whole technology process [64]. Another way public engagement might enhance transparency and credibility of the regulatory process is through inclusion of citizens in deliberations on ethics surrounding technology. This helps to further a more democratic process of regulation. Indeed, Zuboff (2019) advocates for public participation in the regulatory process to be increased because it can enable regulations to be aligned with better value priorities in society [64].

9 Conclusion

Rehabilitation robotics integration in healthcare brings along a lot of significant ethical and regulatory issues that need to be addressed responsibly to prevent irresponsible development and deployment. Ethical issues such as data privacy, informed consent, and algorithmic bias in decision-making processes all come to the forefront. Accountability for actions performed by autonomous systems is also a critical aspect.

Regulatory frameworks need to be matched to the fast track of technological development while safe and effective in real life. It is possible to avoid all of these issues for these stakeholders who include developers, health care providers, and policymakers, by having concerns for clear data practices, actively include multiple points of view in the design and implementation of rehabilitation robots, and a great deal more at the same time. Regular ethical audits and assessments detect and reduce biases and risks, and nurturing a collaborative environment stimulates public involvement in the regulatory process, thereby building trust and ensuring that rehabilitation robotics serve to advance the best values in society. More important, therefore, is the intent to adhere to the aforementioned recommendations in efforts to promote the ethical use of rehabilitation robotics by stakeholders. This will finally lead to achieving improved patient outcomes while protecting both rights at the individual level as well as community interests.

References

1. Beauchamp, T.L., Childress, J.F.: Principles of Biomedical Ethics, 2nd edn. Oxford University Press, New York (2019)
2. Blenman, K.R., Hull, R., Maimela, C., Molefi, T., Khanyile, R., Dlamini, Z.: Society 5.0 healthcare: ethics, legal rights, human rights, safety, and security. In: Society 5.0 and Next Generation Healthcare: Patient-Focused and Technology-Assisted Precision Therapies, pp. 267–291. Springer Nature Switzerland, Cham (2023)
3. Obasa, A.E.: The Ethics of Artificial Intelligence in Healthcare Settings. Stellenbosch University, Stellenbosch, South Africa (2023)

4. Dilsizian, S.E., Siegel, E.L.: Artificial intelligence in medicine and cardiac imaging: harnessing big data and advanced computing to provide personalized medical diagnosis and treatment. Curr. Cardiol. Rep. **16**, 1–8 (2014). https://doi.org/10.1007/s11886-013-0441-8
5. Masiero, S., Poli, P., Rosati, G., Zanotto, D., Iosa, M., Paolucci, S., Morone, G.: The value of robotic systems in stroke rehabilitation. Exp. Rev. Med. Dev. **11**(2), 187–198 (2014). https://doi.org/10.1586/17434440.2014.882766
6. Hong, R., Li, B., Bao, Y., Liu, L., Jin, L.: Therapeutic robots for post-stroke rehabilitation. Med. Rev. **4**(1), 55–67 (2024). https://doi.org/10.1515/mr-2023-0054
7. Elendu, C., Amaechi, D.C., Elendu, T.C., Jingwa, K.A., Okoye, O.K., Okah, M.J., et al.: Ethical implications of AI and robotics in healthcare: a review. Medicine **102**(50), e36671 (2023). https://doi.org/10.1097/MD.0000000000036671
8. Kapeller, A., Felzmann, H., Fosch-Villaronga, E., Hughes, A.M.: A taxonomy of ethical, legal, and social implications of wearable robots: an expert perspective. Sci. Eng. Ethics **26**, 3229–3247 (2020). https://doi.org/10.1007/s11948-020-00268-4
9. Zhang, J., Fang, J., Xu, Y., Si, G.: How AI and robotics will advance interventional radiology: narrative review and future perspectives. Diagnostics **14**(13), 1393 (2024). https://doi.org/10.3390/diagnostics14131393
10. Freyer, O., Wiest, I.C., Kather, J.N., Gilbert, S.: A future role for health applications of large language models depends on regulators enforcing safety standards. Lancet Digit. Health **6**(9), e662–e672 (2024). https://doi.org/10.1016/S2589-7500(24)00124-9
11. Dwivedi, J., Mahgoub, I.: Robotic surgery: a review on recent advances in surgical robotic systems. In: Florida Conference on Recent Advances in Robotics, pp. 1–7 (2012)
12. Garcia-Gonzalez, A., Fuentes-Aguilar, R.Q., Salgado, I., Chairez, I.: A review on the application of autonomous and intelligent robotic devices in medical rehabilitation. J. Braz. Soc. Mech. Sci. Eng. **44**(9), 393 (2022). https://doi.org/10.1007/s40430-022-03692-8
13. Benjamens, S., Dhunnoo, P., Meskó, B.: The state of artificial intelligence-based FDA-approved medical devices and algorithms: an online database. NPJ Digit. Med. **3**(1), 118 (2020). https://doi.org/10.1038/s41746-020-00324-0
14. Iftikhar, M., Saqib, M., Zareen, M., Mumtaz, H.: Artificial intelligence: revolutionizing robotic surgery. Ann. Med. Surg. **86**(9), 5401–5409 (2024). https://doi.org/10.1097/MS9.0000000000002426
15. Marquez-Chin, C., Kapadia-Desai, N., Kalsi-Ryan, S.: Robotic-assisted rehabilitation. In: Brain–Computer Interfaces: Neurorehabilitation of Voluntary Movement after Stroke and Spinal Cord Injury, pp. 27–49. Springer, Cham (2021). https://doi.org/10.1007/978-3-031-01608-0_3
16. Memić, H., Bošnjaković, A., Džemić, Z., Badnjević, A.: Regulations and directives—past, present, future. In: Inspection of Medical Devices: For Regulatory Purposes, pp. 5–31. Springer, Cham (2023)
17. Morioka, H., Hirayama, T., Sugisawa, T., Murata, K., Shibukawa, M., Ebina, J., et al.: Robot-assisted training using hybrid assistive limb ameliorates gait ability in patients with amyotrophic lateral sclerosis. J. Clin. Neurosci. **99**, 158–163 (2022)
18. Shi, J., Chen, X., Hu, H., Ung, C.O.L.: Application of implementation science framework to develop and adopt regulatory science in different national regulatory authorities. Front. Public Health **11**, 1172557 (2023). https://doi.org/10.3389/fpubh.2023.1172557
19. Karthika, B., Vijayakumar, A.R.: ISO 13485: medical devices—quality management systems, requirements for regulatory purposes. In: Medical Device Guidelines and Regulations Handbook, pp. 19–29. Springer, Cham (2022)
20. McDermott, O., Kearney, B.: A review of the literature on the new European Medical Device Regulations requirements for increased clinical evaluation. Int. J. Pharm. Healthc. Mark. (2024). https://doi.org/10.1108/IJPHM-07-2023-0060
21. Iphofen, R., Kritikos, M.: Regulating artificial intelligence and robotics: ethics by design in a digital society. Contemp. Soc. Sci. **16**(2), 170–184 (2021). https://doi.org/10.1080/21582041.2018.1563803

22. Wong, V., Hassan, N., Wong, Y.P., Chua, S.Y.N., Abdul Rahman, S., Mohamad, M.L., Lim, S.: Nurses' adherence to ethical principles—a qualitative study. Nurs. Ethics (2024). https://doi.org/10.1177/09697330241291159
23. Hutler, B., Rieder, T.N., Mathews, D.J., Handelman, D.A., Greenberg, A.M.: Designing robots that do no harm: understanding the challenges of ethics for robots. AI Ethics **4**(2), 463–471 (2024). https://doi.org/10.1007/s43681-023-00283-8
24. Matarić, M.: Socially assistive robotics: human-robot interaction methods for creating robots that care. In: Proceedings of the 2014 ACM/IEEE International Conference on Human-Robot Interaction, pp. 333–333. ACM/IEEE (2014). https://doi.org/10.1145/2559636.2560043
25. Sharkey, A., Sharkey, N.: Granny and the robots: ethical issues in robot care for the elderly. Ethics Inf. Technol. **14**, 27–40 (2012). https://doi.org/10.1007/s10676-010-9234-6
26. Banda, Y.D., Raj, T., Priyadarshini, R.: Applications of robots in assistive technology. In: Robotics and Smart Autonomous Systems, pp. 201–227. CRC Press (2024)
27. Fosch-Villaronga, E., Mahler, T.: Cybersecurity, safety, and robots: strengthening the link between cybersecurity and safety in the context of care robots. Comput. Law Secur. Rev. **41**, 105528 (2021). https://doi.org/10.1016/j.clsr.2021.105528
28. Chan, H.Y., Muralidharan, A.: Care robots for the elderly: legal, ethical considerations, and regulatory strategies. In: Developments in Intellectual Property Strategy: The Impact of Artificial Intelligence, Robotics, and New Technologies, pp. 129–156. Springer International Publishing, Cham (2024). https://doi.org/10.1007/978-3-031-42576-9_6
29. Rajalaxmi, V., Kirthika, R.: Emerging role of artificial intelligence and robotics in physiotherapy: past, present, and future perspective. In: Advancements in Clinical Medicine, pp. 200–214. IGI Global (2024)
30. Tilala, M.H., Chenchala, P.K., Choppadandi, A., Kaur, J., Naguri, S., Saoji, R., Devaguptapu, B.: Ethical considerations in the use of artificial intelligence and machine learning in healthcare: a comprehensive review. Cureus **16**(6), e62443 (2024). https://doi.org/10.7759/cureus.62443
31. Goffin, T., Palmieri, S.: Regulating smart healthcare robots: the European approach. In: Research Handbook on Health, AI, and the Law, pp. 75–91. Edward Elgar Publishing (2024). https://doi.org/10.4337/9781802205657.00011
32. Elahi, B.: Safety Risk Management for Medical Devices. Academic Press, London (2021)
33. De Micco, F., Grassi, S., Tomassini, L., Di Palma, G., Ricchezze, G., Scendoni, R.: Robotics and AI in healthcare from the perspective of European regulation: who is responsible for medical malpractice? Front. Med. **11**, 1428504 (2024)
34. Pons, J.L.: Wearable Robots: Biomechatronic Exoskeletons. Wiley, Hoboken (2008)
35. Zhang, M., Chi, Z., Yang, Z., Mohammed, S., Huang, J.: Prescribed performance control of a human-following surveillance robot with incomplete observation. Cyborg Bionic Syst. **5**, 0085 (2024). https://doi.org/10.34133/cbsystems.0085
36. Mark, T.: Medical electrical equipment Part 1: general requirements for basic safety and essential performance. ADInstruments. https://m-cdn.adinstruments.com/safety-and-quality-standards/7191187326TRF_PSB_ADInstrument_PowerLab.pdf (2014)
37. Vanderborght, B., Albu-Schäffer, A., Bicchi, A., Burdet, E., Caldwell, D.G., Carloni, R., Ijspeert, A.J.: Variable impedance actuators: a review. Robot. Auton. Syst. **61**(12), 1601–1614 (2013). https://doi.org/10.1016/j.robot.2013.06.009
38. Rahman, M.M., Khatun, F., Jahan, I., Devnath, R., Bhuiyan, M.A.A.: Cobotics: the evolving roles and prospects of next-generation collaborative robots in Industry 5.0. J. Robot. **2024**(1), 2918089 (2024). https://doi.org/10.1155/2024/2918089
39. Gogoshin, D.L.: Patient preferences concerning humanoid features in healthcare robots. Sci. Eng. Ethics **30**(6), 49 (2024). https://doi.org/10.1007/s11948-024-00508-x
40. Nozari, H., Ghahremani-Nahr, J., Szmelter-Jarosz, A.: AI and machine learning for real-world problems. In: Advances in Computers, vol. 134, pp. 1–12. Elsevier (2024)
41. Ibrahim, A.M., Abdel-Aziz, H.R., Mohamed, H.A.H., Zaghamir, D.E.F., Wahba, N.M.I., Hassan, G.A., Aboelola, T.H.: Balancing confidentiality and care coordination: challenges in patient privacy. BMC Nurs. **23**(1), 564 (2024). https://doi.org/10.1186/s12912-024-02231-1

42. Manikandan, A., Sanjay, T., Gautham Menon, A.R., Bhaskar, P.B., Mahadev Govind, R., Ramprasad, O.G.: Issues and challenges in security and privacy with e-healthcare. In: Internet of Things Enabled Machine Learning for Biomedical Applications, p. 222 (2024)
43. Gadotti, A., Rocher, L., Houssiau, F., Crețu, A.M., de Montjoye, Y.A.: Anonymization: the imperfect science of using data while preserving privacy. Sci. Adv. **10**(29) (2024). https://doi.org/10.1126/sciadv.adn7053
44. Erickson, M.L.: Legal and ethical considerations for physical therapy documentation. In: Documentation Basics for the Physical Therapist Assistant, pp. 119–126. Routledge (2024)
45. Nash, I., Kennedy-Mayo, D., Antón, A.: Legal issues in reconciling data protection, AI, and cybersecurity under EU law. Missouri Law Rev. **89**(3), 7 (2024)
46. Ebers, M.: Personal care robots under EU data protection law. In: Privacy, Data Protection and Data-driven Technologies, pp. 229–271. Routledge (2024)
47. Banyai, A.D., Brişan, C.: Robotics in physical rehabilitation: systematic review. Healthcare **12**(17), 1720 (2024). https://doi.org/10.3390/healthcare12171720
48. Adams, D.: Leveraging technology and data advances to improve social program coverage and service delivery. (2024). https://doi.org/10.1787/af31746d-en
49. Johnson, M.J., Keyvanian, S., Mendonca, R.J.: Toward inclusive rehabilitation robots. In: Rehabilitation Robots for Neurorehabilitation in High-, Low-, and Middle-Income Countries, pp. 471–498. Academic Press (2024). https://doi.org/10.1016/B978-0-323-91931-9.00032-3
50. Kambur, E.S., Yıldırım, H.H.: Future health technology trends, policy, and governance perspective: the Turkish case. Health Res. Policy Syst. **22**(1), 147 (2024). https://doi.org/10.1186/s12961-024-01217-4
51. Deo, N., Anjankar, A.: Artificial intelligence with robotics in healthcare: a narrative review of its viability in India. Cureus **15**(5), e39416 (2023). https://doi.org/10.7759/cureus.39416
52. Pal, H.: Advancements and limitations in integrating robotics into medicine: a comprehensive review. Multidiscip. Rev. **7**(11), 2024248 (2024)
53. Yakub, F., Khudzari, A.Z.M., Mori, Y.: Recent trends for practical rehabilitation robotics, current challenges and the future. In: Proceedings of the 2014 International Journal of Rehabilitation Research, pp. 9–21 (2014)
54. Dutta, V., Zielińska, T.: Cybersecurity of robotic systems: leading challenges and robotic system design methodology. Electronics **10**(22), 2850 (2021)
55. Kadhim, S., Kadhim, D.A., Hussein, J.S.: A review comparative of lower limb exoskeleton for rehabilitation. Libr. Prog. Int. **44**(2s), 1105–1114 (2024)
56. Colombo, R., Sanguineti, V. (eds.): Rehabilitation Robotics: Technology and Application. Academic Press, Heidelberg (2018)
57. Tseng, K.C., Wang, L., Hsieh, C., Wong, A.M.: Portable robots for upper-limb rehabilitation after stroke: a systematic review and meta-analysis. Ann. Med. **56**(1), 2337735 (2024). https://doi.org/10.1080/07853890.2024.2337735
58. Blom, P.: Assessing the present and future regulatory landscape governing medical devices from the perspective of the Finnish Health Authority. https://www.doria.fi/handle/10024/189788 (2024)
59. Mikolajczyk, T., Ciobanu, I., Badea, D.I., Iliescu, A., Pizzamiglio, S., Schauer, T., et al.: Advanced technology for gait rehabilitation: an overview. Adv. Mech. Eng. **10**(7), 1687814018783627 (2018)
60. Acquisti, A., Brandimarte, L., Loewenstein, G.: Privacy and human behavior in the age of information. Science **347**(6221), 509–514 (2015)
61. Buolamwini, J., Gebru, T.: Gender shades: intersectional accuracy disparities in commercial gender classification. In: Proceedings of the 2018 Conference on Fairness, Accountability, and Transparency, pp. 77–91 (2018)
62. O'Neil, C.: Weapons of Math Destruction: How Big Data Increases Inequality and Threatens Democracy. Crown Publishing Group, New York (2017)
63. Reed, C.: How should we regulate artificial intelligence? Philos. Trans. R. Soc. A Math. Phys. Eng. Sci. **376**(2128), 20170360 (2018)

64. Zuboff, S.: The Age of Surveillance Capitalism: The Fight for a Human Future at the New Frontier of Power. PublicAffairs, New York (2019)
65. Kuner, C., Bygrave, L., Docksey, C., Drechsler, L.: The EU General Data Protection Regulation: A Commentary. Oxford University Press, Oxford (2020)
66. Binns, R.: Fairness in machine learning: lessons from political philosophy. In: Proceedings of the 2018 Conference on Fairness, Accountability, and Transparency, pp. 149–159 (2018)
67. Crawford, K., Calo, R.: There is a blind spot in AI research. Nature **538**(7625), 311–313 (2016)

Advancement and Challenges in Rehabilitation Robotics: Insights from Case Studies

Megha Mudholkar, Pankaj Mudholkar, Santushti Betgeri, and Monisha Mohan

Abstract Emergence of robotic technology in the field of healthcare, helping healthcare professionals in the treatment. One of the advanced Robotic technology is Rehabilitation robotics which is helping patients with physical disabilities in their recovery and rehabilitation. Various rehabilitation robots like Burt Robot, Myopro Robot etc. are available in the market to solve various problems related with physical disabilities. This chapter provides a thorough overview of real-life case studies and the various challenges faced in the development and deployment of such robotic systems for rehabilitation. Further, detailed analysis of different case studies, shedding light on the practical uses, technological progress, and user experiences will be presented in this chapter. Impact of integrating AI, ML, DL and AR/VR like technologies in rehabilitation robot will also be presented. The chapter further delves into future direction and potential solution to the challenges in rehabilitation robotics. This chapter offers a thorough examination of the successes as well as the difficulties encountered, contributing important new information to the continuing conversation about improving the efficiency and accessibility of robotic rehabilitation technologies.

Keywords Healthcare · Rehabilitation · Rehabilitation robot · Robotic rehabilitation technologies

---

M. Mudholkar (✉) · M. Mohan
Department of Computer Engineering, Marwadi University, Rajkot, Gujarat, India
e-mail: meghakunte2000@gmail.com

M. Mohan
e-mail: monisha.mohan@marwadieducation.edu.in

P. Mudholkar
Faculty of Computer Applications, Marwadi University, Rajkot, Gujarat, India
e-mail: mudholkarpankaj@gmail.com

S. Betgeri
Department of Computer Engineering, Vishwakarma Institute of Technology, Pune, Maharashtra, India
e-mail: santushti2014@gmail.com

J. Narayan and H. Gritli (eds.), *Design and Control of Rehabilitation Robots*, Studies in Systems, Decision and Control 585, https://doi.org/10.1007/978-3-031-86977-8_18

1 Introduction to Rehabilitation Robotics in Healthcare

1.1 Overview of Rehabilitation Robotics

Neurological diseases are the major cause of disability and the second cause of death worldwide, therefore being a serious public health problem [1, 2]. The most common neurological disorders include but are not limited to stroke, Parkinson's disease (PD), brain injuries and multiple sclerosis. Every year, about 500,000 people suffer from stroke in the U.S. and approximately in Europe, 1.1 million [1]. In this context, an estimated 6.1 million people suffered from PD worldwide in 2016, and the statistics showed a trend of increasing the burden of the disease [3]. Thus, the number of patients with a need for consultation by doctors who are proficient in neurological disorders is extremely high. As populations are growing and ageing [4], the prevalence of major disabling neurological disorders steeply increases with age, and health providers will face increasing demands for treatment, rehabilitation and support services due to these neurological disorders [5]. These needs usually involve multidimensional features: physical, cognitive, psychological, and medical; they can also be highly complex. Consequences generally relate to impairments of upper, lower, or both limb motions. Traditionally, recovery procedures include the mobilization of extremities and efforts by the patients. It follows that regaining motor function is one of the major goals of neuro rehabilitation, which is also an important prerequisite for the realization of activities of daily living in an autonomous manner.

Rehabilitation robotics has emerged as a rapidly growing field that combines the principles of engineering, medicine, and computer science to develop advanced assistive technologies for individuals with physical disabilities or impairments. In this application domain, the phrase "robotic technology" describes any mechatronic equipment with a certain degree of intelligence that can physically intervene on the behavior of the patient, therefore optimizing and accelerating his/her sensorimotor recuperation. These robots have two main powers: (1) evaluating the human sensorimotor ability; and (2) re-training the human brain to raise the quality of life for the patient.

In traditional rehabilitation, the medical practitioner is responsible for ensuring that patients continue to participate in exercises that are intended to assist injured limbs in regaining their movement function. There is a correlation between an aging population and a rise in severe impairment among stroke survivors [6, 7]. This correlation is made possible by a decline in the ratio of healthcare providers to patients [8]. Over the course of the last decade, there has been a rise in the utilization of robotic rehabilitation technologies. A healthcare provider can manage multiple patients at once in robotic gyms, which is one of the numerous advantages of robotic rehabilitation systems. These robotic systems aim to improve the conventional rehabilitation process by enhancement in the intensity of therapy, adaptive support, and objective measurement of patient progress. One of the major areas of research during the last couple of decades has been the development of rehabilitation robots to treat sensorimotor deficits, especially in the lower limbs.

Designing and controlling rehabilitation robots is a multidisciplinary process, which takes into consideration the underlying neurophysiological mechanisms of sensorimotor function, the specific needs and capabilities of the target patient population, and the technical requirements which allow for safe and effective interaction between humans and robots. This chapter will introduce the field of rehabilitation robotics, pointed out some of the most important considerations in the design and control of such systems then describe several case studies illustrating its clinical applications and possible benefits. Design in rehabilitation robotics has to be informed by an in-depth understanding of the neurophysiological processes underlying sensorimotor function and recovery [1]. Any rehabilitation approach should base its ideas on these insights: the recovery of function is complex and often limited. For example, during locomotion, the activation of certain neural pathways, recruitment of motor units, becomes critical that needs to be considered in developing rehabilitation robot control strategies. Well-known functional tests, such as the Fugl-Meyer Assessment (FMA) and the Action Research Arm Test (ARAT), are used to evaluate the severity of strokes, motor recovery, and rehabilitation.

Rehabilitation robots, through the use of sensors and advanced algorithms, can continuously monitor the patient state and progress and adapt the support provided. For the rehabilitation process to be most effective and stimulate neuroplasticity, the process by which the brain can reorganize itself by forming new neural connections, especially in response to experience and training, a personalized approach is crucial. Designing rehabilitation robots requires a multidisciplinary approach: only by combining very specialized engineering knowledge with a deep understanding of neurophysiology can such systems be developed that will interact safely and effectively with the user and adapt to particular needs and capabilities. One of the main design issues is the development of sophisticated interaction control schemes able to easily merge with user movement patterns, providing the correct degree of support or resistance. The design of such algorithms has to be performed taking into account minimal physical burden for the user while challenging them to actively participate in the rehabilitation process.

Rehabilitation robots may have advantages compared with conventional therapy, including a standardized training environment, adaptive support, and the possibility to increase therapy intensity and dosage while reducing the physical burden for therapists. A famous example of a rehabilitation robot is the Lokomat, which applies weight support and guidance during overground walking practice. It embodies sensors and sophisticated algorithms that enable the real-time monitoring of user performance and adjust, in real time, the level of support during locomotion, stimulating active participation of the user and encouraging neuroplasticity [1]. There exist a number of case studies on rehabilitation robots, some of which will be discussed in the chapter below. These few case studies already illustrate what potentials are opened by rehabilitation robots in complementing and enhancing traditional therapy in an effort to restore improved functional outcomes in patients suffering from sensorimotor deficits. General: Rehabilitation robots design and control for the treatment of sensorimotor deficits is a complex and evolving field, at the frontier between multiple engineering and medical disciplines. One of the case studies that will be presented is rehabilitation robots for upper limb function recovery [2].

A rehabilitation robot is a device designed for therapeutic purposes, which requires a multidisciplinary approach, combining specific knowledge in engineering with profound knowledge about neurophysiology and clinical principles of rehabilitation. Rehabilitation robots should be designed in such a way that they provide a safe, standardized, and adaptable environment for training, one that can be individualized according to specific needs and capabilities of each patient. Rehabilitation robots represent a development that needs to be advanced through a multi-faceted approach, taking into consideration specific needs and capabilities of the targeted patient group. That includes incorporating case studies on the clinical applications and potential benefits of such systems. For instance, case studies have shown that the application of rehabilitation robots is quite useful in improving balance and gait in patients who have either suffered from a stroke, spinal cord injury, or Parkinson's disease. These case studies show how adaptive support and objective feedback provided by rehabilitation robots can enhance the traditional process of rehabilitation toward improved functional outcomes for the patients. Several case studies investigated the application of rehabilitation robots for the treatment of impairments of the upper limb, showing their functionality in enhancing recovery of fine motor skills and hand function. The review with an assortment of case studies will be encompassed in this chapter, broadly covering applications and implications concerning rehabilitation robots in clinical rehabilitation.

Design and control are the most crucial research aspects of rehabilitation robots, which have immense potential to help improve the lives of persons with different forms of physical disabilities or impairments. Based on advanced engineering principles, neurophysiological knowledge, and clinical input, rehabilitation robots can offer a potent complement to traditional approaches in rehabilitation. As rehabilitation robotics is continuously refined, system development will be instrumental in promoting the patient's recovery as well as his quality of life.

1.2 Importance and Impact on Healthcare

Rehabilitation robots can be very important and have much impact on the healthcare system. Advanced technologies may completely change the process of rehabilitation, providing more effective, personalized, and data-driven therapies for people with various physical disabilities or impairments. Specialized knowledge of engineering, neurophysiology, and clinical expertise is applied in rehabilitation robotics. By combining those, rehabilitation robots support traditional rehabilitation methods and help to achieve better functional results among patients. Telerobotics can allow higher patient engagement with more adaptive support, objective feedback, and customizable settings, thereby hastening the pace of neuroplasticity and recovery.

Robot-assisted therapy may also bring to healthcare providers a uniform and quantifiable mode of treatment that would be very valuable in making decisions on and adjusting the personalized rehabilitation plan.

Moreover, as the population further ages and the incidence of sensorimotor deficits continues to rise, rehabilitation robots will become imperative in meeting the healthcare challenges. These systems can help reduce the burden on healthcare, enhance the quality of a patient's life, and in this way, increase efficiency and effectiveness in the model of care provided for rehabilitation. In fact, this is a multidisciplinary area that wishes to include engineering, neuroscience, and clinical rehabilitation in its research into rehabilitation robotics design and control. Rehabilitation robots are designed to assist and augment traditional rehabilitation methods by giving patients with sensorimotor deficits personalized and customized adaptive support.

Neurophysiological studies have greatly contributed to the understanding of the mechanisms that underlie motor learning and neural plasticity, both fundamental to functional recovery in physically impaired patients and to the development of rehabilitation robots with advanced control schemes that include integrated sensors continuously monitoring and adapting support to changing patient needs. These insights have informed the development of rehabilitation robots with advanced control schemes and integrated sensors that can continuously monitor and adapt the support to the patient's changing needs. Robotic technologies for rehabilitation have been around since the turn of the century [1] and are still growing quickly. However, not all pertinent technologies have advanced at the same rate. This field has seen substantial long-term technological advancements that have up to now resulted in advanced clinical and commercial applications, in contrast to wearing exoskeletons [9]. Through focused and repeated exercises for the lower [2] and upper limbs, such devices can promote neuroplasticity. By automating tiresome and difficult therapy [2] and by adjusting to the unique demands of the targeted individuals [1], robotic rehabilitation can lessen the workload for therapists. Rehabilitation robots could then be designed to provide either graded or task-specific assistance, challenging the patients to be more active during the exercise and thus promoting their neural pathway reorganization. Contrary to traditional therapy, rehabilitation robots provide standardized training environments, more intensive therapy, and less physical burden for therapists; however, many challenges still remain regarding the design and control of those systems. Efficient rehabilitation necessitates profound insight into the neurophysiological processes underlying this process and requires the elaboration of special engineering solutions that can easily be integrated into clinical practice.

Based on the previous considerations, below sections highlight recent findings in different fields of rehabilitation robotics that could be efficient and can be easily integrated into clinical practice. In the first section, we present an overview of robotic devices for the elderly, pediatric and stroke survivors with case studies mentioning the objectives functioning and challenges in each case. Each section also exploits accompanying knowledge of other areas, focusing on reconstructive training and valuation of changes induced by robotic rehabilitation on neural and behaviour change. The third section mainly focuses on technological challenges and a thorough analysis is included. We conclude by giving a general perspective on the research in the robotic rehabilitation field by illustrating the future technologies and vision for further enhancement.

2 Case Studies in Rehabilitation Robotics

Rehabilitation robotics is an engaging and developing field where robotics in all their genius helps in the rehabilitation of patients. This chapter gives a number of practical examples and explains practicalities, victories and complications of incorporating robotic devices into rehabilitation treatment schemes in everyday practice. These include patient group such as stroke patients, individuals with spinal cord injuries, patients suffering from protracted conditions, etc., that portrays how innovations in robotic technologies are patient specific in enhancing motor ranges and precision of the rehabilitation regimes over time.

This section features advances in technology, facilitation of collaboration between various disciplines, and the prospects for rehabilitation robotics in clinical practice. Therefore, it appreciates the degree of sophistication in robotic treatment and the changes in associated motor recovery, functional independence, and quality of life of patients. The case studies also portray other important issues such as how well patients can adapt to the intervention, its ease of use, and how it can be incorporated into standard clinical practice. This section also talks about the transformative impact of robots in rehabilitation by examining real-world applications, while also outlining significant challenges and prospects for future research. This detailed overview highlights the convergence of technology and healthcare, wherein clinical objectives are achieved with advanced robotic aid.

2.1 Case Study-1: Robotic-Assisted Gait Training for Stroke Survivors

2.1.1 Background and Objectives

Stroke is a worldwide health problem that affects millions of people every year and usually comes with severe disability such as impairment of walking and balancing functions. Most of the rehabilitation after stroke consists of restoring movement and achieving independence, with particular emphasis placed on gait training. Despite their focus on improving gait, conventional rehabilitation practices such as physical therapy encounter limitations due to factors including patient tiredness, therapist availability and the need for a certain level of intensity of exercise. Robotic-Assisted Gait Training (RAGT) has come in as a new approach to resolve these difficulties because it offers intense therapy to the patient on regular repetitive bases irrespective of the patient's condition.

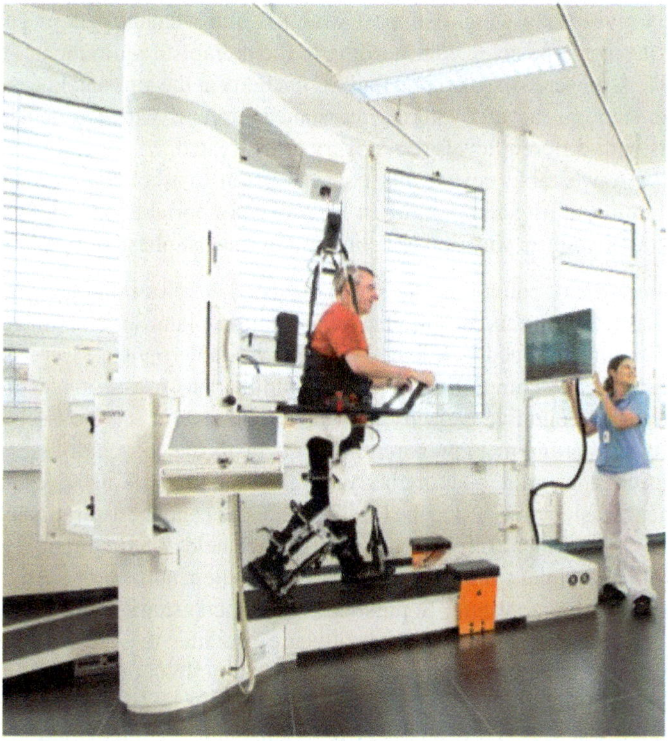

Fig. 1 Lokomat—picture: Hocoma, Switzerland [*Source* https://www.hocoma.com/media-center/media-images/lokomat/]

The primary aim of RAGT training consists of improving patient's walking disability and providing normal gait pattern through external robotic devices such as the Lokomat. Figure 1, this external device assists in motion of the extremities by the patients even in the presence of profound motion dysfunction allowing for neuroplasticity and motor retuning through focused goal-directed practice. The overall aim of RAGT does not only focus on walking retraining but aims at motor recovery and improvement in the quality of life of the client [10].

2.1.2 Implementation and Usage

Robotic-Assisted Gait Training (RAGT) has been successfully implemented in a number of rehabilitation programs around the world, and especially in neuro rehabilitation clinics and hospitals. Many rehabilitation centers exploit robotic systems in clinical conditions; they include [9]:

- Lokomat: This is treadmill confined device with an exoskeleton that facilitates, supports and regulates the movements of the patient's limbs and walking cycle and this gives body weight support which is a movement pattern given by the therapist but can be changed according to the therapist requirements.

- EksoGT: A powered exoskeleton that fits the body at the hips and knees that givers patients standing up and walking assistance support, with the device allowing patients with different levels of recovery to be treated.
- ReWalk: A wearable device that consists of an exoskeleton structure used in rehabilitation of affected gait. Most ReWalks are portable and allow patients to over ground walk as well and are common in the last phase of rehabilitation.

The clinical work pattern in The Robot-Assisted Gait Training pathway (RAGT) includes Assessing a patient, tailored rehabilitation plan, Sessions of robotic gait training, Continuous Feedback and Oversight. Suitable patients were selected for RAGT treatment by a comprehensive multidisciplinary team assessment of spasticity, muscle strength, cognitive capacity and functional state of cardiovascular system. A rehabilitation program is prepared subsequent to the thorough assessment that customizes changes implementing body weight support, walking speed and time of the treatment session in regard to the level of the patient and the aspiration of recovery. Standing, or some other exercises of the patient are held 2–5 times per week for 30–60 min with the patient therapist walking 4 and their adjusting the assistance as needed with a robotic system. Sophisticated robotic systems allow for continuous feedback such as provision for the healthcare provider to gauge progress in the patient, calibrate settings, and modify therapy instructions further which increases the aim of rehabilitation Rehabilitative. In this clinical framework and using these robotic systems, RAGT has been effective in enhancing walking ability and general mobility of individuals suffering from any form of neurological disorders.

2.1.3 Patient Outcomes and Benefits

The use of Robotic-Assisted Gait Training (RAGT) significantly enhances the speed and endurance of patients thereby enabling them to walk longer in shorter time periods. This is linked to the task-oriented and high repetition practice characteristic of RAGT, which builds muscles and improves better control of movements. It also enhances muscle balance and stability by enforcing even distribution of weight and an equal pattern of movement. This feedback helps the users correct the asymmetry in their walking, which is a common concern in post-stroke patients since the risk of falling is high. Moreover, RAGT facilitates neuroplasticity and enhancement of movement re-learning through the strengthening of pathways responsible for walking. Such exhaustive rehabilitation helps patients to recover lost functions and is essential in the process of stroke rehabilitation. RAGT also offers many psychological benefits such as enhanced self-confidence and readiness to have more participation in the rehabilitation process. The course of treatment is well organized and the advantages of the robotic systems are clear to the patients, which guarantees their positivity and motivation during the course of treatment. With the support of such a system, the safety concerns of falling, which is often a huge deterrent when it comes to physical therapy participation, are also addressed. Thus, it has been proven that the rehabilitation process is enhanced with the employment of RAGT, functional recovery of gait, neuroplasticity, and the functional outcomes in stroke survivors have improved as well [11].

2.1.4 Challenges

In spite of several benefits RAGT also has several challenges which are Cost and Market Penetration, Patient Engagement and Motivation, Adaptation to the Individualization, Physical and Cognitive Barriers etc. Availability has been restricted given the costly robotic devices, such as Lokomat, and staff that are trained in gait therapy—which is not easily located in developing or remote regions. The irritant value of such dull, boring and repetitive exercises results in reduced focus and motivation, thus impacting negatively on therapy. Adding elements of gamification or virtual reality increases focus but comes with challenges and costs. It is not easy to change the levels of assistance to each distinct patient's ability levels. Over-assistance leads to passive movement while under-assistance results to irritation or exhaustion. Finding the accurate level of assistance involves fine tuning and reevaluation of levels of assistance very often. RAGT is also less likely to be used for patients suffering from major cognitive difficulties, motor impairment, spasticity, treatment-induced contractures or surgical ones that prevent movement ranges, and/or rule introduction compliance.

Rehabilitation of stroke survivors was enhanced with the introduction of robotic assisted gait training (RAGT) as it provided a more organized and elaborates approach to training gait speed, gait balance, and mobility in general. In as much as RAGT has numerous advantages, there are barriers related to cost, access and patient adherence which make it hard to scale. For the purpose of mitigating those limitations, present-day research and technology are focused on making RAGT easier to use and more user friendly. With an eye towards the future, RAGT has a great potential controlled impact against the stroke that many patients suffer and other similar population of patients who experience difficulty with motion improvement in their activities of daily living, in turn enhances their quality of life and independence [12].

2.2 *Case Study-2: Pediatric Rehabilitation Robots*

2.2.1 Background and Objectives

Robots for Pediatric Rehabilitation are high-end technology that assists children suffering from any impairing physical and neurological conditions. Such conditions include; cerebral palsy, muscular dystrophy, spinal cord injuries, and developmental disorders. Unlike adult rehabilitation, pediatric rehabilitation requires a distinct approach, and these robots are directed to care for children in a more physical and emotional and developmental way. The aim of these rehabilitation robots for children is to customize and enrich therapy session and practices to enable better healing towards restoration of damaged motor functions and encourage regrowth of neural connections. Conventional strategies can prove less effective in enabling the therapist to offer therapy since they can be very demanding in time and resources while manual efforts can be inconsistent. Robotic systems deliver defined, accurate, and adequate repetitive practices while monitoring outcomes' data over time to evaluate patient's progress. Rehabilitation interventions with the robots in children's care the potential is reinforced in the review by Gonzalez Alberto et al., published in the year 2021. The robotics systems are precise, adaptable and include gamified interfaces which are easy to use making the imagined medicine in rehab application more real in pediatric rehabilitation [13].

2.2.2 Implementation and Usage

At rehabilitation centres, clinics as well as children's specialized hospitals, therapy sessions for children with disabilities and other impairments usually incorporate pediatric rehabilitation robots. Depending on a child's needs, one can select a different type of robot however some of the most common include:

- Exoskeletons: wear able devices similar to the Ekso Bionics Pediatric Exoskeleton that assist in limb movements to enable a child walk or entertain extricating limb functions.
- End-Effector Robots: For example MIT-Manus, where the primary concern is the movement of an extremity and enabling a child to perform hand activities through the use of robotic robots or even hand rehabilitation devices.
- Gait Trainers: Which in this case are robots such as the Lokomat that help children with problems of walking patterns by performing actions like body-weight support and guiding legs in the natural movements of walking.
- Interactive Robotic Systems: These include gamified interfaces for example the Armeo Spring Pediatric Robot that uses virtual games in therapy to motivate children engaging them with interactive activities while working on upper limb movements.

Figure 2 shows the custom-designed motorized assembly for providing knee-extension assistance during walking. Most of these robots are equipped with motion analysis and control systems that employ advanced sensors and artificial intelligence to change the level of assistance or resistance during the activity being conducted in real time. Hence, such enables the therapist to assess the performance rate with accuracy and alter the treatment approaches as per one's performance. Besides, different modes of immersive technology such as virtual reality (VR) and augmented reality (AR) can be incorporated into the therapy session to enhance its appeal to kids and cure boredom that may be associated with the repeated performance of the same exercises.

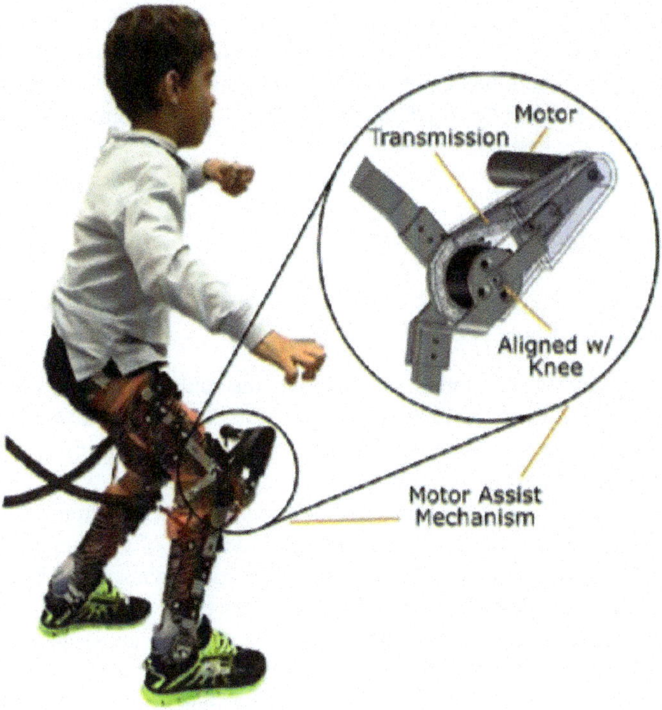

Fig. 2 Exoskeleton "picture: NIH Clinical Center" [*Source* https://irp.nih.gov/catalyst/25/6/robotic-exoskeleton-helps-alleviate-crouch-gait-in-children-with-cerebral]

2.2.3 Patient Outcomes and Benefits

Rehabilitation robots exert a noticeably beneficial influence on the rehabilitation of children, resulting in a marked even greater strides in motor function, balance and cognitive skills. The principal advantages of treatment with robotic devices for children are enhanced motor recovery, enhanced motivation, therapy supported by data, Therapist Load and Productivity, favorable long-term effects etc. Robots help to perform repetitive and accurate. Movements which support the process called neuroplasticity, meaning the abilities of the brain to develop new pathways and regain motor control, are enhanced. Research has indicated that some children with cerebral palsy and use of robotic gait trainers have an accelerated walking speed and a more upright posture as well as better coordination of the limbs than if they were to rely on conventional therapy. The incorporation of features such as gamification and virtual reality lowers the barriers to therapy transactional processes, as people perceive it as play rather than a burden, hence the rate of attrition is minimal. Robots administer and record consistent, guideline movements, furnishing their users with data on their performance levels in order for them to evaluate the effectiveness of the treatment, and modify the course of therapy. Robotics cut the physical work of the therapists allowing them to engage more in complex modalities and patient centered care. Children rehabilitated with the help of robots have been noted to have better motor control and functional independence in the long run. Robotic therapy produces better functional results in very young children suffering from great disabilities when administered as the first line intervention.

2.2.4 Challenges

Pediatric rehabilitation robots demonstrate notable advantages, however, their use is marred by several factors [14]:

- Very Limited Market Potential: Compared to the adult population, which is much more significant, the pediatric market is regarded as less lucrative when it comes to exoskeletons leading to difficulties in garnering support for their enhancement and manufacture.
- Growth and Physical Development: Considering that children are constantly growing, it proves arduous to come up with designs of exoskeletons that tend to such changes in size and proportions. It also complicates the existing problem of designing scalable devices.
- Dimensional Adaptability: The majority of the pediatric exoskeletons have been designed as scaled down adult models, which do not necessarily solve the issues that concern children. More of such approaches entertain the need of the child rather than basing it on the technology are desired.
- Variability in Cognition and Learning: Cognitive capacities and learning baselines differ greatly in children. This makes it challenging when building exoskeletons user-friendly across different cognitive levels.
- Trust and acceptance: Trusting children or parents or caregivers is problematic because it involves the perception of safety, functionality and reliability, which are somewhat conflicting. Safety mistrust, whether it arises from the technology itself or the non-user-friendly design, greatly hinders its use.
- Absence of Guidelines Made for Children: With regard to exoskeletons for children, there is scant development of trends and requirements that could be termed as regulatory, making their safe design and marketing quite challenging.
- Heavy Costs of Development: The design of pediatric exoskeletons is an expensive process focusing on the strong need to assure adaptability, durability, safety and at the same time not having the expectation of high commercial returns.
- Social and Emotional Challenges: Exoskeletons are wearable devices that assist children with movement. On the contrary, they may face difficulties due to emotional and social concerns such as appendages that make them look different especially when in the company of peers or wanting to belong to a certain group.

Pediatric rehabilitation robots provide a revolutionary solution to motor challenges faced by children by offering interesting, accurate, and intensive therapy. This has translated to better results from the patients especially in restoration of motor functions and quality of life after a long period. But some barriers such as cost and children's refusal to accept them existed that limited the adoption. Robotic therapy will, however, require more research and development in order for it to be available and more effective to young children since technology is in constant change.

2.3 Case Study: Robotic Systems for Elderly Care

2.3.1 Background and Objectives

The trends of increasing elderly population provide avenues for enhancement and re-engineering of care for the aged. The populations of the elderly are increasing especially in the continents of Europe and Asia, and thus there is a need for realistic and creative measures to improve the living standards of elderly people. The researchers are investigating the use of robotic solutions to subsidize independent living, alleviate the burden of caregivers, and in general, encourage active aging. Robotic technology in elderly care giving can prove to be useful in a great extent by helping perform certain tasks on daily basis, record some health patterns, and facilitate communication within a social circle. The focus of these systems is to help the elderly remain as active and independent in their homes prior to needing to be institutionalized, which is the idea of "aging in place".

2.3.2 Implementation and Usage

In contemporary society, robotic systems are penetrating various service sectors in the home and institutional environment to ease the process of elderly care. Aid robots which undertake specific activities such as mobility assistance, remembering to take medication, and cleaning all of which help to do away with the need to a large extent the normal care giving routine [15]. Therapy robots that offer movement therapy or rehabilitation in the form of gait training with feedback on the therapeutic process such as the PHAROS system which enables detection and correction of wrong postures in order to keep active rehabilitation in check [6, 15]. Robots such as Paro and Pepper who, by virtue of their design, can be verbally interacted with by seniors, helping them with the sense of having someone around, thereby combating boredom which is a big issue especially for dementia patients taking into consideration the role of socialization towards one's mental and cognitive health (Fig. 3) [7].

Moreover, there are also developed and developing technologies that include, health care management systems that are combined with AI-enabled robots that monitor the patients' physiological signs in real time, collect and store health-related information and alert the caregivers or doctors in cases of emergencies. Most of this system is enhanced by body worn devices to ensure the patient's health is constantly monitored.

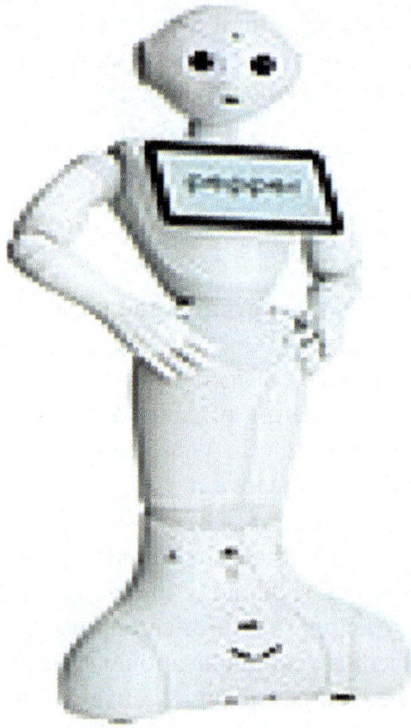

Fig. 3 Pepper robot [*Source* https://commons.wikimedia.org/wiki/File:Pepper_the_Robot.jpg]

2.3.3 Patient Outcomes and Benefits

Research in robotic systems designed for the elderly has revealed many positives. The greatest emotional and physical burden on the caregivers is associated with performance and therefore the robots perform repetitive activities. They also help in better outcomes of the patients through the following edges, medication adherence, health metric monitoring, fall or accident prevention. Even for elderly patients with cognitive deficits, such as dementia, robots like Paro are able to decrease agitation and enhance well-being. In addition, robotic systems are effective in averting crises through the management of physiological parameters, which are monitored and health practitioners notified of any issues in real time in the hope of preventing escalation and serious injuries. Through exercise sessions and cognitive games, certain robots have also been shown to encourage movement and active participation; thus helping to combat the effects of physical and mental deterioration. Robotic systems, one may conclude, stand to change the face of elderly care by offering adaptive assistance tailored to each individual and enhancing their quality of life.

2.3.4 Challenges

Robotic systems have the potential to be very helpful in their application in elderly care, but there are several issues to be resolved [8]. Perceived incapacity of robots for complex understanding of human interaction and emotions or physical integration or flexibility related work. Care of the elderly using robots is not only for the economic and technical solutions. It should be regarded as the eco-social-technical system. Elderly users and caregivers are important in the design process and should be involved from the scratch to the very end of the project, not just for functionality. Elderly users are prone to hesitation due to unfamiliarity of robots and its essence being already pushed and used on them without proper understanding making them feel left out or disregarded. AI in care robots should address the safety concern more i.e. regarding privacy, cognitive and psychosocial safety as well as security. In care robots, a focus on the elderly users' demographics must be integrated inclusively and respect human rights such as access to healthcare services and the right to a standard of living. Where there is limited usage of digital devices in robotic infrastructural systems among the elderly, it predisposes them to increased marginalization. Robots should adhere to UD criteria like the fact that the use shall be available to all persons, easy and with minimum strain associated with both bodily and cyberspace, which robotics has not adequately explored currently. Future initiatives should aim at integrating such values as care and autonomy within technology designed effectively preserving human dignity and enhancing empowerment. Notions such as autonomy and safety may change with the use and acceptance of robots; thus, a different perspective is required that goes beyond physical safety, but elaborates on psychosocial safety and includes the consideration of human agency.

Robotic systems have much promise in the improvement of elderly care from better enhancing patient autonomies to improving patient outcomes and alleviating stress on caregivers. However, factors such as cost, user-friendliness, and privacy issues must be tackled first in order to facilitate usability and success. With advancements in technology, it is expected that the demand for robots especially in caring for older people will increase dramatically as it is a sector likely to be revolutionized due to care giving robots.

2.4 Case Study: RobHand Rehabilitation

2.4.1 Background and Objectives

The RobHand system is a revolutionary robotic exoskeleton designed to aid in the neuromotor rehabilitation of stroke patients, specifically those experiencing impaired hand functions. RobHand exoskeleton is used to assist in the rehabilitation of stroke patients by enabling them to perform repetitive finger flexion and extension exercises. It aims to improve manual function, hand muscle strength, and motor coordination through a structured neuromotor rehabilitation program. This innovative technology seeks to provide a more efficient, consistent, and accessible approach to hand rehabilitation, overcoming the challenges of traditional manual therapy, which can be labor-intensive and time-consuming. The primary goal of the RobHand exoskeleton is to improve hand mobility, promote muscle strengthening, and enhance the overall quality of life for individuals with hand paralysis resulting from neurological conditions such as stroke.

2.4.2 Implementation and Usage

The RobHand rehabilitation system employs a structured therapy protocol, consisting of 16 sessions over a period of eight weeks, with two 60-min sessions conducted weekly. Each session comprises a series of rehabilitation phases, including flexion and extension exercises for the fingers, utilizing the exoskeleton to mimic natural hand movements and simulate motor function through repetitive tasks. The robotic platform provides guidance and assistance as necessary, promoting bilateral rehabilitation. In certain studies, the exoskeleton incorporates EMG-driven control, allowing the patient's healthy hand to control the robotic hand, thereby augmenting the rehabilitation process. The RobHand is an electromechanical exoskeleton device designed to attach to a patient's hand, providing assistance in various finger movement rehabilitation therapies (Fig. 4).

RobHand is used in rehabilitation programs for stroke patients to assist with hand movements. The exoskeleton facilitates passive and active exercises such as hand opening/closing, finger flexion/extension, and various grip movements. It is typically applied in sessions, with patients undergoing two 60-min sessions per week for eight weeks. The device is adjustable, enabling personalized therapy based on the patient's residual motor function. A therapist configures the device for each session, specifying the type of exercise, number of repetitions, and movement velocity [16].

2.4.3 Patient Outcomes and Benefits

Clinical studies evaluating the RobHand system have yielded promising results regarding patient outcomes, with participants reporting high levels of comfort, ease of use, and overall satisfaction. The safety profile of RobHand was exemplary, with no significant adverse effects observed, such as skin irritation or excessive fatigue. Patient satisfaction ratings were consistently high, with an average score of 4 out of 5 on the Quebec User Evaluation of Satisfaction with Assistive Technology. Although improvements in manual function were not statistically significant in all cases, particularly among chronic stroke patients, these findings highlight the need for further research focusing on subacute patients to investigate longer-term benefits and potential greater improvements in manual function.

2.4.4 Challenges

The implementation of RobHand for rehabilitation has encountered several challenges. One key issue is the limited improvement in motor function among chronic stroke patients, highlighting the need to evaluate the system's effectiveness in earlier

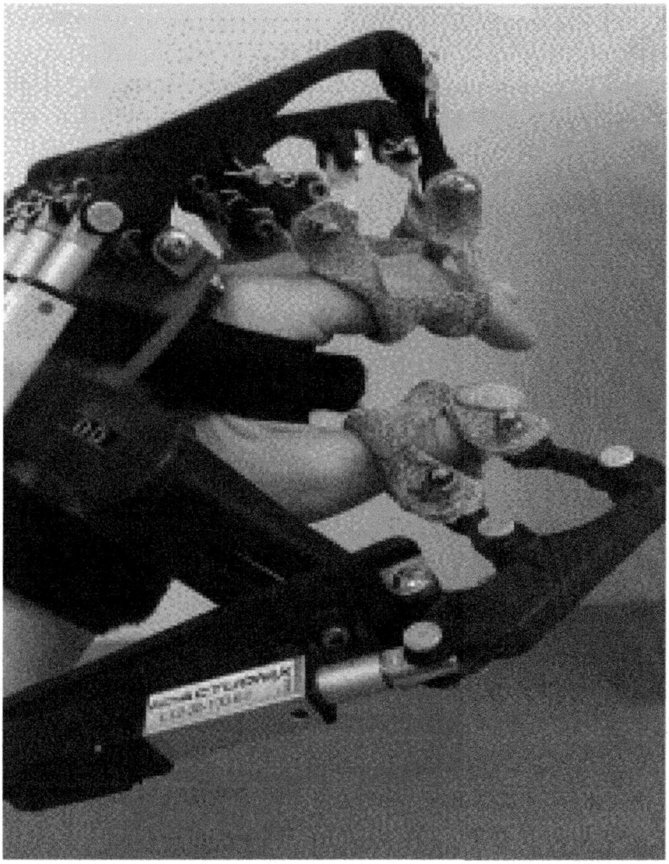

Fig. 4 RobHand exoskeleton [*Source* https://www.frontiersin.org/journals/robotics-and-ai/articles/https://doi.org/10.3389/frobt.2023.1146018/full]

stages of recovery. Additionally, the high cost and complexity of the robotic system pose significant barriers to accessibility, particularly in resource-constrained healthcare settings. Furthermore, while RobHand has demonstrated benefits in user satisfaction and ease of use, additional refinement is necessary to optimize therapeutic outcomes, specifically in enhancing dexterity and hand strength. Some of the key challenges are Limited Functional Improvement, Device Expectations, Stroke Stage Influence, and Evaluation Metrics [16]. The study found no significant changes in manual function after the rehabilitation program, particularly in chronic stroke patients. This is likely due to the long period since their stroke episodes, which may limit further recovery. Some patients expressed lower satisfaction regarding the device's effectiveness, potentially due to high expectations of full recovery, which were not met within the therapy cycle. The effectiveness of RobHand may be limited in chronic stroke patients. Future studies are recommended to focus on subacute patients to evaluate the full potential of RobHand in earlier stages of recovery. Some of the tests used in the study were not suitable for patients with severely impaired motor function, indicating the need for optimizing the evaluation protocols. While RobHand has shown promise as a safe and effective tool for hand rehabilitation, particularly in terms of patient satisfaction and comfort, additional research is necessary to improve its efficacy in promoting motor recovery, particularly among chronic stroke patients. This will enable the optimization of RobHand's therapeutic potential and enhance its impact on improving hand function and overall quality of life for individuals with stroke-related hand impairments.

2.5 *Case Study on Myomo's MyoPro Robot*

2.5.1 Background and Objectives

The MyoPro, a powered upper-limb orthosis developed by Myomo, Inc., is a cutting-edge device designed to facilitate the restoration of mobility and functionality in individuals with neurological disorders or upper-limb paralysis. This device is intended to assist patients who have suffered from strokes, traumatic brain injuries (TBI), or other neuromuscular impairments, including brachial plexus injuries and cerebral palsy, in regaining independence in their daily lives. The primary objective of the MyoPro is to restore lost motor functions, enabling users to perform essential tasks autonomously, such as self-feeding, opening doors, and completing household chores, thereby significantly enhancing their overall quality of life [17].

2.5.2 Implementation and Usage

The MyoPro is meticulously custom-fitted for each patient by certified clinicians, including orthotists and occupational therapists, subsequent to a comprehensive evaluation to ascertain suitability. Upon fitting, the device is utilized in both clinical and domestic settings to support rehabilitation. Its portability facilitates usage beyond the confines of a hospital or clinic, thereby enhancing long-term therapy outcomes. The MyoPro boasts adjustability, with software tailored to each patient's specific needs, enabling the device to adapt in response to changes in muscle strength. For example, in patients with progressive conditions such as ALS, the device settings can be adjusted to maintain functionality. Additionally, a pediatric iteration of the MyoPro is currently in development to cater to children with conditions like cerebral palsy, thereby broadening the device's applicability and potential to positively impact lives.

Figure 5 shows the MyoPro wearable device which can assist patients with brachial plexus injury [17].

Fig. 5 The MyoPro wearable orthotic device [*Source* https://myomo.com/what-is-a-myopro-orthosis/]

2.5.3 Patient Outcomes and Benefits

Numerous studies have consistently demonstrated the efficacy of MyoPro in significantly improving patient outcomes. A notable study revealed that individuals with chronic upper-limb impairment resulting from stroke or traumatic brain injury (TBI) exhibited substantial improvements in motor function, as measured by the Fugl-Meyer scale and other clinical assessments. These improvements were not only achieved in-clinic but also maintained at home, highlighting the device's effectiveness in long-term rehabilitation. Additionally, patients have reported a high level of satisfaction with MyoPro, as it enables them to regain independence in daily activities previously rendered impossible by their conditions. For example, stroke survivors and veterans with long-term disabilities have been able to resume work and improve their overall quality of life, thereby reducing their dependence on caregivers.

2.5.4 Challenges

Despite its considerable potential, the widespread adoption of MyoPro faces challenges. The primary obstacle is insurance reimbursement, as the device's cost can be substantial. Although Myomo is actively working to expand insurance coverage, out-of-pocket expenses may remain high for some patients depending on their insurance plans. Furthermore, ensuring proper training and customization presents an additional challenge. Each device requires precise fitting and adjustment, necessitating extensive coordination between the patient, clinician, and manufacturer. Moreover, both patients and clinicians must navigate a learning curve to optimize device utilization.

2.6 Comparative Analysis of Case Studies

This section examines five case studies, as discussed, on rehabilitation robotics, focusing on their advancements and challenges. The analysis highlights the objectives, usage, challenges, and outcomes of robotic systems in rehabilitation and elderly care. The case studies showcase the potential of robotic systems in improving motor function recovery in stroke survivors and children with neurological impairments, as well as supporting independent living in the elderly. While robotic systems like Lokomat, MyoPro, and pediatric rehabilitation robots demonstrate promise in enhancing mobility, independence, and quality of life, the analysis also identifies significant challenges, including high costs, accessibility barriers, and the need for personalized therapy. This comprehensive review provides a detailed understanding of the benefits and limitations of robotic-assisted rehabilitation in modern healthcare, emphasizing the need for continued innovation and research to fully realize its potential and address the existing challenges (Table 1).

3 Challenges in Rehabilitation Robots

The integration of robotics into rehabilitation practices offers immense potential for improving patient outcomes. However, numerous challenges hinder the widespread adoption and effective utilization of rehabilitation robots. These challenges span technological, human–robot interaction, regulatory, ethical, economic, and logistical domains (Fig. 6).

3.1 Technological Challenges

Rehabilitation robots have to accomplish complex and precise tasks while being tunable to diversified patient needs. However, in any case, the development of technologies to accurately imitate or assist human movement is foreseen as yet remaining challenging. Technologies will have to be developed that will provide advanced sensors, actuators, and control systems for real-time feedback and adaptation based on the patient's physical and cognitive state [18]. Most rehabilitation robots still face some problems, such as low precision in the control of multi-joint exercises, smooth transitions of exercise modes, and diversities among patients. Another technological

Table 1 Comparative analysis of rehabilitation robotics case studies

Sr. No.	Name of case study	Objective	Implementation and usage	Challenges	Remarks
1	Robotic-assisted gait training for stroke survivors	To enhance walking ability and restore functional gait patterns in stroke survivors through robotic exoskeletons, fostering neuroplasticity and motor relearning	Widely used in specialized neurorehabilitation centers using systems like Lokomat, EksoGT, and ReWalk. Involves assessments, personalized therapy plans, and real-time feedback	Cost, accessibility, patient engagement, adaptation to individual needs, physical and cognitive limitations	Promising results in improving gait speed and neuroplasticity, but cost and accessibility remain barriers
2	Pediatric rehabilitation robots	To provide personalized, engaging, and repetitive therapy sessions to foster neuroplasticity and improve motor functions in children with physical and neurological impairments	Commonly used in rehabilitation centers and hospitals. Utilizes devices like exoskeletons, gait trainers, and interactive robotic systems, with gamified therapy options	High costs, limited accessibility, customization challenges, psychological considerations, technological limitations, ethical concerns	Highly beneficial for pediatric motor recovery, with improved engagement through gamification, though cost and access limit widespread adoption
3	Robotic systems for elderly care [8]	To support independent living, reduce caregiver burden, and promote healthy aging through robotic assistance in daily activities, monitoring, and social interaction	Used in homes and institutions to assist with daily tasks, physical therapy, and social engagement. Examples include assistive, rehabilitation, and social robots like Pepper	Technical limitations, high costs, privacy and ethical concerns, user acceptance, training	Can revolutionize elderly care by improving autonomy, but faces technical, ethical, and financial challenges

(continued)

Table 1 (continued)

Sr. No.	Name of case study	Objective	Implementation and usage	Challenges	Remarks
4	RobHand Rehabilitation	To improve hand mobility, promote muscle strengthening, and enhance the quality of life for individuals with hand paralysis from neurological conditions like stroke	A structured therapy protocol involving robotic hand exoskeletons. Sessions involve repetitive exercises for finger movements, using EMG-driven control for better results	Limited improvement in chronic stroke patients, high cost and complexity, need for further refinement in dexterity enhancement	Safe and effective in early-stage recovery, but needs refinement for chronic patients. High cost limits broader use
5	Myomo MyoPro robot	To restore motor functions in individuals with neurological disorders or upper-limb paralysis, allowing them to perform daily tasks independently	Custom-fitted for patients by clinicians, used in clinical and home settings. Provides continuous rehabilitation through EMG sensors detecting muscle signals and amplifying them	Insurance reimbursement, high costs, need for precise fitting and customization, learning curve for both patients and clinicians	Effective in restoring upper-limb motor function but high costs and reimbursement issues hinder accessibility

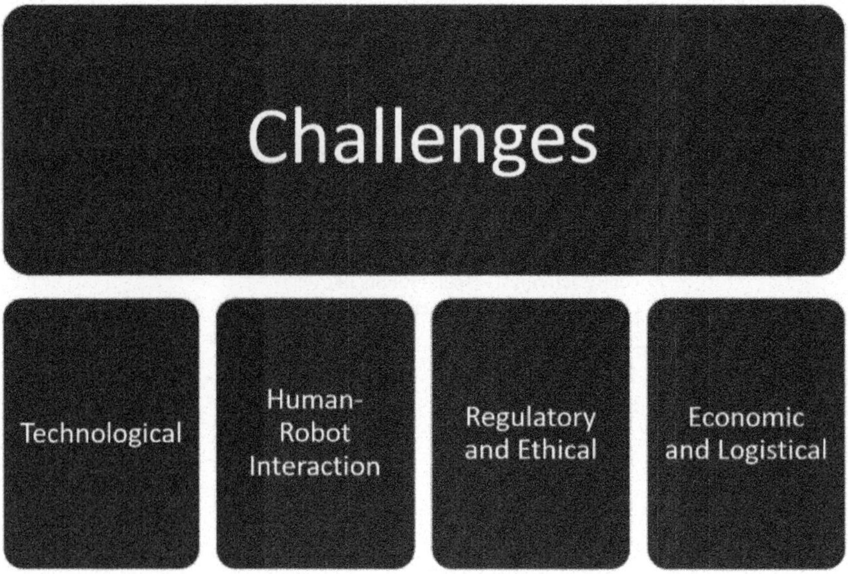

Fig. 6 Challenges in rehabilitation robots

challenge is wearability and portability. Whereas there is a drive towards light and non-invasive exoskeletons and prostheses, many are still bulky and cumbersome; this reduces their practical feasibility in long-term use [19, 20]. Power efficiency, battery life, and miniaturization of components remain significant hurdles. Another important challenge is interoperability among different clinical medical technologies; rehabilitation robots often have to be integrated into the healthcare system, which requires them to be compatible with other clinical devices and software.

3.2 Human–Robot Interaction Challenges

Human–robot interaction is one of the most important features for determining the efficacy of a rehabilitation robot. Robots are never intuitive, safe, and responsive in responding to the needs of various patients. The patients might have different levels of motor functions, cognitive abilities, and emotional reactions, complicated by developing a one-size-fits-all machine. It is also important that the robot strikes a balance between the assistance given and the effort a patient needs to induce to optimize rehabilitation outcomes. Too little, and it might get frustrating; it even could hurt. If it provides too much assistance, it reduces the motivation of the patient to partake in active therapy; if it provides too little, then it could be frustrating and even injure them. Patients also take some time to get used to the use of robotic devices, whereas both the patients and therapists will be presented with steep learning curves. The interaction is designed in such a way that the cognitive load becomes minimum wherein the trust in the technology is enhanced thereof. Resistance from the patients themselves will take place in a few cases because of their feelings of irritation or unease with this technology. Concomitantly with these, the designer has to consider all factors of emotional acceptance and psychological comfort during the design of a robot.

3.3 Regulatory and Ethical Challenges

With the entry of robots into the process of rehabilitation, a host of regulatory and ethical issues are raised. Medical robots are still at the mercy of developing their regulatory frameworks. Such devices have to be strictly checked against safety and efficacy standards. Any malfunction or mistake in the execution of the robot can lead to serious consequences for patient safety, and puts considerable pressure on regulatory bodies to enforce comprehensive testing and validation processes before approval of such systems for clinical use is given [19, 20].

There is an ethical concern of patient autonomy and the place of robots in health care decision-making. Concerns include how much autonomy the robots should have in controlling therapy sessions, depersonalization of patient care for over-reliance on automated systems [19]. Of paramount importance, of course, would be the issue of data security and privacy, since these robots collect sensitive medical and personal data. The whole process requires constant effort in terms of getting proper encryption and privacy protocols to avoid data breaches [20].

3.4 Economic and Logistical Challenges

Probably the single biggest barrier to the further dissemination of rehabilitation robots is their economic feasibility. Many of these systems are very expensive to purchase, maintain, and train personnel to use effectively [20]. In small clinics, or those in low-resource settings, investment in these technologies may be prohibitively expensive, thereby exacerbating inequities in access to advanced rehabilitation services.

Other logistical considerations, besides financial ones, include the integration of these robots within existing healthcare systems. Clinics have to redesign spaces to accommodate the robots and redesign workflows to integrate robotic-assisted therapies without disrupting traditional therapeutic modes. This would take extensive training of the staff and new protocols on how therapists and robots work together to optimize patient care. Another financial barrier involves inconsistent reimbursement policies for robotic rehabilitation services. The treatment covered under many insurance plans is usually not mentioned, and hence it is hard to justify such a cost on behalf of the institutions [20].

The third drawback is that the robotic systems are pretty expensive and are difficult to maintain and upgrade. They require regular maintenance—a lack of which will lead to wastage of time due to their unavailability, something that could negative impinge on therapy sessions. The requirement for highly specialized technicians adds to operational costs. Scaling up the robotic rehabilitation systems also poses a challenge, more so while trying to implement these solutions across multiple healthcare settings [19, 20].

While rehabilitation robots hold a lot of promise for improved therapeutic outcomes, there are several challenges yet to be overcome before these systems become standard in the provision of rehabilitation care. There are technological hurdles in precision, adaptability, and usability that need to be overcome to ensure robots serve as effective recovery tools for patients. Also, human–robot interaction should be refined so that patients and therapists can use such technologies comfortably and effectively. The key to making those many innovations improve the lot of more patients in a variety of settings will lie in the resolutions of regulatory, ethical, and financial challenges. Meeting these challenges means the realization of the full potential of the field of Rehabilitation Robotics to help shape the modern face of rehabilitation therapy.

4 Future Work and Innovation

Rehabilitation robotics is an evolving field focused on enhancing recovery for individuals with motor impairments through innovative technologies. Advances such as soft robotics, artificial intelligence (AI), brain-machine interfaces (BMI), and wearable robotics are pushing the boundaries of what rehabilitation robots can achieve, offering more tailored, efficient, and accessible solutions. These technologies are transforming therapy by improving adaptability, interaction, and real-time feedback, which contribute to better patient outcomes. Additionally, tele-rehabilitation and remote monitoring are increasing the reach of therapy, particularly for those in remote or underserved areas. While challenges related to technology, human–robot interaction, regulation, and cost remain, ongoing advancements and interdisciplinary efforts provide promising solutions. These innovations are set to shape the future of rehabilitation robotics, creating more effective and accessible therapies that improve patient care and quality of life (Fig. 7).

4.1 Emerging Technologies in Rehabilitation Robotics

Rehabilitation robotics is one of the most rapidly evolving fields due to the progress and innovation made in technologies and engineering to ensure better recovery in patients suffering from physical impairments. New, emerging technologies will push the boundaries of what rehabilitation robots can accomplish, creating even more personalized, efficient, and accessible solutions for numerous types of motor impairments. Some of the leading upcoming technologies that will be shaping the future of rehabilitation robotics are hereby highlighted.

4.1.1 Soft Robotics

It is changing the face of rehabilitation robotics by constructing devices that could be flexible, compliant, and capable of emulating the properties of human tissues. Unlike most rigid robots, soft robots will be made of materials that will further accommodate the human body for more comfort and adaptability with the use of the user. It is for this reason that improvements in technology now enable these wearable robotic systems, such as exosuits or exoskeletons, to help patients walk again with much more ease and less pain. In this respect, the soft actuators, sensors, and control systems allow better flexibility for longer use in rehabilitative applications [18, 19].

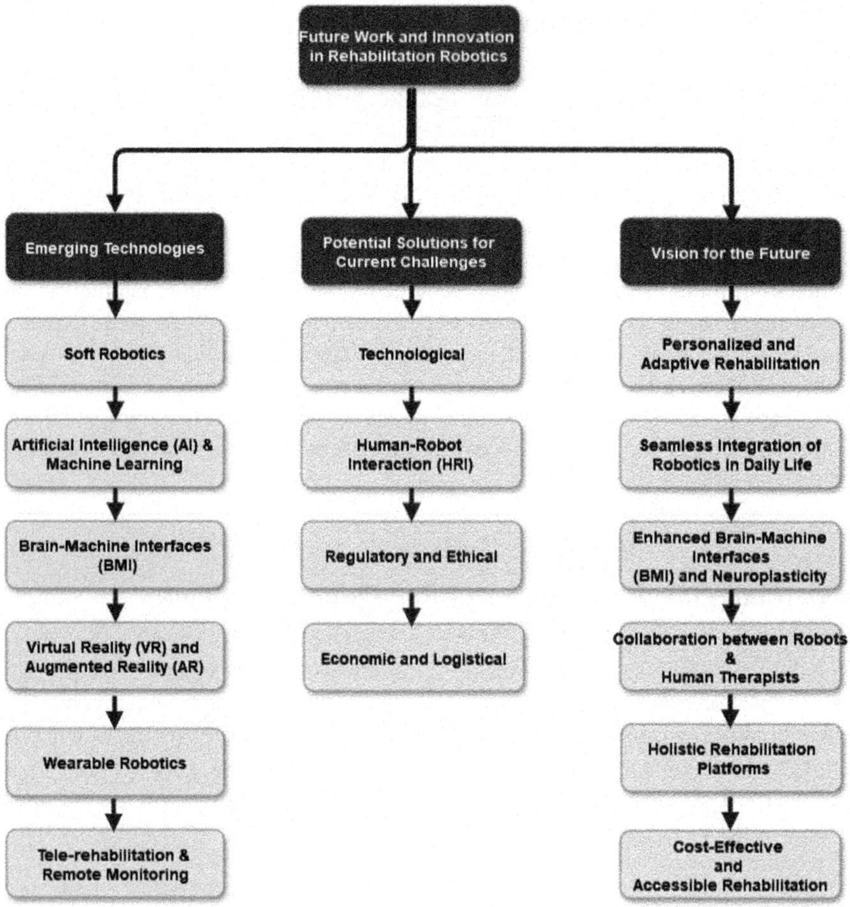

Fig. 7 Future work and innovations in Rehabilitation Robots

4.1.2 Artificial Intelligence (AI) and Machine Learning

AI and ML are currently changing how rehabilitation robots interact with patients. AI-based systems learn from the actions performed by a patient and change the therapy sessions in real time in correspondence to the progress made by them; treatment tends to be more personalized and adaptive, tailored to every individual's needs. Machine learning algorithms can interpret and analyze the patient data for predicting recovery trajectories, optimizing therapy intensities, and offering feedback in real time during various exercises [20, 21]. AI can also be extended to improve rehabilitation robot control strategies. With AI-based controllers, such robots will make more intelligent decisions on the amount of assistance to provide, given a patient's need for sufficient challenge without over-straining. This balancing of assistance and challenge is an important factor in improving neuroplasticity and promotion of motor recovery [20].

4.1.3 Brain-Machine Interfaces (BMI)

Brain-Machine Interfaces are a state-of-the-art technology; it directly connects the brain to robotic systems. BMIs can decode neural signals and transform them into commands, capable of commanding robotic limbs or robotic exoskeletons. New possibilities offered to patients with severe motor impairments-from spinal cord injuries up to stroke-allow the recovering of lost motor activity even when conventional motor pathways are seriously damaged [19]. Recent development of non-invasive BMIs has made this technology much more accessible without the need for surgical implants. In combination with neuro-rehabilitation robots, BMIs allow for a better therapeutic experience by incorporating the brain of the patient directly into the recovery process, thus enhancing neuroplasticity and improving motor function [21].

4.1.4 Virtual Reality (VR) and Augmented Reality (AR)

Virtual reality and augmented reality are fast becoming important parts of modern rehabilitation robotics. Virtual Reality systems create immersive, interactive environments where patients can practice motor skills in a controlled yet engaging way. For instance, VR-based rehabilitation may simulate some real-world tasks-like reaching or walking-in an absolutely safe way that helps patients regain functional ability [21]. On the other hand, AR displays virtual information onto the real world, guiding and providing feedback interactively in real-time during the therapy sessions. It can also enhance the interaction of patients with rehabilitation robots by providing direct visual feedback and instructions right into their field of view, making exercises more intuitive and easier to follow [20].

4.1.5 Wearable Robotics

Wearable robotics, powered exoskeletons, and passive exosuits are becoming lightweight, portable, and accessible in both clinical and home settings. These devices are for assisting walking, standing, and doing activities of daily living; therefore, patients recovering from stroke and spinal cord injuries can be given more independence. Advances in sensor technology and control algorithms have allowed these devices to adapt in real-time to the user's movements, providing the right amount of support exactly when needed [18, 21]. Biologically inspired actuators, on the other hand, such as pneumatic or hydraulic systems, permit wearable devices to more closely approximate human segmental movements. Their wearability and comfort further optimize these systems for practical use in long-term rehabilitation; this extends motion therapy to patients outside of conventional clinical settings at large [19, 21].

4.1.6 Tele-rehabilitation and Remote Monitoring

Tele-rehabilitation, supported by remote monitoring technologies, is expanding access to robotic rehabilitation by letting patients continue therapy from the comfort of their own homes. That involves use of robotic systems which could be controlled or even monitored remotely with healthcare professionals to ensure consistent therapy for the patients even outside a clinic setting [20]. Such systems often combined with AI and IoT technologies, offer real-time remote monitoring of patients' progress to therapists by adjusting therapy regimens and feedback based on data gathered by wearable or stationary robotic devices. The idea of this can be especially appealing in making rehabilitation more flexible and accessible for patients whose residing locations are in remote or underserved areas. New technologies are coming up that are revolutionizing the mode of therapy delivery to patients suffering from motor impairments in rehabilitation robotics. Soft robotics, AI, BMIs, VR/AR, wearable devices, and tele-rehabilitation systems form the driving force for the next generation in robotic-assisted therapies. These innovations are making rehabilitation personalized, more efficient, accessible, helping them achieve better outcomes and improvement in the quality of life. These technologies, constantly evolving, will no doubt play an increasingly active role in the shape of neurorehabilitation to come.

4.2 Potential Solutions for Current Challenges

While identified, the challenges in rehabilitation robotics are not insurmountable. Continuous technological advance, refinement of clinical practice, and true interdisciplinary collaboration could provide possible solutions for many issues faced in this area. Current challenges in rehabilitation robotics call for the following solutions.

4.2.1 Technological Challenges

The development of robust, adaptive, and versatile robots remains one of the primary technological challenges. Solutions to address these challenges include improvement in sensing and control algorithms, developing lightweight and wearable devices, standardization of interoperability. Rehabilitation robots can be designed with highly advanced sensors, together with unparalleled machine learning algorithms for higher precision and adaptability. These adaptive control algorithms may make it possible to have real-time responses from a robot when there are changes in patient performance, thus making therapy personalized and maximally effective throughout the recovery process. AI-driven systems can automatically calibrate movements based on patient feedback and progression. Soft robotics and biologically inspired actuators have emerged as a developing area that will enable the development of lightweight wearable devices without having the challenge of bulky and rigid conventional robots. In fact, such systems have been designed in a manner that they should be able to provide the required support without compromising comfort and flexibility. With further innovation in material sciences, including the use of soft polymers and flexible actuators, there will be fully portable and user-friendly wearable robots. Since the seamless integration of robotic devices with other systems ensures their easy adoption into clinical environments, standardization efforts can amply pave the way for rehabilitation robots. Uniform protocols concerning data exchange and system compatibility will make different devices work more in unison during therapy sessions [20].

4.2.2 Human–Robot Interaction (HRI) Challenges

Human–Robot Interaction (HRI) is a critical area in rehabilitation, and optimizing it is essential for successful therapy. Potential solutions include user-centered design, level s of intelligent assistive, real-time feedback and interaction. The design of robots should be user-friendly for patients and therapists. This will be further helped if the feedback that comes from users during the design process is integrated into the making of robots that are easy to use and intuitive, offering personalized therapy. This will also reduce the cognitive load for the patients and hence make the therapy effective accordingly. Designing intelligent assistive features for the robot include the fact that it could automatically adapt-first, depending on the patient's needs and reduce manual adjustment that a therapist has to make. Using AI and machine learning, robots can detect when to provide more or less assistance appropriately in achieving an ideal situation where most effective therapy is managed in concert with active engagement of the patient [18]. Real-time feedback that comes in forms of visual, auditory, or tactile feedback can increase patient motivation during therapy. Immersive environments through the arising technologies of AR and VR could complement such feedback and make exercises more appealing and motivating for patients [21].

4.2.3 Regulatory and Ethical Challenges

The regulatory and ethical issues in rehabilitation robotics primarily revolve around safety, privacy, and patient autonomy. To overcome these challenges there is a need for regulatory frameworks such as regulatory bodies to come up with clear guidelines and standards stipulating the safety and effectiveness of rehabilitation robots. This means stringent testing protocols that should cover the reliability of the robot, performance under changing conditions, and its capability to handle unforeseen errors during therapy [19]. Also, addressing data privacy and security is important. Most rehabilitation robots collect very private information about their patients. Due to this fact, any device should incorporate robust data encryption and cybersecurity measures. Compliance with global privacy regulations, such as GDPR, will be key to maintaining patient trust, based on application. This can only be ensured if the use is ethnically approved and upholds the will of the patient. The ethical policy directives may be presented on the roles robots play in rehabilitation. In considering ethics, robots have to operate under control, meaning there should be a healthcare professional overseeing the operations to ensure that human therapists are in charge of every procedure in the treatment processes in order to show that patients are not subjected to some decisions made by AI without oversight.

4.2.4 Economic and Logistical Challenges

Accessibility in most cases is limited by the cost and logistic challenges of the rehabilitation robots. The problems have solutions, and they are:

- Lowering the Costs by Modular Design: Designs for modular robotic systems that can easily be adapted depending on specific patient needs can reduce their costs. Upgrades and repairs are much easier in modular robots, hence making them cheaper to use for clinics that cannot afford to purchase a full-scale system. Besides, recent production advances based on 3D printing are foreseen to further facilitate cost reduction in the manufacture of rehabilitation robots [19, 20].
- Tele-rehabilitation and Remote Monitoring: Tele-rehabilitation platforms extend the use of rehabilitation robotics for use in the home environment and reduce the need to actually visit the clinic. Using remote monitoring technologies, patients can carry on with their rehabilitation at home while therapists in turn track their progress in real time and offer guidelines and necessary adjustments. Such can facilitate access to therapy, especially in remote or underserved areas.
- Insurance and Reimbursement Policies: Exerting the influence of advocacy for policy changes specifically to support reimbursement will increase access to these technologies. Healthcare providers themselves can help drive use by showing insurance companies the cost-electiveness of the robotic systems in improving patient outcomes and, thus, decreasing overall healthcare costs in the long run [18].

It is expected that overcoming the challenges of rehabilitation robotics will require a multi-discipline approach: technological innovation, regulatory reform, enhancement in human–robot interaction, and financial solutions. This therefore means that leveraging advancement in AI, machine learning, soft robotics, and tele-rehabilitation can enable the facilitation of higher value and more available, affordable, personalized solutions for the patients. These will complement provisions for strong regulatory frameworks and ethics guidelines that ensure rehabilitation robotics reaches its promise to revolutionize patient care.

4.3 Vision for the Future of Robotic Rehabilitation

The future of robotic rehabilitation holds immense promise for transforming the way patients recover from motor impairments. As technology advances, rehabilitation robots will become more personalized, accessible, and integrated into daily life, offering solutions that are tailored to individual needs. Key innovations such as artificial intelligence (AI), machine learning, brain-machine interfaces (BMI), and wearable robotics will play a pivotal role in creating adaptive rehabilitation programs that adjust in real-time to a patient's progress. These robots will seamlessly integrate into everyday activities, enabling patients to continue their therapy at home or in the community, while remote monitoring and tele-rehabilitation make these services more accessible, especially for those in remote or underserved areas. Enhanced BMIs will bypass damaged motor pathways, allowing patients to control robotic limbs through neural signals, promoting faster recovery through neuroplasticity. Collaborative models where robots and human therapists work together will optimize care by combining the precision and efficiency of robots with the emotional support and strategic insights of therapists. Additionally, future rehabilitation platforms will integrate technologies like VR, AR, and wearable sensors for a more holistic and engaging therapy experience. With innovations in modular design and 3D printing driving down costs, robotic rehabilitation will become more affordable and scalable, making advanced therapies available to a broader population. These advancements will empower individuals to regain their independence and improve their quality of life.

4.3.1 Personalized and Adaptive Rehabilitation

In the future of robotic rehabilitation, therapy will be highly rationed to a specific cause or factor in the patient's ailment. The integration of AI and ML allows the robot to continuously analyze the performances of the patients and automatically adjusts therapy in real time to offer optimal assistance. This adaptability can allow precise tuning of the exercises to match the physical capabilities of the patient and his progress. In time, the robots will be further enhanced to make even better predictions of recovery patterns so that therapies evolve with the rehabilitation process of the patient. For example, more personalized rehabilitation programs will be administered, as robots can currently apply a wide range of therapy modules based on patient data, injury type, and rehabilitation goals. The same kind of injury would no longer be treated uniformly; rehab robots would consider various occurrences, including genetic, biomechanical, or behavioral elements, to come up with optimized paths of recovery that are particular to an individual with different extents of impairment [19].

4.3.2 Seamless Integration of Robotics in Daily Life

With rehabilitation robots becoming lighter and more portable, and their handling easier, they will develop rapidly from purely clinical devices to personal use. Wearable exoskeletons and soft robotic devices will provide the patient with the opportunity to perform therapeutic exercise during daily activity at home, work, or in the community while these robots provide real-time support for walking, reaching, or lifting, thus promoting recovery through everyday movement. It means that this will further be taken with the advancement in tele-rehabilitation and remote monitoring of patients, whereby comfort to patients is given to rehabilitate in their homes while health professionals track the progress, provide feedback, and also alter the course of therapy regimen with ease in real time. Flexibility can further expand access to therapy within rural or underserved populations by minimizing burdensome travel to the clinic; therapy is made convenient and accessible [21].

4.3.3 Enhanced Brain-Machine Interfaces (BMI) and Neuroplasticity

Probably the most promising future development in rehabilitation robotics will be the embedding of advanced brain-machine interfaces. These interfaces will be able to enable the patient's brain directly to communicate with robotic systems, bypassing damaged motor pathways. In this respect, one can confirm that non-invasive BMIs are indeed substances of the future because they become increasingly more precise, affordable, and widely available for patients who will be capable of controlling robotic limbs or exoskeletons according to their thoughts. This will be especially useful for people with certain severe motor impairments dependent on strokes or spinal cord injuries, in that they will restore functional control over movement due to neural activity alone [22]. Future BMIs, combined with neuroplasticity-enhancing techniques, will stimulate those portions of the brain responsible for motor control and accelerate recovery by inducing neural circuit rewiring. This combination of robotic therapy and stimulation of the brain optimizes rehabilitation to recover from ill injury or disease more fully and quickly than is possible today [19].

4.3.4 Collaboration Between Robots and Human Therapists

While robots are going to become increasingly important in rehabilitation, they will not replace human therapists. In their place, collaborative rehabilitation models pepper the future, where both robots and therapists work together in an effort to provide optimal care. Robots perform repetitive and time-consuming tasks, such as leading a limb through precise motions or patient performance metrics, while therapists can then devote more of their energies to higher-order tasks such as motivating and offering emotional support, adjusting long-term rehabilitation strategies to keep patients on track. In time, therapists will be proficient in sensing and customizing the right amount of robotic therapies a patient requires. Moreover, as the AI continues to evolve, robots will be very helpful for therapists by providing necessary data and insights on patient performance, which would definitely go a long way toward data-driven rehabilitation planning that optimizes the results.

4.3.5 Holistic Rehabilitation Platforms

In the coming years, rehabilitation robots will be viewed as part of a much wider integrated holistic rehabilitation platform, combining a number of different digital health technologies. Such platforms combine robotic devices with VR, AR, AI, and wearable sensors to fully engage patients in rehabilitation. Patients can practice functional tasks in immersive VR environments simulating real-world scenarios, thereby making therapy more engaging and effective [21]. The integration of sensors that can be worn and the biofeedback system allows constant monitoring of the patient's health, continuously tracking not only motor recovery but also physiological factors such as heart rate, muscular activity, and oxygen saturation. This information will provide real-time feedback, enhancing the patients' and therapists' capability to monitor progress and make educated adjustments to the therapy regimens accordingly.

4.3.6 Cost-Effective and Accessible Rehabilitation

Probably, one of the main features of robotics rehabilitation in the future will be an affordable price and accessibility of these technologies. The activity related to the development of 3D printing, robotics via open source, and modular design will maintain costs low for enabling small clinics and developing countries to employ the robot rehabilitation systems. By doing so, greater scalability and lower costs of the technologies ensure a future in which patients from all walks of life can access robotic rehabilitation and benefits from advanced therapies. With increased use of tele-rehabilitation, patients anywhere will have better access to therapy services. The remote care models in rehabilitation will further streamline the processes and contribute to solving the global shortage of rehabilitation professionals, which invariably over some time will reduce health care costs.

The future of robotic rehabilitation is brilliant; beginning with advancements in AI, wearable robotics, BMIs, and remote monitoring, this face of rehabilitation is set to change. This will nail into place more personalized, accessible, and effective therapies helpful in the recovery of patients, as they recover within a short duration with better outcomes. When robotics becomes a part of daily life and further integrates into holistic rehabilitation platforms, the future in rehabilitation will be to empower individuals toward looking back at their independence and quality of life. As technology continuously evolves, so does the collaboration between human therapists and robots in setting new standards for rehabilitation care.

5 Conclusion

The field of rehabilitation robotics is rapidly evolving, offering transformative possibilities for enhancing motor recovery and improving the quality of life for individuals with physical impairments. Advances in technologies such as artificial intelligence, machine learning, brain-machine interfaces, and wearable robotics are creating more personalized and adaptive therapy solutions. These technologies allow for real-time adjustments in therapy, ensuring that each session is tailored to the patient's progress and needs. Wearable exoskeletons and soft robotics are making it possible for patients to engage in therapy as part of their daily activities, increasing the potential for recovery outside clinical settings. Tele-rehabilitation and remote monitoring further expand access, particularly for those in underserved or remote areas, by allowing patients to receive therapy at home while their progress is tracked remotely by healthcare professionals. Despite the promising advancements, several challenges persist, including the high cost of robotic systems, accessibility issues, and the need for trained personnel to operate and customize these devices. Additionally, regulatory and ethical considerations must be addressed to ensure safety, effectiveness, and privacy in the use of these technologies. As these hurdles are overcome, the collaboration between human therapists and robotic systems is expected to redefine rehabilitation, combining the precision and efficiency of robots with the insight and emotional support provided by human professionals. This evolving synergy has the potential to set new benchmarks in patient care and recovery.

References

1. Duschau-Wicke, A., Brunsch, T., Lünenburger, L., Riener, R.: Adaptive support for patient-cooperative gait rehabilitation with the Lokomat. (2008). https://doi.org/10.1109/iros.2008.4650578
2. Gassert, R., Dietz, V.: Rehabilitation robots for the treatment of sen-sorimotor deficits: a neurophysiological perspective. BioMed Central **15** (2018). https://doi.org/10.1186/s12984-018-0383-x
3. Miao, M., Gao, X.S., Zhao, J., Zhao, P.: Rehabilitation robot following motion control algorithm based on human behavior intention. Springer Science+Business Media. **53**, 6324–6343 (2022). https://doi.org/10.1007/s10489-022-03823-7
4. Rulik, I., Sunny, M.S.H., Sanjuan, J., Zarif, M.I.I., Brahmi, B., Ahamed, S.I., Schultz, K., Wang, I., Leheng, T., Longxiang, J.P., Rahman, M.H.: Control of a wheelchair-mounted 6DOF assistive robot with chin and finger joysticks. Front. Media **9** (2022). https://doi.org/10.3389/frobt.2022.885610
5. Wu, Q., Wu, H.: Development, dynamic modeling, and multi-modal control of a therapeutic exoskeleton for upper limb rehabilitation training. Sensors **18**, 3611 (2018). https://doi.org/10.3390/s18113611
6. Costa, A., Martinez-Martin, E., Cazorla, M., Julian, V.: PHAROS—PHysical Assistant RObot System. Sensors **18**, 2633 (2018). https://doi.org/10.3390/s18082633
7. Chen, S.-C., Jones, C., Moyle, W.: The impact of engagement with the PARO therapeutic robot on the psychological benefits of older adults with dementia. Clin. Gerontol. **47**, 909–921 (2022). https://doi.org/10.1080/07317115.2022.2117674
8. Khaksar, W., Saplacan, D., Bygrave, L.A., Torresen, J.: Robotics in Elderly Healthcare: A Review of 20 Recent Research Projects. arXiv (Cornell University) (2023). https://doi.org/10.48550/arxiv.2302.04478
9. Tijjani, I., Kumar, S., Boukheddimi, M.: A survey on design and control of lower extremity exoskeletons for bipedal walking. Appl. Sci. **12**, 2395 (2022). https://doi.org/10.3390/app12052395
10. Schicketmueller, A., Lamprecht, J., Hofmann, M., Sailer, M., Rose, G.: Gait event detection for stroke patients during robot-assisted gait training. Sensors **20**, 3399 (2020). https://doi.org/10.3390/s20123399
11. Park, J.M., Kim, Y.W., Lee, S.J., Shin, J.C.: Robot-assisted gait training in individuals with spinal cord injury: a systematic review and meta-analysis of randomized controlled trials. Ann. Rehabil. Med. **48**, 171–191 (2024). https://doi.org/10.5535/arm.230039

12. Morone, G., Paolucci, S., Cherubini, A., De Angelis, D., Venturiero, V., Coiro, P., Iosa, M.: Robot-assisted gait training for stroke patients: current state of the art and perspectives of robotics. Neuropsychiatr. Dis. Treat. **15**(13), 1303–1311 (2017). https://doi.org/10.2147/NDT.S114102.PMID:28553117;PMCID:PMC5440028
13. Gonzalez, A., Garcia, L., Kilby, J., McNair, P.: Robotic devices for pediatric rehabilitation: a review of design features. BioMed. Eng. OnLine **20** (2021). https://doi.org/10.1186/s12938-021-00920-5
14. Fosch-Villaronga, E., Čartolovni, A., Pierce, R.L.: Promoting inclusiveness in exoskeleton robotics: Addressing challenges for pediatric access. Paladyn J. Behav. Robot. **11**, 327–339 (2020). https://doi.org/10.1515/pjbr-2020-0021
15. Zhao, D., Sun, X., Shan, B., Yang, Z., Yang, J., Liu, H., Jiang, Y., Hiroshi, Y.: Research status of elderly-care robots and safe human-robot interaction methods. Front. Neurosci. **17** (2023). https://doi.org/10.3389/fnins.2023.1291682
16. Barria, P., Riquelme, M., Reppich, H., Cisnal, A., Fraile, J.-C., Pérez-Turiel, J., Sierra, D., Aguilar, R., Andrade, A., Nuñez-Espinosa, C.: Hand rehabilitation based on the RobHand exoskeleton in stroke patients: a case series study. Front. Robot. AI **10** (2023). https://doi.org/10.3389/frobt.2023.1146018
17. What Is a MyoPro Orthosis—Myomo. https://myomo.com/what-is-a-myopro-orthosis
18. Morales, R., Somolinos, J.A., Fernández-Caballero A., Ferraresi, C.: Rehabilitation robotics and systems. J. Healthc. Eng. **2018**, Article ID 5370127 (2018)
19. Romero-Sánchez, F., Luporini Menegaldo, L., Font-Llagunes, J.M., Sartori, M.: Rehabilitation robotics: challenges in design, control, and real applications. Front. Neurorobot. **16**, Article ID 957905 (2022)
20. Chua K.S.G., Kuah, C.W.K.: Innovating with rehabilitation technology in the real world: Promises, potentials, and perspectives. Am. J. Phys. Med. Rehabil. **96**(10 (Suppl)), S150–S156 (2017)
21. Nizamis, K., Athanasiou, A., Almpani, S., Dimitrousis, C., Astaras, A.: Converging robotic technologies in targeted neural rehabilitation: a review of emerging solutions and challenges. Sensors **21**, 2084 (2021). https://doi.org/10.3390/s21062084
22. Morales, R., et al.: Brain-machine interfaces in rehabilitation robotics. J. Neural Eng. **13**, 1–17 (2020)

The manufacturer's authorised representative in the EU is Springer Nature Customer Service Centre GmbH, Europaplatz 3, 69115 Heidelberg, Germany. If you have any concerns regarding our products, please contact ProductSafety@springernature.com

Printed and bound by CPI Group (UK) Ltd, Croydon, CR0 4YY

30/09/2026

02225501-0014